THE LOGIC AND METHODOLOGY OF SCIENCE IN EARLY MODERN THOUGHT

FRED WILSON

The Logic and Methodology of Science in Early Modern Thought: Seven Studies

UNIVERSITY OF TORONTO PRESS
Toronto Buffalo London

© University of Toronto Press Incorporated 1999
Toronto Buffalo London
Printed in Canada

ISBN 0-8020-4356-9

Printed on acid-free paper

Toronto Studies in Philosophy
Editors: James R. Brown and Calvin Normore

Canadian Cataloguing in Publication Data

Wilson, Fred, 1937–
    The logic and methodology of science in early modern thought :
seven studies

(Toronto studies in philosophy)
Includes bibliographical references and index.
ISBN 0-8020-4356-9

1. Science – Philosophy – History – 17th century.   2. Science –
Methodology – History – 17th century.   I. Title.   II. Series.

Q174.8.W54 1999   501   C98-932679-9

University of Toronto Press acknowledges the financial assistance to its publishing
program of the Canada Council for the Arts and the Ontario Arts Council.

This book has been published with the help of a grant from the Humanities and Social
Sciences Federation of Canada, using funds provided by the Social Sciences and
Humanities Research Council of Canada.

University of Toronto Press acknowledges the financial support for its publishing
activities of the Government of Canada through the Book Publishing Industry Development Program (BPIDP).

Canadä

To the memory of Rose and Dave Rothman

The longest Tyranny that ever sway'd,
Was that wherein our Ancestors betray'd
Their free-born *Reason* to the *Stagirite*,
and made his Torch their universal Light.
So *Truth*, while onely one suppli'd the State,
Grew scarce, and dear, and yet sophisticate,
Until 'twas bought, like Emp'rique Wares, or Charms,
Hard words seal'd up with *Aristotle's* Armes.

John Dryden, 'Epistle to Charleton'

Contents

x   Contents

xii   Contents

Acknowledgments

The following material has appeared previously. Editors and publishers are thanked for permission to reprint this work.

Study One uses material from 'The Rationalist Response to Aristotle in Descartes and Arnauld,' in E.J. Kremer, ed., *The Great Arnauld* (Toronto: University of Toronto Press 1994), 28–65.

Study Two uses material from 'The Distribution of Terms: A Defence of the Traditional Doctrine,' *Notre Dame Journal of Formal Logic* 28 (1987): 439–54.

Studies Two and Five use material from 'The Lockean Revolution in the Theory of Science,' in S. Tweyman and G. Moyal, eds, *Early Modern Philosophy: Epistemology, Metaphysics and Politics* (New York: Caravan Press 1985), 65–97.

Study Three consists in 'Berkeley's Metaphysics and Ramist Logic,' in Patricia Easton, ed., *Logic and the Workings of the Mind* (Atascadero, Calif.: Ridgeview Publishing Co. 1997), 109–36.

Study Four uses material from 'Hobbes' Inductive Methodology,' *History of Philosophy Quarterly* 13 (1996): 167–86; and from 'Hume's Theory of Mental Activity,' in D.F. Norton, N. Capaldi, and W. Robison, eds, *McGill Hume Studies* (San Diego: Austin Hill Press 1979), 101–20.

Study Six uses material from 'Hume on the Abstract Idea of Existence: Comments on Cummins' "Hume on the Idea of Existence,"' *Hume Studies* 17 (1991): 167–201.

Introduction

In the seventeenth century, those who were developing the new science knew very clearly that what they were doing was very different – different in kind – from what had gone before. Galileo knew when he proposed to ignore Aristotelian forces and search instead for general descriptions of how objects move that he was making a major break with tradition. Robert Boyle knew what he was doing when he attacked appeals to the notion of 'nature' in science: such appeals were at once non-scientific and at the same time a real hindrance to progress in the science that he was advocating and developing. This is not to say that there were no continuities. The practitioners of the new science had studied logic as it was then understood; they had studied Aristotelian patterns of explanation; some were prepared even to go some way with the Aristotelians in speaking of substances and essences. But for all these continuities, there was also major discontinuity: a new way of trying to understand the world was replacing an older way; a new concept of reason and of reasons for things was replacing an older concept of reason and of reasons for things.

Those who adopted the Aristotelian patterns explained the events that we observe in ordinary experience by appeal to the natures of things. These natures were unanalysable powers. Events were explained by citing the exercise of these powers. The natures of things thus constitute the reasons why things are as they are. Reason is the capacity to grasp the reasons of things; it is thus the power to grasp the unanalysable natures of things. We grasp the nature of a thing, that is, have a rational intuition of it, either by means of abstraction, as Aristotle himself argued, or else by means of innate ideas, as was argued by those more in the tradition of Plato. But however we come to have these intuitions, what they are intuitions of, the natures of things, are not given in sense experience.

It was natures in this sense that Molière was to satirize, and that Galileo and

Boyle were to reject: there are no reasons of things in this sense. Rather, when we come to understand something, we place it, as Galileo placed the motions of projectiles, in universal patterns. To understand is to subsume under a general pattern, a pattern of things given in ordinary experience. Reason is the capacity to grasp the reasons of things. But the reasons things happen are now constituted by the general patterns exemplified by these events. Reason has become the capacity to discover these general patterns that describe the ways in which things change and develop. The appropriate method is the experimental method of the new science.

It was not just that the new science involved a shift of cognitive interests, from natures to general patterns, though that certainly happened. What came with the new science was a new metaphysics and a new epistemology, a new notion of what reason is and what is to count as a reason for things happening.

The success of the new science was dramatic. In the relatively few short years that led from Galileo to Newton, its practitioners succeeded in 'demonstrating the frame of the system of the world.' This guaranteed the triumph of the new empirical science over the previous Aristotelian account of natural knowledge. It also secured the success of empirical science over what was, in effect, a variant of the Aristotelian account of reason, namely, the account of reason given by Descartes and the other rationalists. To be sure, there were disagreements between the rationalists, on the one hand, and the Aristotelians, on the other, and on some of these the rationalists were on the same side as the emerging empirical science. This is particularly so with regard to the logic of experiment. None the less, the rationalists shared with the Aristotelians the demand that science begin from apodictic non-empirical first principles, in sharp contrast to the demand of the empirical science of Galileo and Newton that first principles be adopted only after passing the test of experience.

This view of what happened during the scientific revolution was shared by most thinkers until the recent past.[1] For some years more recently, however, there has been a trend not only to emphasize the continuities between the older views and those who were defending and developing the new science, but more strongly to suggest that there was not, after all, a major break, a major discontinuity. Thinkers such as Galileo and Newton, or Bacon and Hume, were just wrong when they argued that they were effecting a sharp break with tradition.

Thus, to take one example, we have I.B. Cohen's *The Birth of a New Physics*. This study explores the development of physics from pre-Aristotelian times through Galileo to Newton. Cohen records the distinction between natural and unnatural or violent motion in Aristotelian thought.[2] He indicates that it has an empirical basis in such patterns as those of stones falling and hot air rising. But he does not go on to indicate that these patterns of behaviour are separated into natural and unnatural not by empirical means but by appeal to a *non-empirical entity*, namely, the nature of the changing thing. A pattern of change which an object undergoes is designated as natural just in case that it is explained by appeal to the nature of the changing object. A pattern of change which an object undergoes is designated as unnatural just in case that it is explained by appeal to the nature of a different object. The natures which ground this distinction are not given in ordinary experience. They are, rather, presented in rational intuition. This *non-empirical grounding* of the distinction between natural and unnatural motion is ignored by Cohen. Nor does Cohen indicate the thoroughly *teleological nature* of Aristotelian explanations. Later, when Cohen comes to compare Aristotle's 'laws' of motion with those of Galileo and Newton, he again fails to notice that the former do while the latter do not refer to the non-empirical notion of natures.[3] There is a continuity between Aristotelian 'theory' and modern physical theory: the former is simply an older version of the latter, replaced as science underwent the improvement which comes with better empirical evidence acquired through patient experiment over the ages. Aristotelian 'theory' is thus assimilated to modern physical theory in a way that overlooks the radical difference between the Aristotelian framework and the framework of modern physics. The former attends to the natures of things in order to understand their empirical behaviour. Modern science, in contrast, as Galileo and Boyle emphasized, concentrates on the patterns of behaviour that we can discover in our empirical investigation of things. Cohen's presentation in effect simply eliminates the need for Boyle to attack the notion of natures as having no place in the new science.

One finds a similar treatment of the history of science in Thomas Kuhn's equally influential work, *The Copernican Revolution*. Kuhn presents much the same story of historical development as does Cohen, though he emphasizes the astronomical side of the story rather than the physics side that Cohen emphasized. Like Cohen, Kuhn presents the basic patterns of the behaviour of things upon which Aristotelian physics focused as rooted in commonsensical empirical observations: rocks move down and the flames of a fire leap up.[4] Kuhn does hint at the teleological nature of Aristotelian explanations,[5] but like Cohen he does not note how the natural/unnatural distinction is rooted in the *metaphysical* and *non-empirical natures* of things. Like Cohen he stresses the continuity of the developments in a way that obscures the fact that part of what was going on in the rise of the new science of Galileo, Boyle, and Newton was the elimination of all attempts to ground explanations in the natures of things. Kuhn's account of the historical development, like that of Cohen, leaves no place for the polemic of Boyle against the then commonly accepted notion of nature.

More recently, S. Shapin has also discussed the scientific revolution in his

book of that title: *The Scientific Revolution.*[6] While Shapin emphasizes the teleological nature of the Aristotelian metaphysics of explanation and its inconsistency with the new mechanical philosophy (163), and while he even notes the
importance of Boyle's polemic against the 'vulgar notion of nature' (151), none
the less he also places Aristotelianism and the system of mechanical hypotheses
as points of view that are on a level with one another. But to the contrary, the
new system of mechanical philosophy as developed by Boyle and as placed on
a firm philosophical foundation by Locke are not on a par. Aristotle's world –
and also the world of Descartes – is a two-level world, as it were: there is sense
experience on the one hand and the world of forms known by reason on the
other. The world of the new science is a one-level world: there is only the world
of sense experience. Mechanical hypotheses are empirical hypotheses about the
world we know by sense experience. They are not claims about forms or
essences or natures that are outside the world we comprehend through sense
and are known only by a reason that innately, or by a mysterious process of
abstraction, grasps or intuits them. In missing the difference between the Aristotelian metaphysics of explanation and the new empirical science, Shapin
misses the deeper point that Boyle is not merely arguing that the Aristotelian
doctrine is wrong but is defending a *new cognitive end* for science and a *new
method* for achieving that end. The new end is knowledge of general matter-of-
fact patterns in the world we know by sense, rather than metaphysical forms
that are outside the world of sense. The new method is the method of experiment, the logic of which was first discerned clearly by Bacon.

Shapin does mention the work of Bacon (*Scientific Revolution*, 65), whose
new method as presented in the *New Organon* is said to have been 'labelled [by
those developing the new science] as a novel method meant to replace the traditional *organon* (Aristotle's body of logical writings) ...' But he does not discuss
how the new method has very different aims from those of Aristotle's organon.
The latter presents a method that is appropriate to the metaphysics of explanation that he develops. Bacon's world has no place for the active powers and teleological forms of Aristotle. What his method aims to do is locate the general
matter-of-fact patterns knowledge of which is the cognitive aim of the new science. Its logic does not aim to present the ontological structure of things in real
definitions appearing in scientific syllogisms. Its logic, rather, aims to proceed
through a body of hypotheses about the general patterns of things and to find
instances that eliminate those that are false, arriving in this way in the end at a
hypothesis that we may reasonably accept as true. Aristotle's method aims to
present the ontological structure of the world in scientific syllogisms; Bacon's
method is that of eliminative induction. By failing to bring out the very different aims of the two methods, Shapin makes them seem to be simple alterna-

tives. In fact, the difference marks a divide that separates an old way of thinking from something very new.

In failing to mark these changes, Cohen, Kuhn, and Shapin all disregard the fundamental point that the Aristotelian metaphysics is accompanied by a doctrine of what reason is, and that this concept of reason is very different from the reason that constitutes the new science. In effect, they are presenting the view that there is but one reason, and that this is common to both Aristotelian science and the new empirical science. They are not the only ones so to obscure the changes. In a recent work, *The Sovereignty of Reason: The Defense of Rationality in the Early English Enlightenment*, F.C. Beiser suggests that we take reason in the 'general sense' taken for granted in the seventeenth century: 'Usually, reason was understood as the business of giving and assessing reasons, as the activity of discovering and evaluating whether we have sufficient evidence for our beliefs.'[7] By taking reason in this generic sense, Beiser tends to miss the change that was taking place in the notion of reason in the seventeenth and eighteenth centuries, the change from the notion of reason that was defended by the Aristotelians and the rationalists, on the one hand, and the notion of reason that was defended by the new empirical science, on the other. This is the contrast between reason as the insight into the metaphysical natures or essences that are the reasons for things and reason as the empirical method of induction and experiment. And in missing this change in the concept of reason, Beiser tends to miss important changes in the terms of the debate concerning the rationality of religion. In fact, as we shall see in the subsequent studies, it was fully reasonable under the Aristotelian and raitonalist concept of reason that there be an appeal in the explanation to a necessarily existent perfect being – God. But in contrast, the new reason of empirical science cannot be made to justify such an appeal. And so the case for God's existence turned from a priori arguments in Descartes, whether ontological or causal, to, as in the case of Boyle, the a posteriori argument from design based on the same principles of inference as those in empirical science. The change in the concept of reason *forces* a change in the arguments used to justify the rationality of religion. Beiser's way of stating his point about reason leads him to overlook the way the debate shifted with the change in the concept of reason. In fact, his way of putting it obscures important aspects of the debate.

In a similar way, Cohen, Kuhn, and Shapin obscure the change in rational methods that occurred as the new science emerged from Aristotelianism and rationalism. Given the very different cognitive ends of the new empirical science, on the one hand, and the older metaphysics, whether Aristotelian or rationalist, on the other, one can expect very different accounts of the method needed to achieve those ends. If reason is the capacity to acquire knowledge of the sorts in which we have cognitive interests, then the different methods define rationality; or, at least, the cognitive ends and the structure of the world determine the rationality or reasonableness of the means adopted to achieve those ends. As Galileo and Boyle, among many, argued, the older methods had not achieved their ends. This suggested that those older Aristotelian and rationalist cognitive ends could not be achieved. They therefore proposed a new cognitive end, one for which they proposed a method that they could establish had reasonable expectation for success. The end was knowledge of matter-of-fact regularity; the method was the empirical experimental method. The work of scientists such as Galileo and Boyle showed that the new method worked, and that the new cognitive end could be achieved. The success of Newton ensured the triumph of the new cognitive end and the new method. At the same time, other thinkers such as Locke and Hume were at work.[8] They developed a metaphysics and epistemology that could be used to defend the new science against attacks by thinkers such as Descartes or the Aristotelian John Sergeant. This new metaphysics and epistemology defended the new science by attacking the metaphysical account of the world that justified the older cognitive end and methods. In particular, it eliminated the old metaphysical necessary connections, whether those of species, genus, and syllogism, as claimed by the Aristotelians, or those of causal structure, as claimed by the rationalists. To say it once again, the pictures of what happened in the scientific revolution given by Cohen, Kuhn, and Shapin obscure these developments.

That at least is the argument which I try to develop in detail in the studies that follow.

I am convinced, then, that something like the traditional story about the emergence of the new science in the early modern period is true. Before that time, there was something that could be called Aristotelianism. This was not science but metaphysics. Then people such as Galileo and Harvey discovered the new science; they set about doing it. Bacon and Descartes proposed methods for the new science. Bacon's inductive method triumphed in the context of the empiricist critique of both the Aristotelians and the rationalists. What emerged was an account of being and an account of reason that were very different indeed from those of the medieval period and from those of the rationalists who continued parts of the older Aristotelian tradition.

This view of things involves the claim that there is a radical discontinuity between the medieval period and what came after. To be sure, the latter – the new empiricist, inductive science – emerged only gradually. But emerge it did: a radically changed account of reason and being. There were, to be sure, continuities. Thus, a logic of consistency carries on, right through, though more highly regarded at some times than at others. But it is the discontinuity that is important, in my view. For that discontinuity marks an advance not only in our understanding of the world but also in our understanding of ourselves.

I have defended the discontinuity thesis in print, and have been reprimanded for not recognizing that there is a continuity of concerns between the medieval period and ours. The present studies are, if you wish, a reply to that reprimand.

These studies, and the thesis that they defend, should, however, also be seen as part of a larger program. To claim that the new science that arose in the early modern period is empiricist is to claim that the views of Hume and of the logical positivists of the 1930s about the nature of causation and of explanation are largely correct. I have defended these claims directly in other works. In *Explanation, Causation and Deduction* (1985), I have defended the empiricist account of explanation defended by Hume and the early positivists, the 'covering law model.' In *Laws and Other Worlds* (1986), I have defended a Humean/positivist account of laws and of causation. In *Empiricism and Darwin's Science* (1991), I have tried to show how the empiricist account of science correctly describes the methodology of Darwin and of biological science. In *Hume's Defence of Causal Inference* (1997), I argue that if Hume stated the problem of induction, he also solved it. At the same time, I attempt to relate Hume's views to more recent positions, such as Bayesian accounts of scientific inference and Kuhn's account of scientific rationality. These books all defend as reasonable the empiricist account of the new science, the account whose success is examined in the more historical studies below. These studies develop other themes, too. Empiricist and positivist accounts of science are often attacked as inadequate through attacks on historically incorrect descriptions of the claims of these philosophies. It is therefore important for one, such as myself, who wishes to defend these philosophies not only as reasonable but as definitory of any reasonable notion of reason, that the historical inaccuracies be removed. Some of this task of correcting the historical record can be found in my *Psychological Analysis and the Philosophy of John Stuart Mill* (1990). The studies below make a similar effort at correcting the pictures of what happened in the past, such as those of Cohen, Kuhn, and Shapin that are now common. These pictures are at once defences of a radical continuity thesis and attacks on the empiricist/positivist account of science and of human reason. It is important for those concerned to defend human reason – the reason of the Enlightenment – that these pictures be corrected. I hope that the studies below are a contribution towards that end.

The story of the emergence of the truly human reason of natural science and of the Enlightenment is a complex one. Several times over the last decade I have tried to sketch out a book that would deal with the relevant interactions of logic, methodology of science, and ontology/metaphysics/epistemology in the early modern period. In the end, I gave up: the themes were simply too interrelated to be organized in a simple manner. The result has been seven essentially independent but none the less interrelated studies. I hope that, taken both individually and together, these studies will help in our understanding of the emergence of the modern mind.

Study One examines the cognitive aims of the new science and the experimental method that it used to try to achieve these ends. The logic of this method was examined by both Bacon and Descartes, and defended by both against the earlier method of the Aristotelians. There is a detailed examination of the method of science that the predecessors of the new science aimed to defend; its purposes and rules are contrasted to those of the new science. At the same time, however, the empiricists and the rationalists proposed different justifications of the new experimental method. That of the rationalists shared many features with the older picture of science deriving from Aristotle. This was criticized by the empiricists, and this empiricist account of the new science was successful against the rationalist alternative. Its success was secured by the success of Newtonian empirical science. It is shown how the Newtonian rules of method were of a piece with the Baconian rules of induction and at variance with the Cartesian rules of method. There was also a radical shift in the notion of force, from that of the Aristotelians and the rationalists to that of the new mechanics of Newton. The former notion in the end pointed towards a God as the cause of order in the universe; the latter was free of this notion. The shift away from the older methods and older accounts of the structure of knowledge involved a radical shift in the view of what was needed in order to explain things.

Study Two examines the changing role of logic in the early modern period. Prior to the new science, syllogistic was not merely a logic of consistency but also a logic of truth. It was criticized in the early modern period, and for the empiricists became a mere logic of consistency. The study begins by looking at what we now call 'traditional logic,' and in particular at the problems of existential import and of the distribution of terms. It is shown that many of the differences have their roots in now-often-forgotten metaphysical theses about the concepts that appear as the terms of categorical syllogisms. It was this metaphysics that was criticized by the empiricists. The formal logic of consistency that the empiricists defended had its roots in the ancient and medieval periods, however, and appeared in particular in the immediate background to the early modern period as part of the educational methods of the schoolmen. As such it was strongly criticized by the new philosophers of the early modern period, rationalist and empiricist alike. The study tries to distinguish these sorts of criticisms from other sorts directed at syllogistic as a logic of truth. The different responses of the rationalists and the empiricists to the notion of demonstrative science is examined in detail.

Study Three picks up a theme from Study Two. Among the criticisms directed at the syllogistic of the schoolmen was that of the humanists who wanted a logic useful for orators and lawyers. The new humanist logic was criticized by both the rationalists and the empiricists, as Study Two makes clear, but the topical logic that the humanists developed was very popular. Study Three argues that one of these new logics, that of Petrus Ramus, should be seen as having certain consequences in the development of empiricism in the work of Bishop Berkeley.

Study Four argues, contrary to most accounts, that Hobbes's methodology of science is not that of demonstration but is rather inductive after the fashion of Bacon. It is argued that the appearance of the demonstrative element enters when Hobbes attempts to give an account of the necessity of natural laws, a problem which, it is argued, was better solved by Hume. Where the new science proposed a new cognitive end, namely, that of knowledge of matter-of-fact regularity, that was not yet to give up the traditional notion of cause. It was the work of Hume, in particular, that argued for rethinking this notion. Hobbes was struggling with the same issue.

Study Five examines in detail the 'rules by which to judge of causes' before Hume. These 'rules,' that is, the rules of eliminative induction, were discovered by Bacon but given their first concise formulation by Hume. Similar rules found in the medievals are seen to serve a very different purpose. Other versions of the rules are located in the early modern period, but none are as carefully crafted as those of Hume. It is suggested that the formulation presented by Hume is his own and represents his genius as a methodologist of empirical science.

Study Six examines the argument for the existence of a necessary being. It is argued that when this argument is located within the context of the substance ontology deriving from Aristotle, then it is valid. The criticism of this argument and of the idea of a necessary being is examined in detail, in particular the criticisms that are found in Hume.

Study Seven provides an examination of the so-called ontological argument for the existence of a necessary being. Again, it is argued that in the context of the rationalist version of the substance ontology that is found in Descartes, the argument is valid. It is further argued that Descartes is correct that there is a sceptical problem at the heart of the traditional metaphysics, and that only something like the appeal to a deity will solve it. Descartes's argument is examined in detail, as are the criticisms that emerged in response to Descartes's defence of the traditional metaphysics. It is shown that the empiricist attacks on the substance ontology proceed naturally from the problems that Descartes located in that ontology and tried, though ultimately failed, to solve. The empiricist critique of the older forms of thought is therefore not merely external; rather, it gains its real strength from the internal problems of the older system, problems that were brought to the fore by the work of Descartes.

In these studies I have tried to enter sympathetically into the spirit of the traditional substance ontology. As I have continued to think about it over many years, it has become more and more clear to me that, if we take it for granted, then much of traditional philosophy makes a great deal of sense – for example, both the ontological argument and the argument a priori for the existence of a necessary being. I think that the empiricists were correct in their criticisms of that metaphysics, and that there was real intellectual progress when their new account of human reason emerged. But the latter does not mean that we should not give the older system the respect that is its due as a remarkable achievement of human thought. I have attempted – with what success the reader will have to judge – to see how the empiricist criticisms of the substance ontology are genuine arguments directed at genuine problems in the heart of the metaphysics. When Descartes asked for objections to his *Meditations on First Philosophy*, he received two sorts. There were, on the one hand, the criticisms of Gassendi and Hobbes. These philosophers refuted Descartes by stating their own positions. Descartes, rightly, was annoyed with these criticisms. Descartes's position was not that of simply rejecting criticism; rather, it was that mere counter-assertion does not constitute criticism. On the other hand, there were the criticisms of Arnauld. The latter makes some devastating points, including the charge of circularity. But Arnauld tries to explore the structure of Descartes's arguments, and provides criticism only when he sees that the very structure of the Cartesian position demands the criticism. Descartes welcomed the views of Arnauld, including his criticisms, which he treated with the respect that they deserve. I would like to hope that the criticisms which I endorse which are directed at the Aristotelians and rationalists are in the style of Arnauld. I do think that the position of the Aristotelians and the rationalists is profoundly wrong, but at the same time I hope that my presentation of the empiricist critique respects the subtlety and depth of the older tradition. Again, the extent to which I have succeeded must be left to the reader to judge.

THE LOGIC AND METHODOLOGY OF SCIENCE
IN EARLY MODERN THOUGHT

Study One

Establishing the New Science:
Rationalist and Empiricist Responses
to Aristotle

Rationalists such as Arnauld followed Descartes in taking Aristotle's philosophy to be a great hindrance to progress in the sciences. Empiricists followed Bacon, who came to the same evaluation of Aristotle and his successors. Indeed, many of the criticisms raised by rationalists and empiricists were identical. Thus, we find Arnauld, in his *The Art of Thinking*, the 'Port Royal Logic,' citing Bacon in support of his criticism of the real definitions given by Aristotle for such things as hot and cold.[1] In some measure these shared criticisms were directed at the versions of Aristotelian logic that had been developed by Agricola and Ramus. These so-called topical logics were designed to be of practical use to lawyers and others who argued cases in a public domain. Both the Cartesians[2] and Bacon[3] were critical of these traditions, as Gaukroger has reminded us recently in respect of the Cartesians[4] and as Jardine has reminded us in respect of Bacon.[5] But besides a logic for lawyers, Aristotle had also offered a method for investigating natural phenomena. This method had abstraction, syllogistic, and real definitions as essential ingredients. Both the Cartesians and the Baconians were also highly critical of this Aristotelian method for investigating natural phenomena. Just as they both rejected the lawyers' idea that the topical logic sufficed for rational thought, both versions of the new philosophy held that the so-called method that Aristotle had proposed was not only wrong but a real impediment to the development of the new science of natural phenomena.

The aim of the present study is to examine this response to Aristotle's method of science. Much of the essay will be devoted to a careful exposition of the cognitive goals proposed by Aristotle and the method that he proposed to achieve those goals. But to begin, we shall examine what both empiricists, Bacon primarily, and rationalists, Descartes and Arnauld primarily, thought the new science was up to, what it aimed at as its cognitive goal, and how both these groups thought, commonly, that the Aristotelian method was totally inadequate as a means to that end. The discussion of Aristotle's method will follow, arguing that the critics were right, and that it fails miserably as a means for satisfying the cognitive goals of the new science. Finally, the essay will argue that the two responses had certain important features in common. In particular, they not only advocated correctly the method of experiment as a means to the cognitive ends of the new science, but also understood the logic of this new method of experiment. At the same time, however, it is also argued that there were important differences. These differences locate the rationalist response in an important way as still resting in the Aristotelian tradition, and at odds with the new science. In the end, there was an adequate understanding of the new science from a methodological point of view only when these remnants of Aristotelianism had been eliminated and a thoroughgoing empiricist account of the method of experiment was at hand.

In my view, a failure to understand the cognitive aims of Aristotle and how they differ from those of the new science – and, correspondingly, a failure to understand the method of Aristotle and how it differs from the method of science – is often a major impediment to a recognition of what was achieved by philosophers and scientists in the early modern period. Too often, in my view, commentators on the history of this period make claims about the continuity between the old and new science that fail to correspond to reality: failing to understand the respective positions save in a superficial way, they fail to recognize the radical break between the old and new ways of thinking.[6] It is the aim of this essay to spell out clearly the differences.

I: New Science: New Methods

What is characteristic of the new science is the *method of experiment*. This is clearly exemplified in the work of Harvey on the circulation of the blood. Harvey's book on this subject is crammed full of experiments. Thus, to take only one, Harvey discusses the issue whether the blood found in the veins is pure or, as held by his opponents, contains vapours. He points out that the quantity of vapours decreases in the presence of cold. He therefore proposes the experiment of ligating an arm, letting the veins swell up with blood, and then immersing it in cold water. If the swelling decreases then there are vapours present; if not, then the blood is pure. When the experiment is performed, there is no decrease in swelling. Harvey concludes that the blood is pure: 'this experiment shows that, below the ligature, the veins swell up not with thinned blood and inflated spirit or vapours (for the immersion into the cold would have depressed such bubbling out) ...'[7]

The logic of the situation is clear. There are two contrary or mutually exclusive hypotheses which together exhaust the alternatives. These are:

> The blood in the veins is pure (that is, a necessary condition for blood in the veins is that it is pure).

and

> The blood in the veins contains vapours (that is, a necessary condition for blood in the veins is that it contains vapours).

We might represent these by:

$$V \supset P$$

$$V \supset {\sim}P.$$

Harvey presents data that falsify the second. The data themselves are derived by inference from other observations. He *takes it on the basis of other data to be true that*

> If the blood in the veins is mixed with vapours then if it is immersed in cold then it decreases in volume

or, in symbols,

$$V \supset [\,{\sim}P \supset (C \supset D)].$$

Since this is true, and since it is given that it is blood in the veins – that is, V is true – it follows that $[\,{\sim}P \supset (C \supset D)]$ is true. Since C is true and D is observed to be false, it follows that $(C \supset D)$ is false. From this it follows that ${\sim}P$ is also false. Hence it is false that

$$V \supset {\sim}P.$$

He then concludes that the first is true.

The logic of the situation is clear: we can see the mechanisms of eliminative induction at work. We are presented with a range of jointly exhaustive and mutually exclusive hypotheses; data are presented which eliminate all but one of the range; it is concluded that the remaining hypothesis is to be affirmed as true. *It is this method that is characteristic of the new science.*

It was this method that the philosophers of the new science were concerned to defend. Bacon's defence and elaboration of this method is well known. Indeed, some argue that it can be traced solely to him. Weinberg, for one, has argued that Bacon was the first to understand clearly the inductive logic of experiment, and that his predecessors had clearly done little more than the 'puerile' method of induction by simple enumeration.[8] But in fact, as I shall argue, recognition, and defence of the method were not limited to empiricists: one finds it also defended by the rationalists. Thus, Descartes also accepted the method as the correct method to science: he explicitly notices Harvey and the experiments that the English physician performed in order to establish the circulation of the blood.[9]

Those who defended this method recognized that it was useful only relative to a certain cognitive goal, one that differs considerably from what the tradition of Aristotle had taken to be our correct cognitive goal.

Descartes proposed the cognitive goal of the new science to be knowledge of nature such that people could intervene effectively in nature.

[A]s soon as I had acquired some general notions concerning physics, and when I began to test them in various particular problems and noticed where they could lead and how much they differed from the principles used up to the present, I believed that I could not keep them secret, without sinning gravely against the law which obliges us to procure, to the best of our ability, the general good of all men. For they made me see that it is possible to arrive at knowledge which is very useful in this life, and that instead of that speculative philosophy taught in the schools, we can discover a practical one, through which, knowing the force and action of fire, water, air, the stars, the heavens, and all the other bodies which surround us, as distinctly as we know the different skills of our artisans, we can use them in the same way for all the purposes to which they are suited, and so make ourselves the masters and possessors, as it were, of nature.[10]

Bacon makes the same point. The old logic has a different goal from the new. He contrasts the 'anticipation' of mind, which is what has hitherto been practised, to the 'interpretation' of nature, which is the task of the new science.[11] The interpretation of nature yields knowledge that is capable of use. 'Human knowledge and human power meet in one; for where the cause is not known the effect cannot be produced. Nature to be commanded must be obeyed; and that which in contemplation is as the cause is in operation as a rule' (*New Organon*, I.iii).[12] Contrasting the logic of the old science with that of the new, Bacon points out that 'the effect of the one [is] to overcome an opponent in argument, of the other to command nature in action.'[13]

Two points are clear. *One*: The cognitive goal is knowledge of matters of fact that can be used to determine the means to be used to achieve human ends: the knowledge must be such as could satisfy our pragmatic cognitive interests – though, of course, the fact that such knowledge satisfies our pragmatic interests does not preclude that it may also be sought out of idle curiosity[14] – Bacon put the point by quoting Solomon to the effect that '[t]he glory of God is to conceal a thing; the glory of a king to search it out' (*New Organon*, I.cxxix). *Two*: The sort of speculative philosophy then taught – the sort that dealt with 'these forms or qualities disputed in the schools,'[15] that is, Aristotelianism – does not satisfy these cognitive interests. Let us deal with these in turn.

1) *The New Cognitive Aims*

In this context, we can ask (a) what more precisely are these new ends? We can ask (b) what method is it that is to be used to achieve this end? And we can ask (c) what is the starting point of this method? As Bacon puts it, both the old and the new logics propose rules to guide the understanding, but his differs from the old 'in the end aimed at, in the order of demonstration, and in the starting point of the inquiry.'[16]

a) The Cognitive Ends of the New Science

It is taken by both empiricists and rationalists that the cognitive goal of the new science is knowledge that serves our pragmatic interests, that is, is capable of being applied. From this it is possible to infer that it has certain characteristics. Thus, if we are to intervene effectively in the new science, then we need to know the *regular patterns* to which particular objects in the universe *conform*. We need to know those conditions that are *sufficient* for what we want to achieve, and those conditions that are *necessary* for what we want to achieve. These laws should be known with as great a degree of *certainty* as possible; and, moreover, these laws *cannot have exceptions*. For if they are uncertain or if there are exceptions, we will not be certain that acting in the appropriate way will actually bring about what we want to achieve.

Descartes leaves these points to be inferred by the reader. Bacon is more explicit in drawing out these points. What science considers is natures, that is, properties of things, and its aim is to discover the *forms* or *nature-engendering natures* of those natures (*New Organon*, II.i). This knowledge is knowledge of causes (II.ii). This knowledge must be certain in the sense that it 'will not deceive him in the result nor fail him in the trial' (II.iv), that is, it must be as certain as possible and without exceptions: 'given the form, the nature invariably follows' (II.iv). The knowledge must thus be of an exceptionless general law. In terms used more recently, what Bacon is arguing is that knowledge that satisfies the cognitive ideal of the new science must be without 'gaps';[17] knowledge that falls short of this ideal has been called 'imperfect.'[18] Specifically, the ideal is knowledge of necessary and sufficient conditions:[19] the form 'is always present when the nature is present, and universally implies it, and is constantly inherent in it. Again, the form is such that if it be taken away the nature infallibly vanishes. Therefore it is always absent when the nature is absent, and implies its absence, and inheres in nothing else' (II.iv). But further, given our cognitive end of knowledge that is capable of satisfying our pragmatic interests, it must be knowledge of the form that causes the nature not in this or that specific sort of substance but causes it in any sort of substance (II.iii). It must, in other words, cover all sufficient conditions; that is, the ideal is a statement of law in which the sufficient condition is the sum of all sufficient conditions. A law of this sort is said by Bacon to be 'free' (II.iv).

This knowledge can be applied in two ways. We can consider a thing to be a collection of simple natures. We then use our knowledge of forms to deduce how to bring about a thing of the required nature (*New Organon*, II.v). Or we can use our knowledge of laws to investigate *processes* that things undergo, including processes through which they grow and develop and processes of generation (II.v). These latter Bacon calls 'latent processes.' He is clear that the *cognitive ideal* in this case includes laws that yield knowledge of (α) all relevant factors that occur within the process, that is, knowledge of all relevant variables, (β) all relevant factors that occur outside the process, that is, knowledge of boundary conditions, and (γ) all details of the process over time, that is, knowledge of what has been called a 'process law,'[20] a law, or rather a set of laws, that gives the necessary and sufficient conditions for any state of the system at any time, where a state of the system is the values of all the relevant variables at a time.

[W]hat I understand by [a latent process] is not certain measures or signs of successive steps of process in bodies, which can be seen; but a process perfectly continuous, which for the most part escapes the sense.

For instance: in all generation and transformation of bodies, we must inquire what is lost and escapes; what remains, what is added; what is expanded, what contracted; what is united, what separated; what is continued, what cut off; what propels, what hinders, what predominates, what yields; and a variety of other particulars.

Again, not only in the generation or transformation of bodies are these points to be ascertained, but also in all other alterations and motions it should in like manner be inquired what goes before, what comes after; what is quicker, what more tardy; what produces, what governs motion; and like points; all which nevertheless in the present state of the sciences (the texture of which is as rude as possible and good for nothing) are unknown and unhandled. For seeing that every natural action depends on things infinitely small, or at least too small to strike the sense, no one can hope to govern or change nature until he has duly comprehended and observed them. (II.vi)

Knowledge of this sort has been called 'process knowledge,'[21] and is quite reasonably taken to be the ideal of scientific explanation of particular facts and processes.[22] After all, with process knowledge one can predict what will happen in a system, given the state of the system at any one time (the initial conditions); one can postdict what has happened in the system previously; one can infer what would happen if one of the variables were to change its value or if one of the boundary conditions were to change; one can infer what one would have to change in order to bring about some state that one desired; and one can infer what would have had to have happened if things were to have been different from what they are. What more could one want to know about the system[23] (in respect of the relevant variables)?[24] The question is rhetorical; the conclusion is that process knowledge is, as Bacon implies, the cognitive ideal for the understanding and explanation of particular things and processes. Bacon gives no example; none was available. Neither does he give much hope that it will be achieved in the state of the sciences as he knew it. Descartes adopts the same model of the ideal explanation, and is more hopeful that it can be attained: 'given that these laws cause matter to assume successively all the forms it is capable of assuming, if we consider these causes in order, we shall finally be able to reach the form which is {at present} that of this world.'[25] In fact Descartes's own work fell far short of this ideal. Neither he nor Bacon could suspect that, before their century was finished, Newton would provide an example of such knowledge in his explanation of the motions of bodies in the solar system.[26]

But of course we have much knowledge of laws that falls short of this ideal: leaving out relevant variables, perhaps or perhaps not specifying all the boundary conditions, or perhaps not being able to supply all the details of how one state develops from or into another, or even perhaps a bit of all of these. Certainly, this was the state of scientific knowledge – that is, knowledge of the sort sought by the new science – in all areas in the early modern period, including such areas as geology and medicine, and also those that already involved the use of mathematics, such as mechanics, astronomy, optics, and music. Such knowledge is 'gappy' – the term is from Mackie[27] – and, relative to the standard of process knowledge, is reasonably called 'imperfect' – the term is from Bergmann[28] – though we must recognize that to call it such is only to compare it to the ideal and not to deny that it can be used as the new science proposed for purposes of explanation and prediction.[29]

b) The Method of the New Science

As Descartes implies and as Bacon makes clear, the cognitive aim of the new science is knowledge of matter-of-fact regularities stating necessary and sufficient conditions. Furthermore, among cases of knowledge of this sort, the ideal one would cognitively prefer is process knowledge, or knowledge of what Bacon calls latent processes: this is knowledge of necessary and sufficient conditions for states of complex processes and, as Bacon puts it, 'must come from those primary and catholic axioms concerning simple natures' (*New Organon*, II.v).

What we need next is a method for achieving these cognitive aims of the new science. This method will be a set of rules conformity to which will constitute a means that is, so far as we can tell, relatively efficient in bringing about the desired ends. The method will provide rules – though not necessarily algorithms – conformity to which will generate knowledge, so far as it can be attained, of matter-of-fact regularities, and, more specifically, rules conformity to which will yield knowledge of regularities that will eliminate the gaps from the imperfect knowledge that we already have, to move us closer to the ideal of process.

Knowledge of these regularities cannot be based on induction by simple enumeration, which is 'puerile'[30] and childish (*New Organon*, I.cv). Such inferences are liable to be upset by contrary instances.[31] This method is highly inefficient as a means for achieving the cognitive goals of the new science. One needs instead a method that will, as it were, take advantage of the negative instances. The correct method should rather 'analyze experience and take it to pieces, and by a due process of exclusion and rejection lead to an inevitable conclusion'[32] (cf I.cv). Knowledge satisfying the cognitive goals of the new science can more efficiently be arrived at by this method – what has come to be known as the method of induction by elimination. Bacon was the first to articulate this method clearly.

These methods aim to discover necessary conditions, sufficient conditions, and necessary and sufficient conditions. Bacon himself was concerned only with the more limited, yet more cognitively desirable, case of necessary and sufficient conditions. (One finds the two more limited cases stated clearly for the first time in Hume.)[33] The method for discovering necessary conditions has been called, following Mill, the *method of agreement*. If one has a pair of hypotheses about necessary conditions

(1)   whenever A then B_1

         whenever A then B_2

then an observation of an individual that is A but not B_2 will eliminate the second alternative as false, enabling one to conclude that the uneliminated hypothesis must be true. As Bacon puts it, we 'proceed at first by negatives, and at last to end in affirmatives after exclusion has been exhausted' (*New Organon*, II.xv). Sufficient conditions are located by the *method of difference*. If we have a pair of hypotheses concerning sufficient conditions:

(2)   whenever C_1 then D

   whenever C_2 then D

then an individual which is C_1 but not D will eliminate the first hypothesis and enable one to conclude that it is the second that is true. Clearly, both methods can be generalized to cover any finite number of hypotheses in the set of alternatives.

   Thus, for example, to illustrate the method of agreement, suppose several people have all become ill upon eating potato salad at a restaurant, but have in other respects had quite different meals, some having meat, some vegetables, some desserts. Being ill and not eating meat eliminates the latter as the cause; being ill and not eating dessert eliminates the latter as cause; and so on. It is the condition in which the individuals who are ill agree that is uneliminated. We therefore conclude that this is the cause or necessary condition for the illness. A simple change will often yield an example of an inference to a sufficient condition. If something changes from C_1 to C_2, and also thereupon changes from not-D to D, one can conclude that C_2, that in respect of which the instances differ, is the cause of D. Thus, Becquerel discovered that burns can be caused by radium – that is, that proximity to radium is a sufficient but not necessary condition for being burned – when he inferred that the radium he carried in a bottle in his pocket was the cause of a burn on his leg by noting that the presence of the radium was the only relevant causal difference between the time when the burn was present and the earlier time when it was not.

   The two methods can be combined in the *joint method of agreement and difference* to yield the discovery of conditions that are both necessary and sufficient. It is this method that Bacon clearly articulates.

   Sometimes it is possible to eliminate an alternative, not on the basis of observation, but on the basis of previously inferred laws. If we know by previous inductions that no C_2 is D, then observation is not needed to eliminate the second hypothesis of (2), and we can infer that what remains, or the residue, gives us the sufficient condition for D. Where an alternative is eliminated on the basis of previous inductions, we are said to use the *method of residues*. This is the form of the method that Harvey used in the experiment cited above.

The methods may be generalized to cover quantitative laws. A cause of Q may be taken not to be a necessary and sufficient condition but a factor P on whose *magnitude* the *magnitude* of Q is functionally dependent. If P varies when Q varies, then one can use methods of elimination to infer that P causes Q. This has been called the *method of concomitant variation*. More complicated methods are needed to infer what precisely is the function that correlates the two magnitudes. Bacon includes in his method a version of the method of concomitant variation (*New Organon*, II.xiii).

It is clear that if we are to conclude that one of (1) is true on the basis of the given data, then we need an additional premise to the effect that

(*)   there is exactly one necessary condition for B and it is among the set consisting of A_1 and A_2.

The existence claim that *there is* a unique cause (necessary condition) that is made here is known as a *principle of determinism*, and the existence of a delimited range of alternatives is known as a *principle of limited variety*. The range will in general be delimited by a generic description rather than by enumeration as in (*). Thus, rather than (*) we will have something like:

(**)   there is exactly one species f of the genus Φ such that whenever B then f.

The assumption is, of course, that

(@) A_1 is Φ and A_2 is Φ.

As Bacon says, the explanatory form must be an instance of a genus (*New Organon*, II.iv). We see immediately that the principle (**) is *generic* where the principles (1) are *specific*. The principle (**) asserts the existence of a law of a certain *form*. The specific laws (1) have *this form in common*; (@) makes this explicit. It is worth noting that (**) involves both existential and universal claims; its structure is, as the logicians now say, mixed-quantificational. This means, on the one hand, that it cannot be conclusively verified – that is a consequence of the universal quantifier – and, on the other hand, that it cannot be conclusively falsified – that is a consequence of the existential quantifier. Similar principles are needed for the other methods. It is evident that the condition of 'exactly one' can be relaxed to 'at least one'; the logic is only slightly more complicated.[34]

Bacon clearly recognizes the need for such principles. Thus, he states that 'in nature nothing really exists besides individual bodies, performing pure individual acts according to a fixed law ...' (*New Organon*, II.ii); that is, the individual or particular events in nature all occur according to 'fixed law.' Thus, we know that *there are* laws there to be discovered. He also assumes a principle of limited variety. Thus, he tells us that in induction (by elimination) we 'must first of all have a muster or presentation before the understanding of all known instances which agree in the same nature, though in substances the most unlike' (II.xi); this sets the problem, which is 'upon a review of the instances, all and each, to find such a nature as is always present or absent with the given nature, and always increases and decreases with it'; this problem is solved with the affirmation of a law 'after exclusion has been exhausted' (II.xv). Clearly, there can be an 'exhaustion' that ends in a definite affirmative only if the variety to be exhausted is indeed limited.

Descartes is less articulate on the details of the method of the new science than is Bacon. Yet, as we have seen, he clearly recognizes that the experimental method of Harvey is a method of exclusion, and that at the end of the exclusion one arrives at an affirmation. He also recognizes that in this process of inference there is an essential role for the principles of determinism and limited variety. Part of the Cartesian method is, of course, deductive. As he puts it, 'I have ... noticed certain laws which God has established in nature, and of which he has imprinted notions in our souls, such that after having reflected sufficiently upon them, we could not doubt that they are exactly observed in everything which occurs in the world. Then in considering the effects of these laws, I seem to have discovered several truths more useful and more important than everything I had learned previously ...'[35] However, there are limits to the success of this method: when the subject matter becomes too complex the human mind is unable to trace out the causal connections from the first principles. 'From describing inanimate bodies and plants, I went on to describe animals, and particularly men. But ... I did not yet have sufficient knowledge of them to speak of them in the same style as of the rest – that is, by demonstrating the effects through the causes, and showing from what sources and in what way nature must produce them ...'[36]

It is in this context that Descartes resorts to the experimental method. Unable to deduce the laws for these complex bodies from first principles, he uses the techniques of Harvey to separate the true from the false and to draw positive conclusions about the regularities to which the actions of the bodies conform. But to do this he must delimit a range of hypotheses. This he does by invoking certain *generic* principles that he (believes that he) has already established. 'I contented myself with assuming that God formed a man's body, entirely similar to one of ours, in the exterior shape of its members as well as in the interior conformation of its organs, without composing it of any matter other than that I had described, and without placing in it, at the beginning, any rational soul, nor any other thing to serve as a vegetative or sensitive soul, except that He excited in its heart one of these fires without light which I have already explained ...'[37] These generic principles of the mechanical philosophy then provide a range of hypotheses for explaining the operation of the human body. It is the job of the experimenter to separate out from this range that hypothesis which is true: where the subject matter is complex, 'I know no other expedient than ... to search for certain experiments which are such that their result is not the same when we explain the effect by one hypothesis, as when we explain it by another.'[38] The method, in other words, is that of the Baconian logic of science: by using observed results to exclude certain hypotheses from a delimited range of possible alternatives, one concludes by affirming the uneliminated hypothesis.

We may conclude that the empiricists and the rationalists recognized that the most efficient method to knowledge of the laws that was the cognitive aim of the new science was the method of experiment, or, what is the same, the method of eliminative induction. The two schools did disagree about the logical status of the principles of determinism and of limited variety, as we shall see. But they were agreed on the cognitive aim of the new science and on the logic of the experimental method that sought to achieve that aim. And in agreeing on these latter they were also agreed that the cognitive goals and methods of Aristotle and the schoolmen were quite wrong.

c) The Starting Point of the Method

The two accounts of the new science also agreed on the starting point of the new science. This was sense, but sense taken as having a role that was different from that which it had for Aristotle and the schoolmen who followed him. As Bacon puts it, in the first instance, the role of sense is limited to deciding the experiment – that is, whether the particular hypothesis being tested is true or false – while the experiment decides the law that governs the actions of the thing – that is, the decision as to which hypothesis it is that is true is determined by the mechanisms of elimination: 'the sense decides touching the experiment only, and the experiment touching the point in nature and the thing itself' (New Organon, I.1). In other words, we must not leap from sense experience to a claim to knowledge of the nature of the thing. We must further recognize that our senses often err (I.lxx). But the new science itself proposes a remedy here. Our senses are part of the natural order, and therefore subject to law. These laws will inform us of the conditions under which we are prone to err, and those where our sense experience is veridical. We can, therefore, appeal to the results of the new science to render our sense experience more reliable: 'certain it is,' Bacon says, 'that the senses deceive; but then at the same time they supply the means of discovering their own errors; only the errors are here, the means of discovery are to seek.'[39]

Where Bacon understands the logic of the situation, he does little to help in practice. Descartes is here more helpful. He not only understands the logic of correcting and aiding our senses as natural instruments to be investigated by the new science, but actually does some of that research and recommends concrete means for implementing that program and acting upon the knowledge that it produces. Descartes makes the same point as Bacon, that our senses err but that the sources of those errors can be discovered. He first notes, 'the stars or other very distant bodies appear much smaller to us than they are,' and other similar errors into which our sense experience leads us;[40] this is raised as an objection, calling into question the possibility of progress towards knowledge. But he then goes on to argue that we can discover the causes of these errors: he attempts to demonstrate 'what changes must occur in the brain in order to cause waking, sleep, and dreams; how light, sounds, odors, tastes, heat, and all the other qualities of external objects can there imprint diverse ideas by means of the senses ...'[41] Then, in the essay on *Optics* appended to the *Discourse on Method*, and intended to illustrate it, he argues that one can understand scientifically, in terms of the laws of perspective, how it is that distant objects appear small, thereby enabling us to correct the errors of sense. 'To sum up, in judging of distance by size, or shape, or color, or light, pictures in perspective sufficiently demonstrate to us how easy it is to be mistaken. For often because the things which are pictured are smaller than we imagine that they should be, and because their outlines are less distinct, and their colors darker or more feeble, they appear to us to be farther away than they are in actuality.'[42] He then goes on in the subsequent (seventh) discourse of the *Optics* to discuss such instruments as the telescope as 'means for perfecting vision.'[43]

In this way the new science provides a 'foundation' or starting point for itself which is, as Bacon puts it, 'deeper and firmer' than that which has hitherto been given to knowledge claims.[44]

2) *What's Wrong with the Old?*

The empiricists and the rationalists both argued that the cognitive aims and methods of the new science were incompatible with the aims and methods of the 'science' proposed by Aristotle and defended by the schoolmen.

The method of Aristotle, Bacon complains, glances at experiments and particulars only in passing, rather than letting the mind dwell among them in an orderly fashion (*New Organon*, I.xxii). The result is notions that are 'confused and overhastily abstracted from the facts ...' (I.xiv). In practice, this amounts to the use of the 'childish' method of induction by simple enumeration to discover first principles (I.cv). But it is more than this; it is to introduce fictions as the explanatory entities rather than regularities, and this proposes in effect a cognitive goal at variance with that of the new science. Thus, Bacon tells us, 'human understanding is of its own nature prone to abstractions and gives a substance and reality to things which are fleeting. But to resolve nature into abstractions is less to our purpose than to dissect her into parts' (I.li). For the new science, as Bacon correctly sees, it is the lawful regularities that determine which forms are explanatory of the actions of things, rather than abstract forms or ideas explaining the actions and thereby the regularities (II.xvii). For the new science, it is things themselves – matter, if you wish – rather than forms that should be attended to; for, 'forms are figments of the human mind, unless you will call those laws of action forms' (I.li). The cognitive goal is knowledge of forms. But this is a chimera; there are no such forms beyond the things themselves and the observable patterns of action of those things. The method proposed for achieving this goal is that of abstraction: the mind proceeds by abstracting the form from the thing. This latter operation proceeds by attending to how the thing is presented in sense experience. Since there are no forms, the method cannot be productive. But attempts to practise this illusory method will interfere with attempts to do the new science, since in practice what it amounts to is attempting to grasp laws by the principle of induction by simple enumeration.

Worse. Once it is supposed that the abstraction has been completed and the form of the thing discovered, the mind can ignore exceptions or explain them away: 'though there be a greater number and weight of instances to be found on the other side, yet these it either neglects and despises, or else by some distinction sets aside and rejects, in order that by this great and pernicious predetermination the authority of its former conclusions may remain inviolate' (*New Organon*, I.xlvi; cf I.xxv, cxxv). Thus, when one accepts abstract forms as explanatory, one ends up accepting as lawful regularities that have exceptions. But this is contrary to the cognitive aims of the new science, which can accept as explanatory only those regularities that are genuinely exceptionless and universal.[45] It is also contrary to the method that the new science uses, that is, the method of experiment or of eliminative induction, which instructs one not to ignore negative instances but to search them out and use them as a tool in the discovery of truth. Aristotelian cognitive goals and methods are thus directly contrary to those of the new science. Not only that, but it is impossible that

Aristotelian 'science' serve our human pragmatic interests: the old science, based on hasty abstraction,[46] is barren of works, remote from practice, and unavailable for action.[47]

The attempt to grasp explanatory forms by abstraction is coupled with another error. This is the 'introduction of abstract forms and final causes' (*New Organon*, I.lxv). That is, it is taken by the tradition that Bacon is opposing that the forms yield teleological explanations of things. This, too, Bacon rejects; final causes are clearly more related to 'the nature of man ... than to the nature of the universe' (I.xlviii).

The traditional method held that once the form was abstracted from things, it yielded self-evident first principles. These were then elaborated by means of the syllogism. These syllogisms were constructed by finding middle terms between extremes. But, while it is no doubt true that things that agree with a middle agree with one another, this is no path to knowledge if the notions that appear as terms are themselves – as they in fact are – 'improperly and overhastily abstracted from facts, vague, and not sufficiently definite, faulty ...'[48] The search for middle terms is, within the traditional scheme, the search for real definitions. But, in fact, the notions that appear in the syllogisms are 'fantastical and ill defined' (*New Organon*, I.xv).

What, in practice, does the traditional method amount to, independently of its claims to discover the metaphysical forms of things? Bacon makes it clear: on this method 'one flies from the sense and particulars to the most general axioms, and from these principles, the truth of which it takes for settled and immovable, proceeds to judgement and to the discovery of middle axioms' (*New Organon*, I.xix). Since this involves only glancing at sense rather than interrogating it; since it involves at best induction by simple enumeration; and since it involves the ignoring of contrary instances; it follows that it is a method not only ill adapted, but positively inimical, to achieving the cognitive aims of the new science. To this he contrasts the experimental method, the method of eliminative induction, which 'derives axioms from the sense and particulars, rising by a gradual and unbroken ascent, so that it arrives at the most general axioms last of all' (I.xix).

Descartes makes the same sort of criticism of the Aristotelian position. Philosophers often examine difficult questions out of order, 'which seems to me [Descartes] to be the same as though they strove to jump from the bottom to the top of some edifice in one leap, either neglecting the staircase which is provided for this purpose or not noticing it ... This is the way ... philosophers [do] who neglect experience and believe that truth will arise from their own heads as Minerva did from that of Jove.'[49] Since formal logic – that is, syllogism – can lead to truth only if its premises are antecedently known to be true, the syllogism is useful to anyone interested in the search after truth.[50] Insofar as the real definitions of the Aristotelians are revealed in the discovery of middle terms of demonstrative syllogisms, this amounts to the rejection of the Aristotelian doctrine of real definitions. As we noted earlier, Arnauld is explicit in rejecting such definitions,[51] citing Bacon as his authority.[52] Moreover, Arnauld, like Bacon, sees the Aristotelian method as proceeding by a sort of induction by simple enumeration, which is in general an unsafe method for discovering regularities in things, since it does not exclude negative instances.[53] The only exception occurs where the induction is complete, but in general this condition remains unfulfilled.[54] The Cartesians likewise rejected all attempts of the sort proposed by Aristotle to explain natural phenomena teleologically, by reference to forms as the active souls of things. Thus, Descartes proposes to explain the workings of the heart and of the sense organs mechanically and without reference to a 'vegetative or sensitive soul.'[55]

What I want now to do is to examine the position of Aristotle in detail. The aim is to show that the criticisms shared by empiricists and rationalists of Aristotle's cognitive goals and of his method that we have just examined are indeed right on. To be sure, the interpretation that I shall offer is no doubt one that will not elicit universal consent. But one of the things that can be said in its favour is that upon it the criticisms advanced by the defenders of the new science in the early modern period do in fact hit home.

II: Aristotelian Science: Aristotelian Methods

Aristotle was perhaps the greatest philosophic mind that the world has seen – though if both the empiricists and the rationalists are correct in their criticisms, he was also profoundly wrong. It is important, I think, to be clear that the defenders of the new science were after a real target, not a straw man. On the one hand, there was the Aristotle of the textbook tradition that Bacon, Descartes, Locke, and so on, had studied in the schools. Thus, Descartes had studied from Eustace of St Paul's *Summa philosophiae quadripartia*, while Locke commented negatively on the physics of Franco Burgersdijck. The new scientists were reacting to this tradition. On the other hand, there was the real Aristotle. It is the latter upon which I shall concentrate, to establish that the target was indeed legitimate, and not merely the artefact of a bastard tradition in late scholastic textbooks. But I shall also suggest that the textbook tradition did not misrepresent this Aristotle, that the picture one could get from the textbooks was not inaccurate.[56]

1) *The Metaphysics of Explanation*

In the final analysis, Aristotle's explanation scheme is of a piece with that of Plato in the *Phaedo*, as Vlastos[57] and Turnbull[58] have argued. What is given in sense are appearances of things. These appearances are, so far as sense is concerned, separable. There is nothing about the sensible appearances, as such, to explain why one is followed by such and such rather than so and so – why, for example, Socrates' imprisonment is followed by his taking the hemlock rather than his going to Thebes. To explain the sequence is to show why one appearance *must* be followed by another; it requires the discovery of a necessary connection. This necessary connection is provided by an underlying activity that determines the sequence or form of sensible events.[59] This activity itself, however, is not given in sense experience. The entity that explains is thus not an entity that we know by sense; it is outside the realm of sense experience.

The activity that explains is understood in purposive terms. 'Nature, like mind, always does whatever it does for the sake of something, which something is its end' (*De anima*, 415b16–17).[60] Thus, the basic thrust of Aristotle's explanation scheme is, like Plato's, teleological, except that it is a teleology which now includes not only the conscious purposive activities of people but also many other sorts of change from those of animal and plants, including their growth, to those of what we should call inanimate objects.[61] Aristotle extends the Platonic analogy to art to include the spontaneous changes of things, and takes the latter to be as goal-directed as the former, and to be explainable by reference to this goal. For, as Aristotle says,[62] 'the same result as is produced by art may occur spontaneously. Spontaneity, for instance, may bring about the restoration of health. The products of art, however, require the pre-existence of an efficient cause homogeneous with themselves, such as the statuary's art, which must necessarily precede the statue; for this cannot possibly be produced spontaneously. Art indeed consists in the conception of the result to be produced before its realization in the material' (*De partibus animalium*, 640a28–32).

Spontaneous growth is caused not by an external *telos* but by an internal *telos*, in which the entity develops so as to instantiate, or instantiate as best it can, a form.[63] Explanation consists of showing how a part or a process contributes to, or permits, the realization of a goal, the relevant goal being the instantiation of a form, the essential characteristics of the thing in question. Aristotle continues:

The fittest mode, then, of treatment is to say, a man has such and such parts, because the conception [form] of a man includes their presence, and because they are necessary conditions of his existence, or, if we cannot quite say this, which would be best of all, then the next thing to it, that it is quite impossible for him to exist without them, or, at any rate, that it is better for him that they should be there; and their existence involves the existence of other antecedents. This we should say, because man is an animal with such and such characters, therefore is the process of development necessarily such as it is; and therefore is it accomplished in such and such an order, this part being formed first, that next, and so on in succession; and after a like fashion should we explain the evolution of all other works of nature. (*Part. an.*, 640a34–640b4)

Thus, we should explain the process by which the backbone gets articulated in the womb by referring to the form of a fully developed man. Man is the sort of creature whose way of life requires that he have the capacity to bend, and so, once the existence of humans is given, including their essential form, then *that* explains the process that produces that form. Empedocles got things the wrong way round when he tried to explain our having an articulated backbone by reference to the accident of the backbone being broken up as the foetus gets contorted in the womb (640a17–23).

The *nature* of a thing is the form that constitutes the final cause; it can also act as an efficient cause (*Part. an.*, 641a25–8). In addition, the nature of an animal is a certain part of its soul (641a28, 641b9–10), and the soul, too, can act, we are told elsewhere, not only as a final but also as an efficient cause (*De an.*, 415b8–12, 21–8). It is, for example, the efficient cause of orderly growth and development of organs that serve essential functions in the organism (416a5) and also the efficient cause of locomotion and of sensation (415b22–5). The nature or soul is also *dynamic*; it is an *active capacity*, *ability*, or *power* or *potency*. Dunamis is defined in the *Metaphysics* as 'a source of movement or change, which is in another thing than the thing moved or in the same thing *qua* other; e.g. the art of building is a potency which is not in the thing built, while the art of healing, which is a potency, may be in the man healed, but not in him *qua* healed' (*Metaph.*, 1019a15–18); *dunamis* or potency is, as the example of an art makes clear, an *active* potency. As such, it is the efficient cause of changes, as elsewhere sculpting is said to be the efficient cause of a statue (*Physics*, 195a5–8). *Dunamis* is the *nature* as the dynamic source of motion, where *nature*, as defined in the *Physics*, is an internal origin of change or stability (192b21–3). Artefacts are subject to change by external influences, but natural objects have an internal origin of change, which is their nature. The nature of a plant is a *dunamis* for assimilating nutriment and using it to grow to the pattern or form of the species and maintain itself there. A plant's nature is a kind of soul; the nature or soul of an animal includes the nutritive *dunamis* along with the *dunamis* for perceiving, as is clear from, for example, *De anima*, Book II, chapter 3, which opens by listing the *dunamis* of the soul as the nutritive, desiderative, perceptive, locomotive, and ratiocinative (*De an.*, 414a29–32). The chapter closes by asserting that the most appropriate account of the soul will be an account of each of these (415a12–13), which, elsewhere in the *De anima*, are described as *parts* of the soul (413b7, 27–32). The soul thus just is this set of *dunamis*. The dynamic soul is thus a set of active potencies which constitute the nature of the thing, in one sense of *nature* (*Ph.*, 192b21–3). This dynamic nature brings it about that the object has the form that the potency aims at actualizing, where this form is *nature* in another sense: 'What is potentially flesh or bone has not yet received its own "nature," and does not yet exist "by nature," until it receives the form specified in the definition, which we name in defining what flesh and bone is. Thus in [this] second sense of "[nature]" it would be the shape or form (not separable except in statement) of things which have in themselves a source of motion. (The combination of the two, e.g. a man, is not "nature," but "by nature" or "natural") ... We also speak of a thing's nature as being exhibited in the process of growth by which its nature is attained ... What grows *qua* growing grows from something into something. Into what then does it grow? Not into that from which it arose but into that to which it tends' (193b1–7, 13–19). One can find these points coming together in the *Generation of Animals*, Book II, chapter 4:[64]

as the products of art are made by means of the tools of the artist, or to put it more truly by means of their movement, and this is the activity of the art, and the art is the form of what is made in something else, so is it with the power [*dunamis*] of the nutritive soul. As later on in the case of mature animals and plants this soul causes growth from the nutriment, using heat and cold as its tools (for in these is the movement of the soul and each comes into being in accordance with a certain formula), so also from the beginning does it form the product of nature. For the material by which this latter grows is the same as that from which it is constituted at first; consequently also the power [*dunamis*] which acts upon it is identical with that at the beginning (but greater than it); thus if it [that is, the *dunamis*] is the nutritive soul, it is also the generative soul, and this [that is, the soul] is the *nature* of every organism, existing in all animals and plants. But the other parts of the soul exist in some living things and not in others. (*Gen. an.*, 740b25–741a3, italics added)

And, be it noted, not only natural objects but also the stuffs of which they and artefacts are made have an internal origin of change, which is their nature. Thus, as the nature of a plant is a *dunamis* for assimilating nutriment and using it to grow into, and maintain itself in, the pattern or form of the species, so the nature of the element earth is a *dunamis* for downward motion (*De caelo*,

310b16–20), and, in fact, in its natural motion moves uniformly in a straight line towards the centre of the universe (*Cael.*, 311b19–20), which is its goal (296b16) and which therefore is its natural place. This natural place is indeed its form, and the natural *dunamis* of earth tends to maintain it in this form – that is, at the centre – once it has arrived there (296b30–297a8; cf *Ph.*, 208b8ff). The dynamic nature of the thing generates the form that the thing becomes; in fact, the dynamic nature *is* the form of the thing, the structured striving that creates the visible or sensible form that the thing presents to us in experience.[65]

We thus see that the traditional explanation schemata of Aristotle include the teleology – explanation in terms of final causes – that was rejected by both the empiricists and the rationalists concerned to replace this old 'science' by the new. Thus, Arnauld objects to making bodies '*knowers*, as is done only too often in the common philosophy, where they claim that brutes know, and that plants choose their nutrition, and that all heavy things seek the center of the earth as their resting place, which they could not do without knowledge (italics added).'[66]

Not surprisingly, the doctrine of explanation in terms of natures connects up with Aristotle's logic, and also with his metaphysics of change. In the *Categories* we are told that the individual things about us – the individual men, trees, statues, rocks, dogs, plants, or planets would all be included – are what exist; 'they are the entities which underlie everything else' (*Cat.* 2b16). These primary substances are 'neither predicable of a subject nor present in a subject' (2a11). There are also 'secondary substances,' that is, the species and genera in which primary substances are included (2a15–17); these 'are predicable of a subject, and are never present in a subject' (1a20). All the categories other than substance are 'present in a subject' (1a24). It is, then, primary substances that undergo processes of development and growth (4a10–12). It is their form which is predicated of them. But this form, we know, also explains. What it explains are the characteristics that things – that is, substances – come to have; these characteristics are the things that are present in substances. Thus the active form or nature that is predicated of a substance explains the characteristics that are present in the substance.[67]

When change occurs, we have a coming-to-be. The *Physics* begins with an analysis of this. When we speak of something coming-to-be, two different sorts of things are said: either 'the man becomes musical' or 'the not-musical becomes musical' (*Ph.*, 189b35). In the former case, that which becomes *endures through the change*, while in the latter case it passes away (190a9). But whether we have 'X becomes Y' or 'not-Y becomes Y,' what we have as the process which makes either, or, rather, both, true is 'X-which-is-not-Y becomes X-which-is-Y.' The product of such a process contains two ingredients both of which are essential to the analysis of change, namely, a subject of change which is a substance (190a14, 33) and a characteristic that is present in the substance (190a33); but the change also presupposes a *privation*, namely, the absence of the characteristic which comes to be (190a4–5). The substance, before the change, was numerically one, but included two distinguishable elements, namely, the substance *per se* on the one hand, and, on the other, that which was to be replaced by its opposite (190b24). Change thus presupposes three things: substance, characteristic, and privation. In the next chapter (Book I, chapter 8), Aristotle shows how earlier thinkers were confused about change, and therefore baffled by it. In particular, that which *is* seemingly cannot be got from that which *is* nor from that which *is not*. Aristotle solves the problem by making it, in a way, both. In the *first* place, nothing comes into being simply from not being. Something does comes into being from its privation which, in itself, is indeed not-being, but it does not come into being from the privation as such. Rather, it comes into being from the privation only incidentally; it comes into being not from a bare privation but only from a privation in a substance. But also, nothing comes into being from being *simpliciter*; it comes into being from something which to be sure is, but not from it as it is but from it as it is not, that is, as not the particular thing that comes to be (191a23–191b26). But *secondly*, with a reference forward to the philosophical lexicon which is included as Book V of the *Metaphysics*, this solution of the *Physics* to the problem of change may also be expressed in terms of the doctrine of potentiality and actuality (*Ph.*, 191b27–9). A thing comes from that which is it potentially but not actually (*Metaph.*, 1017a35–1017b1). This links the analysis of change with which the *Physics* opens to the doctrine that change is explained by art, as 'we say the Hermes is in the stone' (*Metaph.*, 1017b7), or by the active form or power which is predicated of the substance undergoing the change and which moves that substance spontaneously and naturally, as 'we say of that which is not yet ripe that it is corn' (1017b8). And so we find that potentiality is itself explained a few chapters later in the *Metaphysics*, as we have noted, as 'a source of change in another thing or in the same *qua* other' (1019a15, 1020a1–2).

This may be compared with the metaphysics of explanation developed by Plato in the *Phaedo*. (1) Both have sensible characteristics that are present in things. Change consists in such characteristics coming to be present in things where previously they were not present in them. (2) That which the characteristics are present in is *active*. (3) That activity explains teleologically the pattern of sensible characteristics that things present.

But there are *two differences*, which are, however, interrelated. The striving of Aristotle's substances is a *formed* striving. The striving of substances is goal-directed in both Plato and Aristotle, but in Aristotle the striving is itself intrinsically informed by the goal at which it aims whereas in Plato the goal which the striving aims to achieve is determined by the forms which are *separate* from the striving. In Plato, the form that determines the goal is separate from the soul which strives, so that the activity of the latter is, if you wish, in itself unformed and *bare*. Thus, the two differences: (a) while for Aristotle the activity which explains has an intrinsic direction, for Plato it is bare; and (b) while for Aristotle the form which determines the goal of the activity is intrinsic, not separate, for Plato the form is separate from the striving soul. *Where Plato differs from Aristotle there are serious flaws in the teleological metaphysics of explanation, and Aristotle's moves eliminate those flaws to make that metaphysics both plausible and consistent.*

For Aristotle, the active entity is the (primary) substance. It alone exists because it alone can exist separately or independently (*Metaph.*, 1028a34); of the things in the other categories, 'none of them is either self-subsistent or capable of being separated from substance' (1028a23). Now, there is no form apart from substance (1039b18), nor is there substance apart from form (1029a7). Substance as a category is *not* a category of contraries, nor does it admit of variation in degree (*Cat.*, 3b21–4a9). Individual substances thus cannot lose their form and yet remain the individual they are; or, since the form is the active soul, 'soul is inseparable from its body' (*De an.*, 413a3). Thus, if Socrates loses his humanity he ceases to be Socrates (*De generatione et corruptione*, 319b16–21). Aristotle contrasts the sort of unity which a substance has with the sort of unity which a heap has (*Metaph.*, 1031a20); in the latter case, the thing and its essence are two, as is not the case for substance: A heap is formed by mere juxtaposition (1043b5). But the unity of a substance is more than juxtaposition; it involves an inseparability of the parts of its unity, or, conversely, if the parts are separated then the substance ceases to be. And, in fact, the parts themselves change their nature if they cease to be parts of the substantial unity of which they were parts: 'no part of a dead body, such I mean as its eye or its hand, is really an eye or a hand' (*Part. an.*, 641a2–4). *There is thus necessity and inseparability in the unity of a substance.* In particular, there is the unity of the substance and its form, that is, the necessary connection between, or inseparability of, the substance and the form which determines the goal of its activity.

Aristotle also contrasts the unity of a conjunction with that in which something is attributed to a substance (*An. post.*, 93b35). The various senses make us aware of the sensible characteristics of things. Each sense has its own object, and there are certain common sensibles; but the substance of which these are the characteristics is not given in sense, it is only the incidental object of sense (*De an.*, 418a7–26). So far as sense is concerned, then, what is presented to one are separate events. This is all that animals have; men have more, namely, the connection that comes through reason: 'The animals other than men live by appearances and memories, and have but little of connected experience; but the human race lives also by art and reasonings' (*Metaph.*, 980a 25–7). The point is that the senses by themselves do not give us knowledge of the substance which is the cause and support of the characteristics that are given in sense: 'we do not regard any of the senses as wisdom; yet surely these give the most authoritative knowledge of particulars. But they do not tell us the "why" of anything – e.g. why fire is hot; they only say that it is hot' (981b10–13).

Wisdom is defined by Aristotle as founded on the understanding of things' causes rather than on mere recognition or recording of the sensory properties of things (*Metaph.*, 981b27). Elsewhere he tells us that, in a way, knowledge and opinion have the same object, but in another way in the two it differs fundamentally.

Knowledge is the apprehension of, e.g. the attribute 'animal' as incapable of being otherwise, opinion the apprehension of 'animal' as capable of being otherwise – e.g. the apprehension that animal is an element in the essential nature of man is knowledge; the apprehension of animal as predicable of man but not as an element in man's essential nature is opinion: man is the subject in both judgments, but the mode of inherence is different. (*An. post.*, 89a33–8)

Sense, therefore, can yield only opinion; knowledge is reserved to the intellect which grasps the *form* of the object known (*De an.*, 429a15). This form or nature is the active cause of the thing or substance having present in it the characteristics it has; it is that which actively creates and maintains the characteristics not as a heap or a separable and contingent conjunction but as genuinely unified in the necessary structure of the substance.

On Plato's view, too, it was an active entity that *explained* the characteristics of things, and united them into a structured whole. On Plato's view, too, sense could not grasp that unity – for it the characteristics formed a mere conjunction of separable parts; it was reserved to the intellect, reason, to grasp the entities that provided the real explanation by constituting a non-contingent link tying the separable parts together into a genuine unity.

Now, the point is that contingency, or, what is the same, separability, must be explained. It is explained, on Plato's account, by appeal to a soul striving for a form. This striving constitutes a necessary connection that ties the ostensibly separable parts into what is in fact inseparable unity. But now, *the forms are separate from souls*. The connection is therefore contingent, and in need of explanation. That is, if the form is separate from the soul, then it must be explained why the soul strives after the form it does rather than some other form. Why, for example, if the forms are separate from souls, should Socrates not strive after doggie justice rather than human justice? Surely that needs explaining just as much as the sequence of sensible events leading to his drinking hemlock rather than running off to Boeotia.

Separable things, upon Plato's view as upon Aristotle's, require explanation. This is why the appearances given in sense require explanation. The separation is explained, again upon the views of both Plato and Aristotle, by eliminating it. It is eliminated by reference to a causal process of production in which the apparently separate entities come to be embodied in an underlying and unifying matter of some sort. For Plato such production proceeds by reference to a separate form. Here, though, is a separation. Since it is a separation, it requires explanation. Such explanation is by production. So we must say that the form is produced, and that this production yields the embodiment of the form in a matter. But upon the Platonic scheme, any production proceeds by reference to a separate form. So there will be a second form by reference to which the first is produced. But since it is separate, it, too, will require an explanation and this will be by reference to yet another separate form. And so to infinity. Thus, in order to produce the concrete thing the appearances of which are given in sense, there is an unterminating regress of production of necessary conditions.[68] Since all of the necessary conditions are never produced, neither is the concrete thing. Thus, if one grants the Platonic view that the forms are separate, nothing is ever produced, nor, therefore, is anything ever explained.

Note that the same sort of regress will occur if the matter must always be produced.

We find the regress argument in the following passage.

Since anything which is produced is produced by something ... and from something (and let this be taken to be not the privation but the matter ...), and since something is produced (and this is either a sphere or a circle or whatever else it may chance to be), just as we do not make the substratum (the brass), so we do not make the sphere, except incidentally, because the bronze sphere is a sphere and we make the former. For to make a 'this' is to make a 'this' out of the substratum in the full sense of the word [that is, including form as well as matter (*Metaph.*, 1029a3)] ... If, then, we also make the substratum itself, clearly we shall make it in the same way, and the process of making will regress to infinity. Obviously then the form also, or whatever we ought to call the shape present in the sensible thing, is not produced, nor is there any production of it, nor is the essence produced; for this is that which is made to be in something else either by art or by nature or by some faculty [that is, spontaneously]. But that there is a *bronze sphere*, this we make. (*Metaph.*, 1033a23–1033b9)

Aristotle does not hesitate to draw the relevant anti-Platonic conclusion that this argument entails that the forms are not separate.

Is there then a sphere apart from the individual spheres or a house apart from the bricks? Rather we may say that no 'this' would ever have been coming to be, if this had been so, but that the 'form' means the 'such,' and is not a 'this' – a definite [= separate or independent] thing; but the artist makes or the father begets, a 'such' out of a 'this'; and when it has been begotten, it is a 'this such.' And the whole 'this,' Callias or Socrates, is analogous to 'this brazen sphere,' but man and animal to 'brazen sphere' in general. Obviously, then, the cause which consists of the Forms (taken in the sense in which some maintain the existence of the Forms, i.e. if they are something apart from the individuals) is useless, and the Forms need not, for this reason at least, be self-subsistent substances. (*Metaph.*, 1033b14–29)

For our purposes this is the central anti-Platonist argument, though Aristotle advances others, too. For example, if, as the previous argument establishes, the form is not separate, then in the case of production that occurs spontaneously or by nature, the form as *pattern* is not needed: 'it is quite unnecessary to set up a Form as a pattern ... the begotten is adequate to the making of the product' (1034a2–4).

Since forms are never produced, it follows that coming-to-be and passing-away does not apply to forms. They are, as Ross points out,[69] therefore immortal, at least in the sense that they are outside of time. Equally, matter in the sense of potency, that is, the activity that produces, neither comes to be nor passes away; it, too, is timeless or immortal,[70] as Aristotle also says elsewhere:

The matter comes to be and ceases in one sense, while in another it does not. As that which contains the privation, it ceases to be in its own nature, for what ceases to be – the privation – is contained within it. But as potentiality it does not cease to be in its own nature, but is necessarily outside the sphere of becoming and ceasing to be. For if it came to be, something must have existed as a primary substratum from which it should come and which should persist in it; but this is its own special nature, so that it will be before coming to be ... And if it ceases to be it will pass into that at the last, so it will have ceased to be before ceasing to be. (*Ph.*, 192a25–33)

This is of a piece with Plato's argument in the *Phaedo* that since the very being of soul is the activity or life which accounts for change, that activity or life cannot come to be or pass away (*Phd.*, 103a–c), and that therefore the soul as life is in itself deathless (105e) and indestructible (106c), that is, immortal (107c).

This, however, does not imply a later life in the underworld, as Socrates immediately suggests (107a) and takes up in the myth that now follows (107d-115a)
the dialectical – that is, philosophical – discussion; for the tale of the underworld is one where things, even souls, change. All that follows is that, as the
forms are timeless and changeless (78d), so the soul is also, as Socrates earlier
indicates, in itself timeless and changeless (79b–c). Since the forms are immortal (79d), so the soul is thus immortal (80a–b); and since for the Greeks the
mark of divinity is immortality, the soul is like the forms in its divinity (80a–b).
Although there is much mythology involving gods in the *Phaedo*, in the end,
from the philosopher's perspective, it is the forms which are the true gods, and
the soul is immortal, not in having everlasting existence in an underworld, as
the myths have it, but in existing timelessly.

This is not our present theme, however, so let us return to our main concerns.

2) *The Logic of Explanation in Aristotle*

Let us take stock.

Plato and Aristotle both accept that the entities given in sense experience are,
so far as sense experience is concerned, separable and only contingently connected, or, better, associated. For Plato these events are explained by reference
to entities not given in sense experience, namely, souls and forms. Specifically,
the sensible events come to be in the order they do because they are caused by a
soul to be in it through its striving after certain forms; this striving of the soul
after the forms provides the non-contingent connection which explains the
patterns of the sensible events. For Aristotle, too, the separable events are
explained by reference to entities not given in sense experience, namely, a substance. A substance is active, like a Platonic soul, but this activity is in the
direction determined by the form of the substance. The form of a substance is
an inseparable part of it; it is a matter of metaphysical necessity that the active
substance is nothing more than a Platonic soul with the form after which it is
striving collapsed into it. That structural move by Aristotle removes a major
defect of the Platonic metaphysics of explanation. For the rest, the Aristotelian
explanation pattern remains the same as the Platonic: the separable events of
sense experience are explained by appeal to the goal-directed activity of an
entity outside of sense experience. The active substance produces in accordance
with its form the sensible characteristics that come to be present in it. This
active substance constitutes the necessary connection that ties the only apparently separable characteristics into a unity that accounts, as sense experience
alone cannot, for those characteristics' being ordered as they are.

We should be clear: this explanation scheme is anthropomorphic. It is, to be sure, not *crudely* anthropomorphic; Aristotle is not committed to saying, for example, that stones deliberate, and, in fact, he explicitly denies that (*Ph.*, 199a20ff). None the less, he takes *activity* to be the *basic explanatory category*, and the model for activity is clearly volition and the goal-directed activity of men. As it has been put, '[a]ppeal to a *dunamis* is fundamental, for Aristotle, in the way that appeal to a law of nature is taken to be fundamental by modern philosophers of science such as Carl Hempel'[71] – or, for our purposes, Bacon or Descartes. Here the idea of the new science is that one explains events by subsuming them under a covering law or regularity. The regularity is a *general fact* about events in the world of sense experience; it constitutes a timeless pattern among those entities. Thus, where Aristotle explains the events of the world of sense experience by appeal to a *timeless entity outside* the world of sense experience, Bacon and Descartes (and Hempel) explain the events of the world of sense experience by appeal to a *timeless pattern in* the world of sense experience.[72] For the empiricists, since the activity which is explanatory for Aristotle is in itself totally outside the realm of sense experience, explanations in terms of it are vacuous; but such 'explanations' were also held by the rationalists to be vacuous. Indeed, the whole language of capacities is for the empiricists and the rationalists alike to be so reconstructed that it turns out, as Molière also said, that capacities – dormitive powers or whatever – never in themselves explain, and that they seem to only because laws are associated with their use. It has been said that 'No objection ought to be raised to the idea that a capacity can be explanatory.'[73] But both the empiricists and the rationalists raise such an objection, and it is in fact the crucial point of disagreement between them and Aristotle.

To see this disagreement more exactly, let us consider in more detail the patterns Aristotle ascribes to explanation. These are, of course, syllogistic. One has scientific knowledge when one has a syllogism which is also a demonstration (*An. post.*, 71b17–18). A demonstration must yield certainty, for that is what scientific knowledge is,[74] as opposed to opinion, which is concerned with 'that which may be true or false, and can be otherwise' (89a3), and to false claims to such knowledge like those made by the Sophists: 'We suppose ourselves to possess unqualified scientific knowledge of a thing, as opposed to knowing it in the accidental way in which a Sophist knows, when we think that we know the cause on which the fact depends, as the cause of the fact and no other, and, further, that the fact could not be other than it is ... the proper object of unqualified scientific knowledge is something which cannot be other than it is' (71b8–16). So the premises of a demonstrative syllogism are necessary, and this necessity, Aristotle goes on to explain, is a matter of the *form* or *essence*[75] of a thing.

Demonstrative knowledge must rest on necessary basic truths; for the object of scientific

knowledge cannot be other than it is. Now attributes attaching essentially to their subjects attach necessarily to them: for essential attributes are either elements in the essential nature of their subjects, or contain their subjects as elements in their own essential
nature ... It follows from this that premises of the demonstrative syllogism must be connexions essential in the sense explained: for all attributes must inhere essentially or be
accidental, and accidental attributes are not necessary to their subjects. (*An. post.*,
74b5–12)

This form or essence appears as the *middle term* of the demonstrative syllogism: 'Our knowledge of any attribute's connexion with a subject is accidental
unless we know that connexion through the middle term in virtue of which it
inheres ...' (*An. post.*, 76a3–5).

The most perfect syllogistic form is the first of the first figure (*Ph.*, 79a16).
Thus, a scientific explanation of why *C* is *A* looks like this:

(S)
$$B \text{ is } A$$
$$C \text{ is } B$$
$$\therefore C \text{ is } A.$$

Now, 'to know its essential nature is, as we said, the same as to know the cause
of a thing's existence ...' (93a4–5). Hence, the middle term of the syllogism (S),
if it is a demonstrative and scientific syllogism, will be the cause of *C*'s being *A*.
An example which Aristotle gives is this (94a36–94b7): The question is 'Why
did the Athenians become involved in the Persian War?' and this is taken to
mean, 'what cause originated the waging of war against the Athenians.' To this
the answer is, 'Because they raided Sardis with the Eretrians.' Aristotle lets *A* =
war, *B* = unprovoked raiding, *C* = the Athenians. 'Then B, unprovoked raiding,
is true of C, the Athenians, and A is true of B, since men make war on the unjust
aggressor. So A, having war waged upon them is true of B, the initial aggressors, and B is true of C, the Athenians, who were the aggressors. Hence here too
the cause – in this case the efficient cause – is the middle term.'

Conversely, where the middle is not a nature or form or essence, a syllogism
will not explain. For example, if all non-twinkling lights are near, and all planets are non-twinkling lights, we can deduce that the planets are near; but we will
not have *explained* the fact of their being near. The reason is that the middle
term, non-twinkling lights, is not an essence. (No negative term could be an
essence; Plato's argument in the *Phaedo* forces this conclusion.) On the other
hand, if we reverse it to say that all planets are near, and all near things are non-
twinkling, we not only deduce that all planets are non-twinkling, but also

explain it, since nearness is the essence that accounts for non-twinkling (*An. post.*, 78a29–78b11, 98b4–24).

When it comes to explanations in which the conclusion that *C* is *A* involves the attribution of sensible characteristics *A* to *C*, then that fact must be caused by the essence that appears as the middle term; otherwise that fact won't have the necessity that must attach to the conclusion of a demonstrative syllogism (*An. post.*, 75b24). In fact, since the causes which explain are dynamic powers or faculties – for example, the nutritive faculty – the facts to be explained are *patterns or regularities* in observable behaviour: for example, if a thing *x* has a nutritive faculty or soul then it is true that whenever *x* eats food then *x* digests it (*De an.*, 416a20–416b6). Now, a generalization such as this is 'true in every instance' (*An. post.*, 73a27) and this means that there is 'no limitation in respect of time' (*An. pr.*, 34b7). Russell once wrote that, 'one may call a propositional function necessary, when it is always true; possible when it is sometimes true; impossible when it is never true.'[76] Aristotle has this same idea. For him, if some pattern is in fact a real regularity, that is, is always true, then it is necessary. Thus, he tells us that 'if, as we have said, that is possible which does not involve an impossibility, it cannot be true to say that a thing is possible but will never be ...' (*Metaph.*, 1047b3–6), and that 'that which is capable of not existing is not eternal' (108b23). That for which it is possible that it not exist is not eternal; hence, the eternal, that which is always true, is that for which it is not possible that it not exist. But it is a common principle of modal logic that if it is not possible that not *p*, then it is necessary that *p*; and Aristotle accepts this principle when he tells us that 'we say that that which cannot be otherwise is necessarily as it is' (1015a33). Hence, that which is always true is necessary.[77] Aristotle himself says just this: 'a thing is eternal if its "being" is necessary; and if it is eternal, its "being" is necessary. And if, therefore, the "coming-to-be" of a thing is necessary its "coming-to-be" is eternal; and if eternal, necessary' (*Gen. corr.*, 338a1–4; cf *Cael.*, 281b26). Thus, that a pattern of sensible appearances is in fact truly a regularity implies that it is necessary.

Now, this would seem to imply the equivalence of regularity and causal necessity[78] – a thesis that is, in one way at least, precisely the thesis about causation endorsed by Hume.[79] Aristotle is, however, no Humean. For, according to the regularity account of causation, causal necessity *just is* regularity, whereas according to Aristotle, while a general fact holding or being a regularity is due to a *necessity which lies behind it*, 'necessity is *that because of which* a thing cannot be otherwise' (*Metaph.*, 1015b2; italics added).[80] This necessity is *the necessity of the essences which cause things to behave in regular ways.*[81] For Aristotle the covering law pattern of explanation of the new science is not enough.[82] Rather, as others have also pointed out, it is the underlying nature or essence which both explains the observable events and accounts for why the *de facto* regularity of sense is a *law*, that is, why it is a regularity that holds of *necessity*.[83]

Consider a (somewhat simplified) example. Let *C* be some object or, rather, substance; call it Corsicus. Now suppose that Corsicus is such that whenever it is put in water it dissolves. Let *A* be this pattern of sensible events. Then *C* is *A*. The middle term that explains *C* being *A* is the capacity (power, nature, essence, form, soul) of *solubility*. Let *B* be this capacity. Then the explanatory syllogism (S) is:

> Whatever is soluble is such that whenever it is put in water it dissolves (*B* is *A*)
>
> (S')   Corsicus is soluble (*C* is *B*)
> _____
>
> ∴. Corsicus is such that whenever it is put in water it dissolves (*C* is *A*).

Solubility is an active power, like the nutritive power, the activity of which accounts for the pattern of behaviour described in the conclusion. As the nutritive power is one of a set of powers, including the appetitive and the ratiocinative, which make up the soul or form or nature of human beings, so also solubility will be one of the set of powers that constitute the set that makes up the full nature or essence of Corsicus; for example, the nature of Corsicus may be sugar. Thus, the second premise of (S') is necessary because the nature or essence is inseparable from the substance Corsicus. Provided that the major premise of (S') is also necessary, then the conclusion will be necessary, and the syllogism will establish a necessary connection among the events in the pattern mentioned in the conclusion. The necessity of the law derives from the form or essence; thus, explanation of a pattern of behaviour is not obtained by fitting that pattern into a more complex pattern, as in the new science, but by *redescribing* the object whose behaviour is being explained.[84]

Now, as Molière made clear, for the defenders of the new sciences, the major premise of (S') is indeed necessary, but only trivially so, since it is true by nominal definition. For that reason, for the critics of Aristotle (S') is not more explanatory than

> Whatever is a bachelor is an unmarried male
>
> (S")   Callias is a bachelor
> _____
>
> ∴. Callias is an unmarried male.

(S″) doesn't explain since the minor premise merely *restates* what the conclusion says. As for the major premise in (S″), that is in fact redundant; for the rule of language

'Bachelor' is short for 'unmarried male'

which licenses its assertion as a necessary truth equally licenses the rewriting of the minor as the conclusion and therefore also licenses the inference of the conclusion from the minor premise alone. It is for this reason that Arnauld excluded attempts to explain by appeal to substantial forms:[85] he gives it as a rule of method 'not to multiply beings without necessity, as is so often done in ordinary philosophy, as when, for example, people do not agree that the diverse arrangements and configurations of the parts of matter suffice to make a stone, some gold, some lead, some fire or some water, unless there is in addition a substantial form of stone, gold, lead, fire, or water, really distinct from all conceivable arrangements and configurations of the parts of matter.'[86]

On the other hand, for Aristotle (S′) *is* explanatory. The major premise cannot, therefore, be true by definition. It must be a substantive truth. Thus, for the empiricist, capacities are analysable in the sense that the term 'soluble' in the major premise of (S′) can be defined by the right hand side using only (the principles of logic and) terms that refer to observable characteristics of things; while for Aristotle, since the major premise of (S′) cannot be a definitional truth, capacities (dispositions, tendencies, powers) cannot be analysed into (patterns of) observable characteristics of things. *Aristotelian explanations are in terms of the unanalysable active dispositions and powers of things.*

If the major premise of (S′) is not definitional, it is also not contingent; Aristotle holds, as we have seen, that the premises of a scientific syllogism are *necessary*. The empiricist divides propositions, as Hume does,[87] into two mutually exclusive and jointly exhaustive classes. There are, on the one hand, those propositions which state relations of ideas, and which are therefore necessary in the sense of being tautological, devoid of factual import, and, on the other hand, those propositions which state matters of observable fact and which are therefore contingent. If the former propositions are, as they are often called, analytic, and the latter synthetic, then for Aristotle, in contrast to the empiricist, there is a third category, that of propositions which are both synthetic and necessary. (Kant's term, 'synthetic a priori,' is not quite appropriate for Aristotle, but the point is much the same.)[88]

Let us suppose for the moment that the essence of Corsicus is exhausted by the single disposition of solubility (as the essence of the element earth is exhausted by its disposition to move in a straight line towards the centre of the universe). Soluble things will have a variety of sensible characteristics, such as colour, as well as the sensible characteristic of dissolving in water. Suppose that all soluble things in the Aegean are, simply as a matter of accident, blue, while soluble things elsewhere are a variety of other colours. An inhabitant of the Aegean who has not travelled might be tempted through observation to list *two* laws about such things:

(P$_1$)   Whenever in water then dissolves

(P$_2$)   Always blue

and might be tempted to offer the following definition for the kind of thing Corsicus is:

(1) $K^1 =_{Df} P_1 \& P_2$

In fact, however, only (P$_1$) is a lawful regularity, and the *correct* definition is

(2) $K =_{Df} P_1$

As *nominal* definitions, of course, both (1) and (2) are equally correct. But they are not equally correct as attempts to reflect the dynamic essence the activity of which explains the observable properties of the things we are trying to define. *The essence of a substance provides a criterion for determining the correctness of a definition in terms of observable characteristics of the kind to which the substance belongs.* Statements like the major premise of (S′):

> Whatever is soluble is such that whenever it is put into water it dissolves

make just this sort of connection between the essence or nature as an unanalysable disposition, on the one hand, and, on the other, the essence or nature as defined by observable characteristics. Since such statements pick out one nominal definition among many as *the* correct definition, such statements have reasonably been called *real definitions*. Like nominal definitions, statements of real definition are necessary; but unlike nominal definitions, statements of real definition are synthetic, substantive truths with genuine explanatory power.

The nature or real definition of a thing determines what will happen to that thing in the appropriate circumstances. It also excludes. Thus, if it is (part of) the nature of a thing that it is soluble, then if that thing is in water it will *not* retain its shape, and, moreover, since this is true by virtue of the nature or essence which the thing necessarily has, it is *necessary* that the thing will *not* retain its shape. That property is *impossible*. Conversely, dissolving when in water is *not impossible*. But what is not impossible is *possible*. Hence, while it may in some sense be logically possible that some thing which is naturally soluble not dissolve when in water, none the less the latter is not *really possible* given the *nature* of the thing. The nature or real definition thus determines the *real possibilities* concerning what a thing can be. But a nature or essence is active. Thus, the notion of possibility in Aristotle is tied to the notion of activity: a thing being in a certain way is really possible only if the thing is actively disposed to be in that way.[89]

In the Aristotelian framework, then, one may look at the nature of a thing in two ways. There is the *dynamic nature* on the one hand. On the other hand, there is its *definitional nature*, the list of observable characteristics which are caused by the dynamic nature. It is the dynamic nature which determines the correctness of any proposed definition of the definitional nature.

Moreover, it is evident that *the real task in defining any natural kind does not end with observation; what is crucial is that one go beyond the level of sense experience to grasp the essence of the thing.* Thus, an account of what the name 'goat-stag' signifies does not qualify in Aristotle for the honorific title of *definition (horismos)* since goat-stags do not exist:

Since ... to define is to prove either a thing's essential nature or the meaning of its name, we may conclude that definition, if it in no sense proves essential nature, is a set of words signifying precisely what a name signifies. But that were a strange consequence: for (1) both what is not substance and what does not exist at all would be definable, since even non-existents can be signified by a name: (2) all sets of words or sentences would be definitions, since any kind of sentence could be given a name ... (3) no demonstration can prove that any particular thing: neither, therefore, do definitions, in addition to revealing the meaning of a name, also reveal that the name has *this* meaning. (*An. post.*, 92b26–35)

Thus, just as Plato's forms provided a standard of real meaning to judge the correctness of definitions, so too do *Aristotle's forms or natures provide a standard of real meaning to judge the correctness of definitions.*

We may conclude, then, subject to qualifications about what Aristotle calls violent motion to which we shall come directly, that in order to understand the observable behaviour of a thing – in order to explain it – one must give the essence (nature, form) of the thing. That is, this is so insofar as the observable behaviour *can* come to be understood; for some of the characteristics may simply be quite accidental, and therefore incapable of being explained.[90] However, while this gives an *understanding* of the *full behaviour* of a thing, insofar as it can be understood, a *fuller understanding* of that behaviour can yet be achieved.

The essence or nature of a thing is not a single power (save in the case of the most elementary things); it is, rather, a set of powers, as we have seen. One should add, perhaps, that it is not a *mere* set, in the sense of a heap or a conjunction (*An. post.*, 92a30), but a set with a structure that provides a real unity (*Metaph.*, 1037b25, 1045a8). It is through this unity and structure which the essence has that it *confers* unity and structure on the individual substance of which it is the essence (1052a33). For Aristotle, *the logical structure of an essence or form is made explicit when it is defined as a species in terms of genus and difference.* The question of 'how essential nature is revealed' is answered by 'definition' (*An. post.*, 90a36–7, 91a1), and definition of an essence consists in treating it as a species and giving its definition in terms of genus and specific difference (96b15ff). Thus, to use the standard example, the essence *man* is revealed in its definition as *rational animal*, where *animal* is the genus and *rational* is the specific difference.

The search for definitions in this sense is also in a sense the search for middle terms of syllogisms.[91] Thus, the definition of *man* as a *rational animal* is 'exhibited' in the demonstrative syllogism:

> All rational is animal (*B* is *A*)
> All man is rational (*C* is *B*)
>
> ∴ All man is animal (*C* is *A*).

This demonstration shows how an attribute 'attaches' to a given subject (*An. post.*, 91a2), but the conclusion cannot itself be man's essential nature, in the sense of fully revealing it, since it omits the middle term of the premises which mediates or connects the two, thereby functioning as an inseparable item in the logical structure of the essence of man (91a27–32). Essential definitions thus cannot be demonstrated (91b10). The point is, in fact, that to know the essential definition and to know the syllogism that supposedly demonstrates it are of a piece; to know the one is to know the other (93a17–19). The essential definition packs into a single sentence the contents of the syllogism which is to demonstrate it; or, equally, the syllogism unpacks that sentence. So one should say that the syllogism does not demonstrate but rather 'exhibits' the essential definition (93b18).[92] The point is that the logical structure of both is in effect equivalent, with a middle mediating between extremes according to the rules of class inclusion.

An essential definition is, as we have said, a real unity (*Metaph.*, 1037b25). It is therefore a necessary truth; but, again, it cannot be *merely* definitional since, quite clearly, it provides a standard for the correctness of our nominal definitions of the definitional natures of things. Essential definitions are thus statements of synthetic necessity. And equally, the premises and conclusion of the syllogisms that exhibit such definitions must be synthetic necessities. Thus, sensible characteristics present in things are explained by appeal to the dynamic nature which necessitates them, and they receive a fuller explanation when the internal structure of that necessitating essence is revealed in the premises of the syllogism (or series of syllogisms) that exhibits that structure. And in this fuller understanding one arrives at last at basic premises which are not only necessary but indemonstrable: 'the peculiar basic truths of each inhering attribute are indemonstrable' (*An. post.*, 76a17).

That explanation, and now fuller explanation, is in terms of synthetic necessary truths we have already stressed. Another point is worth emphasizing. Full understanding consists in exhibiting a structure of genera and specific differences. Thus, *to fully understand a thing is to grasp the ways in which, in its essence, it is the same as, and different from, other things.* It is to recognize the analogies and disanalogies between its behaviour, in its essentials, and the behaviour of other things.

Now, this being so, it is evident that it immediately implies that simile, and even metaphor, is fully a part of explanation. It is not at all surprising, then, that, as one author says, 'in order to carry out the study of "necessary" causes successfully, Aristotle is going to appeal to the ancient method of explanation by image and analogy ...':[93] that, if one may so speak, is the very essence of Aristotle's idea of explanation. This is in sharp contrast to the new science of both the empiricists and the rationalists. For the latter, as Bacon and Descartes both recognized, as we have seen, analogies and disanalogies among laws have an important place in the research process – these analogies are given by the principles of determinism and limited variety that are essential to the working of the logic of the experimental method, as both Bacon and Descartes recognized; but to note such analogies is *not*, as for Aristotle it was, to increase understanding of individual events. Rather, for empirical science such analogies *suggest specific hypotheses which, if they survive testing, can be used as laws in covering-law explanations of individual events, and which, being specific, provide explanatory detail and therefore understanding of individual events which is beyond what the suggestive but more abstract analogies could ever yield.*[94] To put it in a formula, for the new science, it is laws that explain, not relations of sameness and difference. In the end Aristotle does not distinguish science from poetry.

Before turning to the issue of how we know or grasp the forms or natures or essences of things, we must cash in a promissory note about explanation that was issued a short while ago.[95]

3) *Laws of Nature in Aristotle's Philosophy of Explanation*[96]

We have said that for Aristotle what happens *always* is *necessary*. That means that for Aristotle laws of nature are true generalizations, and as such are exceptionless. In fact, this is not Aristotle's position. The necessary, rather, is what happens *always or for the most part*. Thus, when discussing certain events which happen in nature and which can be explained by citing causes as middle terms of syllogisms so that these natural events are necessary, we are told that '[s]uch occurrences are universal (for they are, or come-to-be what they are, always and in every case); others again are not always what they are but only as a general rule; for instance, not every man can grow a beard but it is the general rule. In the case of such connexions the middle term too must be a general rule ... connexions which embody a general rule – i.e. which exist or come to be as a general rule – will also derive from immediate basic premises. (*An. post.*, 96a8–19). The same point is made elsewhere (cf *An. post.*, 87b9–27; *Gen. an.*, 727b29, 770b9–13, 772a35, 777a19–21; *Part. an.*, 663b28; *An. pr.*, 25b14, 32b4–13). It is important to note that what happens for the most part does not cease to be necessary when it ceases to be strictly universal. The explanations of the relevant patterns are therefore scientific, with necessary conclusions, rather than examples with contingent conclusions that constitute examples which show that not all scientific understanding in Aristotle is of the necessary.[97] The question remains, however, how the necessary can have exceptions.

Now, Aristotle distinguishes two senses of 'necessary.' We have thus far largely concerned ourselves with events the explanation of which consists of their being produced by the natural activity of the substance in which they occur. This is the case of *immanent causation*. But events can also occur in a substance which are caused by another substance acting on the first. This is *transeunt causation*. Thus, for example, all art in which an artist modifies a second substance involves transeunt causation. The two sense of 'necessary' are related to these two kinds of causation.

Any substance will, normally, act in accordance with its nature; the real potentialities constituting the nature of the substance will, normally, be actualized. But sometimes they are not actualized: the actions of a second substance may prevent this. That is, a natural tendency will be actualized unless the actions of a second substance prevent it. When the natural motion is prevented, the cause is an external substance. It is, in other words, an instance of transeunt causation. Thus, some cases at least of transeunt causation involve one substance bringing about in a second a motion or change that is contrary to the natural motion of the latter. Such 'unnatural' motion is sometimes called 'violent.' Thus, each of the simple bodies called elements has a natural motion; for example, the earth moves naturally in a straight line towards the centre of the universe. But such substances may also exemplify upward motion; in such a case it is not so much that the substance moves, that is, moves itself, but that it is moved, that is, is moved by another (cf *Cael.*, 300a20ff). Such motion is unnatural or violent, or, as Aristotle also says, 'constrained': 'the constrained is the same as the unnatural' (300a24). It is in this context that Aristotle introduces his two senses of 'necessary': 'Necessity ... is of two kinds. It may work in accordance with a thing's natural tendency, or by constraint and in opposition to it; as, for instance, by necessity a stone is borne upwards and downwards, but not by the same necessity' (*An. post.*, 94b38–95a3; cf *Metaph.*, 1015a20ff).

The sense of 'necessity' ties in with Aristotle's view of 'possibility.' For something to be possible, it must be more than merely 'conceivable' or 'logically possible': it must have the *capacity* to exist. That is, a 'possibility' is a 'natural tendency' or a 'natural power,' that is, one of the tendencies, powers, or capacities that is part of the nature or form of a thing. Aristotle conceives these strivings as successful unless prevented by some external force.[98] Thus, Aristotle writes of a man who potentially knows a certain subject: 'When he is in this condition, if something does not prevent him, he actively exercises his knowledge; otherwise he would be in a contradictory state of not knowing'; and he goes on immediately to make a similar point about ordinary physical objects: 'In regard to natural bodies also the case is similar. Thus what is cold is potentially hot: then a change takes place and it is fire, and it burns, unless something prevents and hinders it. So, too, with heavy and light: light is generated from heavy, e.g. air from water (for water is the first thing that is potentially light), and air is actually light, and will at once realize its proper activity as such unless something prevents it' (*Ph.*, 255b3–11). He puts it more abstractly elsewhere, that where there is a capacity for something to occur, that is, where it *can* occur, then it *will* occur: 'he who says of that which is incapable of happening either that it is or that it will be will say what is untrue; for this is what incapacity meant' (*Metaph.*, 1047a12–14). For, he asks, what is there to prevent a possibility from coming to pass, unless it is impossible? (*De motu animalium*, 699b29).[99]

Thus, if *p* is an unnatural event brought about by an external force, it is necessary, in the second sense of necessary. But if *p* is unnatural then it is contrary to some natural state *q* which is prevented by the external force from happening. So not-*q* is also necessary. But if it is necessary that not-*q*, then, since the necessary is that which it is not possible that not, it follows that it is not possible that *q* – that is, that *q* is impossible.

In the other sense of 'necessary,' however, it is the natural motion that is necessary. For it flows from the nature of the substance, and the nature or form of a substance is inseparable from it; that is, the substance has its form necessarily, and therefore has necessarily the characteristics which that form or nature causes to come to be in the substance.

Motion which is natural is free, unconstrained motion. In one sense, therefore, motion which is free is necessary: it is free because it is caused by the unconstrained striving of the very substance which is moving; in such motion, the substance is a *self-mover* and in moving itself unconstrained by anything external its motion is free. Thus, when an earthy object is unsupported, it moves in a straight line downwards towards the centre of the universe; such motion is *free fall*.

Conversely, motion which is unnatural is forced, constrained motion; that is why it can be said to be violent. In this sense, then, unnatural motion is necessitated, forced upon the substance and contrary to the goal which the substance itself, through its active nature, is striving to achieve: in such motion the substance does not move itself but is *moved by another*.

In modern science, the notion of force is a dead metaphor. It means mass × acceleration, and any suggestion of coercion plays no role; nor does the notion of a self-mover, as in the case, still spoken about in this way, of the free fall of heavy objects. For Aristotle, however, there is nothing metaphorical at all about these ways of speaking: the basic anthropomorphism of dynamic natures ensures that the talk of freedom and spontaneity in the case of natural motion, on the one hand, and force, constraint, and violence in the case of unnatural motion, on the other hand, has a perfectly legitimate place. *Aristotelian science must be contrasted to, and not confused with, modern science.*[100]

The central point of contrast, however, is perhaps the fact that the regularities that are to be reckoned lawful by the Aristotelian criterion of being regularities necessitated by the nature of the thing, that is, regularities caused by the striving of a thing in accordance with its nature, *can have exceptions*. Violent motion is always such an exception.[101] It is this category of violent motion that establishes the contrast to modern science and to empiricism. For what modern science aims to do is to explain matters of empirical fact by appeal to *exceptionless regularities*. What Aristotelian *scientia* aims at, in contrast, is knowledge of those regularities that are accounted for by the natures of things. Violent motion does not exhibit such regularities. To be sure, the motion of the substance that causes the violent motion itself flows from the nature of that substance; it is natural. But the violent motion that it produces will in general be the accidental outcome of activities that do not have the violent motion as their aim. In that sense, the violent motions will be exceptions to natural laws. Thus, as Aristotle says, 'There is no knowledge by demonstration of chance conjunctions; for chance conjunctions exist neither by necessity nor as general connexions but comprise what comes to be as something distinct from these' (*An. post.*, 87b19–22). The point of difference, then, between Aristotelian *scientia* and the new science is not merely one in the philosophy of science, about whether explanation is by mere matter-of-fact regularities or by regularities that are grounded in natural necessities. From the perspective of Aristotle, *there is no point* to the general concern of modern science to discover exceptionless regularities. From the perspective of Aristotle, where explanation consists in appeals to natures, the central aim has to be the discovery of those natures, and not the discovery of exceptionless regularities. Thus, given the aim that the Aristotelian metaphysics lays down for science, much of the activity of modern scientific research is *useless*. To be sure, it may serve other aims – for example, manipulative – *but it does not yield scientific knowledge and therefore does not serve the natural cognitive goal of man of coming to know the natures of things*. From the viewpoint of the empiricist, it is, of course, Aristotle's anthropomorphism that opens up the way to this divergence between Aristotelian explanations and those of modern science.

It is this same anthropomorphic content that opens the way to the legitimacy of the use of the legal analogy in the case of natural laws. In modern science this, too, is a dead metaphor: what science seeks in order to explain are *exceptionless regularities about matters of empirical fact*. Since they are exceptionless, they are not lawful in the Aristotelian sense, for, as we have just seen, when he introduces the notion of violent motion Aristotle allows that regularities which he would reckon necessary and therefore lawful can have exceptions. Moreover, scientific laws are *mere* regularities; they have no normative implications. For Aristotle, in contrast, a regularity which is a natural law, that is, one caused by the activity of the nature of the thing, is *also* normative, describing how the thing *ought* to behave.

Greek legal terminology was already applied to nature by Anaximander, who said of the things of the world that they come to be from the Unlimited and that they return to it when they perish, this process taking place 'by moral necessity, giving satisfaction to one another and making reparations for their injustice, according to the order of time' (frag. 1).[102] Here a certain general rule is implied – encroachments by one element will be counteracted. But, in the first place, this rule is not said to be exceptionless, nor, in the second place, is there any attempt to represent nature in all its detail as subject to law. The Sophists, however, were later to *contrast nomos* (law or convention) and *phusis* (nature); the former is subject to cultural variation, and is therefore less regular than the latter. When, still later, Plato combined the words *nomos* and *phusis* to speak of a law of nature (*nomos tes phuseas*), he was speaking with no doubt deliberate paradox. In the *Gorgias* (483e), Callicles argues that when the stronger subjugate the weaker, though they are not following a law made by us, they are none the less following a law, namely, a law of nature. And in the *Timaeus* (83e), Plato remarks that sometimes our blood is replenished not from food and drink, as happens for the most part, but otherwise in a way contrary to the laws of nature.

It is the Aristotelian notion of nature that can make sense of this. The second Platonic example would simply be a case of violent motion. In the first case, the appeal would be to a *human nature* – except, of course, both Plato and Aristotle would argue that Callicles, like Thrasymachus, has a false view of human nature, that there are more impulses to human nature than the urge to acquire power in order to be able to satisfy as many of one's lower desires as possible. This example makes clear how the link-up between law in the normative sense and nature is effected: it depends upon nature in the Aristotelian sense being *teleological*, that is, *an activity intrinsically directed towards a goal*. For, 'the good [is] ... that at which all things aim' (*Nicomachean Ethics*, 1094a2–3; cf 1076a3).

Now, there is for persons more than one end, and we choose some of these for the sake of others, the final ends. It is the latter that define the 'chief good' (*Eth. Nic.*, 1097a27). This chief good is sought for its own sake (10974a34), and happiness consists in the achievement of this final end, or, what is the same, the achievement of what is desired for its own sake (1097b1). Human activities serve a certain function; ultimately, they serve the final ends which man has (1097b25ff). These ends include the nutritive (in common with plants and animals), sensitive (in common with animals), and rational, which is peculiar to man (it is the difference which defines the species). In man the other ends function to serve the differentiating end, which regulates them (1098a1ff). One may serve these ends poorly or excellently; virtue consists in serving them excellently (1098a15–17; 1106a15–16). Thus, 'the virtue of man ... [is] the state of character which makes a man good and which makes him do his work well' (1106a22–3). Since in regard to human appetites one can have an excess, a defect, and a correct intermediate (1106b15–17), virtue is the habit, or state of character, of acting for the good, and where this is a matter of choice, it is the choosing of the means which is determined by rational deliberation to be the one which best satisfies our ends – that is, as it turns out, the mean or intermediate option in the case of each appetite (1106b36–1107a2). As for deliberation, 'We deliberate not about ends but about means' (1112b13). We may, of course, deliberate about intermediate ends, but if in our search for means to bring these about we find one cannot be achieved, we give it up: 'if we come upon an impossibility, we give up the search, e.g. if we need money and this cannot be got; but if a thing appears possible we try to do it. By "possible" things I mean things that might be brought about by our own efforts ...' (1112b24–7). The reason for this is simple enough: 'choice cannot relate to impossibles, and if any one said he chose them he would be thought silly ...' (1111b21–2).

Now, the final ends we desire for their own sake. But, as G.E. Moore taught us to ask, are these ends desirable, not in the sense of being capable of being desired but in the sense of being *worthy* of desire? Ought we to desire those ends? Ought they to constitute the good of man? To raise this question is to begin to deliberate about these ends, that is, to treat them as intermediate ends relative to some further standard which is final. It is also to treat them as things which it is possible, through our efforts, to change. But the final ends for which man strives, that is, the ends at which the human *telos* is directed, are determined by human nature itself. And this nature or form is something that each of us has as a matter of metaphysical necessity. It is therefore not possible to change the final ends of man. And if it is impossible, we give up trying to find out how to achieve it. It is simply pointless or silly to pretend to have ends beyond the final ends determined by human nature. So the question: Are these desirable, in the sense of being worthy of desire? is a question that would be silly to answer in any way but the affirmative.

Kant has made us aware that '*ought* implies *can*'; if we *can't* be something it is *pointless* to say that we *ought not* to be it. The converse of this principle is that '*must* implies *ought*.'[103] Its point, not surprisingly, is the same as that of Kant's principle: it is *pointless* to say about what we *must* be anything other than that we *ought* so to be. Aristotle's claim that man's nature determines man's good is, in effect, a conclusion based upon this '*must* implies *ought*' principle. As a matter of metaphysical necessity, we have the nature we have. That is how we must be: we must be the sort of person who aims at the ends which, by our nature, we desire for their own sake. Since we *must* be the sort of person who has those final ends, it is pointless to hold other than that we *ought* to be that sort of person, that is, that the ends that we naturally desire for their own sake are also desirable for their own sake in the sense of that they are *worthy* of desire.

If virtue is, as Aristotle says, the habit or state of character of acting for the good, then striving as we must for our natural ends is virtuous. In effect, then, the '*must* implies *ought*' principle is none other than the piece of common wisdom that it is only reasonable to make *a virtue of necessity*, and that it is only reasonable in the face of necessity to make of it a virtue because not to do so – to say that either what must be is forbidden or an alternative is permitted – would be at best pointless and at worst frustrating and painful, as trying to do the impossible so often is.

It should be emphasized that what we do of necessity, in the sense in which we are now speaking, is not done under compulsion – the *other* sense of 'necessity' – but *freely*. As we emphasized before, moving oneself to achieve one's natural ends is a necessary motion – since it is a motion that derives from the nature or from that which is necessarily and inseparably part of us – but it is also a free motion – since it is not constrained by any external force.

It thus turns out that *natural patterns* of behaviour are also patterns of behaviour that we *ought* to exemplify. Regularities which have *natural necessity* are also patterns of how we *ought* to behave. Regularities which are *natural laws in the scientific sense* relevant to explanation are also *natural laws in the legal sense* relevant to moral evaluation. Aristotle's metaphysics of explanation thus not only justifies the legal metaphor but in fact shows that the evaluative terminology is more than metaphor. The metaphysics not only justifies certain patterns of explanation but also provides the metaphysical foundations for the natural law position in ethics.

For our purposes we need not pursue Aristotle's ethics further. Since most virtues are acquired habits, the social conditions for their development cannot be ignored. This the *Nichomachean Ethics* notes at its beginning (Book I, chapter 2) and it ends (Book X, chapter 9) with a transition to the *Politics*; Aristotle's ethics is, as is Plato's, in the end inseparable from his political science. But this is surely true of any moral philosopher. In this respect, Spinoza and Hume surprise us no more than Aristotle or Plato.

It is perhaps worth noting that Aristotle's teleological vision of the universe, in which each entity in it has a soul with its own necessary *telos*, enables the legal terminology to be extended quite reasonably from the human case through animals and plants to even the elements. The hierarchy of souls is thus also a hierarchy of values.

4) *Our Knowledge of the Forms of Things*

It is sense which provides us with our first contact with the individual substances external to us; but what knowledge apprehends is the form, the universal, which is the cause of the individual's having, insofar as it exists naturally, the sensible characteristics it in fact presents to one (*De an.*, 417b22). Sense alone can give rise only to opinion, which apprehends the characteristics of things as separable, that is, as capable of being otherwise. But in knowledge they are apprehended as incapable of being otherwise (*An. post.*, 39a33–5; cf *Eth. Nic.*, 113b20). For, knowledge apprehends the form or nature which causes their coming to be present in the substance, unifying them into a whole and thereby constituting a necessary connection among them. Sense may, in a way, by recognizing the definitional nature of a thing, recognize what a thing is. But that remains at the level of opinion, and in that sense one's knowledge of *what* the thing is is only accidental. One's grasp of *what* a thing is is not accidental when one grasps its dynamic nature, the form which causes the definitional nature, and, indeed, is what transforms the latter into a genuine and not a quasi-unity or mere conjunction. It is the province of knowledge to grasp that form. This knowledge of forms is incorrigible: 'of the thinking states by which we grasp truth, some are unfailingly true, others admit of error – opinion, for instance, and calculation, whereas scientific knowledge and intuition are always true ...' (*An. post.*, 100b5–8). Or, again, 'it is not possible to be in *error* regarding the question of what a thing is, save in an accidental sense ... About the things ... which are essences and actualities, it is not possible to be in error, but only to know them or not to know them. But we do inquire what they are, viz., whether they are of such and such a nature or not' (*Metaph.*, 1051b25–33; cf *De an.*, 430a26–88). Like knowledge of the forms in Plato, as described in the *Meno*, so also for Aristotle, knowledge of forms, only now inseparable forms, is tied down and absolutely certain; it is, as he says, not *possible* that it be in error. This does not, as Aristotle also says, preclude inquiry, but it does mean that inquiry, if successful, has as its product a state that necessarily excludes the possibility of a need for any further inquiry.

The mind grasps some necessary truths as the conclusions of demonstrative syllogisms; but, of course, first principles cannot be arrived at, or justified, by such inferences. Rather, '... it is *intuitive reason* that grasps first principles' (*Eth. Nic.*, 1141a7–8), or, equivalently, the indemonstrable knowledge which is the grasp of immediate premises is *rational intuition* (*An. post.*, 88b35–89a1). *What inquiry aims at, then, is the rational intuition of the natures of things.*

In what does such rational intuition consist? What is the ontology of the knowing situation? Already in Aristotle's time it was an old dictum to say that like knows like (*De an.*, 427a27), but in Aristotle this dictum takes on a fairly specific content. For Aristotle, the likeness amounts to an *identity*: in knowledge the mind is identical with its object (429a16, 429b20, 431a1). Since the object of knowledge is the form or nature, it follows that in knowledge the form that is in the thing known is also in the mind that knows it (431b30). But the form is not in the mind as a characteristic of it; for if it were, then the mind would *be* Socrates or Corsicus, that is, a man or a lump of earth, let us say, or in any case whatever substance it is that is known; but the mind is clearly not any of those. Thus, it is not the material object, only its form, which is in the mind. As the medievals were later to say, the form of the substance known is in that substance substantially but in the mind only intentionally. Thus, *rational intuition of a form consists in the form itself being literally in the mind.*

If the form itself is literally in the mind, as Aristotle says, then so is its logical structure. If a specific nature is in the mind, then so is its genus and difference, since the specific form simply *is* the unity of genus and difference (*Metaph.*, 1037b25–7). Let us call the form of a thing known *qua* in the mind a 'concept.' Then, when the mind analyses a specific concept into genus and difference, this *logical structure of the concept* is *eo ipso* the *ontological structure* of the thing known.[104] But the definition in terms of species and difference is exhibited in the syllogism. Thus, the movement of thought which occurs when the mind understands a proposition as the conclusion of a demonstration which exhibits a definition is not only a movement of thought as it knows, but also the logical structure of a concept and the ontological structure of the thing known. *Syllogism is thus not only a logic but also an onto-logic and an epistemo-logic.* The core to this is the notion of substance; for its nature is at once the cause of the being of its substance and also that which is, in its logical structure, exhibited in syllogism: 'as in syllogisms, substance [=essence] is the starting-point of everything. It is from "what a thing is" that syllogisms start; and from it also ... processes of production start' (*Metaph.*, 1034a30–2). Equally, 'scientific knowledge ... is a state of capacity to demonstrate' (*Eth. Nic.*, 1139b31), and science is the mental disposition or capacity to demonstrate.

The new science, too, insists upon deductive logic as playing a central role in explanation. According to the defenders of the new science, both empiricist and rationalist, the explanation of an event consists in the deduction of that event from another event (the initial conditions) and an exceptionless matter-of-empirical-fact regularity (the covering law).[105] But the deduction is purely tautological, only unpacking what is already implicit in the premises. On Aristotle's account of explanation, however, *demonstrative syllogisms of science* reveal the underlying necessities of the world and the internal structure of those necessities. Thus, for Aristotle, in contrast to the new science of both the empiricist and the rationalist, syllogism has a substantive role to play in explanation.[106] This substantive role is, of course, dependent upon the basic Aristotelian idea that explanation is always in terms of the unanalysable active powers or dispositions of the natures of things. But in any case, both the rationalists and the empiricists were to the point when they criticized not merely the topical logics of Agricola and Ramus but also the *syllogistic logic of explanation* which they attributed to Aristotle.

Syllogism is not, however, according to Aristotle, a tool of discovery. To construct a scientific syllogism is to discover its middle term. Syllogism exhibits this knowledge, but it does not discover it. Since sense perception is only of sensible appearances, it does not yield scientific knowledge (*An. post.*, 87b28).

None the less, according to Aristotle, perception is of central importance; it is from our sense perceptions that knowledge of the universal is elicited:

if we were on the moon, and saw the earth shutting out the sun's light, we should not know the cause of the eclipse; we should perceive the present fact of the eclipse, but not the reasoned fact at all, since the act of perception is not of the commensurate universal [which is that which 'makes clear the cause' (88a5)]. I do not, of course, deny that by watching the frequent recurrence of this event we might, after tracking the commensurate universal, possess a demonstration, for the commensurate universal is elicited from the several groups of singulars. (87b39–88a4)

Where sense is lacking, a science will also be lacking, since the universals from which science proceeds are got from sense. Although we do not know the reasons of things by sense, we learn them from sense. These reasons for things, their forms or natures, we then put into syllogisms to exhibit their internal logical structure.[107] After a certain number of experiences of a fact, the universal dawns upon us in an act of intuitive reason. The man who has the 'faculty of hitting upon the middle term instantaneously' is the man of 'quick wit' (*An. post.*, 89b10).

There is a two-stage process here. The first stage consists in scanning the particulars. The second consists in *abstracting* the form from the particulars so scanned. This whole process is called *induction*. The product of this process is a rational intuition of the form of the substance that is the cause of the sensible appearances from which the process has begun. For the power of abstraction is simply the power to lift the forms from the sensible appearances they cause and put them (the forms) in the mind. It is by induction that one arrives at the starting points of scientific knowledge: 'induction is the starting point which knowledge even of the universal presupposes, while syllogism proceeds *from* universals. There are therefore starting points from which syllogism proceeds, which are not reached by syllogism: it is therefore by induction that they are acquired' (*Eth. Nic.*, 1039b27–30). In fact, since demonstration is about forms, it follows that 'intuitive reason is both the beginning and the end [of inquiry]: for demonstrations are from these and about these' (1143b10–11). The faculty for grasping first principles is what Aristotle calls *nous*; through this capacity the mind 'cognitively grasps the first principles in an immediate and non-discursive way.'[108]

Complete induction has the force of demonstration, Aristotle points out (*An. pr.*, Book II, chapter 22), but the induction which arrives at first principles is seldom complete. Aristotle describes the process in this way: 'states of knowledge are ... developed from sense perception. It is like a rout in battle stopped by first one man making a stand and then another, until the original formation has been restored. The soul is so constituted to be capable of this process.' The rout is, of course, the flux of sensory experience. We determine that the rout has been stopped when we recognize that a certain structured formation has been achieved. The recognition of this structure is the intuition of the form. 'When one of a number of logically indiscriminable particulars has made a stand, the earliest universal is present in the soul: for though the act of sense perception is of the particular, its content is universal – is man, for example, not the man Callias. A fresh stand is made among these rudimentary universals, and the process does not cease until the indivisible concepts, the true universals are established: e.g. such and such a species of animal is a step towards the genus animal, which by the same process is a step towards a further generalization' (*An. post.*, 100a9–b4). An example from the *Topics* makes it equally clear that induction is normally imperfect, and also clear that it is not the 'puerile' and 'childish' induction of the naïve empiricist which consists in registering and enumerating sensory observations but has instead as its product the rational insight into form: 'e.g., the argument that supposing the skilled pilot is the most effective, and likewise the skilled charioteer, then in general the skilled man is the best at his particular task' (*Top.*, 105a14–16).

Plato had argued in the *Meno* that our knowledge of forms was innate. But, while that argument established that the knowledge said to be innate could not have been taught in the way Greek youth were taught Homer, it none the less hardly followed, as Plato suggested, that it was innate. The argument of the *Phaedo* strengthened the case of the *Meno* by arguing that we have concepts of things – for example, perfect equality or perfect justice – which we could not get by means of sense observation. Aristotle suggests that intuitive knowledge, in all its certainty and infallibility is not 'innate in a determinate form,' without our consciously knowing it from the beginning; and it certainly could not be 'developed from other higher states of knowledge' since that would introduce circularity or a regress (*An. post.*, 100a9–11). But it is equally hard to believe that it comes from nothing at all (99b–30). Rather, then, than it coming from a higher form of knowledge, Aristotle proposes that it develops from a humbler source, a lower faculty. This faculty is perception, the discriminative capacity already present in animals. The first stage in the development from sense to knowledge is memory, when one 'continue[s] to retain the sense impression in the soul' when perception as such is over. Out of frequently repeated memories develops experience, 'for a number of memories constitute a single experience.' In experience 'the universal [is] now stabilized in its entirety within the soul, the one beside the many which is a single identity within them all.' Out of experience originate both art or the skill of the craftsman, insofar as our concern is with becoming, and the knowledge of the scientist, insofar as our concern is with being. At this point the mind is in a position to abstract the universal (99b35–100a13).

This transition is made possible because of the fact that perception has the universal implicitly within it: 'though the act of sense perception is of the particular, its content is universal' (*An. post.*, 100a17–100b1). For Plato, the forms are separate from the souls, and therefore from the sensible characters that come to be in souls as the latter strive to imitate the forms. There is no question of their being tied to sense experience, nor, therefore, of their being discoverable by the mind in sense experience. Given the separation, Plato has little option but innateness. However, the separation of the forms is a serious error, and once this separation of the forms is eliminated they can, as active powers, be recognized as tied inseparably to the sense impressions which they cause and provide a necessary connection among. Thus, when the impressions are given in sense, they bring with them to the mind the nature or form that ties them together and is their cause. Sense itself has not the capacity to grasp this form, so it is only the incidental object of sense; but it is there, tied to sense impressions, ready for the mind to abstract. As for the *Phaedo* argument, that forms are only imperfectly embodied in experience, Aristotle can agree. For all that substances strive to embody their form, there are always in fact, in the sublunary world, contingent circumstances that introduce forces that create unnatural events in any substance, and these prevent the perfect embodiment of the form. Still, once one has observed a regular but not exceptionless pattern of behaviour and captured it in experience, it will be possible to abstract from this imperfect regularity the form that causes it and would cause it with perfect regularity were it not for external forces that sometimes constrain it and prevent its natural fulfilment. From the imperfect embodiment of forms one cannot infer their separation; nor from the fact that sense does not grasp perfect embodiments of the forms does it follow that the mind cannot learn of the forms by beginning from sense.

The 'rout' of sensory experience is stopped, then, when the mind grasps the form. Thus, the rout is stopped when experience comes to be structured in the mind by the form, now a concept existing intentionally in the mind. This structure on the side of the mind is identical to the structure on the side of the substance known. For the form that, in the substance, causes the sensible appearances is identical to the form that comes to organize the experiences in the mind. As Aristotle puts it, as we saw, the 'rout' is stopped when 'the original position has been restored' (*An. post.*, 100a12).

This coming to be of the form in the mind is the result of the mind's own *activity*. In one sense the mind is passive, a mere potentiality, capable of having all forms impressed upon it (*De an.*, 429a15). But the *power or capacity of discovering the form in sense* is one which distinguishes man from animals (414a19–20). This capacity of the mind to discover the form in sense experience, and to put it not just implicitly but actually in the mind, is the power of abstraction. Through it the mind rises above mere experience, mere repetition, which is the best that sense can achieve, and grasps real connection, that is, the reasons or forms which account causally for those patterns in sense experience. In abstraction the mind comes to have in it the form of the substance whose sensible appearances are experienced, and in thus becoming identical with the cause of those experiences the mind thinks that cause. In the exercise of this capacity to abstract forms, mind is active (430a15–16); it moves itself. Indeed, since man is in his essence a *rational* animal, that is, one capable of grasping the reasons of things – the forms – it follows that the thinking faculty which is that ratiocinative part, as opposed to the nutritive and appetitive parts, of the soul, must be 'in its essential nature activity' (430a18). *The form of man, his essential activity which differentiates him from all other things, is to inform himself of the being or reasons of things.* As Aristotle puts it elsewhere, in the opening sentence of the *Metaphysics*, 'All men by nature desire to know' (980a22).

Thus, for Aristotle, as Ross has put it, 'induction is ... a process not of reasoning but of direct insight, mediated psychologically by a review of particular instances.'[109] Its aim is to discover the middle terms of syllogisms. It begins in experience, but unlike induction for the empiricist, does not end there in the discovery of a pattern *in* experience. Rather, it goes beyond the world of experience. How much experience it takes to trigger the insight depends upon the particular person; only some have 'quick wit,' the 'faculty of hitting upon the middle term instantaneously' (*An. post.*, 89b10). It is misleading in the extreme, then, to suggest that Aristotle did not overlook the need for empirical investigation since he held that 'to discover the full essence of a natural kind takes empirical investigation.'[110] While experience does play a role for Aristotle, it is far indeed from the sort of empirical investigation into exceptionless matter-of-fact regularities which the new science demands. The latter Aristotle *did* overlook; indeed, he did not even conceive that such a program of research was worthwhile.

We can see this sort of thing in the views of Duns Scotus concerning experiential knowledge of ordinary things. Scotus was arguably the first to take with some seriousness the issue of our inductive knowledge of things,[111] and so an examination of those views will be particularly revealing. In particular, it will illustrate with considerable clarity the great difference between the concerns of the scholastics on the one hand and the empirical investigators who defended the new science.

Truth, according to Scotus,[112] involves the relation of a thing to an exemplar.[113] Of the exemplars, there are two: one created, one uncreated. The uncreated exemplar is the form in the mind of God, the eternal pattern for the thing or substance that He has created. The created exemplar is the form *in* the created substance.[114] The truth of the thing is in the first instance the form in the mind of God. It is, in the second instance, the created form in the substance. For, the form in the thing is the created image of the idea in God's mind. We can therefore come to know the truth of the thing by having an intuition of the form in the thing. This we obtain by abstracting that form from the thing as it is given to us in sense experience.[115] The form *qua* in the mind is the intelligible species.[116] So far as the thing is known by sense alone, it is merely a collection of properties. It is the form in the thing that creates the unity out of these properties or accidents, and makes a substance out of them.[117] These forms or natures of things produce regular patterns of behaviour. Nature in this sense is uniform, according to Scotus. This is the meaning of the scholastic formula that 'Natura determinatur ad unum': 'The proposition that nature is determined towards a single end is to be understood not in the sense of being ordained to produce one and only one singular effect, but in the sense that it is ordained to a definite pattern of production – natural agents in this respect differing from voluntary agents which are not ordained to a definite pattern but rather determine themselves to one or the other of two opposite actions.'[118] Nature, or the nature of a thing, is the ultimate source of being, or the being of the thing, and the power that is the source of all its becomings, doings, and beings. Hence the principle that '[q]uidquid evenit ut in pluribus ab aliqua causa non libera, est effectus naturalis illius causae ...':

(#)   Whatever occurs in a great many instances by a cause that is not free is the natural effect of that cause.[119]

The effect which frequently follows a non-free cause is the natural effect of that cause. The regularity that we observe in the world we know by sense is thus rooted in the natures of things. Thus, even if a seeker after causes has observed only a restricted number of instances of a set of sensible appearances, he or she can still conclude that these have the same cause, and that this cause will produce the same effects always and everywhere (*semper et in omnibus*): 'As for what is known by experience, I have this to say. Even though a person does not experience every single individual, but only a great many, nor does he experience them at all times, but only frequently, still he knows infallibly that it is always this way and holds for all instances.'[120] We know this by virtue of (#), which is a self-evident proposition 'reposing in the soul.'[121] Observation of the world generates knowledge of this proposition – though, to be sure, 'the senses are not a cause but merely an occasion of the intellect's knowledge ....'[122] Thus, 'it is recognized through experience and is certain that for the most part nature acts uniformly and in an orderly way.'[123]

Scotus gives an example of how a fact that is known first by sense experience, as contingent and probable only, can come to be known scientifically, with demonstrative certainty.

[A]t times we experience [the truth] of a conclusion, such as: 'The moon is frequently ecplised' ... Sometimes, beginning with a conclusion thus experienced, a person arrives at self-evident principles. In such a case, the conclusion which at first was known only by experience now is known by reason of such a principle with even greater certainty ... for it has been deduced from a self-evident principle. Thus, for instance, it is a self-evident principle that when an opaque body is placed between a visible object and the source of light, the transmission of light to such an object is prevented. Now, if a person discovers by way of division that the earth is such an opaque body interposed between sun and moon, our conclusion will no longer be known merely by experience as was the case before we discovered the principle. It will be now known most certainly by a demonstration of the reasoned fact, for it is known through its cause.[124]

What is important, however, is that even causal facts that have demonstrative certainty are not invariable. Even though it is self-evident that

(#)   Whatever occurs in a great many instances by a cause that is not free is
         the natural effect of that cause

it is not only not self-evident that

         As often as the cause is posited, its proper effect must also be posited

it is simply false. In this sense there is no absolute necessity to created or imperfect causes. 'To be able to produce an absolutely necessary effect does not pertain to the perfection of a secondary cause; in fact, there is no such thing (as a necessary effect) in a secondary cause ... for (the concept of) an absolutely necessary causation includes a contradiction.'[125] Natural laws, therefore, need not hold universally. To be sure, there are some such laws that do – for example, the laws that describe the motions of heavenly objects. But even these can be described universally as happening 'per causam naturalem ordinatam' – according to ordained natural causes, that is, predictably – only on the supposition 'quod causae naturales sibi dimittantur' – that is, only on the condition that it is left to itself – which is to say only on the supposition 'quod per virtutem divinam non impediantur' – that is, the supposition that it is not interfered with by the divine omnipotent power.[126] God is not bound by the laws for the creatures He has brought into being; He may at any time suspend them. But this would of course be a miracle. More importantly for the scientist interested in the explanation of things as they appear in the ordinary course of events, there are *chance* causes. 'Natural causes, normally productive of certain effects, can sometimes have their normal operations prevented due to extrinsic interfering factors.'[127] It is true that when there is no interference, a natural cause will produce the effect at which it aims: 'Omnis causa sibi dimissa ... producit effectum cujus est per se' (Any natural cause, if left to itself, will produce its proper effect).[128] The production of the proper effect is an exercise of power, and the cause will exercise that power to its fullest so long as it is not prevented from doing so by some stronger power: 'Causa naturalis agit ad effectum suum secundum ultimum potentiae suae quando non est impedita' (A natural cause produces its effect to the utmost of its power so long as it is not prevented from doing so).[129] Thus, if we have

(#)   Whatever occurs in a great many instances by a cause that is not free is the natural effect of that cause

as a principle, then we also have

(&)   Every natural cause, unless prevented from doing so by some interfering factor, produces its effect necessarily.

That is why all observed effects attributable to secondary causes can be so attributed with only conditional necessity, not absolute necessity: a cause whose operation can be impeded, however unlikely such interference might be, is not a necessary cause.[130]

Scotus tells us that when we know things only by experience, we have no notion of a necessary tie among events, a connection that is grounded in the natures of things. He points out that sometimes

we must be satisfied with a principle whose terms are known by experience to be frequently united, for example, that a certain species of herb is hot. Neither do we find any ... prior means of demonstrating just why this attribute belongs to this particular subject, but must content ourselves with this as a first principle known from experience. Now even though the uncertainty and fallibility in such a case may be removed by the proposition 'What occurs in most instances by means of a cause that is not free is the natural effect of such a cause,' still this is the very lowest degree of scientific knowledge – and perhaps we have here no knowledge of the actual union of the terms but only a knowledge of what is apt to be the case. For if an attribute is an absolute entity other than the subject, it could be separated from its subject without involving any contradiction. Hence, the person whose knowledge is based on experience would not know whether such a thing is actually so or not, but only that by its nature is it apt to be so.[131]

Experience produces only probability, although we can raise this probability higher by invoking the principle (#): this principle will make it certain that an object that for the most part produces certain effects is by its nature apt to do so. Even so, there is still the further principle (&): this principle allows that there might well be chance causes that are operative, preventing a thing from producing the effect that it is naturally apt to produce. *Natural laws have exceptions.*

It has been suggested that this puts Scotus's views close to those of the empiricists in making it reasonable to conclude that induction is always infirm. Thus, it has been remarked that '[e]xceptional occurrences ... do not invalidate the method of induction. For they cannot shake our certitude of the principle of induction which remains self-evident and true. Exceptions, however, serve to show the occasional inapplicability of the method, and to render the observer more cautious in attributing a particular effect to a particular cause.'[132] But *the method of induction of the empiricists aims at finding* true *matter-of-fact regularities.* If there are exceptions, then one does not have a *true* matter-of-fact regularity, and the method has not (yet) achieved its goal. For Scotus, in contrast, the method aims at finding necessary connections based in the natures of things. Knowing such a connection is quite compatible with there being exceptions to the natural law which it implies. The differences are two. (1) The empiricists affirm and Scotus denies the need to find true matter-of-fact regularities. (2) Scotus affirms and the empiricists deny the need to pass beyond the realm of sense experience to a realm where we can rationally intuit objective necessary connections. *Given the very different cognitive ends of the empiricists, on the one hand, and Scotus, on the other, it is misleading indeed to suggest that their views of science and of scientific method are equally 'inductive.'*

To be sure, for both the method – let it be commonly called 'inductive,' if you wish – that yields knowledge does not in the end yield unqualified certainty. The point is that the source of the uncertainty is very different, resulting from very different cognitive concerns. For the empiricists, the problem of certainty is *the problem of induction* – what is commonly called 'Hume's problem.' For these thinkers, the problem of certainty has its roots in the fact that an inductive inference is always from a sample to a generalization, from an observed sample to a total population. Since, in the absence of objective necessary connections, there is nothing in a sample that guarantees that what regularly happens there will regularly happen in the population, it follows that the inference from sample to population, the inductive inference, is always uncertain. For Scotus, in contrast, the problem of certainty is *the problem of abstraction*. Here the problem consists in the fact that what we know in sense are contingent patterns of separable events, and that it is necessary to decide that an observed pattern is due to nature, that is, is rooted in the nature of the changing thing, or is due to chance, to some factor that interferes with, and impedes, the normal operations of a thing. To be sure, one abstracts a form from observed events, *but how does one know for certain that one has abstracted the correct form, the form that grounds the natural course of events, the natural laws that account for the observed structure of experience?* In the end, one does not know, or, at least, know for sure, and there is therefore an uncertainty in our knowledge that is just as problematic for Scotus and the Aristotelians in general as the uncertainty due to 'Hume's problem' is problematic for the empiricist.

The problem is that of 'warranting universals,' as Peter Dear has expressed it.[133] Dear points out quite correctly that 'within the Western philosophical tradition, it [Hume's problem of induction] appeared only after its constitution in the practice of seventeenth century natural philosophy ... Before the seventeenth century the modern problem of induction did not exist.'[134] Dear does not, however, locate the issue in the context of the new cognitive goals that had emerged: the goal of coming to know true matter-of-fact regularities, *exceptionless* regularities. Only in this context does 'Hume's problem' begin to make sense. The universal propositions that the inheritors of Aristotle sought were quite different: they were patterns that were supposed to be rooted in the metaphysical natures of things. Within the latter context, the existence of already-observed exceptions, let alone unobserved exceptions, was not a serious problem: all laws had exceptions! Thus, 'unlike Hume ... [Scotus and] scholastic-Aristotelians throughout the preceding century and a half had not treated the issue as one of particular philosophical significance for their project. Instead, they had treated it as a minor practical matter.'[135] At the same time, within the context of the experimental philosophy, exceptions were especially revealing of the truth: the logic of experiment as laid out by Bacon and Descartes required the use of negative instances in a progressive move towards the truth about matter-of-fact regularity. Again, as Dear puts it, 'Aristotelian "monsters," unlike Baconian ones, were not illuminating rarities that served to reveal nature's workings, but were instead nature's mistakes, in which regular processes had become spoiled through adventitious causes.'[136] Dear's discussion is valuable, but, still, he fails to locate the use of negative instances in the context of the *logic* of the experimental method. Only then do we recognize the very different *logical* and *ontological* roles of our ordinary sensible experience of things.

The major point that Dear makes, however, is certainly correct: observation by means of the senses plays very different roles in Aristotelian science on the one hand and the new experimental science on the other. On the scholastic-Aristotelian view, what is needed is nothing more than firm insight into the natures of things, insight that can be gained from a repetition of positive instances of the supposed natural law. Dear quotes Franciscus Aguilonius, an Antwerp-based Jesuit mathematician and pedagogue, who wrote in a 1613 text-book on optics:

If the senses always cheated the mind's acuteness, there could be no science, which [is what] the Academics [= Academic sceptics] strive to assert; and if they never failed, the most certain experience would be had by a single act. Now, the matter is, however, in the middle [of these extremes]: for although the senses are sometimes deceived, usually however they do not err. Hence it is, that whenever an experience is certain the first time, it is confirmed by the repetition of many acts agreeing with it.[137]

Sense of course, as Scotus put it, is the occasion and not the cause of the certain knowledge that we obtain: 'the senses are not a cause but merely an occasion of the intellect's knowledge ...'[138] It is not the changing sensible appearances of the object that cause the knowledge but rather the necessary and immutable nature of the thing: 'it is not precisely this mutability in the object that causes the knowledge; it is the nature of this mutable object that does so, and this nature is immutable.'[139] For this reason, a necessary ontological proposition such as

(#)   Whatever occurs in a great many instances by a cause that is not free is the natural effect of that cause

is known infallibly. For, the knowledge is caused by the object known, and (#) applies to this cause too.

This proposition is known to the intellect even if the terms are derived from erring senses, because a cause that does not act freely cannot in most instances produce an effect that is the very opposite of what it is ordained by its form to produce ... if the effect occurs frequently it is not produced by chance and its cause therefore will be a natural cause if it is not a free agent. But this effect occurs through such a cause. Therefore, since it is not a chance cause, it is the natural cause of the effect that it produces.[140]

But this simply won't do: how much experience is needed to distinguish a chance cause from a non-chance or natural cause? Even if we glomb onto, or have a rational intuition of, a nature, how do we know that we have abstracted the correct one? If our knowledge of the operation of natures is restricted to what they are 'apt' to produce, then our knowledge is not reliable until we have a criterion for discerning when they are acting reliably and when not. But if the basis for relying upon our knowledge of natures is the fact that natures for the most part are reliable in their operations, then we have no clear way to separate chance from natural implantations of propositions in the intellect.

These are clear problems, but the root problem is that the whole doctrine of natures was open to criticism. Where the defenders of the scholastic-Aristotelian tradition were busy repeating the old dogmas, the defenders of the new science, in contrast, argued that the whole Aristotelian program was not worthwhile. For Aristotle there is a two-stage process: surveying the particulars, and then abstracting the form. The defenders of the new science argued that explanations in terms of forms were vacuous. That eliminates the second stage of the Aristotelian process. There remains the first stage. What does this look like? Without the second stage it is a simple-minded survey of particulars, at once the 'childish' induction by simple enumeration and at the same time a use of data simplistically acquired without the use of instruments and other techniques for supplementing unaided sense. It is clear that, seen thusly, Bacon's rejection of Aristotelian processes as based on notions 'improperly and overhastily abstracted from facts, vague, not sufficiently definite, faulty'[141] is just. Some defenders of Aristotle and the scholastics object to the claim that 'the scholastic logicians made the essence of induction to consist in enumeration.' It is argued that 'from the very fact that Aristotle and the Scholastics considered it possible to reach a truth about '*all*,' actual and possible, known and unknown, by an acquaintance with '*some*,' they must have recognized a method of ascent to the '*all*,' other than enumeration.'[142] Indeed they did recognize such a method. Unfortunately, that method, the method of abstracting natures, was criticized by their successors, those who developed the new inductive methods. That criticism consisted in calling into question the whole metaphysics of natures. It called into question, in other words, the very possibility of there being any method of the sort that Aristotle proposed. Moreover, the ascent from 'some' to 'all' for the new science was different from that ascent in Aristotle. For the new science, the 'all' was a strictly universal matter-of-fact regularity. For Aristotle and his scholastic successors, the 'all' was a 'natural law,' a pattern that was supposed to be rooted in the metaphysical natures of things but which allowed for exceptions. But the latter could be defended only so long as the notion of a 'nature' escaped criticism. It did not so escape. And so, from the viewpoint of the new cognitive end and the new experimental method of eliminative induction, the Aristotelian method, or, rather, what seemed to be left of it, could quite reasonably be characterized as 'induction by simple enumeration.' With the natures eliminated as metaphysical vacuities, this is what was left, and, relative to the new cognitive end, it is indeed 'puerile,' worse than useless.

It is worse than useless because it leads people to maintain as 'self evidently true' propositions that are, from the viewpoint of the new science, simply false. Thus, Charles Schmitt comments as follows on Zabarella's appeal to experience:

> The limitations of Zabarella's use of experience ... lie in his reluctance to abandon certain deeply ingrained Aristotelian notions, even in the face of contrary evidence. Although many Aristotelian teachings on natural philosophy just did not seem to agree with everyday experience, every attempt was made to retain as much as possible from the peripatetic system and if a certain concept could feasibly be retained by making minor adjustments elsewhere in the system this was usually done. Major discrepancies in the laws of motion had been noticed several times previously – e.g., by Philoponous and Simplicius in antiquity, by Buridan and Oresme in the fourteenth century, and by Benedetti and Tartaglia in the sixteenth – but the framework of the Aristotelian system was still generally accepted by Zabarella.[143]

It is precisely for this reason that Schmitt is wrong to suggest later that 'Zabarella ... clearly points the way toward seventeenth-century experimental observational science.'[144] So are others, such as J.H. Randall, Jr, who argue for a continuity between late scholastic and Renaissance Aristotelianism and the methods of the new science.[145] This is wrong because it misses two central differences between the old and the new methods. The one is that the two differ with regard to their cognitive ends. The other is that the methods themselves are very different, the one aiming to use experience to eliminate falsehood and move towards truth and the other aiming to abstract natures from experience. Both these differences turn upon the acceptance by the Aristotelians of the metaphysics of natures and its rejection by the defenders of the new experimental science. There is a discontinuity in cognitive ends, in methodology, and in ontology and epistemology. To be sure, there are continuities in language, in examples, in topics of concern, and so on. But we must not let these obscure the discontinuities. The search for precursors must not mislead us into thinking there are no discontinuities. Nor should we let our sense of continuity obscure the fact that, from the point of view of the new experimental science, the older methods were in fact unreasonable, unjustified, and worse than useless. How else could one describe them, when, upon critical evaluation, they turned out to offer nothing more than induction by simple enumeration?

If this criticism is just, so is Bacon's critical rejection of the Aristotelian process of induction and subsequent elaboration of principles syllogistically, what he refers to as the 'anticipation of the mind,' and which he contrasts to the 'interpretation of nature' – that is, eliminative induction – that he proposes.[146] Bacon's reasonable criticism is that the Aristotelian inference 'flies from the senses and particulars to the most general axioms, and from these principles, the truth of which it takes for settled and immovable, proceeds to judgment and to the discovery of middle axioms,'[147] a process which takes the most general propositions as 'certain fixed poles for the argument to turn upon, and from these to derive the rest by middle terms ...'[148]

Descartes similarly rejects Aristotelian claims in his famous discussion of the wax example in Meditation II: the constantly changing nature of the wax, the infinity of forms that it takes on, prevents the mind from being able to abstract the essence from one's sense experience of the wax.[149] The result is that, far from attaining certainty, the Aristotelians are left with only 'these very convenient weapons of debate, the probable syllogisms of the Scholastics.'[150] From this perspective, that Aristotelian induction is inefficacious and incapable of providing the claimed content for demonstrative syllogism, the criticisms of Aristotle that were shared on these points by the rationalists and the empiricists were entirely reasonable.

For Aristotle and the scholastics such as Scotus, there was a guarantee that one could acquire knowledge in the sense of incorrigible certainty. This guarantee was provided by the notion that there are certain 'natural tendencies' of the human mind towards the acquisition of knowledge. But for the defenders of the new science, whether empiricist or rationalist, the whole talk of powers is vacuous. In particular, therefore, talk of the power of abstraction is vacuous. But it is the 'naturalness' of this power of abstraction, the supposed fact that it is a feature of the human 'nature' or 'form' or 'essence,' that, in the end, is supposed to provide the guarantee for the whole of the Aristotelian system. 'By nature' all persons aim to know, the system claims. Coming to know is thus a *natural motion* which, like all natural motions, will be successful unless it is constrained. But such unnatural interventions – that is, being caused to abstract the wrong form – happen only incidentally. For what is natural is the norm; it is what happens *always or for the most part*. Hence, error, while it can occur, is unnatural, and the natural powers of the human mind guarantee that success – knowledge of laws – is the usual outcome of induction. But what sort of guarantee is *this*, if, as both the empiricists and the rationalists argue, the whole thing is vacuous? Invoking the natural power of the mind to know forms will not, contrary to suggestions of some such as Gaukroger,[151] enable the Aristotelian to evade the rationalist and empiricist critique, which cuts much more deeply indeed; Descartes and Arnauld both see this clearly, as I shall argue below.

It was Galileo who shifted cognitive interest from natures or forms to general regularities. His concern was with projectile motion. The tradition deriving from Aristotle had it, as Aquinas put it, that 'the movement of heavy bodies towards the center' is their natural motion (*Summa*, FS, Q54, A2, RObj 2).[152] This is a matter of intuiting the nature of such objects as objects that gravitate. The motion of projectiles is contrary to this natural motion. This motion is therefore violent, and must be explained in terms of external forces, and in particular in terms of the entity that set the projectile in motion. Aquinas therefore states that 'the necessity whereby an arrow is moved so as to fly towards a certain point is an impression from the archer, and not from the arrow' and tells us that 'the violent necessity in the movement of the arrow shows the action of the archer' (FP, Q103, A1, RObj 3). The end of its movement derives not from its own nature but from the nature of the external object that is causing the violent motion: 'an arrow tends to a determinate end through being moved by the archer who directs his action to the end' (FS, Q1, A2). The object of Aquinas's science, so far as it concerns projectile motion, is a matter of locating more precisely the external force that accounts for this violent motion. As is well known, this in fact proved a difficult task. For, once the projectile leaves the hand of the archer or whoever or whatever sets it in motion, there seems to be no external force capable of maintaining that violent motion. And if there is no such force, then it is as if the obstacle to natural motion was removed: the violent motion should cease and the projectile should immediately return to a natural motion towards its natural end. But the projectile does not cease its apparently violent motion. Where, then, is the external force that maintains that violent motion? We need not go into attempts to solve this problem. Suffice it to say that it is a problem only so long as we maintain both that there are forms or natures of things in terms of which we must explain the motions of objects, and that the forms of heavy objects determine that their natural motion is straight downwards towards the centre of the earth or, what was for many the same, the centre of the universe. What Galileo did was reject the idea that the main task of science was to search for these forms that determine natural motion and violent motion, and in particular that the main task for the physicist with regard to projectiles was to discover the external force accounting for the violent motion. As Stillman Drake has argued, Galileo gives up this old task of science (*scientia*) and replaces it with a new task.[153] The old cognitive interest was a metaphysical interest, non-empirical; it is replaced by a new cognitive interest, an interest that can, unlike the old, be fulfilled by empirical research.

In Galileo's dialogue on *Two New Sciences*, Salviati represents Galileo. Sagredo provides wise and intelligent commentary. It is the latter who introduces 'the question agitated among philosophers as to the possible cause of acceleration of the natural motion of heavy bodies.' He then proposes an answer in terms of an external force impressed upon the object by the substance that initiated the non-natural or violent motion that is contrary to its natural motion as a heavy (or gravitating) object straight downwards.

[L]et us consider that in the heavy body hurled upwards, the force impressed upon it by the thrower is continually diminishing, and that this is the force that drives it upward as long as this remains greater than the contrary force of its heaviness; then when these two [forces] reach equilibrium, the moveable stops rising and passes through a state of rest. Here the impressed impetus is [still] not annihilated, but merely that excess has been consumed that it previously had over the heaviness of the moveable, by which [excess] it prevailed over this [heaviness] and drove [the body] upward. (*Two New Sciences*, 158)

And so on. The traditional philosopher Simplicio immediately raises some objections to this account in terms of active forces. As Stillman Drake points out in his note to this text (158n), this reasoning of Sagredo is precisely that which Galileo himself had used in his earlier thinking concerning free fall and projectile motion. But Galileo's spokesman, Salviati, instead of replying to the points raised by Simplicio, now suggests a very different move. He suggests that this old problem simply be abandoned, as a search after fantasies, and that they search instead for a simple description of the patterns in which these objects move.

The present does not seem to me to be an opportune time to enter into the investigation of the cause of the acceleration of natural motion, concerning which various philosophers have produced various opinions ... Such fantasies ... would have to be examined and resolved, with little gain. For the present, it suffices ... to investigate and demonstrate some attributes of a motion so accelerated (whatever be the cause of its acceleration) that the momenta of its speed go on increasing, after its departure from the rest, in that simple ratio with which the continuation of time increases, which is the same as to say that in equal times, equal additions of speed are made. And if it shall be found that the events that then shall have been demonstrated are verified in the motion of naturally falling and accelerated bodies, we may deem that the definition assumed includes that motion of heavy things, and that it is true that their acceleration goes on increasing as the time and the duration of the motion increases. (158–9)

What we see here is Galileo changing the question that the physicist was to ask: instead of forces grounded in the metaphysical natures of things, he proposed, instead, to discover the *exceptionless patterns* or *regularities* of motion. Move, Galileo is saying, from a metaphysical problem that admits of no solution to an empirical problem that can be solved.[154] In fact, as we know, when Galileo gave up the quest for natures and natural forces and searched instead for empirical regularities, he was remarkably successful in discovering regularities in the motions of things. Not only did he discover the precise form of the motion of objects in, as one says anachronistically, 'free fall,' but he also discovered the regularities that describe the motions of all projectiles. The latter, as he discovered, always moved along curves having the form of parabolas.

In this context, mathematics did not attempt to describe a deeper underlying metaphysical essence as the syllogistic described the deeper ontological structure of the essences of things; rather, mathematics was for Galileo merely a tool to describe the observable motions of things, a tool to record the regularities in a convenient language.[155]

As Stillman Drake has convincingly argued in, for example, his *Galileo: Pioneer Scientist*,[156] Galileo was able to make his great breakthrough with regard to the motions of objects precisely because he gave up the search after metaphysical forces of the sort that Aquinas thought were there. The very notion of what science was up to had changed. But, it is equally clear, the notion of cause had not. To insist that science search after regularities *rather than* causes is not yet to replace the notion of cause as it had come down in the philosophical tradition with the notion of cause as regularity. It is not yet to replace the Aristotelian notion of cause that we find in Aquinas by the notion of cause as regularity that found its first clear statement in Hume. In order for this to happen, the whole notion of Aristotelian causes as active powers, forms, natures, and essences had to be subjected to criticism and eliminated. Only upon that elimination could the discovery of causes in the sense of regularities be proposed as the goal of science.

Now, as Popkin has argued, the early modern period was characterized by a widespread development of sceptical arguments and positions. These derived from Montaigne, of course, and found their greatest development perhaps in Bayle. But in the background there were such figures as the academic sceptic Simon Foucher and the Bishop of Rouen, Pierre Daniel Huet. There were many motives for this development, not least of which was the attempt to argue for tolerance on the basis of the fact that no one really knew anything for certain. Many of those who were active in pursuing the new science of Galileo were moved by just these motives. But, at the same time, these same arguments were an attack on the Aristotelian forms or natures or essences that were the central explanatory entities of the older science that was defended by the scholastics, such as Aquinas. Galileo's practical success in actually discovering laws that satisfied the new cognitive interest in matter-of-fact regularities undoubtedly had an impact on his contemporaries that cannot be underestimated. Still, the philosophical and sceptical arguments against the Aristotelian position were of great importance in the eventual success of the new science.

For the defenders of the new science, the Aristotelian patterns are not only wrong in suggesting the search for powers, but are positively inimical to a correct pursuit of knowledge via the methods of the new science. For Aristotle it is not universal regularities for which we are enjoined to search but patterns in sensible experience that are caused by natures. The goal of the new science is thus ignored. Which regularities are to be counted as lawful is, for the defenders of the new science, a matter of inductive inference, controlled by the data of sense experience; for Aristotle, what counts as a law is determined by insight into forms which comes from sense experience but is not controlled by it. For the new science, deduction subsumes events in experience under patterns discovered in experience; that is what explanation *is* for the new science. For Aristotle, in contrast, the deductive structure of the syllogism, while not a tool of discovery – it is a tool for teaching (*An. post.*, Book I, chapter 1), and the middles are discovered by induction – does play a substantive role in explanation in exhibiting the *ontological structure* of reality. Even if the empiricist and rationalist objections to this view of syllogism as revealing the ontological structure of reality were waived, one would still have to say, as both Bacon and Descartes said, that it is an impoverished logic. To explain reality one needed, Descartes argued, not just syllogism and definition by genus and difference, but the deductive richness of geometry – which, moreover, had not only a method of teaching, namely (as with Aristotle), the deductive or synthetic method, but also a systematic method of investigation which makes the inductive method as described and practised by Aristotle seem as if it proceeded by whim and fancy, metaphor and the imagination of poets, rather than the deliberate and systematic processes any investigation that pretends to be rational should surely demand. (For more on the deductive method in Descartes, see Study Seven, below.)

Furthermore, while Aristotle scorns mere empiricism, he has a need of it. He does regularly connect his philosophy of nature to the art of the craftsman. He tells us, for example, in the *Metaphysics* that 'experience is knowledge of particulars, art of universals ... knowledge and understanding belong to art rather than to experience ... men of experience know *that* a thing is so, but not why, while the others [that is, those who possess knowledge and understanding] know the why and the cause' (*Metaph.*, 981a15ff). For this reason, artists can teach, while men of experience – mere experience – cannot. The point claimed by Aristotle here is that if we know natures, we know causes, and if we know causes we can intervene systematically as the artist does to produce ends that we desire. In fact, however, the introduction by Aristotle of unnatural motions defeats this idea. If unnatural outcomes are always possible, then we cannot know that the outcome in a given case will conform to regularity or be an exception. Knowledge of Aristotelian laws is therefore useless for prediction – contrary to those who defended the new science, for whom explanation is by exceptionless regularity and is therefore symmetrical with prediction.[157] We can know whether an Aristotelian natural cause is operating, rather than some external force, only *ex post facto*, after we note the outcome. Aristotelian science is therefore useless for the craftsman who requires prediction in order to achieve control. In other words, even if there is such a thing as an Aristotelian science, it is, contrary to what Aristotle claims, useless for the craftsman, and must be supplemented by the second-best but far more useful matter-of-fact inductive science that both the empiricists and the rationalists of the early modern period were concerned to defend.

But if the Aristotelian science is useless, that is something that is, in a way, of only incidental concern to the philosopher – though a concern with human well-being must object to a philosophy which directly conflicts with the practice of empirical science which *is* useful. More fundamentally, the concern of the philosopher must be whether or not it is true. It would become the empiricist position, developed systematically by Hume, that the whole philosophy of natures, substances, unanalysed powers, and objective necessary connections is fundamentally false. On these points it turns out that there is in fact a certain disagreement between the rationalist and the empiricist, the former defending objective necessary connections while the latter rejects them. While both the rationalists and the empiricists defended the new science against Aristotle and his scholastic successors, none the less they differed on this point: here we find their responses to Aristotle to be quite different. It is to this difference that we must now turn.

III: Rationalist versus Empiricist Accounts of the New Science

The empiricists and the rationalists were agreed on the cognitive ends of the new science. They were agreed that the proper method was that of experiment, and they were agreed that the logic of experiment was the logic of eliminative induction. Finally, they were agreed that the cognitive goals and methods of Aristotle were not only different from those of the new science but were in fact a positive hindrance to progress in the latter. But there is also a crucial difference between the rationalists and the empiricists. As it turns out, it is the latter whose views are in conformity with the practice of the new science, while the rationalists adopt a position that is in most important aspects indistinguishable from the viewpoint of Aristotle.

The experimental method of the new science is an eliminative method. If the logical mechanisms of falsification are to succeed not only in eliminating hypotheses but to result in the acceptance of an uneliminated hypothesis, then one must have reason to believe antecedently that there is only a limited range of possible hypotheses and that one of the hypotheses in this range is true. These principles that must be antecedently accepted are the principles of limited variety and of determinism. Both Bacon and Descartes saw clearly the need for such principles. The difference between the two responses to Aristotle consists in the different status each assigns to the principles of determinism and limited variety.

The most spectacular instance of the new science was the work of Newton.[158] Newton's achievement was to 'demonstrate the frame of the System of the World,' as he described it (page 397) at the start of Book III of the *Mathematical Principles of Natural Philosophy*. What Newton demonstrated was *a process law for the solar system*:[159] it was this that captured the imagination of the world, and made the triumph of the new science secure. At the same time, he had eliminated, in Book II, section ix, the Cartesian rival explanation.

As the rejection of Descartes's position indicates, Newton's science proceeds by a process of elimination. This shows its connection to the method defended by Bacon and Descartes. In fact, the method used by Newton is logically of a piece with the methods of eliminative induction described by Bacon and Descartes. But the mathematical formulations used by Newton, since they were permitted by his subject matter, enabled him to employ the eliminative methods in a highly sophisticated way.

The mathematical formulations also permitted Newton to represent the process law for the system in a highly compact way. Bacon spoke of the process law as describing the latent processes of a system. The laws that Bacon is talking about are laws that describe necessary and sufficient conditions for all states of the system. The processes were said to be latent because they were largely unknown in detail. The problem is to grasp that detail. It is often not possible.[160] Newton discovered that this detail could be represented mathematically in a highly compact way in the case of the solar system, and, more generally, in the case of n-body mechanical systems.

What Newton discovered here was that motions depended upon forces and that the forces could be represented as the product of mass and acceleration, where acceleration is the rate of change of velocity and distance the rate of change of distance.[161] That is, acceleration is the second derivative of distance with respect to time. We have the ordinary differential equation of the second order, which we would represent as

$$(@)\qquad F = m \times [d^2s/dt^2].$$

When this is integrated, we obtain a family of curves for the path or orbit of the object through time. Initial conditions provide values for the constants of integration, and pick out which among this family is the actual orbit of the object. Different possible values for the constants of integration give the necessary and sufficient conditions for any particular state at any particular time. There will be one such equation for each object in the system. Solving them jointly yields the development of the whole system over time. The family of differential equations provides, in other words, the process law for the system. The fact that the family of equations can be solved analytically only for the simple case of the two-body system is a minor point. It simply means that for more complex systems the solution must be represented as the limit of a convergent infinite series. This means that we can never actually reach the real solution though we can always approximate it as closely as we want. But this is a technicality that need not detain us.

The problem in practice is to find the force function F to insert in ($@$). That is: What specifically is the force that governs the motions of objects in any given system? Newton succeeded in solving this problem for the solar system.

Newton makes two assumptions. For one, he assumes that *there are* forces satisfying condition ($@$). For two, he assumes that these forces *satisfy certain conditions*, including the action-reaction law and the law for the vector addition of forces.

Next, he takes for granted the observations that confirm Kepler's three laws. (In the third edition of the *Principia* these data are listed at the beginning of Book II as 'Phenomena.')

Assume the simplest case of a two-body system. The assumption is that *there is* a force that satisfies ($@$). But there are an infinite number of possibilities for this: for example, it varies inversely with the distance, inversely with the distance squared, inversely with the distance cubed, and so on, independently of the masses, directly with the product of the masses, directly as the product of the squares of the masses, and so on. What Newton is able to deduce is that if Kepler's first two laws are satisfied, then the force varies inversely as the square of the distance (Book I, sections ii and iii). The assumptions determine that a certain range of hypotheses as to the force function is acceptable. How is this range to be narrowed? How are we to eliminate all hypotheses but one? What Newton was able to show was that, if we accept Kepler's laws, then that leads demonstratively to the *exclusion of all possibilities but one*, namely, the force function that we have come to know as the gravitational (Book I, Proposition XI). Since we assume that there is such a force, and since we now know that this is the only possibility compatible with Kepler's laws, accepting the latter requires us to accept that the force that is, as one says, acting is the gravitational. Insofar, then, as we accept Kepler's laws on the basis of observation, we are to that extent required to accept that the force function is gravitational. And once we have the force function we have the required differential equations (@) that we can solve to obtain the process law for the system.

The logic is essentially that of eliminative induction. Observational data lead us to accept as correct the one uneliminated hypothesis from a previously given range within which we previously accepted that there is one that is true. What Newton uses in addition, however, is a sophisticated mathematical apparatus that enables him to eliminate all alternatives but one by deducing that, if we accept certain well-supported inferences from the observational data – namely, Kepler's laws – then we must accept one specific hypothesis from the infinite range and reject all others as false. Newton does not proceed to collect data that exclude *seriatim* one hypothesis after another from the range; rather, he uses data to draw inferences that then enable him to deduce the one hypothesis that must be accepted and whose acceptance entails the rejection of all others in the range. Given the deduction, the one set of data serves simultaneously to determine the acceptable hypothesis and to reject the remainder. The logic is that of eliminative induction; the mathematics makes the processes of elimination proceed in a highly efficient manner. But the point remains: the logic is the Baconian and Cartesian logic of eliminative induction. *Newton's great inference conforms to the rules of the experimental method of the new science, that is, the rules of eliminative induction; and it has as its product a process law, that is, a law that provides the best explanation of the sort that forms the cognitive aim of the new science.*

Newton's actual inference is, in fact, more complicated than this, and still more impressive. He recognizes that Kepler's laws are not exactly true, and that deviations from them are caused by the fact that the objects in the solar system do not form a two-body system but rather (as he thought) a seven-body system (see Book III, Proposition III, and the Scholium to Proposition XIV). Thus, we must take into account not only the gravitational forces between a planet and the sun, but also the forces of other planets that act on the one in which we are interested. Thus, in calculating the orbit of Mars one needs to take into account not only the gravitational force of the sun but also that of Jupiter. When we move from two-body systems to these more complicated one, then the mathematics becomes more complex. But Newton is able to show not that there is exactly one hypothesis about forces that is true, namely the gravitational, but only the somewhat weaker conclusion that the true hypothesis is in a very narrow interval around the gravitational hypothesis. The eliminative methods are at work again, but this time the mathematics does not allow Newton to conclude on the basis of the data that exactly one hypothesis is true. He can conclude only that the true hypothesis is in a certain very narrow range – sufficiently narrow that one can for all practical purposes assume that the force function is the one that is at the centre of the range, namely the gravitational. In any case, for our purposes the major point is that the method that Newton is using, for all its mathematical sophistication, is in its logic that of eliminative induction.

In these inferences one has to assume what in effect are principles of determinism and limited variety. These are the assumptions, first, that *there are* forces, and, second, that these forces *satisfy certain conditions. These assumptions are given by the basic laws or axioms of mechanics.* In effect, they are given by what we have come to know as Newton's three laws, together with the law for the vector addition of forces.

The first law states: 'Every body continues in its state of rest, or of uniform motion in a right line, unless it is compelled to change that state by forces impressed upon it.' Newton has defined the notion of 'motion' in his 'Definition II': 'The quantity of motion is the measure of the same, arising from the velocity and quantity of matter conjointly.' Motion is thus defined as what we now call 'momentum':

$$m \times v$$

The 'change of motion' mentioned in the first law, then, is the product of the mass and the change of velocity:

$$m \times dv = m \times d^2s.$$

It is clear that the 'forces' mentioned are the momentary forces that are required to introduce the instantaneous change of state, that is, change of motion (= momentum). This is clear also from 'Definition IV' where we learn that 'An impressed force is an action exerted upon a body, in order to change its state, either of rest, or of uniform motion in a right line.' The reference, therefore, is to what we would represent by

$$F \times dt.$$

Thus, what the first law, the 'law of inertia,' asserts is that

> Whenever there is a change of state $m \times dv$ then *there is* an external force $F \times dt$ which provides a necessary and sufficient condition for that change

though it does *not* say *specifically* exactly *how* the external force determines that change, that is, exactly what is the functional relation between the force and the change of motion. The first law thus asserts that *there are certain laws, without asserting specifically what those laws are.* But it does lay down certain restrictions on these laws: these laws *satisfy the condition* of relating accelerations to circumstances. The first law states: for any object *there is* in the external circumstances in which it is situated a necessary and sufficient condition for any change of motion. Contrary to what Aristotle had held, it is accelerations and not velocities that are correlated to circumstances.

The second law states: 'The change of motion is proportional to the motive force impressed; and is made in the direction of the right line in which that force is impressed,' or, in symbols,

$$F \times dt = m \times dv$$

or, what is the same, the more familiar

$$F = m \times [dv/dt]$$
$$= m \times [d^2s/dt^2]$$

that is, (@).[162] This law lays down a condition that *further limits the logical form* of the laws that the first law asserts to exist: the functional relationship is one of direct proportionality and determines the acceleration to be in a certain direction. The third law states: 'To every action there is always opposed an equal reaction: or, the mutual actions of two bodies upon each other are always equal, and directed to contrary parts.' This law *still further limits the range of possible hypotheses.*

Finally, as a further limitation, there is the law of the vector addition of forces, the 'composition law' for mechanics: 'A body, acted on by two forces simultaneously, will describe the diagonal of a parallelogram in the same time as it would describe the sides by those forces separately.' Newton refers to this as 'Corollary I,' which he then generalizes to any number of forces in 'Corollary II.' In fact, this law is not a 'corollary' but is an independent law. For our purposes, however, we need not speculate about what leads Newton into this logical error.

We should note that if we obtain specific laws of the form (@) for each of the two bodies in a two-body system, then we have a system of differential equations that can be solved, up to constants of integration, to describe how the system changes over time. In other words, what we have in the specific laws that the first law asserts to obtain is a *process law* for the system. The composition law shows how we can extend this to the *n*-body case.

Newton uses these laws to assure himself that there are forces that explain the motions of the planets, and he uses them to pick out a range of possible hypotheses. Then, using data obtained by observation, he eliminates all but one of these hypotheses, which he then accepts as the law that correctly describes the motions of the solar system. What he deduces is that the relevant force is gravitational. Then, knowing this specific law for interactions in the solar system, he can obtain the process law that was so to impress the world that subsequently no one was ever able to challenge the notion that, so far at least as the inanimate world is concerned, it is the new science alone that yields knowledge.

We have seen that Newton's method is that of eliminative induction, aided by some sophisticated mathematics. We now see that what we now call *Newton's laws of motion constitute the principles of determinism and limited variety that are essential if the eliminative mechanisms are to work.*

In eliminating alternatives, Newton eliminates the conjectures of Descartes, though he also devotes a special argument in Book II to show that the Cartesian suppositions must be false if we accept, on the one hand, the Newtonian laws of motion, and, on the other hand, the data that require the orbits of planets to be elliptical.

This method conforms to that proposed by Descartes himself. He, too, aims to give a description of the motion of the heavens.[163] He takes for granted the 'principal natural phenomena [of the heavens] to be investigated ...' (*Principles*, III, ¶ 5). He appeals to these, and, more specifically, to such then-recently discovered phenomena as the phases of Venus to reject the Ptolemaic description of the solar system (III, ¶ 16). He accepts the mathematics of Copernicus and Tycho as correctly describing the system; these two systems are equivalent as mathematical devices, differing in effect only as to the system of coordinates that are used: earth-centred (Tycho) or sun-centred (Copernicus) (III, ¶ 17). These motions suffice, Descartes believes, to explain the phenomena that he has listed: 'all this is easy for those who have some knowledge of Astronomy' (III, ¶ 37). But he must, as it were, fill in the details of the process, what Bacon called the latent process.

The framework is established by certain basic laws. In particular, like Newton he introduces the law of inertia. This he has introduced in Book II of the *Principles*. Here he states '[t]he first law of nature: that each thing, as far as is in its power, always remains in the same state; and that consequently, when it is once moved, it always continues to move' (II, ¶ 37); and '[t]he second law of nature: that all movement is, of itself, along straight lines; and consequently, bodies which are moving in a circle always tend to move away from the center of the circle which they are describing' (II, ¶ 39). These laws, and others that Descartes proposes, provide the mechanical framework. He appealed to this framework in his discussion of Harvey and the circulation of the blood. He appeals to it also in the case of the heavens.

Descartes, in fact, does not know the details of the process: 'we have not been able to determine ... the size of the parts into which this matter is divided, nor at what speed they move, nor what circles they describe' (*Prin.*, III, ¶ 46). What he proposes instead is to tell a 'likely story' to show how it is that the facts of astronomy, including the mathematical descriptions of Tycho and Copernicus, are compatible with the mechanical framework established by the law of inertia and the other basic principles. 'Let us suppose,' he says, 'if you please, that God, in the beginning, divided all the matter of which He formed the visible world into parts as equal as possible and of medium size ...' (ibid.). A series of further hypotheses follows. These enable Descartes to explain, so he claims, the accepted phenomena: 'These few {suppositions} seem to me sufficient for all the effects of this world to result from them in accordance with the laws of nature explained previously, as if they were [the] causes [of these effects]' (III, ¶ 47).

The story that results is, because the eliminative mechanisms have not done their work of eliminating all alternatives, not known to be true:

although perhaps in this way it may be understood how all natural things could have been created, it should not therefore be concluded that they were in fact so created. For just as the same artisan can make two clocks which indicate the hours equally well and are exactly similar externally, but are internally composed of an entirely dissimilar combination of small wheels: so there is no doubt that the greatest Artificer of things could have made all those things which we see in many diverse ways. (*Prin.*, IV, ¶ 204)

This will do, however: 'it suffices if I have explained what imperceptible things may be like, even if perhaps they are not so' (ibid.). This is because, so long as the likely story does actually describe the regularities among observed facts, then it matters not for practical purposes whether the supposed mechanisms actually obtain or not.

because Medicine and Mechanics, and all the other arts which can be perfected with the help of Physics, have as their goal only those effects which are perceptible and which accordingly ought to be numbered among the phenomena of nature. {And if these [desired] phenomena are produced by considering the consequences of some causes thus imagined, although false; we shall do as well as if these were the true causes, since the result is assumed similar as far as the perceptible effects are concerned}. (ibid.)

Others, like Newton and the other practitioners of the new science who came after Descartes, were not to be so easily satisfied: they wanted to go beyond hypotheses so far as the details of the process was concerned; they wanted the eliminative mechanisms to be employed to find which among the various 'likely stories' was the one that was actually true.

But for all the cognitive desirability of actually knowing the details of the process, it is none the less true that science must often settle for less. At earlier stages, such as those represented by Descartes, the development of 'likely stories' or 'how possibly' explanations as they have been called,[164] is an important feature of science.[165] Such a story, like the story that Descartes attempted to tell about the solar system, is 'likely' because it will be compatible with certain basic axioms. But it remains a 'story,' a 'mere' hypothesis, because the mechanisms of elimination have not shown that it is the only possible hypothesis. Descartes saw this clearly:

I wish what I shall write ... to be taken only as an hypothesis {which is perhaps very far from the truth}. But, even though these things may be thought to be false, I shall consider that I have achieved a great deal if all the things which are deduced from them are entirely in conformity with the phenomena: or, if this comes about, my hypothesis will be as useful to life as if it were true, {because we will be able to use it in the same way to dispose natural causes to produce the effects which we desire}. (*Prin.*, III, ¶ 45)

What was to happen, however, as we know, was that the later research of Newton was to establish that, though the Cartesian story was 'likely' in the way in which Aristotle's was not, because it was compatible, as Aristotle's was not, with the law of inertia, it was in fact false: one could not assume the basic laws of motion and the data that established elliptical orbits, and consistently suppose the Cartesian story to be true.

The point I want to emphasize here, however, is the basic compatibility that we have seen thus far between the Newtonian and Cartesian programs as parts of the new science. In particular, both understand the role of the eliminative mechanisms and the role of the basic laws of mechanics as constituting the relevant principles of determinism and limited variety that must be taken for granted if those mechanisms are to work.

The difference lies in the grounds which each uses to establish these basic principles. Here is Newton on the law of inertia, his 'Law I':

Projectiles continue in their motions, so far as they are not retarded by the resistance of the air, or impelled downwards by the force of gravity. A top, whose parts by their cohesion are continually drawn aside from their rectilinear motions, does not cease its rotation, otherwise than as it is retarded by the air.

Galileo assumed the law of inertia; under this assumption he was led to form a hypothesis concerning projectile motion; this hypothesis was confirmed by data that also eliminated alternatives, for example, circularity. Thus, discovering a law while guided in the research by the assumption of the law of inertia tends to confirm the law of inertia, and renders its use elsewhere reasonable. *For Newton the axioms of mechanics are acceptable because they have led to the discovery of specific laws.* The laws, of course, have the mixed quantificational structure characteristic of principles of determinism and limited variety. That means they are not falsifiable. Thus, failure to find a law that the theory asserts to be there does not falsify the theory; it may equally be true that one has simply not searched hard enough. This is the lesson about method in science and about the logical structure of theories that Kuhn has made abundantly clear to us.[166] But from this it does not follow, as Popper would foolishly have it, that these laws are therefore somehow non-empirical. To the contrary, there is no reason in logic to reckon that a statement is non-empirical simply because it involves an existential quantifier. A statement is to be reckoned empirical just in case that matter-of-fact evidence is used to argue for and against it. Newton so uses matter-of-fact evidence. Thus, *for Newton and therefore for the new science that he represents, the basic axioms that constitute the principles of determinism and limited variety essential for the working of the experimental method are factual claims, a posteriori, to be defended on the basis of empirical and, ultimately, observational evidence.*

The same point can be made with respect to the third law, that 'To every action there is always opposed an equal reaction ...' Here is one illustration:

If a horse draws a stone tied to a rope, the horse (if I may to say) will be equally drawn back towards the stone; for the distended rope, by the same endeavour to relax or unbend itself, will draw the horse as much towards the stone as it does the stone towards the horse, and will obstruct the progress of the one as much as it advances that of the other.

Here is another, the case of impact, that is, of what we have come to know as impulsive forces:

If a body impinge upon another, and by its force change the motion of the other, that body also (because of the equality of the mutual pressure) will undergo an equal change, in its own motion, towards the contrary part.

Note what Newton does here. He in fact provides us with two *very specific* laws. He then, in effect, points out that they have a certain *generic form in common*, to wit, that mentioned by the second law. Newton then *generalizes* from these specific laws, to the claim of the third law that for every specific sort of system falling under the genus of mechanical system the specific force law that applies in that system will share this same generic form. This generalization is the third law. What it does is place generic constraints on the specific laws that we may expect to hold in mechanical systems of specific sorts that we have not yet examined.

Newton's procedure is the same with respect to the second law: 'The change of motion is proportional to the motive force impressed ...' The 'change of motion' refers, as we have seen, to 'change of momentum.' The 'forces' that Newton has in mind are, again, clearly impulsive. As the example used in the case of the third law makes clear, it is this sort that he has in mind when he provides the empirical illustration of the second law:

If any force generates a motion, a double force will generate double the motion, a triple force triple the motion, whether that force be impressed altogether and at once, or gradually and successively.

Again note what Newton does here. He first takes the case of purely impulsive forces, that is, those that are 'impressed altogether and at once.' He then considers those that are 'impressed ... gradually and successively.' The 'successively' covers the case of a sequence of impulsive forces. The 'gradually,' however, adds the notion that the forces may compound incrementally. The law itself *generalizes* from these specific cases to a *generic form* that applies to all cases, including those where the incremental steps become infinitesimally small and we reach the limit of continuously applied forces.

Cohen is misleading when he asserts that

[t]he projected revisions of the Second Law and the basis for using it to derive a curved trajectory for the motion of projectiles have enabled us to see how Newton justified the application of the Second Law in the *Principia* to continuously acting forces of gravity. We have seen Newton state more than once in the *Principia* that such a result was implied by the Second Law, although in point of fact it seemed to follow only from the particular nature of Def. VIII and was not ever proved to be a valid extension of Newton's Second Law save as an intuitive extension of the law from impulses to continuous forces, a step which he never justified by rigorous logic or by experiment. But Newton's intuition served him well, for it enabled him to apply correctly a law which originated in

the seventeenth-century problem of impact to the universal problem of the gravitational forces governing the motions of the heavenly bodies. Newton's genius enabled him to transfer the concept of real forces, seen in impact, and in terrestrial weight, to imagined forces such as forces on the moons, tides, planets, and comets, and thereby to construct the first satisfactory dynamics of the world-system. This was the decisive step in constructing the physical system of the *Principia*.[167]

Cohen's claim is that Newton justified the second law by reference to the case of impulsive forces and that he 'never justified by rigorous logic or by experiment' the generalization to the case of infinitesimally short impulses, that is, continuous forces, even though Newton successfully applied the second law in the latter sort of case in a way that led to the discovery of certain laws. Cohen is correct: the general case provides Newton with the rule that leads him to the discovery that objects that obey Kepler's laws require a central force of the inverse square form. This is proved for the circumjovial moons in Proposition I, Theorem I, of Book III of the *Principia*, and is generalized to other planetary objects in later theorems. The central theoretical principles are proved in Book I, and the proofs make clear that Newton conceives of central forces as impulsive, but that the impulses are applied instantaneously and successively, that is, the law is taken to apply to continuously operating forces. This extension of the second law from the illustrative example of impulse is not a matter of 'rigorous logic': to the contrary, it is a matter of empirical generalization. But it *is* justified by 'experiment.' For, in the first place, the law is a generic law obtained by generalizing from certain specific laws, in particular, those concerning impulsive forces. It is this sort of data, the successful application to known cases, that provides the initial justification for Newton's accepting the second law. This sort of data is, of course, 'experimental.' Secondly, this generic law is then applied to new systems, specifically such systems as the systems of the circumjovial moons, the circumsaturnal moons, the circumsolar planets, and so on, and the result is the discovery of specific laws that describe the motions in those systems. In each of these cases there is the delimitation of a range of hypotheses by the generic laws, including the second law; these hypotheses share a certain common form determined by those generic laws; the latter assert that exactly one of these specific hypotheses is true; subsequently, the observational data confirm one of these and eliminate the rest as false. The discovery of these laws that the generic laws, including the second law, assert to be there, to be discovered, in turn confirms the generic laws of the theory. It is this *success* in guiding research in hitherto unexplored areas, including that of continuously acting forces, that provides the subsequent justification by 'experiment' for Newton's further application of his theory to cases to which it originally had not been applied. Thus, contrary to Cohen, it was not just intuition but solid scientific practice that justified Newton's applying the second law to the wide range of cases in which he tested it and found it to be successful. Intuition played a role only in helping him to pick out among the possible generic hypotheses that fit the original body of laws the one that could be successfully applied in the discovery of new laws. This sort of 'intuition' plays a key role, as Kuhn has emphasized,[168] in creating new generic theories in periods of revolutionary science.[169] It does not follow from this, however, that acceptance of a generic theory like Newton's is somehow founded on something other than 'experiment' and observational data.[170]

How, then, does Newton justify accepting the generic theory that he uses as a guide in experiment, that is, as principles of determinism and limited variety? *By observational data.* The principles that he cites are accepted initially because they generalize from known to unknown cases, from known laws that exemplify a certain logical form to the law about laws that any specific system of a generic sort will exemplify a specific law having that same generic form.

The method that Newton uses, we should emphasize, is that proposed by Bacon. Bacon criticizes the Aristotelians, who fly 'from particulars to axioms remote and of almost the highest generality ... and taking stand upon them as truths that cannot be shaken, proceed to prove and frame the middle axioms by reference to them ...'[171] For Bacon, in contrast, 'the middle are the true and solid and living axioms, on which depend the affairs and fortunes of men ...' (*New Organon*, I.civ). The middle axioms are, clearly, what we have called the specific laws. Bacon proposes that the reason derives 'axioms from the senses and particulars, rising by a gradual and unbroken ascent, so that it arrives at the most general axioms last of all' (I.xix). The propositions that we should take as axioms 'are not [the] abstract' principles of the Aristotelians (I.civ) but generalizations of the middle axioms; the axioms should be such that 'those intermediate axioms are really limitations' (ibid.), that is, they should be generic laws of which the specific laws are instantiations. The relation between general axioms and specific laws as characterized by Bacon is thus the relation between the axioms of mechanics and the specific laws investigated in practice by Newton. Further, according to Bacon, the middle axioms or specific laws are established by a process of 'rejections and exclusions' (I.cv), that is, by the method of eliminative induction, the method that Newton used to discover the laws for the circumjovial planets and for the solar system. In addition, according to Bacon, 'in establishing axioms by this kind of induction, we must also examine and try whether the axiom so established be framed to the measure of those particulars only from which it is derived, or whether it be larger and wider. And if it be larger and wider, we must observe whether by indicating to us new particulars it confirm that wideness and largeness as by a collateral security ...' (I.cvi). The general or generic axioms apply not only to the specific systems from which they are first abstracted but to further systems hitherto unexamined; they delimit ranges of hypotheses for these further systems, and if the method of 'proper rejections and exclusions' leads to the affirmation of specific laws in these new areas, then that constitutes *empirical evidence further confirming the general axioms*; it provides, as Bacon says, '[c]ollateral security.'[172] And this, of course, is Newton's procedure. He obtains his axioms by abstracting from specific cases such as Galileo's law and the action of impulsive forces; he then applies these generic axioms to further cases, such as the solar system and the circumjovial planets; by a process of elimination he obtains laws for these systems that are themselves instantiations or limitations of the generic axioms; the discovery of these further laws provides additional evidence confirmatory of the axioms. *Newton's method in practice is the empirical method outlined by Bacon.*

To this empirical method of Bacon and Newton we may now contrast the procedure of Descartes. This will bring out the differences between the empiricist response to Aristotle and the rationalist response.

Descartes first argues that

[w]e ... understand that it is one of God's perfections to be not only immutable in His nature, but also immutable and completely constant in the way He acts ... From this it follows that it is completely consistent with reason for us to think that, solely because God moved the parts of matter in diverse ways when He first created them, and still maintains all this matter exactly as it was at its creation, and subject to the same law as at that time; He also always maintains in it an equal quantity of motion. (*Prin.*, II, ¶ 36)

He uses this to derive the law of inertia:

Furthermore, from this same immutability of God, we can obtain knowledge of the rules or laws of nature, which are the secondary and particular causes of the diverse movements which we notice in individual bodies. The first of these laws is that each thing, provided that it is simple and undivided, always remains in the same state as far as is in its power, and never changes except by external causes. (II, ¶ 37)

These laws do apply to particular things; we can observe that the actions of the latter conform to these laws. But, unlike the case of Newton, it is not to this observational evidence from the phenomena that Descartes appeals when he offers reasons for accepting the law of inertia. Rather, he *deduces* it from the immutability of God, something which is in turn established by *reason alone*. In short, Descartes holds that the basic axioms are a priori truths, *clear and distinct, or self-evident, truths*.

This marks the basic difference between the empiricist and rationalist responses to Aristotle. Both accept that the method of eliminative induction is the method of the new science that they are both concerned to defend. This means in particular that they both accept the necessity of assuming certain principles of determinism and of limited variety as guiding the use of empirical data with respect to the elimination of false hypotheses. But for the empiricists these principles are themselves empirical, acceptance of which is justified by the discovery of laws that they assert are there to be discovered. For the rationalists, in contrast, accepting the basic axioms is justified only if they are either themselves self-evident or can be deduced by self-evident steps from other propositions that are self-evident. Thus, Arnauld tells us that '[a]ll that is contained in the clear and distinct idea of a thing can be truly affirmed of the idea of that thing,'[173] and then establishes as his 'First Rule' for axioms that '[i]f moderate attention to the subject-idea and to the attribute suffices to show that the attribute is truly contained in the subject-idea, then we have a right to take as an axiom the proposition joining the attribute with the subject-idea.'[174] Conversely, where there is no self-evidence, then we cannot have an axiom. This is the 'Second Rule': 'When simple consideration of the subject-idea and the attribute is insufficient for our seeing clearly that the attribute belongs to the subject-idea, the proposition joining the two ideas must not be taken as an axiom.'[175] The rule, in short, is none other than the first of Descartes's rules of method: 'never to accept anything as true that I did not know evidently to be such; that is to say, carefully to avoid haste and bias, and to include nothing more in my judgments than that which presented itself to my mind so clearly and distinctly that I have no occasion to place it in doubt.'[176]

To this we contrast Newton's *Rules of Reasoning in Philosophy*, the first of which asserts that '[w]e are to admit no more causes of natural things than such as are both true and sufficient to explain their appearances.' What counts is the *capacity to explain and predict*. There is no requirement that the alternatives to the axioms be inconceivable; it is factual evidence that counts, not self-evidence.

Moreover, prediction is to be based on exceptionless laws; causes, in other words, are to be understood as implying generalizations. Newton makes this point when he states his second 'Rule,' which, it should be noted, he takes to be a straightforward inference from the first: 'Therefore to the same natural effects we must, as far as possible, assign the same causes.' He gives several examples. One is 'the descent of stones in Europe and in America.' The first rule tells us that the assigned cause must be 'true,' that is, true as a matter of fact, and a sufficient condition for the effect. The first rule tells us, in other words, that when we assign a cause we need inductive evidence that supports the claim that the cause is in fact sufficient for the effect. But Galileo did provide such evidence. The second rule tells one to assume that these have the same cause unless one has reason, *empirical reason*, to suppose something to the contrary. That is, one will take it to be *not possible* to assign the same cause to the same effect only when observational or inductive evidence establishes that it is not possible. The second rule thus instructs us to base our inferences to causes or, what is the same, to laws on empirical or inductive evidence. Thus, the second rule tells one to assign the same cause to all motions directed towards a massive central body: descending stones towards the earth, the moon towards the earth, the circumjovial planets towards Jupiter, and the solar planets, including the earth, towards the sun: the force of gravity that moves the stone towards the earth should be assumed to be the relevant cause in each case, as Newton argues in the Scholium to Proposition IV, Theorem IV, of Book III.

What Newton means to allow as evidence and what he means to exclude is further explained in Rule IV: 'In experimental philosophy we are to look upon propositions inferred by general induction from phenomena as accurately or very nearly true, notwithstanding any contrary hypotheses that may be imagined, till such time as other phenomena occur, or by which they may either be made more accurate or liable to exceptions.' It may well be that contrary hypotheses can be *imagined*. Since they are imaginable, they are clearly possible, not contrary to reason; and since they are not contrary to reason, it follows that the proposition that is accepted is not self-evident. A general proposition – for example, Newton's axioms – may thus be accepted on the basis of inductive evidence even if it does not conform to the rationalist rule of Descartes and Arnauld that requires axioms to be self-evident.

Bacon accepts the same point. He is critical of the Aristotelians for flying from sense particulars to first principles and 'taking stand upon them as truths that cannot be shaken ...'[177] This he contrasts to his own method, which moves from particulars to middle axioms to more general axioms 'of which those intermediate axioms are really limitations' (*New Organon*, I.civ). Such a more general axiom, since it is 'larger and wider, we must confirm whether by indicating to us new particulars it confirm that wideness and largeness ...' (I.cvi). Where such confirmation is possible, so is disconfirmation: it is thus possible to conceive or imagine phenomena by which, as Newton puts it, the axioms of which Bacon speaks 'may either be made more accurate or liable to exceptions.' Where contrary evidence is imaginable, so, too, is the falsity of the proposition. In short, for Bacon, axioms are not self-evident, in conformity to

Newton's practice but contrary to the position of Descartes and Arnauld.[178] To be sure, the axioms, if arrived at by proper inductions will be certain – after all, one of Bacon's criticisms of Aristotle is that the method of the latter yields no certainty! The point is that the certainty at which Bacon aims is what later thinkers called 'moral certainty' as opposed to the 'infallible certainty' of self-evidence.[179]

If Newton in 'Rule IV' emphasizes once again the relevance of observational inductive evidence, he also excludes certain ways of challenging the results of inductive inference: we are to accept for purposes of explanation and prediction the results of inductive inference 'notwithstanding any contrary hypotheses that may be imagined.' This rule appeared only in the third edition of the *Principia*, and was apparently prompted by the suggestion of his correspondents that Newton's inductive inference to the law that motions in the solar system can be explained by appeal to the law of gravity could be challenged by the hypothesis that the real cause of those motions is not gravity but an unknown power of the deity. Newton's point is that the suggestion that induction reveals only the apparent cause whereas something else is the *real* cause is a suggestion *not supported by any evidence whatsoever*: it is a *mere* hypothesis. But in the absence of matter-of-fact evidence, Newton is saying, there is no reason to accept that hypothesis, nor, therefore, any reason to reject the conclusion of the inductive inference.

The hypothesis could be taken to provide grounds for rejecting the induction only if it is accepted as true, that is, as *determining* what is to be explained rather than as suggesting further experiments that might be performed. Newton had made this point already in a letter to Oldenburg intended for the French Jesuit Ignatius Pardies, who had raised objections to Newton's account.[180] Pardies had suggested two grounds for questioning Newton's account. One was based on certain misunderstandings of Newton's experiments (Newton, *Correspondence*, 170); these matter-of-fact issues are not our present concern; they can be replied to by matter-of-fact arguments in any case (169). What matters for us, rather, is the methodological point implicit in Pardies's claim that Newton's results could be challenged because there are explanations – including those of the Cartesians, as well as those of Grimaldi and Hooke – that account as well for the facts as does Newton's explanation. Newton argues, correctly, that Pardies offers *hypotheses* – *mere* hypotheses – to argue against his (Newton's) experimental results. This, Newton holds, is methodologically illegitimate. As he puts it: 'the best and safest way of philosophizing seems to be this: first to search carefully for the properties of things, establishing them by experiments, and then more warily to assert any explanatory hypotheses. For hypotheses should be fitted to the properties which call for explanation and not be made use of for determining them, except in so far as they can furnish experiments' (169). Propositions are to be accepted only on the basis of observational inductive support, and they can be used to *determine* phenomena only *after* they have acquired such support. To suppose otherwise is to allow caprice to enter into science, and to destroy all certainty: 'if anyone makes a guess at the truth of things by starting from the mere possibility of the hypothesis, Newton does not see how to determine any certainty in any science; if indeed it be permissible to think up more and more hypotheses, which will be seen to raise new difficulties' (169).

For Newton, then, non-inductive support of the sort claimed by the Cartesian for advancing certain hypotheses as a priori acceptable does not exist; it is a figment of the imagination. The only hypotheses that are acceptable are those that have acquired inductive support and have, therefore, ceased to be hypotheses.

In any case, Newton's strictures against 'hypotheses,' including 'Rule IV,' make clear, as do the other rules, that Newton allows in agreement with Bacon and with his own practice that a proposition can function legitimately as an axiom for a theory, and be used to guide research, even if, contrary to the Cartesian methodology, it is not self-evident. More strongly, as his remarks to Pardies make clear, he thinks that the resort to hypotheses of the sort that one finds in the rationalists leads to arbitrariness in science: Cartesian methodology is contrary to the requirements of the cognitive goals of the new science to find, so far as possible, true matter-of-fact generalities.

Newton's strictures on hypotheses here echo, with little doubt non-accidentally, the criticisms that Bacon made of the Aristotelians. Aristotle had, Bacon argued, 'come to his conclusion before; he did not consult experience, as he should have done, for the purpose of framing his decisions and axioms, but having first determined the question according to his will, he then resorts to experience, and bending her into conformity with his placets, leads her about like a captive in a procession.' Bacon can't resist a dig: 'So that even on this count he is more guilty than his modern followers, the schoolmen, who have abandoned experience altogether.'[181] In effect, Newton is accusing the Cartesians of the same faults as those of which Bacon is accusing the Aristotelians.

As we shall now see, there is a real point to this charge.

IV: The Downfall of Rationalist Accounts of the New Science

One finds the difference between Newton and Descartes repeated in the textbook traditions that followed upon their work. Thus, consider J. Rohault's *System of Natural Philosophy*, a textbook in Cartesian science.[182] Rohault agreed, of course, with the critics of Aristotle, that the explanations proposed by Aristotelians in terms of forms were vacuous (*System*, I, 104); that in order to say something about the nature of objects, 'if we would succeed herein, and say something more than ordinary, we must descend to Particulars, notwithstanding the Custom of Philosophers, who seldom do so, but for the most part content themselves with proposing abundance of loose Questions, which we may look upon as superfluous, and from which we gain no advantage' (I, 102–3); and that the opinion of Aristotle in the case of motion in particular was a weak induction from imperfect experience (I, 47). He argued, against Aristotle, that the law of inertia could in fact be found to be exemplified in experience; he cited, like Newton, among other things, the example of Galileo's law for projectiles (I, 47). But this was his argument *against* Aristotle. Unlike Newton, Rohault did not conceive this as providing his argument *for* the law of inertia. Rather, Rohault argued, following Descartes, that the law of inertia was to be derived a priori from the self-evident truth of God's immutability (I, 46–8). Rohault's book had been translated by John Clarke, and notes of a critical nature were added by his brother, the Newtonian Samuel Clarke. These notes directed Newtonian criticisms at the Cartesian system that Rohault presented. Clarke argues that the preservation of motion of which Rohault speaks is illustrated in only some empirical cases, those involving perfectly elastic objects; for other objects, the Cartesian law does not hold (*System*, I, 46n, 48n), arguing in effect that the relevant evidence is empirical, not a priori (cf I, 83n). Clarke, in other words, challenges the claimed self-evidence of the Cartesian law. Moreover, in his notes he advances the Newtonian refutation of the Cartesian hypothesis concerning vortices (II, 73n–74n) and sketches Newton's argument for explaining Kepler's laws by means of an inverse square force (II, 74n–76n, 96n–98n).

To Rohault we may contrast William Whiston, whose *Sir Isaac Newton's Mathematick Philosophy More Easily Demonstrated*,[183] provided a fairly elementary textbook for the Newtonians. He repeats the Newtonian refutation of Descartes's vortex hypothesis (*Newton's Mathematick Philosophy*, Lect. 31), and outlines the demonstration of the frame of the system of the world (Lects 33, 34). These, of course, are based, ultimately, on Newton's laws of motion, which are presented early on in Lecture 5, which begins with a statement of the law of inertia. In contrast to Rohault, Whiston has no a priori arguments in favour of the law; instead, he offers as justification for accepting the law of inertia and the other laws of Newton's theory the same sort of empirical support, based on the shared form of known specific laws, that Newton himself offered, namely, such things as projectile motion, the motion of a top, and so on. There is in effect by way of omission a criticism of the Cartesian claim that the fundamental laws had to be established a priori.

John Clarke's Newtonian textbook, *A Demonstration of Some of the Principal Sections of Sir Isaac Newton's Principles of Natural Philosophy*,[184] repeated the standard Newtonian justifications for accepting the basic laws of motion. He added, in commenting on Newton's 'Rule IV' against hypotheses, that 'the real Nature and Constitution of Things is a Matter-of-fact, and my be come at by constant Observations and repeated Trials. It is impossible by abstract Reasoning ... to find out what the particular Qualities of that Matter [in the planetary system], and what Laws of Motion the several Parts or Particles of it are subject to' (*Demonstration*, 100). Newton's method, he went on, 'must surely be preferable to meer Dreams and Chimaera's which have no Ground in Nature, but are directly contrary to the whole Procedure by Experiments ...' (101). This is directly an argument against there being any role in (the new) science for the self-evident principles that the rationalists deemed to be essential to the experimental method.

Newtonian commentators on the Cartesian tradition were in these ways, explicitly or by omission, critical of the a priori starting point of the latter. What we find, of course, is that these arguments were successful: the Cartesian tradition gradually died out. The immediate consequence was that *the new science became wholly empirical*. The question is: What brought about this change?

In part, of course, we have given the answer: it was the Newtonian argument that a priori justifications were not needed because empirical, inductive justifications sufficed. But this is not by itself complete. For of course it was open to any rationalist to reply that although the moral certainty achievable by inductive argument sufficed for practical purposes, for purposes of philosophy, in contrast, one needed the infallible certainty of self-evidence. To proceed as the Newtonians did, and as Bacon did, was to settle for the second best. What was needed was an argument that the second best is the best we can do, and is therefore, *faute de mieux*, in fact the best.

This argument was in fact developed by the empiricist critics of the rationalists. There were several aspects to this argument. One: The concepts used by the rationalists, the notion of force in particular, became in the works even of the rationalists devoid of their non-empirical content; it gradually became obvious that there was no need for the non-empirical concepts which the rationalists none the less insisted were needed if we were to attain *scientia*. Two: Infallible certainty is unattainable, even on Cartesian grounds, in most areas. Three: Such certainty is in any case unattainable in principle. And four: An empirical investigation of the mind's own workings reveals the absence of any of those sorts of ideas that would have to be present were the Cartesian ideal of infallible certainty to be attainable.

1) *Disappearing Powers*

As we have seen, the cognitive goals of the new science were incompatible with the sorts of knowledge of matters of fact that the Aristotelian patterns required one to aim at. On Aristotle's account of the patterns that we discover in sense experience, one can have both natural and unnatural motions. The former are the laws of nature; the latter are exceptions to those laws. The aim of Aristotelian science is to know forms and, therefore, natural laws. But this means, then, as we have seen, that Aristotle aims at knowing regularities that, since they are not exceptionless, do not satisfy the cognitive goals of the new science. What Descartes and the Cartesians do is remove this fault: as they develop the Aristotelian position, there is no possibility of violent change. That being so, there is no possibility for there being exceptions to natural laws. This means that in the search for causes, the aim of the Cartesians turns out in an important way to be the same as that of the empiricist scientists who took Bacon to be their model.

Consider projectile motion once again. The natural motion of massy objects is to fall straight down, vertically towards the centre of the Earth. The projectile has a horizontal component to its motion. That motion is therefore unnatural. The natural motion is explained in terms of the exercise by the changing substance of its natural power. The unnatural motion is explained in terms of the exercise of a power by an external substance. The problem with projectile motion is that no one can discover the external force that is being applied.

Descartes makes several important points.

In the first place, he insists that when a substance moves there are no *powers* in the substance that are being exercised to bring about the motion. Thus, in the *Principles* he advances the issue, '[w]hat ought to be understood about the striving of inanimate objects towards motion,' and then argues: 'When I say that these little globules strive, {or have some inclination}, to recede from the centers around which they revolve, I do not intend that there be attributed to them any thought from which the striving might derive; I mean only that they are so situated, and so disposed to move, that they will in fact recede if they are not restrained by any other causes.'[185] There is no directive thought that moves the objects, no exercise of a power towards an end. The teleology of Aristotle has gone, at least so far as concerns the motions of ordinary bodies. Thus, *like Galileo, Descartes eliminates the appeal to Aristotelian forces*: in explaining motion in general and projectile motion in particular, there is no claim that one must appeal to the exercise of Aristotelian powers by the changing objects, whether internally or externally.

In this case, actually, Descartes is talking not about projectiles near the surface of the earth but rather of the motion of objects in the celestial regions. But the point is the same, since he has earlier argued that the Aristotelian distinction wherein the natural motions of terrestrial objects are in straight lines whereas those in the heavens are circular is spurious: both kinds of matter are the same, and both involve objects for which motion in a straight line continues unless impeded. He has earlier made the point '[t]hat space does not in fact differ from material substance,'[186] from which he concludes that 'this shows ... that the matter of the heaven and the earth is one and the same; and that there cannot be a plurality of worlds.'[187] That is, he concludes that the Aristotelian distinction between straight-line terrestrial motions and circular celestial motions is spurious.

But a second point follows from the fact that 'space does not in fact differ from material substance.' This is the point that upon the Cartesian scheme, *there is only one material substance*, namely, the whole indefinitely large thing called space. What are ordinarily called substances – for example, stones, projectiles, planets, human bodies – are merely *parts of* this one material substance. Now, the distinction between natural and unnatural motions depends upon there being two different substances, one acted upon, which is moving unnaturally, and one acting upon the former, causing the unnatural motion. Since there is only one material substance, it follows that it is no longer possible to draw the distinction between natural and unnatural motion. What we have is one material substance. This changes in accordance with its essence – though, to repeat the point just made, this does *not* involve the exercise of an active power on the part of the substance. All change in material substance is simply motion in conformity with the essence, with no possibility that another substance will force that substance to behave in some other way that is contrary to its essence. Hence, within the Cartesian scheme, *there is no unnatural motion*, motion in which a substance is forced to act contrary to its essence.

Upon the Aristotelian scheme, motion is in conformity to a natural law if it is motion in conformity with the nature or essence of the changing substance. Motion is contrary to natural laws if it is unnatural or violent. It is thus the possibility of unnatural motion that allows the Aristotelians to hold that there are violations of natural laws. Conversely, *when the possibility of unnatural motion is eliminated, there are no violations of natural laws*.

Of course, one can distinguish what happens when one cause operates from what happens when several causes operate. If there are several causes present and operating, then what happens is not the same as what happens when one alone is operating. In that sense, the presence of other causes impedes the effects that would occur were only one of them operating. As Descartes puts it, 'inasmuch as it often happens that several different causes act simultaneously against the same body and some impede the effect of others; depending on whether we consider the former or the latter, we can say that this body strives or tends to move in {several} different directions at the same time.'[188] Although we can speak in terms of impeding motions, and therefore as if there were such things as unnatural motions, in fact what we have is the action of several causes acting simultaneously, and there is nothing 'unnatural' about the motion that results: it is precisely that which is required by the essence of the material world of which the several causes and the body being acted upon are all parts.

The patterns that one aims to discover in the material world are therefore *patterns which have no exceptions.*

(To be sure, we must allow for miracles, when the all-powerful God intervenes directly in the workings of the material world to bring about a change that is not determined by the essence of material things. But physics, the science of material things, cannot be expected to take account of miracles. Having said this, one should also note that the possibility of miraculous intervention by a powerful substance does have a role to play in Cartesian thought. This is what makes room for the Evil Genius of the first of Descartes's *Meditations.* We shall have to examine this possibility below, and again in greater detail in Study Seven.)

Because the cognitive aim of Cartesian science thus becomes the discovery of exceptionless patterns, it is clear that it has the same aim as that proposed by Galileo and by the empiricists, such as Boyle and Newton, who followed Bacon. The rationalists and the empiricists thus stand jointly against the Aristotelians in making the cognitive aim of the new science the knowledge of exceptionless matter-of-fact regularities.

Having said this, it must also be said that when one takes into account the basic framework, then the Aristotelians and the rationalists are of a piece, and are together to be contrasted to the empiricists. The latter aim to know matter-of-fact regularities on the basis of empirical evidence. In contrast, the rationalists and the Aristotelians both aim at *scientia*, that is, at an incorrigible knowledge of *essential truths.* For the empiricists, incorrigible knowledge of matter-of-fact regularities is simply not possible: the inference to the regularity is an inference to a general or universal truth on the basis of evidence drawn only from a sample. So, of course, the knowledge is always fallible, corrigible, and therefore not *scientia.* Moreover, they have their doubts not only about whether essences can be known, but even about whether there are such entities. The empiricists, in other words, reject the framework that enables both the Aristotelians and the rationalists to claim that they can go beyond the merely probable inferences from sample to population to the incorrigible truths of *scientia.* What makes the rationalist or Cartesian position with regard to cognitive aims seem to coincide with that of the empiricists is not that the former abandoned substance metaphysics, as the latter in effect did. Rather, it has to do with some specific moves with regard to the essence of material things. On this point the Cartesians distinguished themselves from the Aristotelians by eliminating that feature of the latter's view that made it possible to speak of natural laws as having exceptions. The Cartesian view of material substance eliminated for practical purposes all possibility of unnatural motions. When that possibility went, so did the possibility of there being exceptions to natural laws. It is this which brought the rationalists close to the empiricist position on the cognitive aims of the new science, in spite of the fact that the two disagreed on whether *scientia* of the sort at which the rationalists, with the Aristotelians, aimed was even possible.

We should note, moreover, the point made previously, that for Descartes change in the material world is not explained in terms of the exercise of active powers by the changing bodies. In this respect, too, the Cartesians disagree with the Aristotelians and side with the empiricists: *in the material world there are no active powers*. The empiricists and the rationalists are also in this respect one with regard to the patterns they aim to discover in the material world.

This feature of the new mechanics was widely accepted. Thus, consider the small tract 'A Discourse of Local Motion,'[189] by the otherwise anonymous 'A.M.' A.M. attempts to lay out the basic regularities describing the motion of ordinary physical objects. In particular, he disagrees with the laws of motion that Descartes proposed in the *Principles*. Descartes got them wrong, but so did A.M. That, however, is not the present point. Rather, what is interesting is the way in which A.M. insists that '[a] Body is in it self indifferent to Rest or Motion' ('Discourse,' 1), rejecting the idea, as did Descartes, that motion always requires a mover. A.M. rejects the version of the Aristotelian account of the causes of motion according to which the external force moving a body 'impresseth therein a certain Quality, call'd *Impetuosity*, and that as long as this Quality lasts, the Motion lasts also' (8). Whatever such a quality may be, it certainly is not a cause. Even if the quality were to disappear, the motion none the less does not stop 'because there is no new Cause, producing Rest, and Motion cannot cease, but Rest must be produced instead thereof' (13). He goes on to indicate that '[t]he Bodies which we move, do cease to move, because they are impeded' (13). And like Descartes, he holds that while an object will continue to move in a straight line unless impeded, curved motion requires a cause (17).

The notion, common to the Cartesians and the empiricists, that in causation in physics or mechanics we find only regularity, not activity, was 'in the air.'

This makes clear how misleading it is to suggest, as some have done, that for Descartes rectilinear motion is natural.[190] The thought is that where the cause is an external impeding cause, the motion is unnatural. Since non-rectilinear motion is, according to Descartes's second law of motion, always explained in terms of an external, impeding cause, it must be unnatural. We must recognize, however, that the natural/unnatural distinction makes sense only in a context in which natures are active powers and explanation is in terms of these powers. But the world of material bodies contains no such natures. This means that the distinction makes no sense in the Cartesian context, at least at the 'practical' level. Instead of explaining things at the 'practical' level in two different ways, in terms, on the one hand, of natural and, on the other hand, of unnatural or violent motions, Descartes simply searches for *patterns*. Rectilinear motion is one such pattern. Motion brought about by external, impeding causes is another. They are both explanatory in the same way, as patterns. In effect, to suggest that for Descartes rectilinear motion is natural because it does not involve an external, impeding cause is to try to impose (or reimpose) Aristotelian categories on the patterns Descartes claims to be laws, even as Descartes has just finished eliminating those categories from mechanics at the 'practical' level. It is to misread what Descartes is trying to do.[191]

At the same time, however, as we insist that there are in material objects no active powers in the Aristotelian sense, we must also recognize that for the Cartesians, *activity does not entirely disappear.*

Descartes distinguishes 'the general cause of all the movements in the world' from 'the particular [causes], by which individual parts of matter acquire movements which they did not previously have.'[192] With regard to the latter, there is no activity, only pattern or regularity. However, '[a]s far as the general {and first} cause is concerned, it seems obvious to me that this is none other than God Himself, who, {being all-powerful} in the beginning created matter with both movement and rest; and now maintains in the sum total of matter, by His normal participation, the same quantity of motion and rest as He placed it in at that time.'[193] Two things are to be noticed.

First, God is an active force. Through His (or Her) activity changes occur in material things in accordance with the essence of that substance. So far as particular material things are concerned, there is no activity: there are only patterns, just as for the empiricist. But, none the less, there really are activities that connect the events that occur in material things. In this respect, Descartes agrees with the Aristotelians against the empiricists: for both the former, in contrast to the latter, causation is more than regularity, except that, for Descartes, the relevant activities are the activities of God. Where the Aristotelians locate causal activity not only in God but also in individual particular things, *Descartes locates the activities that connect events in material objects wholly and exclusively in God.*[194]

Second, from the nature of this activity, its immutability, Descartes deduces, or at least thinks that he can deduce, a priori the patterns that bodies must conform to in the material world. As he puts it, '[f]urthermore, from this same immutability of God, we can obtain knowledge of the rules or laws of nature, which are the secondary and particular causes of the diverse movements which we notice in individual bodies.' He then infers that '[t]he first of these laws is that each thing, provided that it is simple and undivided, always remains in the same state as far as in its power, and never changes except by external causes.'[195] Several other rules for the behaviour of material things are also deduced from the same premises. As is well known, Descartes stated a number of rules that do not in fact hold. For what we are about, that need not detain us.[196] What is important is that, for Descartes, *God plays a causal role in the explanatory scheme.* For Descartes as for Aquinas, the search for causal explanation leads in the end to a God, an all-powerful, active, necessary being. We shall be exploring this notion that God is ultimately involved in all explanations in Studies Six and Seven, below. For the present, the point simply is that *for Descartes, Aristotelian forces, located in God, continue to play a role in physical explanations.*

This is in contrast to the new physics that was developed by the empiricists, and by Newton in particular. Leibniz came to distinguish kinematics from dynamics. Kinematics provides the rules that describe all possible motions, while dynamics provides the rules that describe all actual motions. The laws of dynamics are justified by appeal to the active forces of substances. Thus, Leibniz, like Descartes, appeals to active forces to justify the claims he makes concerning the laws of mechanics. The Newtonians also insisted that one needed to introduce the notion of force into physics, and that it was necessary to separate dynamics from kinematics. But it is important to recognize that Newtonian dynamics is very different from the dynamics of Leibniz. Where Leibniz located the laws of dynamics in the activities of metaphysical substances, the Newtonians separated the notion of force from that of substances. As Leibniz put his case at one point, 'although all particular phenomena of nature can be explained mathematically or mechanically by those who understand them, it becomes more and more apparent that the general principles of corporeal nature and of mechanics themselves are nevertheless metaphysical rather than geometrical and pertain to certain forms or indivisible natures as the causes of what appears rather than to the corporeal or extended mass.'[197] Leibnizian forces, in short, are Aristotelian exercises of active powers of substances.

In contrast, for the mechanics of Newton, this notion of force as the exercise of an active power has disappeared. The Newtonians, unlike Leibniz, took seriously Galileo's move with respect to projectile motion: give up the search for Aristotelian forces and search instead for matter-of-fact regularities that correctly describe the motions in question. The concept of force in Newtonian mechanics is simply a useful notion for stating the regularities that we find hold for material things. It is a concept that can be measured and understood in terms of the observable characteristics of things. In contrast, the Aristotelian forces, whether they were understood to exist as Aristotle claimed, or as Descartes claimed, or as Leibniz claimed, are never given in sense. Pierre Costabel has emphasized this point. Noting the special sense in which the term 'dynamics' was used by Leibniz, he took up the issue whether one could use this term to describe anything of the mechanics contained in Newton's *Principia mathematica*. His central conclusion was that

[i]n this 'method of philosophy' [that is, that of the *Principia*], *huic philosophandi modo*, the question of knowing if force is real or not, if it is a primary notion or not, is not formally posed nor does it emerge. The question is one of 'deriving' a way of arguing from mechanical phenomena translated into mathematical terms, and in which the 'demonstrative' scheme should be applicable throughout. The mechanics of Newton is truly a *Mechanica rationalis*, but it is not dynamics in the sense of Leibniz.[198]

This is just: Newton's pragmatic way of treating the concept of 'force,' keeping it as a concept that could be used to state matter-of-fact regularities, while divorcing it from the metaphysical context it had had in both the rationalists and the Aristotelians, meant that there was a radical difference between the dynamics of Leibniz and that of Newton. It marks precisely that transition from the older ways of explaining to the new ways of explaining in the empirical science of Bacon and Newton.

As for Descartes's mechanics, there is, as we have argued, a clear sense in which the forces in the strict sense, that is, in the Aristotelian sense of activities, are to be located in God. This reading is of a piece with that which has been developed by Gary Hatfield, who puts the point this way: 'In the end, then, the view that assigns to Descartes a wholly geometrical treatment of matter and motion ... stands; Descartes removed causal agency from the material world and placed it in the hands of God and created minds, that is, in the hands of immaterial substances endowed with the power to act on nature.'[199] But there have been those who disagree with this reading of Descartes.

Thus, Guéroult has allowed that all force is ultimately to be traced to God; but, none the less, he proposes that 'the characteristic of these forces [that is, force of rest and force of motion], in contradistinction to the Divine will that they manifest, is that they are immanent in "nature" or extension and ... can be calculated at each instant for each body, according to the formula *mv*.'[200] Now, it is certainly true that Descartes has something like '*mv*' as a concept in his physics. Thus, here is his third law for the motions of material bodies: 'This is the third law of nature: when a moving body meets another, if it has less force to continue to move in a straight line than the other has to resist it, it is turned aside in another direction, retaining its quantity of motion and changing only in the direction of that motion.'[201]

Here we have a notion of force, one clearly related to '*mv*.' Of course, it cannot literally be the *mv* or momentum of later physics, since Descartes has clear notions of neither *mass* nor *velocity*. As Miller and Miller point out in their translation of the *Principles*, Descartes means by 'quantity of motion' not the product of mass and velocity but rather the product of size (that is, volume) and speed.[202] Physicists were soon to discover that this purely geometrical notion of mass would not do; this is the point that Leibniz was making when he insisted that we must distinguish dynamics from kinematics. This development makes it clear that Descartes could not have got as far as he did without *some concept of mass, however vague that might be, and therefore some concept something like the later concept of momentum.* In some way, then, as Guéroult has said, Descartes needs some notion of *force* that can more or less *like Newton's concept* be calculated from the observed data concerning moving bodies.[203] What Guéroult is arguing, then, is the not-very-surprising thesis that Descartes, in stating his rules as matter-of-fact laws, must introduce an empirical concept of force that plays a role close to that of Newton's concept of momentum.

What is illegitimate is Guéroult's further inference that this (semi-)*empirical concept of force* is to be identified with *force in the Aristotelian sense.* Here is that inference: 'In reality, force, duration, and existence are one and the same thing (*conatus*) under three different aspects, and the three notions are identified in the instantaneous action in virtue of which corporeal substance *exists* and endures, that is, possesses the force which puts it into existence and duration.'[204] Alan Gabbey has given a subtle version of Guéroult's inference, making use of a scholastic distinction between *causae secundum esse* and *causae secundum fieri.* The forces that move things and maintain them at rest, considered in themselves or *per se*, are the divine activities that sustain and ultimately explain processes among material bodies. As such, they are *causae secundum esse.*[205] However, when these forces are viewed not *per se* 'but as quantifiable causes of change in the corporeal world, or as reasons ... explaining absence of change of a certain kind in particular instances, they are *causae secundum fieri*, and are the causal agents at work in the ... three Laws of Nature.'[206] On this basis, Gabbey argues that, '[i]n this subtle and complex ontology of force Descartes does not exclude force from body or its actions.' He continues: 'Strictly speaking God is the ultimate real cause and the only true substance, but speaking at the "practical" level of physical investigation, forces – whether of motion or of rest – are real causes in their own right and distinct from motion and rest.[207] This is no doubt just: it is possible to distinguish the forces present from motion and rest. This should not surprise us: after all, force for Descartes is given by the product of size (volume) and speed (or, as it was to become, *mv*), which is clearly not the same as either motion or rest. This concept appears, moreover, in the statements of the rules or laws that describe the behaviour of material objects. This is not the issue. Rather, what is at issue is whether *force at the 'practical' level*, as Gabbey puts it, is an instance of *activity in the Aristotelian sense* in the way that God's causal activities are activities in the Aristotelian sense.

It is clear that Leibniz's *forces* are activities in the Aristotelian sense. For him, the *primitive force*, which through its limitations in interacting material objects gives rise to what he calls the 'derivative force,' is, he tells us, 'nothing but the first entelechy [*entelechia prima*], [and] corresponds to the *soul* or *substantial form*.'[208] This is the force that picks out among all kinematically or geometrically possible motions of material bodies those that are actual; that is, this is the force that defines *dynamics*. '[L]ike time, motion taken in an exact sense never exists, because a whole does not exist if it has no coexisting parts. Thus there is nothing real in motion itself except that momentaneous state which must consist of a force striving towards change. Whatever there is in corporal nature besides the object of geometry, or extension, must be reduced to this force.'[209] The point is that the Leibnizian force *is*, quite clearly, an Aristotelian activity, an exercise of an active teleological power.

It is precisely this that Descartes denies is present in material objects: for him, as we have seen, '[w]hen I say that these little globules strive, {or have some inclination}, to recede from the centers around which they revolve, I do not intend that there be attributed to them any thought from which the striving might derive ....'[210]

Gabbey is correct in suggesting that there is a place in Cartesian mechanics for the concept of force as predicated of material bodies. But from this it does not follow that force as predicated of material objects – force at this 'practical' level – is an Aristotelian activity. To the contrary, while Descartes does allow, as Gabbey correctly argues, that there are forces predicated – at the 'practical' level – of material objects, these simply are *not* in any sense activities: activity in the Aristotelian sense is, for Descartes, restricted to God.

It is important to recognize that forces as predicated of material objects, that is, forces at the 'practical' level, are in fact *truly predicated* of material objects. Garber has suggested otherwise, proposing that they are really located 'nowhere at all.'[211] This suggests that it is false to predicate forces in the 'practical' sense of material objects. But that amounts to the suggestion that we cannot predicate the concept of size times speed of material objects, which is absurd. What, then, does Garber have in mind? Explaining this thought, Garber points to Descartes's first law of motion. The law states that bodies continue in motion or in a state of rest. This law is explained by reference to God's conserving activities. Thus, the first law states in effect that God will act on the world in such a way that moving bodies will keep moving and bodies at rest will stay at rest. Garber then suggests that '[t]his can be *described* by saying that bodies, *as it were*, have a force to continue their motion, or exert a force to maintain their rest. But this is not to attribute anything real to bodies over and above the fact that God maintains their motion and as a consequence they obey a law of persistence of motion.'[212] This says: take the legitimate concept of force at the 'practical' level; interpret it as an Aristotelian force – which it *is not*; then you can *speak as if* bodies were moved by or kept at rest by *as it were (Aristotelian)* forces. This is no doubt so. Moreover, it no doubt represents why we always have a temptation to read Aristotelian concepts into the Cartesian ones, as Guéroult and Gabbey do (though Leibniz did not). And when this is done, the (Aristotelian) forces are indeed imaginary and therefore quite strictly speaking 'nowhere.'

Hume noted an important tendency of the human mind. 'It is a common observation,' he points out, 'that the mind has a great propensity to spread itself on external objects, and to conjoin with them any internal impressions which they occasion, and which always make their appearance at the same time that these objects discover themselves to the senses.' It is this, he argues, that accounts for the tendency to think that causation involves the exercise of active powers. It is this propensity, he proposes, that 'is the reason why we suppose necessity and power to lie in the objects we consider, not in our mind, that considers them; notwithstanding it is not possible for us to form the most distant idea of that quality, when it is not taken for the determination of the mind, to pass from the idea of an object to that of its usual attendant.'[213] Garber is showing us how this tendency works in the case of Descartes. But Descartes, like Hume, was able to resist the propensity.

In any case, at the 'practical' level, that is, the level of ordinary material objects, Descartes's argument is that there are no Aristotelian forces, no unanalysable active powers. In this respect, his views are of a piece with those of Hobbes. Hobbes defined the notion of cause in terms of necessary and sufficient conditions: 'a CAUSE simply, or an entire cause, is the aggregate of all the accidents both of the agents how many soever they be, and of the patient, put together; which when they are all supposed to be present, it cannot be understood but that the effect is produced at the same instant; and if any one of them be wanting, it cannot be understood but that the effect is not produced.'[214]

Power is not different from cause, according to Hobbes. Power simply refers to what will or would happen if certain conditions are met. As he puts it, '*cause* respects the past, *power* the future time.'[215] And so, 'whensoever any patient has all those accidents which it is requisite it should have, for the production of some effect in it, we say it is in the *power* of that patient to produce that effect, if it be applied to a fitting agent.'[216] Thus, being sugar is sufficient for dissolving if put in water. It therefore has the power of dissolving if in water. To say that if sugar has the power of solubility is to say that if sugar is put in the appropriate context, that is, in water, then dissolving results. What is important is the causal regularity that

(*)   For any x, if x is sugar then if x is in water then x dissolves.

To say when we assert

(**)   Sugar is soluble

that *solubility* is a *power* which is present in things which are sugar is to say just what (*) says. Thus, (**) simply asserts in other terms what (*) asserts. This is *all there is* to powers. So what Hobbes is arguing is that

(+) x is soluble

simply means

(++)   if x is in water then x dissolves.

His point, then, is no different from Molière's about the dormitive power of opium.

It is the same point that Hume makes when he states that '[t]he distinction, which we often make betwixt power and the exercise of it, is equally without foundation';[217] or, as he also put it elsewhere, '[i]t has been observed, in treating of the understanding, that the distinction which we sometimes make betwixt a power and the exercise of it, is entirely frivolous ...' He continues, however, arguing that 'though this be strictly true in a just and philosophical way of thinking, it is certain it is not the philosophy of our passions, but that many things operate upon them by means of the idea and supposition of power, independent of its actual exercise.'[218] He concludes, with Hobbes, that we have a power when we have a pattern or regularity: 'power has always a reference to its exercise, either actual or probable, and that we consider a person as endowed with any ability when we find, from past experience, that it is probable, or at least possible, he may exert it ... power consists in the possibility or probability of any action, as discovered by experience and the practice of the world.'[219]

To insist, with Hobbes and Hume, that (+) simply is *defined by* (++) is to insist that

($)   Whatever is soluble is such that if it is in water then it dissolves

is true *ex vi terminorum*, that is, by virtue of nominal definition. In contrast, of course, for the Aristotelians, ($) is not true by nominal definition; it is, rather, a substantive truth. This means, as we have seen, that for the Aristotelian, (+) does not have the same meaning as (++); they are distinct concepts. As Hume would put it, for the Aristotelian there is a distinction between the power and its exercise. Once one abandons that distinction, as Hobbes and Hume – and Molière – do, then what explains is not the power as such but the pattern, the regularity.

What holds for solubility holds for force. Hobbes provides the following account: 'I define FORCE to be the *impetus* or quickness of motion multiplied either into itself, or into the magnitude of the mover, by means whereof the said movement works more or less upon the body that resists it.'[220] We need not go into the details. The point is that the concept of force is defined in terms of the motions of things, and is explanatory only in the sense in which the concept of power is explanatory. That is, what counts is not the concept but the regularity. In that sense, as Gabbey correctly points out, 'forces in general have evaporated completely from his system ...'[221] That is, what has evaporated is the notion of force as an active unanalysable Aristotelian power that explains. This is not to say that forces cannot truly be predicated of bodies. To the contrary, that is the whole point of Hobbes's definitions, to ensure that the concept of force does indeed have an empirical content – to ensure, in other words, that it is useful at the 'practical level.' In just this sense, for Newton one does not explain simply by appeal to the concept of force; rather, for Newton what is explanatory are *force laws*, that is, the patterns that describe the interactions of objects under different conditions, and specifically, those patterns that connect the *ma* (mass times acceleration) of an object to the circumstances in which the object is located.

In precisely this same sense, force for Descartes has also evaporated at the 'practical' level. But where this is all there is to force for Hobbes, it is not so for Descartes. For, besides force at the 'practical' level, there is also the activity of God, the only real force that operates. For Hobbes, causation just is the pattern. For Descartes, this is all there is to causation at the 'practical' level. That is why the science defended by the Cartesians can be seen as having the same cognitive goals as the science defended by the empiricists. But where the pattern *is* explanatory for Hobbes – and for Hume – this is not true of Descartes. For Descartes, the pattern by itself is not explanatory. Rather, what, ultimately, is explanatory is, as it was for Aristotle, *activity*. Descartes might have got rid of the powers of finite substances, but he none the less retains the category of active power in a way in which Hobbes – and Hume – do not. In this sense, the cognitive aims of Descartes are of a piece with those of the Aristotelians, and are opposed to those of the new science of Galileo and Bacon.

It has been argued that Descartes's world is a 'world without ends.'[222] It is certainly true that the *created world* is a world without ends. It is straightforwardly denied that ordinary material objects have any Aristotelian active powers. This means that Aristotelian teleology is equally straightforwardly denied of material objects. None the less, there is still *activity* in Descartes's world. The fact that it has, as it were, evaporated from the material world and is now located wholly within God does not detract from the fact that, in the last analysis, for Descartes, as for Aristotle, explanation is in terms of *active powers*. In that sense, it is *not* a world without ends; it is just that the ends are now wholly located within God. To suggest that Descartes's world is 'without ends' is therefore seriously misleading, implying as it does that Descartes has made a decisive break with the Aristotelian tradition. To the contrary, Descartes accepts the basic Aristotelian framework of substances, essences, and activities. To imply that these are absent from Descartes is to move him closer than he really is to the empiricist tradition.[223]

If it is true, as I have just been arguing, that the science of Descartes was the *scientia* of the Aristotelians, it is also true that the world of ordinary things was stripped of Aristotelian powers. In this respect the Cartesian picture of the world of ordinary things is very much of a piece with that of the empiricists. It did not take very much to make the world fully empiricist: all one had to do was deny activity even of God. Hume saw this clearly.

He gives a correct characterization of the Cartesian position:

Matter, say they, is in itself entirely unactive and deprived of any power by which it may produce, or continue, or communicate motion: but since these effects are evident to our senses, and since the power that produces them must be placed somewhere, it must lie in the Deity, or that Divine Being who contains in his nature all excellency and perfection. It is the Deity, therefore, who is the prime mover of the universe, and who not only first created matter, and gave it its original impulse, but likewise, by a continued exertion of omnipotence, supports its existence, and successively bestows on it all those motions, and configurations, and qualities, with which it is endowed.[224]

He then gives the argument that concludes that God does not have any powers of the Aristotelian sort.

We have established it as a principle, that as all ideas are derived from impressions, or some precedent perceptions, it is impossible we can have any idea of power and efficacy, unless some instances can be produced, wherein this power is perceived to exert itself. Now, as these instances can never be discovered in body, the Cartesians, proceeding upon their principle of innate ideas, have had recourse to a Supreme Spirit or Deity, whom they consider as the only active being in the universe, and as the immediate cause of every alteration in matter. But the principle of innate ideas being allowed to be false, it follows, that the supposition of a Deity can serve us in no stead, in accounting for that idea of agency, which we search for in vain in all the objects which are presented to our senses, or which we are internally conscious of in our own minds. For if every idea be derived from an impression, the idea of a Deity proceeds from the same origin; and if no impression, either of sensation or reflection, implies any force or efficacy, it is equally impossible to discover or even imagine any such active principle in the Deity. Since these philosophers, therefore, have concluded that matter cannot be endowed with any efficacious principle, because it is impossible to discover in it such a principle, the same course of reasoning should determine them to exclude it from the Supreme Being. Or, if they esteem that opinion absurd and impious, as it really is, I shall tell them how they may avoid it; and that is, by concluding from the very first, that they have no adequate idea of power or efficacy in any object; since neither in body nor spirit, neither in superior nor inferior natures are they able to discover one single instance of it.[225]

The step that was required to move from the Cartesian world to the world of the new empirical science was therefore not very large. Descartes made the move very easy. For all that, it was a momentous move. For it meant that the new science no longer needed God in any explanatory role.

Still, as Hume indicates, what was crucial was the argument against the Cartesian account of ideas. This argument was more complex than Hume's bare statement suggests it to be. It is to this argument that we must now turn.

2) Cartesian Ideas

It has of late been recognized that Descartes's views on ideas do not fit the standard picture, with us since Thomas Reid, of representationalism.[226] This was, of course, the position of Arnauld, who argued against Malebranche's representationalism and against his representationalist reading of Descartes.[227] There is much to be said for this non-standard reading.

Descartes takes perception to fall under the category of thought.[228] Thoughts in general and perceptions in particular are modifications of the mind; they are acts of the mind. But thoughts and perceptions are also about, or as Brentano taught us to say, they 'intend,' certain objects. They do this by virtue of *ideas* which inform them. An idea, then, is the form of an act of thought or perception; as Descartes puts it in his replies to the Third Set of Objections, 'by an idea I mean whatever is the form of a given perception.'[229] But an idea in this sense is not itself a content of thought that is apprehended by the mental act of which that idea is the form; we are *aware of* the idea, but that idea is not the *object of perception*. Ideas, in other words, are not representative entities that intervene, as it were, between the act of the mind which is the perceiving and the object perceived.[230] For, Descartes distinguishes two senses of the term 'idea.' In one sense, it is a modification of the mind, and as such is something of which the mind is immediately aware; this is what Arnauld had in mind when he pointed out that 'our *thought or perception* is essentially reflexive upon itself ... for I never think without knowing that I think.'[231] However, an act of thought or perception is also intrinsically connected to the object or body which it intends, by virtue of which connection the mind is aware of that object or body. As Arnauld puts it, 'the *idea* of an object and the perception of an object [are] the same thing.'[232] He continues: 'an object is present to our mind when our mind perceives and knows it'; when I so conceive of a thing, that 'thing is *objectively* in my mind ...'[233] Descartes draws the relevant distinction between two senses of the term 'idea' in the 'Preface' to the *Meditations*: 'in this word "idea" there is here an equivocation. First it can be taken materially, as an operation of my intellect ... or it can be taken objectively for the body which is represented by this operation.'[234] This he elaborates in his reply to the Second Set of Objections, a passage which Arnauld also quotes:[235]

II. *Idea.* I understand this term to mean the form of any given thought, immediate perception of which makes me aware of the thought ...
III. *Objective reality of an idea.* By this I mean the being of the thing which is represented by an idea, in so far as this exists in the idea ... whatever we perceive as being in the objects of our ideas exists objectively in the ideas themselves.[236]

Crucial here to the position of Descartes and Arnauld is the notion of 'objective being.' This notion was derived by Descartes from sources in late Scholastic philosophy, Suarez in particular.[237] According to Suarez, creatures, that is, created beings, have essences. A creature insofar as it is real instantiates this essence. The reality or true being of the creature consists of the properties that are predicated of it; these properties jointly express the essence of the creature. A creature *qua* essence is a potential or possible being, not an actual being; it is, in this sense, not a reality. Thus, 'the essence of a creature, or the creature itself prior to God's creative act, has in itself no true real being; and in this sense, that of the sort of being involved in existence, the essence is not a thing, but is simply nothing ...'[238] Prior to creation the creature, or what is the same, the creature *qua* essence, has no *esse reale intrinsecum*, no real being intrinsic to it (Suarez, *Disputationes*, 31.2.1) or *verum esse real*, true real being. An essence becomes real through its 'receiving true being from its cause [that is, God].'[239] However, there is a sort of reality that attaches to the essences of creatures that is absent from those things that are purely fictitious: they, unlike the latter, are *capable of real intrinsic being*, that is, of being created by God. 'It is this [the capacity for true real being] that distinguishes the essences of creatures from things fictitious or impossible, such as a chimera; and in this sense creatures are said to have real essences, even if they do not exist.'[240] We thus distinguish the potential being of essences both from true real or real intrinsic being and from nonbeing. 'The being which belongs to essences before the divine act creating them is only a *potential or objective being*' (italics added).[241] Connections among the essences are ground of the necessity of eternal truths. 'For God to have had true scientific knowledge from eternity that man is a rational animal, it is not required that the essence of man have from eternity some real actual being, since "is" does not here signify real and actual being, but only an intrinsic connection between the extreme terms of the proposition; this connection is not grounded in actual being, but in potential being.'[242] The eternal truths are connections among essences that hold independently of true real being, what actually exists – that is, what makes them eternal and necessary; in this sense the connections they represent need have no basis in an actual reality. It does not follow, however, that those necessary connections have no being whatsoever. To the contrary, they exist in the mind of God. The essences are known by God eternally (*Disputationes*, 31.2.1). In this sense, the essences which have potential or objective being exist in the mind of God. However, 'this being known is not some sort of real being intrinsic to them.'[243] Similarly, the necessary truths are objects of God's knowledge; indeed, to say that they have objective being simply *is* to say that they are known, or, what is the same, have mental existence: 'From time eternal there has been no truth to these propositions [the essential truths] save for their being objectively in the divine mind, not being in themselves subjects or realities.'[244] Suarez makes the same point in a variety of ways: 'the essences of creatures only have a being in their cause, or an objective being in the mind,'[245] and '[t]hey do not exist in themselves; they exist only objectively in the mind.'[246]

Descartes does not adopt the Suarezian position completely – he argues for a different relationship between the eternal truths and God[247] – but these points are not important for what we are about, which is, rather, the fact that *the objective existence of things in the mind is constituted by the presence in the mind of the essence of the thing.*

Descartes refers to actual being as 'formal' being or reality (*esse formale*): the formal being of a thing is constituted by its properties or modifications. This he distinguishes from the objective being of a thing, that 'mode of being, by which a thing exists objectively or is represented by a concept of it in the understanding.'[248] This is 'the way in which its [the intellect's] objects are normally there.'[249] When a thing has objective being in the mind, that thing is not a mode or property of the mind, that is, it is not 'formally' in the mind; otherwise the mind would actually *be* the thing. None the less, though the thing that is objectively in the mind lacks, as such, formal reality, it is still not nothing[250] and is to be contrasted to ideas which are merely chimerical.[251] Thus, as Descartes put it to Caterus in the replies to the First Set of Objections,

the idea of the sun is the sun itself existing in the intellect – not of course formally existing, as it does in the heavens, but objectively existing, i.e. in the way in which objects are normally in the intellect. Now this mode of being is of course much less perfect than that possessed by things which exist outside the intellect; but, as I did explain [in Meditation III], it is not therefore simply nothing.[252]

Arnauld also makes the object perceived to be itself in the mind that knows it, there in the mind only objectively: '*the perception* of a square indicates more directly my soul as perceiving a square and *the idea* of a square indicates more directly the square insofar as it is *objectively* in my mind.'[253]

One must, of course, distinguish the thing from the thing *qua* formally existing, or, in Suarez's terms, the thing *qua* having true real being. It is the thing *qua* formally existing that is *there* in the world, existing as the entity with which other real objects, including ourselves, interact. In Suarez's terms, when we distinguish the thing from the thing *qua* true real being, what we are distinguishing is the essence of the thing from the thing as a true real being. Thus, when Descartes and Arnauld distinguish the thing – the sun, say – from the sun *qua* formally existing, the thing that is said to be in the mind, that is, the sun insofar as it does not formally exist, must be the essence of the thing. In other words, what is present to the mind when the thing is objectively in it is not this real object but its essence.[254] We therefore see that for Descartes and Arnauld, as for Suarez, *the objective existence of things in the mind is constituted by the presence in the mind of the essence of the thing.* But once we recognize this we also recognize that *the Cartesian account of perception is structurally that of Aristotle, in which the mind knows the object by virtue of having the form or essence of that object literally in it.*

To be sure, Descartes and Arnauld reject the mechanisms by which Aristotle explained the presence of the form or essence in the mind. For Aristotle there was the process by which the form was transported from the object known to the knowing mind. Descartes rejected this story, as is well known, to replace it by an atomistic account that fit with the new science.[255] This, though, is relatively trivial. Aristotelians were soon able to adapt Aristotle's account of cognition to an atomistic picture of reality.[256] But Aristotle also required the form to be abstracted from the objects given in sense. Descartes and Arnauld rejected this mechanism, too; that, after all, was the point of the wax example in Meditation II. This mechanism is replaced by the doctrine of innate ideas: all the forms or essences of things are native to the mind, and all we must do is retrieve them. But for Aristotle what is crucial to having a rational intuition of the thing known is for the form or essence of the thing to be present in the mind. *This central claim of the Aristotelian position is retained by Descartes and Arnauld. The rationalists reject not the central thesis of the Aristotelian account of knowledge in terms of rational intuition but only inessential claims about the mechanisms by which rational intuition is attained.*

We further recognize the source of the disagreement between the empiricists and the rationalists concerning the method appropriate to the new science, the rationalists insisting upon an a priori justification for the relevant principles of determinism and limited variety and the empiricists arguing that for these general axioms, as for more specific laws, nothing more or less than inductive support was needed. The rationalists Descartes and Arnauld held that a priori knowledge of infallible certainty was possible for the same reason that Aristotle held that it was possible: there are in things forms that we know apodictically and which provide necessary connections among events in the world. And such knowledge was needed in the science they defended because only through a knowledge of essences and therefore of the necessary connections among things could one know which regularities were genuinely laws and which were not. *The rationalists held that a priori first principles are needed and attainable because they retained certain central features of the Aristotelian framework.*

The rationalists did indeed reject the idea that syllogism revealed the ontological structure of reality. They replaced syllogisms by geometry: the ontologic of the world was to be found in Euclid rather than in the textbooks of the logicians. But from the perspective of the basic metaphysics, this was a detail: the rationalists in fact retained the basic structure of the traditional Aristotelian metaphysics. In particular, they retained the Aristotelian notion that science was, *ideally, scientia* in the Aristotelian sense, that is, in the sense that at its best science consists of *demonstrations proceeding from self-evident propositions concerning the natures or essences of things.*

The rationalists gave up little more than some of the pretensions of the Aristotelians that demonstrative knowledge was available outside areas that could be treated mathematically, that is, the so-called 'high sciences' of astronomy, mechanics, optics, and music.[257] Thus, while the rationalists did agree, as we have argued, with the empiricists on many criticisms of the Aristotelians, they did so on very different grounds: the empiricists held that demonstrative knowledge and *scientia* were impossible, and on this basis objected to Aristotelianism; the rationalists held that such knowledge was possible, and objected only to claims by the Aristotelians to have obtained it in areas where they (the rationalists) held (reasonably, agreeing with the empiricists) that it was absent.

We thus see that in the last analysis the rationalists did not reject Aristotelianism. If they objected to parts of the traditional method to science, they did not object to the basic claims concerning the ontological structure of the world – *objective necessary connections do exist* – and the consequent methodological claim – *the ideal of explanation is that of demonstration based on self-evident principles*. The empiricists, in contrast, held that explanation did not require self-evident principles. This was the theory of Bacon and the practice of Newton.

But this theory could not be sustained as long as there was no criticism of the traditional metaphysics with its ontology of necessary connections. So long as that went unchallenged, the traditional ideal shared by the Aristotelians and rationalists such as Descartes and Arnauld would remain as an alternative to the empiricist account of science. Bacon had stated the empiricist alternative to the tradition, but he had not articulated a metaphysics in opposition to the rationalist-Aristotelian metaphysics of necessary connections. It was Locke who would state the basic outlines of the empiricist metaphysics that would succeed in eliminating the rationalist account of the methodology of the new science.

However, if the rationalists did accept the central metaphysical and epistemological principles of the Aristotelians, it is also true, as we have noticed, that they overcame one defect of the Aristotelian position. It turns out that this is as important for the issue of coming to know scientific laws as it is for defining the aims of science.

The relevant defect of the Aristotelian position is that the cognitive goals of the new science were incompatible with the sorts of knowledge of matters of facts that the Aristotelian patterns required one to aim at. The point is, to repeat, that on Aristotle's own account of the patterns we discover in sense experience, one can have both natural and unnatural motions. The former are the laws of nature; the latter are exceptions to those laws. The aim of Aristotelian science is to know forms and therefore natural laws. But this means, then, as we have seen, that Aristotle aims at knowing regularities that, since they are not exceptionless, do not satisfy the cognitive goals of the new science. What the rationalists do is remove this fault: as they develop the Aristotelian position, there is no possibility of violent change. That being so, there is no possibility for there being exceptions to natural laws. Such an exception arises when it is allowed that one can have an essence that is predicated of a substance the properties of which are other than those that correspond to this essence, that is, in Cartesian terms *a substance the formal reality of which does not correspond to its essence.*

On the Aristotelian account of knowledge, perceptual error will occur when one abstracts from what is given in sense the 'wrong' form. We experience the patterns of sense. From these we abstract a form. If the form that we abstract is not the same as the form of the substance that presented us with the sensible impressions of which we were aware, then we have a case of perceptual error. What makes such error possible is the possibility of violent or unnatural motion, either in the thing perceived or in the mind that knows. *The issue, then, for the Aristotelian is this: By what criterion do you know that the process of abstracting has been successful?* Aristotle himself avoids this issue by holding, as we saw, that perception is itself a natural process that tends, generally, 'for the most part,' to produce the outcome of knowledge. But that merely raises the further question, *How do you know that the process of abstracting currently taking place is natural?* To this Aristotle provides no answer.

The rationalists solve this problem in part by accepting the account of rational intuition, in which the form or essence is in the mind ('objectively'), while rejecting the doctrine of abstraction: clearly, there can be no question of the process of abstraction proceeding in some unnatural way; there is no such process. The issue for the Cartesian, therefore, is this: *Is the essence that is objectively in the mind instantiated in a substance the formal reality of which exhibits that essence?* The answer that the Cartesians give is that every essence that is in the mind is instantiated in a substance whose formal reality does exhibit that essence. In this sense, for the Cartesians, *ideas are never false.* Error arises, they hold, not through our ideas but through our judgment. (How this is possible is the main point treated in Meditation IV.) To be sure, there are certain ideas that are obscure, not clear and distinct, and these in fact mislead our judgment, causing us to fall into error. Such ideas may be said to be 'materially false,' but this is quite compatible with the basic thesis that all our ideas are true. '[E]ven though ... true and formal falsity can characterize judgments only, there can exist nevertheless a certain material falsity in ideas, as when they represent that which is nothing as though it were something. For example, my ideas of cold and heat are so little clear and distinct that I cannot determine from them whether cold is only the absence of heat or heat the absence of cold, or whether both of them are real qualities, or whether neither is such.'[258]

If all our ideas are true, and error lies solely in the judgment, then the proper procedure is to discover a method that will control the judgment so that it does not affirm where its ideas do not justify, through their clarity and distinctness, such affirmation. In the Aristotelian scheme, whether a perception or idea was true could be challenged by challenging the claim that the abstraction had been successful. The Cartesians solved this problem of error by dispensing with abstraction and holding that all our ideas are true. Yet the latter still needs justification: we still need to know the answer to the question asked before: Is the essence that is objectively in the mind instantiated in a substance the formal reality of which exhibits that essence? After all, could not an external substance, an evil genius, perhaps, be able to act with sufficient force as to impose upon the substance instantiating the essence that is objectively in our minds a formal reality – that is, modifications and properties – contrary to that essence?

The answer that the Cartesians give is clear. The essences are ideas in God's mind, and God does not deceive. This, in fact, is the basic thrust of the argument of the *Meditations*. Arnauld is succinct: it is only by our own reasoning that we are assured that the essences which have objective reality in our minds also have formal reality outside us. 'I am entirely sure that there is truly an earth, sun and stars outside my mind *only by reasoning* ...' (italics added).[259] Moreover, not only is it reason alone that assures us that the essences in our mind have any formal reality at all outside the mind, it is also reason alone that assures that the formal reality of the substances of which these essences are predicated actually does express those essences. Reason assures us of this, the truth of all our (clear and distinct) ideas, because it assures us that these substances *cannot* exist unnaturally, in ways contrary to their essences. God is sufficiently powerful that He can guarantee that there is no evil genius who can force things to be in ways that are contrary to their essential natures, and, moreover, He is no deceiver; from which it follows immediately that we ought to accept as true, that is, as correctly representing their formal reality, all ideas that are clear and distinct. For this argument to be successful, we must, of course, accept the basic premise that the omnipotent God would not mislead us: He in His power ensures that substances appear, or are formally, as they are essentially: '*only God could represent the assumed appearances to the mind, and God is not a deceiver*' (italics added).[260] The relevant premise is the principle that 'we ought to accept as true what could not be false without our being forced to admit in God things completely contrary to the divine nature, such as to be a deceiver, or subject to other imperfections which the natural light shows us evidently cannot be in God.'[261]

In effect, what the Cartesians do is adopt an Aristotelian metaphysics in which there are no unnatural or violent motions. This eliminates the problem of error. However, it can do so only if it can *guarantee* that there is *no* unnatural existence, that is, *no* substance the formal reality or being of which is contrary to the essence that is properly to be predicated of it. The Cartesians do, indeed, solve the problem of error, but only at the cost of then requiring a proof that God exists, and that He is no deceiver. Arnauld has likely been alone in rejecting the notion that Descartes had attempted to prove much more than the evidence would bear. Certainly, few have been convinced. In any case, we shall explore these themes more fully in Study Seven, below.

3) *Locke's Challenge to Aristotelianism and Rationalism*

Historians of science have long recognized a distinction valid in the early modern period between the 'low sciences' and the 'high sciences'; the 'low sciences' – geology, alchemy, astrology, and medicine – could not produce the sort of mathematical demonstrations that could be produced by the 'high sciences' of mechanics, optics, and astronomy.[262] Brown suggests that the difference consists in the fact that the high sciences had some reasonable theories, whereas these were absent in the low sciences, where '[t]he practitioners ... were confined to inferring from one particular occurrence to another with the aid of those generalisations that the history of their craft suggested to one.'[263] Hacking argues that because the known laws in the low sciences were grossly imperfect, in the sense discussed above,[264] these sciences needed a notion of evidence that allowed one to talk in terms of degrees, in a way in which the mathematical or higher sciences needed no notion of evidence beyond the self-evidence and infallible certainty of the traditional *scientia* or demonstration. Hence, Hacking suggests, the notion of probability tended to emerge in connection with the low sciences.[265] Brown argues that this thesis does not stand scrutiny, since there were arguments among the practitioners of the high sciences about whether their theories actually fit the empirical data.[266]

Now, I think there is some force to this argument of Brown's: if one's interest is in empirical fit, that is, in the capacity to explain and predict in terms of matter-of-fact regularities, then there were indeed similar problems in both the high and the low sciences.

But matters are worse than that: not only were there similar problems, but there was no impulse to fix those problems, in either area, high or low. For, on the whole, reflective thought of a pre-scientific sort did not entertain empirical fit as a worthy cognitive aim, nor did it provide any efficient methodological rules for generating knowledge of the required sort. In the first place, given the notion of knowledge, that is, *scientia* in the Aristotelian sense, there is no particular motive to be interested in securing a better empirical fit. What was important was the grasp of forms or natures or essences and not merely the empirical fit, that is, the regularities in the patterns of behaviour. Moreover, in the second place, given the possibility of unnatural or violent change, which was a part of the traditional notions, one could reasonably expect that there would be *no* neat fit to reality, no exceptionless patterns of action. This was true at least in the terrestrial regions described by physics, though the Aristotelians were harder pressed to account for the lack of empirical fit in the celestial regions, which were taken to be perfect and therefore not subject to violent change. But, in general, it was true to say that the Aristotelian patterns were, therefore, as the empiricist and rationalist defenders of the new science argued, a positive hindrance in any search for unconditional regularities insofar as the explanation scheme was prepared to accept as laws patterns that admitted exceptions. Finally, in the third place, the method that Aristotle and his successors advocated – namely, as Brown points out,[267] the Aristotelian method of searching for demonstrative syllogisms with premises known with certainty – was not efficient in the discovery of exceptionless regularities, as we have seen.

None the less, the mathematical sciences gave the *illusion* of demonstration, whereas, as Brown says, 'the low sciences could provide relatively few logical demonstrations,'[268] that is, not even the illusion of *scientia*. Having said that, however, one must also point out that, even if one did not have *scientia*, the Aristotelian ideal, one *did* have knowledge that was useful. *What the low sciences made clear was that useful knowledge of the sort that was the cognitive goal of the new science could be attained in the absence of knowledge of any necessary connections or a priori truths.* Such, for example, was the implication of Harvey's researches on the circulation of the blood. Such, too, was the implication of the new knowledge of diseases that resulted when Locke's friend, the physician Thomas Sydenham, applied the scientific method – what Locke was to call the 'historical plain method'[269] – in the low science of medicine.

Locke was to emphasize this point in his discussion of the role of syllogism in reasoning, that is, in effect, whether science had any need of the Aristotelian ideal of demonstrative knowledge. Brown fails to recognize this role that the low sciences played in establishing the new science in place of the old ideal of *scientia*. And he fails to recognize this role because he fails to notice that there is a serious distinction to be made between the cognitive goals of the new science and those of the old science. Both Bacon and Descartes knew better than Brown on this point; he would have done well to have read them.

Locke took up Bacon's point that one need not identify the norms of causal inference with demonstrative syllogisms. He made the point characteristically, in his down-to-earth way:

Tell a country gentlewoman, that the wind is south-west, and the weather lowering, and like to rain, and she will easily understand, it is not safe for her to go abroad thin clad, in such a day, after a fever: She clearly sees the probable connexion of all these, viz. south-west wind, and clouds, rain, wetting, taking cold, relapse, and danger of death, without tying them together in those artificial and cumbersome fetters of several syllogisms, that clog and hinder the mind, which proceeds from one part to another quicker and clearer without them: and the probability which she easily perceives in things thus in their native state, would be quite lost, if this argument were managed learnedly, and proposed in mode and figure. (*Essay*, IV.xvii.4)

In this way, Locke gives a place to the low sciences (now including meteorology) in a realm where one could reason about causal connections in the absence of demonstrative syllogisms. *But what legitimates this place*?

Brown, for one, has not asked this question. Neither, it is clear, did Bacon. It was Locke's achievement among those who defended the empiricist account of the new science to have done so.

What was essential was to *argue philosophically* for the absence of the necessary connections that had to be present were the Aristotelian account of causal inference to be acceptable. Bacon had *taken it for granted* without argument that the basic principles of science were not necessary; Locke provides a *philosophical case* that they were not. It was this development that was crucial to the elimination of the Aristotelian account of causal inference *and the rationalist, Cartesian account of the new science.*

Both the Aristotelians and the rationalists argued that there are objective necessary connections that bind simple ideas together to yield the real definitions of things. It is evident, Locke proposes, that we do not know the forms or essences that yield the necessary connections required for an Aristotelian or rationalist understanding of why parts of things cohere (*Essay*, IV.iii.26). But even if we knew why the minute parts of things cohere, we still would not know everything that was necessary for knowledge of the sort required by both the Aristotelians and the rationalists for demonstrative knowledge of causes. For, the form or essence must account for all the actions of a substance – all its properties – that are neither unnatural (a category excluded by the Cartesians) nor accidental or occasional. Now, in fact, the *regular* activities of material substances include the production of our sensations of colour, sound, and so on – that is, the ideas of secondary qualities. If these activities of the substances are to be known scientifically in the sense required by both the Aristotelians and the rationalists, the regularities concerning observable properties that are revealed in sense experience must be demonstrable from necessary truths grounded in the essences of the substances. But, for that to be possible, we must discern necessary connections among our sensations of colour, sound, taste, and so on; we must be able to recognize the ontological ties among these that are grounded in the essential natures of the things. These necessary connections must be both ontological, in the things themselves and connecting objectively the properties of those things, and epistemological, giving us, when in the mind, the rational intuitions of those entities that is necessary for knowledge of them in the sense of *scientia*. But, Locke argues, *we grasp no such connections.*

'Tis evident that the bulk, figure, and motion of several bodies about us, produce in us several sensations, as of colours, sounds, tastes, smells, pleasure and pain, etc. These mechanical affections of bodies, having no affinity at all with those ideas they produce in us (there being no conceivable connexion between any impulse of any sort of body, and any perception of a colour, or smell, which we find in our minds) we can have no distinct knowledge of such operations beyond our experience ... (IV.iii.28)

What Locke is doing here is providing the required argument against our having to suppose that there is an ontology of real essences and necessary connections that we must turn to if we are to genuinely understand the world. To the contrary, because we can admit no objective necessary connections, we must not aspire to the now-inapplicable Aristotelian and rationalist ideal of demonstrative science. *Since we are not acquainted with real essences and necessary connections, such entities can have no place in accounting for most important features of the world, and the knowledge that we have of those features can be based only on contingent matter-of-fact propositions.* As Locke continues, we 'can reason no otherwise about them, than as effects produced by the appointment of an infinitely wise agent, which perfectly surpasses our comprehension ...' (*Essay*, II.iii.25; see also IV.iv.10).

Locke is here appealing to what has become known as the principle of acquaintance to exclude any claim that our ontology, at least in the areas of the 'low sciences,' must include real essences or necessary connections of the sort required to justify the Aristotelian and rationalist ideal of demonstrative knowledge. By means of this argument Locke was able to justify the rejection of that ideal, removing forever whatever plausibility it had previously had.

4) *The Sceptical Response to the Rationalists: Huet*

For all that Locke insisted both that infallible Cartesian certainty was impossible in many areas and also that this did not in fact matter for the purposes of the new science so far as these areas were concerned, none the less he also allowed that the Cartesian cognitive standard of self-evident truth had a role to play.

Locke was for the most part within the empiricist camp, and insisted that, in the absence of real essences and necessary connections, one could not attain the infallible certainty that was the cognitive goal of Aristotelian *scientia* and Cartesian science alike. Often enough, however, he qualified his judgments concerning the probable nature of most of our knowledge in ways that make clear that he had not yet divorced himself from the ideal of *scientia*. Thus, he says that, given the fact that we are acquainted with no real essences, no necessary connections, in most areas of life, we must rest content with fallible claims based on inferences from past experience.

We are not therefore to wonder, if certainty be to be found in very few general propositions made concerning substances: our knowledge of their qualities and properties goes very seldom further than our senses reach and inform us. Possibly inquisitive and observing men may, by strength of judgment, penetrate further, and, on probabilities taken from wary observation, and hints well laid together, often guess right at what experience has not yet discovered to them. But this is but guessing still; it amounts only to opinion and has not that certainty which is requisite to knowledge. (*Essay*, IV.vi.13)

Yet knowledge in the traditional sense of *scientia* is also possible, though only in limited areas: 'I doubt not but that it will be easily granted, that the knowledge we have of mathematical truths is not only certain, but real knowledge ...' (IV.iv.6).

Locke was thus still willing to admit a place for the Cartesian standard of self-evidence as the ultimate criterion for the rational acceptability of propositions. If proofs and probabilities falling short of the demonstrations of traditional rationalist and Aristotelian *scientia* had to be used, they were none the less clearly second best relative to the Cartesian standard.

However, if Locke was not willing in the end to challenge the traditional doctrine that the only legitimate form of knowledge was defined by infallible certainty, there were others who were not willing to allow even this much, and were prepared to mount a sceptical attack on the rationalist cognitive standard of infallible certainty and self-evidence.

Of course, the challenge was old: it had already begun in the ancient world. We find it in Sextus.

The major premise of an Aristotelian syllogism is a universal statement. That means that it is about a whole population. But our experience is always of a limited sample, and, since events in sense experience are separable, there is nothing in one or several individuals – that is, in a sample – that will guarantee anything about either the next individual or the remainder of the population. On the basis of sense experience alone, one cannot claim to *know* the major premise of a syllogism that purports to be explanatory. Sense cannot yield scientific syllogisms – that is, scientific in the sense of *scientia*.[270] For Locke, in most areas we cannot have knowledge of the essences that we must know if we are to discover necessary connections beyond the separable impressions of sense experience. The conclusion is inescapable: we have no *knowledge* of laws. The best that we can do is belief or opinion. And if we take as given the Cartesian norm of assent accepted by Descartes and Arnauld, that we assent to no proposition that could conceivably be false, then it follows that, if we are limited to the world of sense experience, we can assent to nothing: the Cartesian norm leads to a scepticism as radical as any ever imputed to a Pyrrhonist. In effect, the response of Locke is not to embrace scepticism but to reject the Cartesian norm for assent; it is to reject the idea of knowledge as *scientia* and to redefine the notion of knowledge to allow for, what is self-contradictory on the Aristotelian and rationalist scheme, *fallible knowledge.*

Sextus's point is that no induction from particulars is complete, and knowledge, therefore, of universal premises, if it exists, is not achieved through examining a total population.

This, of course, Aristotle grants. The major premises of scientific syllogisms are not generalizations from particulars; if they were, they could not have the necessity they do have. Moreover, the Aristotelian process of induction is not a matter of generalizing from knowledge obtained by sense of a sample; it is, rather, a matter of the mind's abstracting the universal from a set of particulars. This does not eliminate the sceptical problem, however. If the universal is derived from a particular, then one still cannot use the syllogism to provide understanding of the particular. If the particular in question was among those studied to obtain the universal, then the inference is otiose; and if it was not, then there is still no guarantee that what has held in the past will hold in the future.[271] The Aristotelian can reply that the form is inseparable from the substance, but that itself is a claim that requires proof – how do you *know* it to be true? – and, moreover, even in the next instance of the form, how do you know that its causal power will in fact have its natural outcome? Might there not be some external force that causes an unnatural result?

In any case, there is still, as we have pointed out, the question of how you know that the process of Aristotelian induction was successful in the sense of yielding the form that actually causes the appearances that are presented to one? Might not the appearances be so confused that we err? For example, might we not be caused to err by unnatural effects among the appearances or by external substances that cause our own processes of perception to work unnaturally and lead us in either case to abstract the wrong form? How can the possibility of illusion or delusion ever be sufficiently controlled for so as to guarantee that abstraction is always successful? These possibilities show that even propositions which make no greater reference to forms than the crucial minor premise of the Aristotelian syllogism 'This C is B' (say, 'Corsicus is soluble') or 'These A's are B's' inevitably lack certitude, and can therefore never be classed as knowledge.[272]

These Pyrrhonian themes were used by Montaigne; and Descartes himself, in attempting to respond to Montaigne, was to employ these same patterns, as we all know, when he applied the 'method of doubt' in Meditation I, and then again in Meditation II, when he used the wax example to argue from the alterability of appearances to the impossibility of having obtained the form or essence of the substance by abstraction. This critique of the Aristotelian theory of knowledge made it possible to reintroduce the Platonic theme of innate ideas. None the less, we should again emphasize, the *basic* Aristotelian categories of substance, nature or essence, and so on, remained unchallenged. In spite of the Platonic theory of knowledge, Descartes remained, as we have seen, an Aristotelian in his basic metaphysics. In particular, *he retained the basic idea that knowledge was incorrigible, infallibly certain.*

Descartes uses the Aristotelian metaphysics itself to deepen the sceptical critique beyond where the ancient sceptics had taken it. In the end, as we have seen, Aristotle's reply to the sceptics is that persons by nature aim to know, that what is natural is necessary, and that what is necessary happens always or for the most part. It follows that erroneous perception is simply not what one normally encounters, and it can, therefore, be dismissed as not a problem, that is, dismissed as one problem that cannot cut to the basis of the metaphysics. More generally, for Aristotle, all possibilities are at some time actual. If they are not actual, it is because something external is causing them to be in an unnatural state or process that prevents their actualization. But unnatural processes are relatively rare since, once again, the natural is the necessary and the necessary happens always or for the most part. Thus, error can exist in Aristotle's metaphysics alongside the claim that all knowledge is infallible. It turns out, however, as Descartes recognized, that Aristotle has in fact not eliminated the possibility of *systematic* error.

Descartes asks the simple question: What guarantee is there that the natural *does* happen normally? Perhaps there could be a substance with great power the nature of which is to *prevent the best* from happening. That is, since the natural is what any substance ought to be, it is the nature of this powerful substance to prevent the natures of other substances from being actualized. If such a substance exists, then there will be every reason to suppose that the natural, however necessary, will in fact never occur. Now, Descartes agrees that it is natural for mind to know. Indeed, all one must do is clear away by means of the method of doubt the rubbish that blocks this natural movement, and the latter will occur, the mind moving, as in the *Meditations*, in a series of reasonable steps from ignorance to knowledge. Knowledge is the human good. But perhaps there is a powerful substance – an evil genius – whose nature it is to prevent the good. If so, he will in particular so act as to prevent the mind from coming to know.

The problem of the evil genius, we now see – we shall explore the notion in greater detail in Study Seven, below – is a general problem for Aristotle's metaphysics; it is the problem of unnatural motion applied to the case of knowing. Given the possibility of the evil genius – given this gap in Aristotle's metaphysics – then that metaphysics can itself undercut any claims that it might be the object of knowledge (*scientia*).

Aristotle simply assumes that what is natural is, since necessary, what happens always or for the most part; that is, he simply assumes that what is best for things is what is normal. His scholastic followers, such as Aquinas, make the same simple assumption.

Aquinas[273] makes the distinction made previously by Aristotle between things which are natural and things which are unnatural or violent. '[A] thing is called natural because it is according to the inclination of nature,' he tells us, and 'we call that violent which is against the inclination of a thing' (*Summa*, FP, Q82, A1). Violent motion always involves a hindrance of a natural inclination to be in the natural way: 'that which has a form actually, is sometimes unable to act according to that form on account of some hindrance, as a light thing may be hindered from moving upwards' (FP, Q84, A3). Furthermore, violent motion needs an active cause to hinder the natural inclination: 'nothing violent can occur, except there be some active cause thereof. But tendency to not-being is unnatural and violent to any creature, since all creatures naturally desire to be. Therefore no creature can tend to not-being, except through some active cause of corruption' (FP, Q104, A1, Obj 3). Now, sense organs are not deceived when it comes to their proper objects – or, rather, *not normally* deceived: they work satisfactorily unless something comes along to interfere with their normal operation. '[S]ense is not deceived in its proper object, as sight in regard to color; unless accidentally through some hindrance occurring to the sensile organ – for example, the taste of a fever-stricken person judges a sweet thing to be bitter, through his tongue being vitiated by ill humors' (FP, Q85, A6). Aquinas makes this remark in the context of an argument why the intellect does not err. The conclusion to be drawn is that the intellect does not err unless accidentally through some hindrance. But the intellect knows things by virtue of those things being in the mind: 'intelligent beings are distinguished from non-intelligent beings in that the latter possess only their own form; whereas the intelligent being is naturally adapted to have also the form of some other thing; for the idea of the thing known is in the knower' (FP, Q14, A1). Thus, the intellect does not err with regard to the truth of things – unless accidentally through some hindrance. The question is: How do we know whether or not there is any such hindrance? The mind, after all, can be directly affected by God, as when he gives us dreams that foretell the truth. But in the same way the mind can be directly affected by demons.

[T]he outward cause of dreams is twofold, corporal and spiritual. It is corporal in so far as the sleeper's imagination is affected either by the surrounding air, or through an impression of a heavenly body, so that certain images appear to the sleeper, in keeping with the disposition of the heavenly bodies. The spiritual cause is sometimes referable to God, Who reveals certain things to men in their dreams by the ministry of the angels, according Num. 12:6, 'If there be among you a prophet of the Lord, I will appear to him in a vision, or I will speak to him in a dream.' Sometimes, however, it is due to the action of the demons that certain images appear to persons in their sleep, and by this means they, at times, reveal certain future things to those who have entered into an unlawful compact with them. (*Summa*, SS, Q95, A6)

No doubt such interventions in the natural workings of the world are in a way miraculous, where a miracle is understood as a systematic violation of everything that nature requires, or, rather, that is required by the totality of the natures of things. 'A miracle properly so called is when something is done outside the order of nature. But it is not enough for a miracle if something is done outside the order of any particular nature; for otherwise anyone would perform a miracle by throwing a stone upwards, as such a thing is outside the order of the stone's nature. So for a miracle is required that it be against the order of the whole created nature' (*Summa*, FP, Q110, A4). A systematic hindrance of the workings of the intellect could prevent it from grasping the truth of things. Such spiritual interference would violently prevent the intellect's natural working, and would prevent it from naturally achieving its natural cognitive end of the truth. 'God alone can do this [cause a miracle],' Aquinas concludes (FP, Q110, A4). No doubt it would require a very powerful spiritual being systematically to deceive us. But why could it not be a demon? So long as it was powerful enough it could cause our intellects to fall into error, to grasp forms that are not the truth of the world – as sense can sometimes fall into error even with respect to its proper objects and present to us images that incorrectly represent their causes. However, we recognize that natured things naturally achieve their ends, ends that are determined by the supreme being, God. As Aquinas puts it in the 'fifth way':

We see that things which lack intelligence, such as natural bodies, act for an end, and this is evident from their acting always, or nearly always, in the same way, so as to obtain the best result. Hence it is plain that not fortuitously, but designedly, do they achieve their end. Now whatever lacks intelligence cannot move towards an end, unless it be directed by some being endowed with knowledge and intelligence; as the arrow is shot to its mark by the archer. Therefore some intelligent being exists by whom all natural things are directed to their end; and this being we call God. (FP, Q2, A3)

Now, 'a natural desire cannot be in vain' (*Summa*, FP, Q75, A6), and, therefore, since 'the rational creature naturally desires to know all things' (FP, Q12, A8, Obj 4), it follows directly that the natural desire to know things will in fact be fulfilled: given God's goodness, we cannot systematically err. And, as the 'fifth way' indicates, we are systematically aware of the goodness of God's operations. So we know that we cannot systematically err: we are beings such that coming to know is our natural end; our natural ends are such that they must for the most part be fulfilled; we know through experience that God arranges things for the best; and we know, consequently, that we are beings who can attain to *scientia*.

But all this presupposes that we can somehow see that God arranges everything for the best, including all the operations of our faculty of reason. Unfortunately, it also does not allow that there is some powerful being who presents to our intellect reasons that are not in fact the reasons for things, or, what amounts to the same, a being who is powerful enough to ensure that the reasons (essences, natures, forms) in the mind that seem to explain things as they appear to us are in fact not the real reasons for those things. If all the reasons that appear to us to be adequate are in fact false and inadequate, then it is also false that things are arranged for the best. Which is to say: There is no guarantee that our natural inclination to know things will, whatever the appearances are, be fulfilled. Aquinas, however, following Aristotle, simply does assume that we can see that everything is arranged for the best, including the arrangements for our coming to know things: he simply assumes that in this best of all possible worlds, our rational faculty is on the whole in good working order, and that we can therefore simply take for granted that *knowledge* = scientia *is possible*.

But this, of course, is *to assume that there is no evil genius whose aim is to ensure that we are systematically deceived about the reasons for things, about the natural laws that obtain in the world*.

What we must recognize, however, is that simply to assume that there is no evil genius is to leave open the possibility that such a substance exists. And if it is possible that she exists, then our knowledge claims are all, possibly, erroneous. Some of our ideas are sufficiently clear and distinct in themselves that they are intrinsically indubitable. But no matter how intrinsically indubitable such claims seem to be, and indeed are, it is none the less possible that they are wrong. Thus, even if we cannot avoid assenting to them, they are, by the Cartesian ethics of belief, unworthy of assent, and we must proceed in our search after knowledge as if we did not assent to them.

We have already noted the strategy that Descartes, and, following him, Arnauld, used to escape this problem at the very core of the Aristotelian metaphysics. Descartes discovers among our ideas the idea of God. This idea is of a being that is self-existent; that is, it is the idea of a being that is necessary in the sense of being so infinitely powerful as to be able to maintain Herself in existence no matter what. No evil genius could therefore prevent Her from existing according to Her nature, that is, the idea that we have of Her. But this idea includes the idea of Her goodness, that is, of Her aim to secure for finite beings the best that is, in a finite world, possible. Thus, though She cannot prevent some unnatural events, that being a consequence of finitude, and, where it is present, the free will of finite beings, none the less She can in Her infinite power guarantee that what is natural, that is, the best for things, does happen, if not always, then for the most part. In particular, in Her goodness, She will be no deceiver and will, moreover, guarantee that we are not systematically deceived, that is, will guarantee that, properly cleansed, the mind will achieve knowledge (*scientia*). The idea of an evil genius who systematically deceives is logically impossible, given our idea of God. Thus, the idea of God excludes the possibility that challenged the worthiness of the indubitable for our assent. The idea of God not only eliminates the possibility of the evil genius, but, since She, who must, in Her infinite power, exist, is not only no deceiver but also one who acts for the best, the idea that we have of Her thus also guarantees that our clear and distinct or intrinsically indubitable ideas will turn out to be true. The idea of God eliminates not only the evil genius but, by infallibly guaranteeing the truth, eliminates that and any other possibility of error. One can now claim legitimately to know, in the sense of *scientia*, the things of the world, that is, to know that our clear and distinct grasp of essences is true in the sense that things of which these essences are predicated exist in accordance with, and not unnaturally or contrary to, those essences.

Descartes, of course, allows for the possibility of error: error is the result of judgment running ahead of reasons. Were we able to control our judgment and orient it constantly towards the best, that is, clear and distinct reasons, we would not err. That we do err, that we do not always control our intellect for the best, is the result of our free will, rather than God's causation. Descartes thus solves the problem of error along the lines of the traditional solution of the general problem of evil.

Descartes thus shares with Aristotle the notion that knowledge of causes is *scientia* and infallibly certain while at the same allowing that error is possible. They differ only as to the sources of error. To this we may contrast the empiricist position which holds that knowledge of causal propositions is knowledge of matter-of-fact regularities. This knowledge is based on inductive inferences from a sample to a population. It is, therefore, intrinsically uncertain. One can never achieve the Aristotelian and rationalist goal of infallible certainty, though, to be sure, moral certainty is often attainable.

There are at least three sources for possible error according to this scheme that was in crucial respects shared by the rationalists and the medieval Aristotelians: (1) The necessary propositions that define laws of nature hold only always *or for the most part*. Hence, laws of nature may have exceptions, and inferences based on them from present to future may turn out to be wrong. (2) The inference from particulars to a universal proposition is never fully certain because of the possibility that either the process of sense experience or the process of abstraction will misfire. (3) God in Her power is capable, according to some, including Ockham and Descartes, of changing the essences of things at any time, and to that extent they cannot be taken categorically to be necessary.

R. Brown ('History,' 666–7) mentions these three things without, however, clearly distinguishing them. On the basis of these points he claims that the medievals recognized that causal judgments are fallible, that there is a feature about them that enables one to hesitate and hold that there is something probabilistic attaching to them. He then connects this to the concern, present at least since Hume, within the empiricist camp, that an inference from an observed sample to a full population is always fallible. Brown concludes that both the medievals and the later empiricists had a concern about causal judgments being fallible, and somehow probabilistic. Their concerns, he concludes, are essentially the same: there is no discontinuity in the history of science between the *scientia* of the medievals and the science that emerged in the early modern period.

The conclusion simply does not follow from the premises. It is true that both the medievals – and the rationalists – thought that there was a probabilistic element in causal reasoning. But it does not follow that these concerns are the same. Moreover, it is absolutely essential to notice that, within the medieval–Aristotelian–rationalist metaphysics, probabilistic reasonings fit in with the possibility of infallible certainty, while also noticing that this is impossible within the empiricist ontology.

For Aristotle and the scholastics such as Scotus, there was a guarantee that one could acquire knowledge in the sense of incorrigible certainty. This guarantee was provided by the notion that there are certain 'natural tendencies' of the human mind towards the acquisition of knowledge. But for the defenders of the new science, whether empiricist or rationalist, the whole talk of powers is vacuous. In particular, therefore, talk of the power of abstraction is vacuous. But it is this 'naturalness' of this power of abstraction, the supposed fact that it is a feature of the human 'nature' or 'form' or 'essence,' that, in the end, is supposed to provide the guarantee for the whole of the Aristotelian system. 'By nature' all persons aim to know, the system claims. Coming to know is thus a *natural motion* which, like all natural motions, will be successful – unless, that is, it is constrained. As Scotus puts it, '[a] faculty does not err in regard to an object that is properly proportioned to it unless the said faculty is indisposed.'[274] Unnatural interventions, such as being caused to abstract the wrong form, happen only incidentally. For what is natural is the norm; it is what happens *always or for the most part*. Hence, error, while it can occur, is unnatural, and the natural powers of the human mind guarantee that success – that is, knowledge of laws – is the usual outcome of induction. But what sort of guarantee is *this*? Even if we glomb onto, or have a rational intuition of, a nature, how do we know that we have abstracted the correct one? If our knowledge of the operation of natures is restricted to what they are 'apt' to produce, then our knowledge is not reliable until we have a criterion for discerning when they are acting reliably and when not. But if the basis for relying upon our knowledge of natures is the fact that natures for the most part are reliable in their operations, then we have no clear way to separate chance from natural implantations of propositions in the intellect.

Scotus cannot even argue that he knows by the nature of the intellect that the intellect by its nature is such that it *never* is indisposed. That would imply that there are some natural laws that are not subject to exceptions, that there are no powers that could interfere with, or impede these natural activities of the substance. Now, to be sure, there are some such laws that never have exceptions, such as the laws that describe the motions of heavenly objects. But even these can be described universally as happening 'per causam naturalem ordinatam' – according to ordained natural causes, that is, predictably – only on the supposition 'quod causae naturales sibi dimittantur' – that is, only on the condition that it is left to itself – which is to say only on the supposition 'quod per virtutem divinam non impediantur' – that is, the supposition that it is not interfered with by the divine omnipotent power.[275] God is not bound by the laws for the creatures He has brought into being, and may at any time suspend them. This would, of course, be a miracle. But, as Descartes was later to ask, might there not be a powerful being who could work the miracle of deceiving us even with respect to what seem like the most proper and certain of the objects of the intellect?

Scotus, however, does not undertake these Cartesian reflections. Nor is that the point in which we are at present interested. Rather, there are three central points: (A) Scotus argues that certain and infallible knowledge of natural laws is – in some cases, at least – actual. (B) Because natural laws have exceptions, many of them can be known only through sense experience and by inference, and to this extent they are fallible and merely probable. The problem of knowledge is thus the problem of how we can use our self-evident knowledge of principles to overcome the fallibility and probability of these judgments. (C) Even where we cannot, for whatever reason, overcome the fallibility of various judgments with regard to causes, the *real possibility* of such knowledge remains, both as an ideal and as an ideal that we could reasonably expect to attain if we undertake various additional reflections.

Those who propose to emphasize the supposed continuity between the medievals and the empiricist defenders of early modern science focus upon (B), the probability of many causal judgments. Scotus, as we see, recognizes at least three sources for possible error, fitting the patterns that we have suggested hold in the metaphysical framework that was, in these crucial respects, shared by the rationalists and the medieval Aristotelians: (1) The necessary propositions that define laws of nature hold only always *or for the most part.* Hence, laws of nature may have exceptions, and inferences based on them from present to future may turn out to be wrong. (2) The inference from particulars to a universal proposition is never fully certain because of the possibility that either the process of sense experience or the process of abstraction will misfire. (3) God in Her power is capable, according to some, including Scotus and Descartes, of changing the essences of things at any time, and to that extent they cannot be taken categorically to be necessary.

R. Brown ('History,' 666–7), as we have noted, mentions these three things in his defence of the continuity thesis; but, at the same time, he fails to notice that the three are all compatible with (C), the notion that there is something about the human intellect in terms of which it makes sense to say that infallible certainty is possible, both generally and in particular cases. What makes this possible is the doctrine, shared by the rationalists and the Aristotelians, that there are forms or natures or essences of things which, when known, reveal with certainty the necessary logical structure of things. It may be difficult to know when we have grasped the correct form, but, for all that, the metaphysical framework allows that there are such forms: the idea is taken for granted. And so, within the Aristotelian-rationalist framework, the notion of infallible certainty makes sense. What Brown fails to notice is that when the empiricists reject these notions of forms and essences they give up any possibility that the human intellect can achieve something more than moral certainty. Brown sees that both the medievals and the empiricists recognized the possibility of error, and therefore infers a continuity of concerns. So much the worse, he concludes (668–9), for the theses of Hacking and me that there is a radical break between the old *scientia* and the new science and that this break consists in the disavowal by the new science, in contrast to *scientia*, of infallible certainty and its admission of only probable knowledge. In particular, I would argue, Brown doesn't understand the basic structure of Aristotelian metaphysics and how it radically differs from the empiricist philosophy of the new science. He therefore fails to see that the radical break between the older traditions and the new science *understood and practised in empiricist terms* consists precisely in the rejection of the doctrine of forms and essences that makes it reasonable for philosophers to claim that infallible *scientia* is possible and must play a role in science.[276]

Descartes continued to defend the older tradition of *scientia*. Like Scotus, he defends thesis (A) that certain and infallible knowledge of natural laws is, in some cases at least, actual. He allows also with regard to certain natural laws, such as those of physiology, that something like proposition (B) obtains – that they can be known only through sense experience and by inference, and that to this extent they are fallible and merely probable. The problem of knowledge is thus the problem of how we can use our self-evident knowledge of principles to overcome the fallibility and probability of these judgments. And, clearly, he continues to defend proposition (C), which states that even where we cannot, for whatever reason, overcome the fallibility of various judgments with regard to causes, the *real possibility* of such knowledge remains, both as an ideal and as an ideal that we could reasonably expect to attain if we undertake various additional reflections.

Locke for the most part rejects (A). But he still accepts (C), and therefore still accepts a central point of the Aristotelian-rationalist metaphysics. Although he did not argue, alongside Descartes and Arnauld, that we need a priori assumptions about God in order to validate our claim to have knowledge in the sense of *scientia*, he none the less left a place in his discussion of knowledge for knowledge of this sort. Locke did argue that one could do 'low' science without reference to any knowledge of essences. His picture of this sort of science therefore conformed to Bacon's theory and Newton's practice. None the less, however attenuated the doctrine of real essences was to become in Locke's works, it still remained, and so long as it was present it remained possible for the rationalists to argue that these sciences remained 'low' and that the science of Newton, insofar as it did not aim at infallibly certain premises, fell short of the ideal, viz., *scientia*, to which science at its best should strive.

In order for the new science to emerge as the *fully empiricist* enterprise that it was to become in theory and practice, the pursuit of infallible certainty had to be eliminated as a possible cognitive goal. The illusion of infallible certainty was, in fact, challenged by various thinkers who attacked the Cartesian restatement of the traditional position. Among these were Simon Foucher[277] and René Rapin.[278]

Perhaps most interesting, however, was Pierre-Daniel Huet.[279] He apparently started off as a Cartesian of sorts, but, after serving with Bossuet as preceptor to the Dauphin and becoming Bishop of Soissons and, later, of Avranches, he became a sceptical critic of Cartesianism.[280] In his *Censura philosophiae Cartesiana*[281] he criticized Descartes's famous *cogito, ergo sum* along the lines that had been suggested by Hobbes and Gassendi in their 'Objections' to the *Meditations*, questioning the validity of the inference. More originally, however, he noted that mentally to assert the propositions of the *cogito* was to produce a thought sequence extended in time. By the time one comes to the end of the inference, the mental assertion of the proposition which was the starting point of the process has now become a remembered mental event. But memory is fallible, so it might be that after all one did not think. Moreover, as one makes the inference, the conclusion is still in the future and one does not yet recognize the truth that one is. Huet concludes that Descartes's 'I think, therefore I am' should more properly be stated as 'I may have thought, therefore perhaps I may yet be' (*Censura*, ch. i–ii). Supposedly self-evident truths such as the *cogito* turn out to be radically contingent. It follows that where the rationalists thought they could find necessary connections there are no necessary connections. The notion that certitude is attainable, in principle if not in practice, requires us to have the idea of a necessary connection, the idea of a form or essence of a thing. In order to have such an idea, we need to have at least one instance of such an entity. In taking apart the *cogito*, in showing that even here we have nothing but separable parts, Huet is challenging the claim that we have therein an instance of a form or nature or essence, of an entity that could provide an objective necessary connection among the separable events of ordinary experience.

With equal care Huet dissected the Cartesian criterion of truth. The latter critique was extended even more devastatingly in Huet's *Traité philosophique de la foiblesse de l'esprit humain*.[282] In this posthumous work, Huet defended academic scepticism (which he argued was not significantly different from Pyrrhonism [*Traité*, 139–50]). He advanced a wide battery of sceptical arguments designed to show that 'l'Homme ne peut connoître la Vérité par le secours de la Raison avec une parfaite & entiere Certitude' (22). There are many degrees of certitude, and faith alone leads to perfect or infallible certitude (16ff). In the task of living, the correct rule for apportioning our assent to propositions that we discover is the only rule that turns out to be available to us, to wit, the rule of probability. 'Il faut suivre dans l'usage de la view les choses probables, comme si elles étoient véritables' (207ff). It is the senses that provide the rule of truth (208). From these alone do we obtain information on the basis of which we can form judgments about the world. But, that being so, it is impossible to achieve evidence that admits of perfect certainty. We have left only probability, with which we must make do, *faute de mieux*. In particular he recommends the sceptical and non-dogmatic cognitive attitude of the empiricists of the Royal Society, 'cette nouvelle Societé de Philosophes Anglois, qui a élevé tant d'excellens Esprits, condamne l'arrogance des Dogmatiques ...' (221).

The effect of critiques such as those of Huet was to force people to recognize that the claims made by the rationalists – and the Aristotelians – that *scientia*, infallible certainty, was possible were claims that reasonable men could dismiss. Once they were dismissed, however, one could also dismiss criticisms of the 'low' sciences and the science of Newton for falling short of the ideal of *scientia*: an impossible ideal is not one for which one might reasonably aim. The critique of the claim that *scientia* was possible by thinkers like Huet freed the new science from the impossible cognitive aspirations of the rationalists and from the other dross that remained of Aristotelianism.

5) *The Empirical Science of the Human Mind*

There is another aspect of the critique of Cartesianism that it is essential that we not overlook.

Descartes's argument from the claim that God is no deceiver succeeds in defending the notion that knowledge is incorrigible and infallibly certain. It does so, however, only by taking for granted that we have such ideas as those of substance, of essence, of God as a necessary existent, and so on. Locke succeeds in making a place of sorts for the empiricist science of the 'low' sciences and the science of Newton, but he none the less leaves a remnant of rationalism untouched: however attenuated, the real essences remain to define a standard of certainty to which rationalists could argue science ought to aspire. That dross had yet to be completely eliminated.

That complete elimination would be possible only with the elimination of the claims that we have notions of the sort that could make infallible knowledge possible. That is, the very ideas of substance, of form or essence, and of objective necessary connection which alone make possible the notion that we can attain infallible certainty must be subjected to a critique that shows them to be illusory.

Now, these concepts are concepts of entities that transcend the world we know by sense experience. *What must be shown is that we have no concepts that transcend the world of sense experience.* Once this was achieved, the new science would be free of rationalism and Aristotelianism and empiricist throughout in both philosophy and practice.

The crucial move can be found in its origins in both Huet and Locke, when they both insisted that the methods of the new science apply to the mind itself. The result was a world which, unlike that of Aristotle, Descartes, and Arnauld, is all on one level, the level of sense. With the elimination of the higher level, the need to introduce rational intuition is gone. No longer are there two ways of knowing – sense and intuition: all knowing now begins and ends in sense. For Aristotle and Descartes, a person's reason is his or her spiritual element: it is his or her point of contact with God. For them, humankind is semi-divine, split between a heavenly rational part and a sensible corporeal part. Locke and Huet attack this idea of humankind, insisting that humans are thoroughly of *this* world. The rational powers that supposedly lift persons out of this world are simply not part of our experience; there are other explanations of the activities for which they purport to account. The reason that we actually use, and which is the reason we ought to use, is as natural as sensation. There is nothing about science and knowing that marks a break in humans between a sensible level and some higher level. Any claims that reason may do more than begin and end in sensible knowledge is reasonably dismissed. This marks the dismissal of the inhuman view of humankind (that we are partly not of this world in which we live and breathe), and the adoption of the truly humane Enlightenment view of humankind as human and no more than human.

It is Locke's position that '[i]f we can find out how far the understanding can extend its view, how far it has faculties to attain certainty, and in what cases it can only judge and guess, we may learn to content ourselves with what is attainable by us in this state' (*Essay*, Intro., sec. 4). His purpose is 'to inquire into the original, certainty and extent of human knowledge, together with the grounds and degrees of belief, opinion and assent' (Intro., sec. 2). The method that he proposed was the 'historical, plain method' (ibid.), which, given that 'historical' is understood to mean 'experimental' or 'observational,' is simply the empirical Baconian and Newtonian method of the new science. The result was an attack on the metaphysics that from Plato and Aristotle through Descartes had supported the view that each person is a being divided from him- or herself, a being with two ways of knowing. The 'real essences' that were the core of that metaphysics Locke declared unknowable. Since the method of empirical science relies not on rational intuition for knowing particulars but on sense observation alone, Locke declares that 'real essences' are not a proper subject for science. For Aristotle and for Descartes, method moves from sense to intuition and concentrates on the latter, valuing the former only as a trigger for the latter. Method, for such a philosopher, is not unitary. For Locke, on the other hand, method *is* unitary. He succeeds in making it unitary by distinguishing empirical science from metaphysics: the latter is a 'science' of essences, or natures, or forms, or what-have-you, lying, as Aristotle and Descartes would have it, beyond and above the world of sense. For Locke that world is unavailable, and any problems with somehow understanding it are not those of science.[283]

We have the method of empirical investigation. This constitutes the fallible science of the world of sense experience. This method is being, as it were, turned upon itself to be used to investigate its own practice. This reveals that we *do* make do without any concepts that refer to entities that transcend the world to which the method is being applied and which contains its practice. More strongly, it reveals, in its investigation of the genetic origins of the concepts that we do use, that all these concepts in fact derive from sense experience. *We have no concepts that refer to the transcendent entities knowledge of which Aristotle and Descartes claimed was essential to the practice of science: we now see that those claims are simply illusory.* That is the achievement of Locke and the others who insisted that the new method of science must be used to investigate mind and thought as well as body. But further, the whole issue of concept formation is redirected. What turns out to be important is not the way words reflect ideas, as if we have, as both the Aristotelians and the rationalists claim, ideas that are somehow prior to language. Rather, what is important is the way we use language to reflect the way the world is.

In Locke, the latter issues are not yet fully clarified.

To be sure, Locke exorcises real essences from playing any significant role in ontology or epistemology in the area of the 'low' sciences and, in fact, in the area of most causal judgments. The necessary connections based on real essences that the Aristotelians and the rationalists claim must be there if we are to be making genuine causal judgments are simply not there. The judgments are, therefore, not genuinely causal in the traditional sense, that is, they are not instances of *scientia*. But, Locke concludes, so much the worse for the claim that causal judgments are a matter of *scientia*: they *are* genuine causal judgments, and that shows that we have no need for the notion that genuine causal judgments are a matter of discerning the real essences that yield necessary connections.

Locke establishes this through an examination of the experience from which our concepts of things in these areas are supposedly derived. It is Locke's claim, which he succeeds in establishing in some detail, that the 'historical plain method' reveals no other genetic antecedents besides sense experience, that this experience contains no necessary connections, and therefore that the concepts we form to explain this experience involve, contrary to the claims of the Aristotelians and the rationalists, no necessary connections. This, on the one hand.

On the other hand, he also suggests a specific account of the formation of our ideas of species of things. This is by 'separation.' In holding this, Locke simply takes over views from his predecessors.

For Locke, forming ideas by abstraction consists in 'separating them from all other ideas that accompany them in their real existence' (*Essay*, II.xii,1). This sort of separation is the mechanism by which 'the mind makes the particular ideas, received from particular objects, to become general' (II.xi.9). Arnauld, in the Port Royal Logic, made the same point: abstraction occurs with respect to 'choses ... composées' when one 'les considérant par parties, et comme par les diverses faces qu'elles peuvent recevoir.' In abstraction what is thus known is considered separately from the whole; either one considers 'les parties séparément' in the case of 'parties intégrantes,' or one 'peut séparer les choses en divers modes,' or, finally, 'quand une même chose ayant divers attributs, on pense à l'un sans penser à l'autre' (*La Logique*, I.v).[284] The former occurs when, for example, we separate in thought the parts of the body. This is the simplest form of abstraction. Berkeley has no problems with this notion: 'I will not deny I can abstract, if that may properly be called *abstraction*, which extends only to the conceiving separately such objects, as it is possible may really exist or be actually perceived asunder' (*Principles*, sec. 5).[285] Neither does Hume have problems with this notion; it is the basis of his – and Berkeley's – view that ideas are either simple or complex and that some of the complex ideas are a result of our combining simple parts in ways that are not found in reality, as, for example, in our idea of the New Jerusalem (T, I.1.I, 2–3).

It is the other two sorts of abstraction that are more problematic. Consider the first of these two forms of abstraction. We begin with simple ideas of sensation, such as white (*Essay*, II.iii.1). One forms the abstract idea of white by separating the property 'white' from the other qualities with which it is conjoined in the white particulars, such as snow or milk (II.xxi.73), that are given in sense experience. The resulting idea is distinct from other ideas, such as those of blue or heat (II.xii.1). The example given by Arnauld is that of the geometers who explore the idea of three-dimensional space by considering first length alone, then area, and finally volumes or solids (which is the same, for these Cartesians as 'corps,' body) (*La Logique*, I.v). This is one form of separation. By this means one forms abstract ideas of species. But whatever is red is coloured. Here we have not only the abstract idea of a species but also an abstract idea of a genus. The method of forming the abstract ideas of genera is rather different. The mind, 'so to make other yet more general ideas, that may comprehend dif-

ferent sorts [species] ... leaves out those qualities that distinguish them ...'
(Locke, *Essay*, III.vi.32). That is, one forms the abstract idea of a genus from
the ideas of several qualities by separating the genus from the species. Thus,
one forms the idea of being coloured by separating the property of being
coloured from such properties as white, red, and so on, which are, of course,
themselves abstract ideas. One forms the abstract idea of an equilateral triangle
by separating this property from particular equilateral triangles.

Que si je passe plus avant, et que ne m'arrêtant plus à cette égalité de lignes, je considère
seulement que c'est une figure terminée par trois lignes droites, je me formerai une idée
qui peut représenter toutes sortes des triangles. Si ensuite, ne m'arrêtant point au nombre
des lignes, je considère seulement que c'est une surface plate, bornée par des lignes
droites, l'idée que je me formerai pourra représenter toutes les figures rectilignes, et ainsi
je puis monter de degré en degré jusqu'à l'extension. (Arnauld, *La Logique*, I.v)

Thus, one forms the abstract idea of extension, or of being extended, by separating the property of extension from the property of triangle or circle or some
other figure – that is, from these abstract ideas. These abstract general or universal ideas signify many particulars by signifying properties in those particulars. But these properties do not exist independently of those particulars. These
ideas do not therefore represent or signify things that can exist as such, independently of particulars. 'general and universal ... belong not to the real existence
of things ... their general nature being nothing but the capacity they are put into
by the understanding, of signifying or representing many particulars' (Locke,
Essay, III.iii.11). Or again,

Quoique toutes les choses qui existent soient singulières, néanmoins, par le moyen des
abstractions que nous venons d'expliquer, nous ne laissons pas d'avoir tous plusieurs
sorts d'idées, dont les unes ne nous représentent qu'une seule chose ... et les autres en
peuvent également représenter plusieurs, comme lorsque quelqu'un conçoit un triangle
sans y considérer autre chose, sinon que c'est une figure à trois lignes et à trois angles;
l'idée qu'il en a formée peut lui servir à concevoir tous les autres triangles. (Arnauld,
La Logique, I.vi)

In general, one forms the abstract idea of a specific property by separating that specific property from the concrete particulars presented to one; and
one forms generic abstract ideas by separating the generic property from the
specific.

The problem with this view is that it permits the reintroduction of all the traditional notions. According to the traditional doctrine, taken over, of course,

by rationalists like Descartes and Arnauld, necessary connections are obtained when we grasp the essential structures of things. These essences are given by species and genera; as Arnauld puts it, 'A definition ... identifies the nature of a thing by identifying the essential characteristics of the thing. The defined word is defined in terms of words expressing a genus restricted by a specific difference ...'[286] Once Locke allows that we can grasp species and genera apart from the things that instantiate them, he is clearly allowing the Aristotelians and rationalists to claim that this cognition is of the essential structure that constitutes the necessary connections of things.

Berkeley fully recognized the dangers implicit in Locke's claim that we form abstract ideas by separation. In his defence of idealism, he points out that the defender of material objects may retreat from the claim that the external cause of our sensations is a material substance to the weaker claim that it is a 'unknown *Somewhat*' devoid of positive characteristics of the sort we know, something that is 'inert, thoughtless, indivisible, immovable, unextended, existing in no place.' Berkeley replies that if this be the concept of matter, then it is no different from the concept of nothing, and can have no more explanatory power than the latter (*Principles*, sec. 80). Berkeley then goes on to consider the possibility that his opponent will claim that the concept of matter that he advances is not that of nothing, for 'in the foresaid definition is included what doth sufficiently distinguish it from nothing – the positive abstract idea of *quiddity, entity,* or *existence*' (sec. 81). If existence were an entity that non-separately accompanied all other entities that one experienced but could be separated in thought, then the world could be populated with all sorts of unacceptable entities. Berkeley thus recognizes the dangers of having abstract ideas as separable parts of concrete particular ideas: they allow the reintroduction of all the entities which, the empiricist was claiming, on grounds provided by the 'historical plain method,' were not in fact to be found ingredient in our causal judgments and other mental processes.

We do, of course, have abstract ideas in the sense that we have ideas that are general, representing in some way not just particular things but species and genera. The issue is whether they are formed by separation. The issue is whether Locke's genetic account is accurate. It was disputed by both Berkeley, who provided the philosophical case, and by Huet, who provided the case in terms of a psychogenetic account. These points of view were synthesized by Hume.

Berkeley challenged Locke's account of abstract ideas in his *Principles* (Introduction, secs 8–10). The three propositions

(a) (Necessarily) whatever exists is particular,
(b) Whatever is possible in thought is possible in reality,

(c) Abstract ideas are formed by separating specific and generic properties from existing particulars,[287]

are inconsistent. The same argument was expressed by Hume in this way:

'tis a principle generally receiv'd in philosophy, that every thing in nature is individual, and that 'tis utterly absurd to suppose a triangle really existent, which has no precise proportion of sides and angles. If this therefore be absurd in *fact and reality*, it must also be absurd *in idea*; since nothing of which we can form a clear and distinct idea is absurd and impossible. (T, I.I.vii, 19–20)

Berkeley and Hume conclude: so much the worse for the traditional doctrine of abstract ideas. Again as Hume puts it,

Now as 'tis impossible to form an idea of an object, that is possest of quantity and qual-ity, and yet is possest of no precise degree of either; it follows, that there is an equal impossibility of forming an idea, that is not limited or confin'd in both these particulars. Abstract ideas are therefore in themselves individual, however they may be general in their representation. (ibid., 20)

Berkeley did not himself go beyond this negative claim to propose an empir-ical account of the causal origins of our abstract ideas. Huet does rather better.

Huet notes that Descartes bases his claim to have infallibly certain knowl-edge on the claim that he has ideas that are innate, and not acquired through sense experience. *This claim he challenges on empirical grounds.* 'Mais quelque diligence que j'aye apportée à cette recherche, je n'ai trouvé en moi aucune Idée, qui ne m'ait paru très-clairement être venue du dehors; & dont je n'aye reconnu la source dans les objets exterieurs d'où elle étoit partie, & la voye même par où elle a trouvé entrée dans mon Entendement.'[288] This, he holds, argues inductively that for others the origins of ideas are similar: 'J'ai cru ensuite pouvoir juger de l'Entendement des autres par la mien' (*Traité*, 197). What, then, is Huet's abstract idea of a triangle? It is, he argues on *phenomeno-logical, that is, empirical, grounds*, a sensuous image fuzzy in its details but resembling the various impressions of triangles that he has experienced over time: 'Ceux qui sont dans une opinion contraire demandent, d'où m'est venue l'Idée d'un Triangle. Je répons qu'elle m'est venue d'une infinité de Triangles que j'ai vûs, d'où je me fais fait un Idée obscure & confuse de Triangle, qui n'est point déterminée, ni circonscrite par des bornes certaines' (198). Huet then proceeds to give similar accounts of our ideas of number and of God (198–201).

This leaves open, however, the hook-up between our ideas and language. If our ideas are, in fact, as Huet, Berkeley, and Hume claim, then how is it that an idea, that is, an image, that is particular becomes abstract, or, in Locke's terms, how does it acquire 'the capacity ... of signifying or representing many particulars' (*Essay*, III.iii.11). It cannot signify them by their all being present to the mind when the particular idea that represents them all is present, since, as Hume says, 'the capacity of the mind be not infinite' (T 18). Hume's solution turns on arguing that the ideas and impressions that an abstract idea signifies 'are not really and in fact present to the mind, but only in power' (T 20). This capacity is one that is acquired or learned through a process of association. Thus, Hume's positive account of abstract ideas depends upon his associationist psychology.[289] Just as our notion of causation is understood in terms of association based on the relation of contiguity, so abstract ideas are understood in terms of association based on the relation of resemblance.[290] Ideas (images) and impressions that resemble each other in a certain respect, say in being red or in being coloured or in being extended, will through the mechanisms of association become so associated in thought that any one of them can introduce another member of the set. At the same time, a word (that is, a sound or mark) comes to be associated with ideas and impressions insofar as they are associated with one another via some resemblance relation. Thus, 'red' comes to be associated with all members of the resemblance class of red impressions and ideas, 'coloured' comes to be associated with all members of the resemblance class of coloured impressions and ideas. And so on. When we encounter an impression or contemplate an idea (image) which is a member of the same resemblance class, that idea or impression introduces the word – the general term – that refers indifferently to each and every member of the resemblance class.

When we have a resemblance among several objects, that often occur to us, we apply the same name to all of them, whatever differences we may observe in the degrees of quantity and quality, and whatever other differences may appear among them. After we have acquired a custom of this kind, the hearing of that name revives the idea of one of these objects, and makes the imagination conceive it with all its particular circumstances and proportions. But as the same work is suppos'd to have been frequently applied to other individuals, that are different in many respects from that idea, which is immediately present to the mind; the word not being able to revive the idea of all these individuals, only touches the soul, if I may be allow'd so to speak, and revives that custom, which we have acquir'd by surveying them. They are not really and in fact present to the mind, but only in power; nor do we draw them all out distinctly in the imagination, but keep ourselves in a readiness to survey any of them, as we may be prompted by a present design or necessity. The word raises up an individual idea, along with a certain custom;

and that custom produces any other individual one, for which we may have occasion. (T 20–1)

The mechanisms of association provide the explanation of how a word becomes general. The abstract idea just *is* the members of the resemblance class *qua* habitually associated with each other, or, more accurately, the associated resemblance class insofar as the general term has become associated with it. The parts of the idea, those impressions and ideas that fall under it, are the members of that class – though, to be sure, when the idea is before the mind, instancing a use of the general term, those parts are present not actually but *only potentially or dispositionally*, ready to be recovered through association as the occasion permits or requires.

A similar account of abstract ideas, in terms of resemblance and the mechanisms of association, is given by David Hartley in his *Observations on Man*.[291] If Hume develops the position in a more philosophically interesting way, Hartley's was certainly more influential as a psychological theory. For, Hartley's is certainly the first systematically worked out psychological theory of the processes of the human mind. As part of that theory, he makes clear that there is no need to introduce necessary connections in order to account for scientific inference.[292]

What Hume and Hartley work out in detail is the notion that human beings are thoroughly this-worldly. They do this by taking seriously the proposal of Locke that one use the method of science, the 'historical plain method,' to describe and explain not only the entities of physics but also the workings of the human mind. What they discover is that *as a matter of empirical fact* there is nothing in the human mind whose origins cannot be traced to ordinary experience, either sense experience in the case of external objects, or internal sense in the case of such things as the passions. In particular, Hume and Hartley argue on the basis of matter-of-fact observations that language has its origins in the ordinary world of sense experience – taken as including, of course, the fact of other human beings and the fact that oneself and others interact socially.

In this picture of humankind, *language* is understood not as something that is merely *expressive* of thought, that is, thought that is prior to and independent of language. Rather, language is taken to be *constitutive of* thought; it enters as the very structure of thought: for Hume, abstract ideas – that is, the medium in which we think – consist of ideas and impressions such that (i) they are associated one with the others on the basis of some relation of resemblance, and such that (ii) a general term, that is, a sound or mark, has become associated with things that resemble one another in the relevant respect. Our habits of thought 'follow the words' (T 23).

The patterns here are both causal, that is, the result of association, and semantic, that is normative. The patterns are artificial, in the sense of being, on the one hand, learned and, on the other, of serving fundamental human needs with respect to communication. But conventions that serve fundamental human needs are converted through the mechanism of sympathy into norms. So the learned linguistic patterns that are abstract ideas, the syntax of language, and the standards of causal inference ('the rules by which to judge of causes and effects') all become converted into shared rules of language and inference, exactly as the norms of property and contract become converted into the shared rules of artificial virtue.[293]

All these conventions can be altered through deliberate choice in order to satisfy interests that we have. Some changes in syntax and semantics occur slowly, and unconsciously. But others are deliberate. Poets regularly make syntactical innovations to achieve new effects. Semantic rules are changed quite regularly. If one has a cognitive interest in matter-of-fact truth, then that will lead one to discipline one's thought to conform to the rules by which to judge of causes and effects. Or so Hume argues, and so Baier correctly reads him. What the need to communicate generates is the general interest that there be conventions of language. The fact that one is born into a linguistic community gives the accepted rules of that community an advantage at the gate over possible alternatives; but since we learn those rules so early, they become so deeply engrained that one can only think of modifying them piecemeal, never of their total replacement. Neither political constitutions, as Livingston has argued,[294] nor our ethics of belief, nor even the logical syntax of language can be evaluated from an a priori perspective and the possibility of revolutionary change contemplated. At least, they ought not to be so evaluated: for, if Hume is correct, there is no Plato's heaven where we can find either an ideal republic, or an objective necessary causal tie, or any other sort of object for our transcendental reason to grasp.

The crucial move consists in construing thought as something essentially linguistic. Hume saw this himself when he wrote in his correspondence that 'in much of our own thinking, there will be found some species of association. 'Tis certain we always think in language, viz. in that which is most familiar to us; and 'tis but too frequent to substitute words instead of ideas.'[295] It is this theme that Baier understands fully, and in this she may be contrasted to many other commentators, such as Peter Jones. Jones has remarked on the passage just quoted that '[u]nfortunately, the view that we think in language is left unsupported; unless he means by it that the *expression* of thought requires language, Hume did not mention any such view in his earlier philosophical work, and it may reflect his reading subsequent to it.'[296] But, of course, Hume *did* support the view that thought *is* language: the argument is constituted by his doctrine of abstract ideas on the one hand and his account of linguistic conventions on the other. Perhaps because these occur respectively in Books I and III of the *Treatise*, Jones has failed to make the connection. Jones locates Hume's views on language in the context of French thinkers such as Lamy, du Tremblay, and du Marsais, who held that linguistic conventions serve human needs.[297] These thinkers were, however, of a piece with the old tradition, also found in earlier thinkers such as Descartes and Locke, in which language derives its signification from its expressing thoughts which are both non- and pre-linguistic. Always eager to locate Hume within a set of precursors, Jones misses Hume's break with the precursors when he makes the radical innovation that thought does not precede but *is* language, and instead attributes to Hume the view of the old tradition with which Hume broke, that the role of language is merely that of expressing thought, and that language acquires its signification through expressing thoughts, that is, thoughts which must be prior to the language that merely expresses but does not constitute them. Jones, in fact, attributes to Hume the position of Descartes and Locke, that we clarify speech by turning to the thoughts or ideas which lie behind it, and give it its signification: 'No matter how many problems Hume leaves unexplored, his own position is moderately clear: talking is distinct from thinking, and most talk expresses thought; when confused by talk, we have to struggle to identify the thought behind it, and in the rarefied regions of philosophy we often find that such thought is itself confused or incoherent.'[298] Now, of course, thought is different from talking, that is, overt verbal behaviour or, if you wish, speech acts. Hume agrees: for a thought to be expressed in overt behaviour, an act of volition is, often at least, required. It does not follow, however, that since thought is not talk – that is, speech acts – therefore it is not linguistic. It is also true that when we are attempting to communicate with others, and we are presented by another with a speech act the meaning of which is unclear to us, then the *first step* in attempting to understand that event consists in attempting to *identify its immediate causes*, that is, the belief and the desire or intention that triggered the volition that produced the speech act. For Hume, however, there is a *second step* that is required if we are to grasp the signification of language: *we must trace out the conventional or habitual links which it has to entities in the world that is given to us in sense experience.* ''Tis impossible to reason justly, without understanding perfectly the idea concerning which we reason; and 'tis impossible perfectly to understand any idea, without tracing it up to its origin, and examining that primary impression, from which it arises' (T 74–5).

What this means is that where the Cartesian program, endorsed on this point by Locke, requires one to *turn inward* to discover the signification of language, the new scientific view of humankind developed by Hume and Hartley requires one instead to *turn outward*, to the world that language, through its syntactical and semantical conventions, tries to describe. For Hume, we must examine not the fit of our language to our ideas but the fit of our language to the world. Jones refers to the passage just quoted,[299] but fails to recognize its significance. As a consequence, he ignores the second step that Hume prescribes for grasping the signification of what is said, that is, the tracing out of the habitual or conventional ties of semantics that link terms to what they designate. Instead, Jones attributes to Hume the Cartesian and Lockean view that to grasp the sense of what a person says it suffices to turn to the 'thought behind it.' But when Jones locates Hume in this tradition of Locke, Descartes, Aristotle, and Plato, in which the meaning of language derives from thought which is non- and pre-linguistic, he simply fails to recognize that in Hume the transition from the other-world of Plato and Aristotle, of Descartes and Locke, to naturalism is finally complete.

For Hume, in contrast, there has been a major leap beyond the tradition inherited from the Aristotelians and the rationalists, and which Locke inherited, the notion that there are ideas that are prior to language; as Annette Baier has put it: 'without language ... there would be no generality, no abstraction ...'[300] These conventions are part of the natural causal order, resulting by means of the associative mechanisms from natural causes. At the same time, these conventions are normative, where this normativity also has a naturalistic explanation. Baier emphasizes this naturalism of Hume in her discussion of Hume's account of causation (*A Progress*, 89ff). According to this, there are two definitions of *cause*. The first idea of cause is constant conjunction. The second is that of a determination of the mind to pass from the idea of the cause to that of the effect. Hume offers these definitions after an elaborate causal analysis of the origin of our idea(s) of cause. The causal analysis itself exemplifies the patterns mentioned in the two definitions. 'Causal inferences are what enable us to trace and recognize constant conjunctions, and their effect on us. Hume's double definition of the causal relation itself displays a meta-causal relation, and that is what gives it its authority' (91). Hume's definition of the concept of *cause* is *not* confused, unlike those of the rationalists, precisely because he has traced out the causal ties, which are also semantic rules, which link the concept to things of which we actually have experience.

In fact the causal analysis not only exemplifies the patterns stated in Hume's two definitions, but it also exemplifies more specific patterns, to wit, those identified by Hume as the 'rules by which to judge of causes and effects,' that is, in effect, the rules of eliminative induction. Hume examines the many *causes* of true belief and error, and concludes that, on the whole, true beliefs are the causal upshot of inferences that conform to these rules, while other patterns of inference, those of 'unphilosophical probability,' have error as their upshot.[301] Causal reasoning in conformity to the rules shows us that conformity to those rules is the best means we have for achieving our goal of truth, the passion of curiosity. The rules are 'reflexivity-tested human norms' (Baier, *A Progress*, 100). They are 'products of our reflection' (ibid.), not a priori standards deriving from God or a world of forms.

Our cognitive standards are here being judged *practically*: they are, in fact, conducive to satisfying our cognitive interest in truth, the passion of curiosity, and are, therefore, rationally acceptable. There is no *pure reason*, that is, reason that is somehow self-evident and self-justifying. One can justify a set of cognitive norms as rationally acceptable only if conforming to them satisfies some passion or other. 'Where reason is lively, and mixes itself with some propensity, it ought to be assented to' (Hume, T 270). Specifically, of course, when we are concerned with truth, the relevant propensity is that of curiosity, the love of truth. The philosophical enterprise, the search after truth, is a way of indulging the sentiments[302]: 'I am uneasy to think that I approve of one object, and disapprove of another; call one thing beautiful, and another deform'd; decide concerning truth and falsehood, reason and folly, without knowing upon what principles I proceed,' and if I do not indulge these sentiments, then 'I *feel* I should be a loser in point of pleasure; and this is the origin of my philosophy' (T 271).[303] The search after truth takes its place in the larger search after happiness that provides the full structure of one's life. And the epistemic norms that a 'just' philosophy adopts are those conformity to which, experience tells us, will in general lead in a human way to my satisfying my human curiosity.

For as superstition arises naturally and easily from the popular opinions of mankind, it seizes more strongly on the mind, and is often able to disturb us in the conduct of our lives and actions. Philosophy on the contrary, if just, can present us only with mild and moderate sentiments; and if false and extravagant, its opinions are merely the objects of a cold and general speculation, and seldom go so far as to interrupt the course of our natural propensities. (T 271–2)

Reason thus comes to take its place among the (natural) *virtues*:[304] ''Tis impossible to execute any design with success, where it is not conducted with prudence and discretion; nor will the goodness of our intentions alone suffice to procure us a happy issue of our enterprizes. Men are superior to beasts principally by the superiority of their reason ... All the advantages of art are owing to human reason' (T 610).[305]

By focusing on human ends and human means of achieving those ends, *including our cognitive ends*, Hume eliminates the non-human, or rather inhuman, standards of traditional epistemology. Hume abandons the chase after self-evident principles and instead contextualizes the norms relative to human interests and human wants, and to *human* means, tested in *human experience*, for satisfying those interests and wants.

It is not my intention to argue that Hume, however important he was, succeeded in working out alone the claims that secured the empiricist account of the new science. He was, rather, part of the movement that began with Locke, included others such as Huet, and continued on through Hartley into the mainstream of scientific psychology. What this tradition, represented at its best in Hume, had succeeded in doing was to ensure that one could understand the new science as using no concepts other than those that were tied securely to the world that we know primarily by means of our senses. The natural scientific approach to language and thought itself establishes, at least as securely as empirical inductive arguments establish anything, that this is indeed the case.

And in this way disappeared any illusion that we need for causal judgments in general or science in particular any concepts that transcend the world we know by ordinary sense experience. The empiricist philosophy of the new science has now been made fully secure; it can continue its empiricist practice unbothered by reproaches from the rationalist camp that it is somehow falling short of a significant cognitive ideal: to the contrary, empiricist science can reply, that ideal, and, what is the same, the ideal of the Aristotelians, is an illusion, and may be dismissed as such.

But there is more to the story than this. The rejection of Aristotelianism and rationalism required a variety of other moves. We shall examine a number of these in the studies that follow.

Study Two

Logic under Attack:
The Early Modern Period

I can remember conversations at the University of Toronto in the 1960s in which Thomist philosophers such as A.C. Pegis distinguished their logic, rooted in a realistic metaphysics, from the logic of philosophers like Bertrand Russell, which was rooted in a nominalistic metaphysics. This view can be found stated in Jacques Maritain's *Formal Logic*.[1] It can also be found stated clearly, though in a more pragmatic context, in P. Coffey's *Science of Logic*.[2] It seemed clear to me, however, that one could adopt Russell's logic and also hold that the properties of things were universals, that is, that they could be present in more than one individual. This seemed to imply that one could accept both Russell's logic and also universals, that is, accept Russell's logic and be a realist. I did not know what the philosophers such as Pegis had in mind, and attributed their view to the failure of Thomists to have any sensitivity to developments in philosophy that happened subsequent to the Middle Ages. I subsequently discovered that I was not alone in these sentiments; Wittgenstein wrote a scathing review of Coffey's logic text.[3] No doubt there was in my own case the added sense that these guys were old fogeys. However, now that I too have become an old fogey, I can feel some of the exasperation they must have felt at the incapacity of youth to grasp the obvious. In any case, I now think that there was a real point to what they argued. What I hope to do in this essay is make my amends to Professor Pegis and the others and to try to state the real difference between traditional logic as they understood it and Russell's logic. Whether they would agree with what follows is another thing, one that, unfortunately, we cannot check out.

At the same time, I hope also to discover the real disagreement between them and Russell. As I will argue, the differences are not so much logical as ontological, though I will emphasize that in fact there are two different logics, logics appropriate to two different ontologies. I shall try to bring out the arguments that were used in the early modern period to establish that the logic of the Aristotelian tradition ought to be rejected in favour of what is in effect the logic of

Russell. In fact, the traditional logic was criticized from two perspectives, that of the rationalists such as Descartes and Arnauld, on the one hand, and that of the empiricists, on the other. As it turns out, the logic of the rationalists did not, in fact, differ in essentials, at least not in the ontological essentials, from that of Aristotle. It was the empiricists, like Locke, using an appeal to the ontological principle of acquaintance (PA), who argued for something like Russell's logic and Russell's ontology.

But as we shall also see, all along there was a version of logic that was used in speech making and argument, and whose core was the notion of keeping things consistent, rather than displaying the ontological structure of the universe. This sort of logic is closely connected with the logic of Russell. As things happened, the ontological logic of Aristotle was closely connected with this logic of consistency in time but separate from it in its essential purpose. Moreover, the logic of consistency was worked out in the Middle Ages in a peculiar context, that of scholastic disputation. Both the rationalists and the empiricists reacted against, and criticized, the scholastic art of disputation. The rationalists, however, opted into the Aristotelian notion that logic was rooted in an ontology of necessary connections. The empiricists, in contrast, rejected this notion, and the metaphysics which it defends by appeal to PA incorporates a logic that is nothing more than a logic of consistency.

In short, there are a number of strands concerning logic and metaphysics that we must disentangle if we are to become clear on the new logic that replaced the traditional logic of Aristotle.

I: Traditional Logic

1) The Problem of Existential Import

The difference between traditional logic and modern logic is often located in the textbook tradition in the doctrine of existential import:[4] traditional logic does, where modern logic does not, allow propositions of the form

A: All S are P $(= SaP)$

to have what is called 'existential import.' In the traditional logic, the inference from A propositions to propositions of the form

I: Some S are P $(= SiP)$

is taken to be valid. Similarly, inferences from propositions of the form

E: No S are P $(= SeP)$

to propositions of the form

O: Some S are not P $(= SoP)$

are taken to be valid. However, if we have the usual translations into the formalism of modern logic, namely,

(Af) $(x)(Sx \supset Px)$

(Ef) $(x)(Sx \supset \sim Px)$

(If) $(\exists x)(Sx \, \& \, Px)$

(Of) $(\exists x)(Sx \, \& \sim Px)$

then the inferences from A to I and from E to O are both invalid.

In fact, of course, there are a series of other inferences traditionally taken to be valid but which turn out to be invalid when the forms are translated into the symbolism of modern logic. Most importantly, there are the inferences on the traditional square of opposition, most of which turn out to be invalid when rendered in the symbolism of formal logic.

> SaP and SeP are contraries (cannot both be true, but can both be false).
>
> SiP and SoP are subcontraries (cannot both be false, but can both be true).
>
> SaP is superalternate to SiP (if the former is true so is the latter).
>
> SeP is superalternate to SoP.
>
> SaP and SoP are contradictory (cannot both be false and cannot both be true).
>
> SeP and SoP are contradictory.

Of these, only the last two remain valid upon translation into the symbolism of modern logic. This can be seen using the standard Venn diagrams.

A proposition of the form

$(x)(Fx)$

is true in the universe U if and only if for every interpretation of x into individuals in U, Fx turns out to be true. As usual,

$p \supset q$

is false if and only if p is false and q is true, and is otherwise true. That is, it is true if p is true and q is true, and is true if p is false. Hence, (Af) = $(x)(Sx \supset Px)$ will be false just in case that there is at least one interpretation of x into individuals of U such that Sx is true and Px is false. If there is no such individual, then (Af) is not false, and is, therefore, true. That is, (Af) is true just in case that the class of S's which are not P is empty. This gives the standard Venn diagram of two overlapping circles in a square, the square representing the universe U and the two circles representing S and P. That (Af) is true is represented by shading the portion of the S circle which lies outside the P circle. The shading represents that the class of S's outside P is empty. The diagram thus represents the truth conditions for (Af).

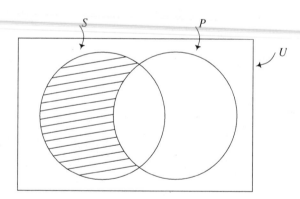

Another way of stating the truth conditions for (Af) = $(x)(Sx \supset Px)$ is this: (Af) is true if and only if for all interpretations of x into individuals of U *either* (i) every interpretation of Sx is false – that is, when there are no S's, or, as one says, when S is empty – *or* (ii) every interpretation of x which makes Sx true also makes Px true. In the above diagram, condition (i) will obtain if the overlap of S and P is empty, while (ii) will obtain if there are individuals in the overlap. The diagram asserts nothing about whether there are or are not individuals in the overlap of S and P; it leaves open whether (Af) is true because (i) obtains or is true because (ii) obtains.

The proposition (If) = ($\exists x$)(Sx & Px) is true if and only if for at least one interpretation of x into individuals of U, Sx & Px is true, that is, if and only if there is at least one individual in the overlap of S and P. We represent this in the Venn diagram by putting an x in the overlap.

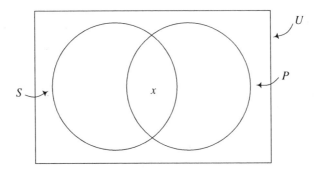

Now, an inference is valid if and only if the form is such that it guarantees that the set of conditions which make the premises true contains a set of conditions which make the conclusion true. Or, in more detail: An inference is 'valid in a universe U' just in case that for all possible interpretations of its predicate constants into subsets of U, there is no case where the premises turn out true and the conclusion false. Then, an inference is 'valid' just in case that it is valid in all non-empty universes.

In terms of Venn diagrams, what this amounts to is this: The diagram for the premises must already contain the diagram for the conclusion. If this is so, then the truth conditions for the premises will already contain the truth conditions for the conclusion. Since S and P have been interpreted into arbitrarily chosen subsets of U, it follows that what holds for this interpretation will hold for all possible interpretations of S and P into subsets of U. It follows that the inference is valid in U. But U itself is an arbitrarily chosen non-empty universe. So what holds for U holds for all non-empty universes. It follows that the inference is valid in all non-empty universes and therefore is valid. It is clear, therefore, from the Venn diagrams that the inference

(Af), so (If)

is invalid: the diagram representing the truth conditions for (Af) does not contain the diagram that represents the truth conditions for (If).

Similarly, consider one of the traditional inferences with regard to the contraries *SaP* and *SeP*. Take, for example,

SaP, so ~ *SeP*

This will translate into symbols as (Af), so ~ (Ef):

$(x)(Sx \supset Px)$, so ~ $(x)(Sx \supset \sim Px)$.

The Venn diagram for (Ef) is this:

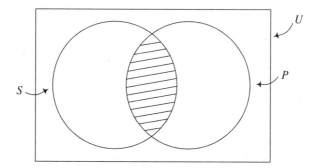

which indicates that the overlap of *S* and *P* is empty. Since ~ (Ef) will be true if and only if (Ef) is false, we can diagram the truth conditions for ~ (Ef) by diagramming the *falsity* conditions for (Ef). Since (Ef) will be true just in case that the overlap of *S* and *P* is empty, it will be false just in case that the overlap is not empty, that is, if and only if *there are* individuals in the overlap. We therefore diagram the falsity conditions for (Ef) by putting an *x* in the overlap of *S* and *P*.

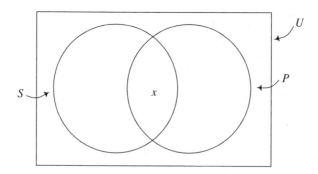

This diagram will also represent the truth conditions for ∼ (Ef).

But now we see that the inference from (Af) to ∼ (Ef) is invalid: the diagram giving the truth conditions for the former does not contain the diagram for the truth conditions for the latter.

In contrast, we do see that the inferences with regard to contradictories are all valid. Consider, for example, the inference

(If), so ∼ (Ef).

The diagram representing the truth conditions for the premise already contains the diagram representing the truth conditions for the conclusion. The inference is therefore valid.

One may use the Venn diagrams to show that all the traditional inferences are invalid when translated into symbols, except for the contradictories.

Interestingly enough, if we consider the propositions

[a] $(x)(Sx)$

[e] $(x)(\sim Sx)$

[i] $(\exists x)(Sx)$

[o] $(\exists x)(\sim Sx)$

it turns out that all the traditional patterns of the square of opposition hold, provided only that we assume that the universe U is non-empty:

[a] and [e] are contraries

[i] and [o] are subcontraries

[a] and [e] are superalternate to [i] and [o] respectively

[a] and [o] are contradictories

[e] and [i] are contradictories.

Consider the inference from [a] to [i]. [a] is true in U if and only if everything in U is S, that is, if and only if U outside S is empty. The Venn diagram for [a] will therefore consist of a square with a single circle inside, the circle representing the things that are *S*. Since [a] is true just in case that *U* outside *S* is empty, we must shade in the area within the square that is outside *S*. But *U* is non-empty. In order to present the presence of individuals within *U*, we must put an *x* within the circle representing *S*.

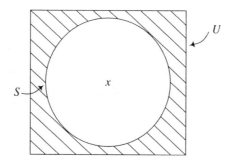

Now, [i] is true if and only if at least one individual in *U* is *S*. We represent this by putting an *x* in the circle representing *S*:

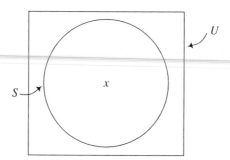

But we now see that the diagram that represents the truth conditions for [a] already contains the diagram for the truth conditions for [i]. Hence, the inference of subalternation from [a] to [i] is valid. The other inferences on this square of opposition also all turn out to be valid.

It is clear why the inference from [a] to [i] is valid while that from (Af) to (If) is not. In a non-empty universe, the truth of [a] implies that there must be some *S*, and therefore that [i] is true. In contrast, (Af) can be true even if *S* is empty; indeed, it is true if *S* is empty. But, if *S* is empty, then (If) is false. So, the truth of (Af) is compatible with the falsity of (If), and the inference from the former to the latter is therefore invalid: it can have a true premise and a false conclusion.

Traditional inferences other than those on the square of opposition are also invalidated when they are translated in the symbols of modern logic. There are various forms of conversion, for example. Simple conversion

SeP, so PeS

SiP, so PiS

remains valid, but conversion by limitation

SaP, so PiS

SeP, so PoS

turns out to be invalid upon the usual translation. All instances of obversion, however, remain valid. Thus, if nS is used to represent the things that are not S, then obversions such as

SeP, so SanP

SiP, so SonP

remain valid upon translation into symbols.

The move that is customarily made to save the inferences on the traditional square of opposition is to try to adjust the translations of the four categorical forms. Since the inference from A to I turns out to be invalid because (Af) can be true when there are no S, in which case (If) is false, the first thought is simply to stipulate that we add to the symbolic translation a clause to the effect that there are S:

(Af1) $(x)(Sx \supset Px) \ \& \ (\exists x)(Sx)$

(Ef1) $(x)(Sx \supset \sim Px) \ \& \ (\exists x)(Sx)$

(If1) $(\exists x)(Sx \ \& \ Px)$

(Of1) $(\exists x)(Sx \ \& \sim Px).$

This is, in effect, the suggestion of I. Copi.[5] But it will not by itself do. We have now made sure that subalternation is valid. But it turns out that (Af1) and (Of1) are no longer contradictories. So, to ensure that these inferences hold we have also to modify the translations of *I* and *O*. The following will solve the immediate problem:

(Af2) $(x)(Sx \supset Px) \& (\exists x)(Sx)$

(Ef2) $(x)(Sx \supset \sim Px) \& (\exists x)(Sx)$

(If2) $(\exists x)(Sx \& Px) \vee (x)(\sim Sx)$

(Of2) $(\exists x)(Sx \& \sim Px) \vee (x)(\sim Sx).$

But this will not do overall either. We will still be able to move from true propositions such as

No swine are mermaids,

which is true, to conclusions such as

Some mermaids are not swine,

which is false because there are no mermaids. The inference would be

SeP premise

$\therefore PeS$ conversion

$\therefore PoS$ subalternation.

The problem is that we have been able by the conversion to move an empty term from the predicate place to the subject place. In order to exclude invalid inferences of this sort, it is necessary to exclude the possibility not only of subject terms being empty but also of predicate terms being empty. We therefore need the following translations:

(Af3) $(x)(Sx \supset Px) \& (\exists x)(Sx) \& (\exists x)(Px)$

(Ef3) $(x)(Sx \supset \sim Px) \& (\exists x)(Sx) \& (\exists x(Px)$

(If3) $(\exists x)(Sx \& Px) \vee (x)(\sim Sx) \vee (x)(\sim Px)$

(Of3) $(\exists x)(Sx \ \& \sim Px) \ \lor \ (x)(\sim Sx) \ \lor \ (x)(\sim Px)$.

This is, in effect, the suggestion of P.F. Strawson.[6] But it will not do either.[7] Suppose that you are a materialist, so that the class of material things is a universal class: everything that exists belongs to it. In that case you could infer from the true proposition that

>All bodies are material

the false proposition that

>Some immaterial things are not bodies.

The inference would go like this:

>SaP premise
>
>$\therefore SenP$ obversion
>
>$\therefore nPeS$ conversion
>
>$\therefore nPoS$ subalternation.

Here the problem is that P is a universal class, so that its complement nP is an empty class. Once the empty classes have been thus reintroduced, the traditional inferences once again become invalid. In order to preclude this possibility we have once again to re-adjust our translations so as to preclude this possibility. What we need is this:

(Af4) $(x)(Sx \supset Px) \ \& \ (\exists x)(Sx) \ \& \ (\exists x)(Px) \ \& \ (\exists x)(\sim Sx) \ \& \ (\exists x)\sim(Px)$

(Ef4) $(x)(Sx \supset \sim Px) \ \& \ (\exists x)(Sx) \ \& \ (\exists x(Px) \ \& \ (\exists x)(\sim Sx) \ \& \ (\exists x)(\sim Px)$

(If4) $(\exists x)(Sx \ \& \ Px) \ \lor \ (x)(\sim Sx) \ \lor \ (x)(\sim Px) \ \lor \ (x)(Sx) \ \lor \ (x)(Px)$

(Of4) $(\exists x)(Sx \ \& \sim Px) \ \lor \ (x)(\sim Sx) \ \lor \ (x)(\sim Px) \ \lor \ (x)(Sx) \ \lor \ (x)(Px)$.

This has, unfortunately, become somewhat ridiculous: we have managed to save all the traditional inferences but only at the cost of requiring symbolic translations that are hardly credible. It is in fact the standard translations

SaP	All *S* are *P*	$(x)(Sx \supset Px)$
SeP	No *S* are *P*	$(x)(Sx \supset \sim Px)$
SiP	Some *S* are *P*	$(\exists x)(Sx \,\&\, Px)$
SoP	Some *S* are not *P*	$(\exists x)(Sx \,\&\, \sim Px)$

that are plausible: the more we fiddle with these the farther we get from a plausible set of truth conditions for the four traditional categorical forms.[8]

There is, in fact, an alternative upon which we can keep both the usual – and plausible – translations into symbols, on the one hand, and, on the other hand, also keep as valid all the traditional inferences of the square of opposition.
    The notion of validity with which we have been working is this:

> An inference is valid if and only if the form is such that it guarantees that the set of conditions which make the premises true contains a set of conditions which make the conclusion true.

Or, in more detail:

> An inference is 'valid in a universe *U*' just in case that for all possible interpretations of its predicate constants into subsets of *U*, there is no case where the premises turn out true and the conclusion false. Then, an inference is 'valid' just in case that it is valid in all non-empty universes.

Let us call this notion of validity 'B-validity,' after George Boole, who was the first person who defended in a serious way the standard translation of the four traditional forms and who was the first to conclude that the traditional inferences of the square of opposition were for the most part invalid.[9]
    It is possible to define another notion of validity, however. We could have:

> An inference is 'valid in a universe *U*' just in case that for all possible interpretations of its predicate constants into *non-empty non-universal* subsets of *U*, there is no case where the premises turn out true and the conclusion false. Then, an inference is 'valid' just in case that it is valid in all non-empty universes.

Let us call this notion of validity 'A-validity,' after Aristotle.

Now, the traditional inferences were invalidated by the fact that, where we allow empty classes or universal classes to replace the constants *S* and *P* in the categorical forms, it was possible to go from true premises to false conclusions. What the notion of A-validity does is insist that no substitutions of empty or universal classes shall be made for *S* and *P*. Without the problematic cases to worry about, the traditional inferences all turn out to be valid, that is, A-valid, while we are also able to retain the normal translations into symbols.[10]

The same point can be put in a slightly different way. Let us define the notion of an 'A-tautology' in a way parallel to the notion of 'A-validity':

> A proposition 'holds in a universe *U*' just in case that for all possible interpretations of its predicate constants into *non-empty non-universal* subsets of *U*, there is no case where it turns out to be false. Then, a proposition is a 'tautology' just in case that it holds in all non-empty universes.

An A-tautology will be vacuous simply because it will be true by virtue of its form; that is, its form is such that in all possible cases it will turn out to be true. If *T* is a tautology, then to conjoin it to a proposition will add no additional information; it will be a vacuous addition. Thus, if *S* is some proposition, then *S* will be logically equivalent to *S* & *T*. It turns out that each of the conjuncts of

$$(\exists x)(Sx) \,\&\, (\exists x)(Px) \,\&\, (\exists x)(\sim Sx) \,\&\, (\exists x)\sim(Px)$$

is an A-tautology: for all interpretations of *S* and *P* into non-empty non-universal subsets of non-empty universes, there is no case where any of these turn out to be false. Since each conjunct is a tautology (A-tautology) so is the conjunction. It follows that

(Af4) $(x)(Sx \supset Px) \,\&\, (\exists x)(Sx) \,\&\, (\exists x)(Px) \,\&\, (\exists x)(\sim Sx) \,\&\, (\exists x)\sim(Px)$

is logically equivalent (A-equivalent) to the standard translational form

$(x)(Sx \supset Px).$

Similarly for the other three categorical forms.

The form

$SaS \equiv (x)(Sx \supset Sx)$

is of course B-tautological, and therefore also A-tautological. In contrast,

$$SiS \equiv (\exists x)(Sx \ \& \ Sx) \equiv (\exists x)(Sx)$$

is not B-tautological. This is because the notion of B-tautology allows that there are empty classes. But *SiS is* A-tautological, since this notion does not admit of empty classes. Not surprisingly, in a systematic or axiomatic presentation of traditional syllogistic, *SiS* plays a crucial role; in effect, it secures the absence of empty classes.[11]

Just as the notion of B-validity is captured by the technique of Venn diagrams, so the notion of A-validity is captured by the Euler diagrams for testing the validity of inferences.

The Euler diagrams represent the ways in which non-empty non-universal classes can be related to each other. As these are customarily presented,[12] there are six possible cases. (α) *S* and *P* coincide; (β) *S* wholly contained in *P* but *P* not wholly contained in *S*; (γ) *P* wholly contained in *S* but *S* not wholly contained in *P*; (δ) *S* partially contained in *P* and *P* partially contained in *S*; and (ε) *S* and *P* are disjoint. These relationships, which are mutually exclusive and jointly exhaustive appear in the following diagrams.

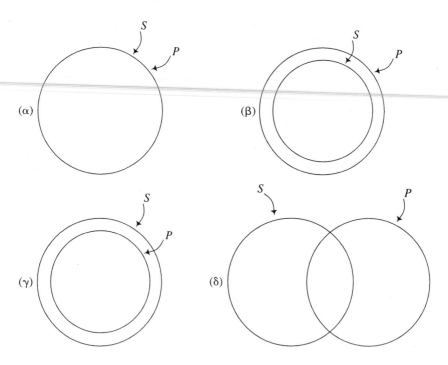

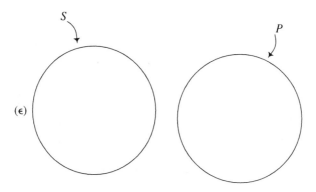

The A proposition *SaP* is true in circumstances α and β, and is otherwise false; the I proposition *SiP* is true in circumstances α, β, γ, and δ, and is otherwise false; in all possible circumstances in which the A proposition is true, the I proposition is true; the inference of subalternation from A to I is therefore valid, that is, A-valid. Again, the E proposition is true in circumstance α, and is otherwise false. Hence, in any circumstance in which A is true E is false, and in any circumstance in which E is true A is false; there are no possible circumstances in which they can be true together. However, both A and E are false in circumstance δ; so there are possible circumstances in which they can both be false. A and E thus turn out to be contraries, on the criterion determined by the concept of A-validity

Since *nX* is whatever is outside *X*, the proposition *SanP* is true in circumstances ε and is otherwise false. But this is the same set of truth conditions as for *SeP*. Hence *SanP* and *SeP* are logically equivalent, on the criterion for being an A-tautology. Again, *SenP* is true in circumstances α and β and is otherwise false. But these are the same truth conditions as for *SaP*. So, on the definition of A-tautology, *SenP* and *SaP* are logically equivalent. It follows from this that obversion of A and E propositions is A-valid.

And so on. Using the Euler diagrams it is clear that all the traditional inferences are A-valid.

We thus see that there is a simple way to retain as valid all the traditional inferences while also retaining the usual translations into the symbolization of modern formal logic. This consists in using a narrower conception of validity, A-validity, rather than the usual B-validity.

If we look at the matter in simple pragmatic terms, then, in all likelihood we will prefer B-validity to A-validity. After all, we want to allow statements of the E form such as

(x)   No mermaids live in Lake Ontario

to be true and therefore to be assertible. But if we adopt A-validity, then we have excluded propositions of this sort: all classes have members. But the point of (x) is that it is true precisely because there are no mermaids at all: there are none in Lake Ontario because there are none. We want, therefore, to allow that there are empty classes. It seems reasonable to allow universal classes also. That being so, we will then prefer B-validity to A-validity. This is the sort of reasoning that moved philosophers such as Russell to defend B-validity, and to suggest that the traditional inferences such as subalternation were simply invalid.

Why would one prefer A-validity, as all the traditional logicians clearly did? Why would one exclude both empty classes and universal classes?

2) The Distribution of Terms

Before taking up the issue of A-validity, however, it will pay to look at another traditional doctrine, that of distribution.

This doctrine provides one of the rules to which commonly appeal is made when judging the validity of immediate inferences and syllogisms. The relevant rule is this:

(RD)   No term may be distributed in the conclusion which was not distrib-
         uted in one of the premises.[13]

Keynes explained this rule, in a section of his text headed 'The Distribution of Terms in a Proposition,' in the following way: 'A term is said to be distributed when reference is made to *all* the individuals denoted by it; it is said to be undistributed when they are only referred to partially, i.e. information is given with regard to a portion of the class denoted by the term, but we are left in ignorance with regard to the remainder of the class' (68). This doctrine has been criticized in several ways. Geach[14] and Barker[15] have attacked Keynes's formulation directly, while Katz and Martinich[16] propose to replace Keynes's criterion with another, quite different, criterion. In fact, however, the criticisms of Geach and Barker are unsound, while the criterion proposed by Katz and Martinich presupposes that of Keynes.

The objection that Geach raises is that there are two semantical relations that a term has according to the statement of the rule of distribution, namely, that of *denoting* and that of *referring*. Thus, 'man' always denotes each and every man, but refers to some men only in one context (when it is not distributed) and in another refers to all men (when it is distributed). The problem is the semantical relation of *referring*, and how it is to be distinguished from denoting. 'The whole doctrine hinges on this distinction, but neither Keynes's nor any later exposition tells us what this distinction is' (*Reference and Generality*, 6). Since the notion of 'referring' remains unexplained, this doctrine of distribution is worthless, and the rule (RD) for evaluating inferences that is based on it, that a term distributed in the conclusion must be distributed in the premises, is worthless.

Now, it is true that Keynes does not explicitly explain what he has in mind. Yet it is not so unclear as Geach suggests. Indeed, it is sufficiently clear for one to recognize that Keynes does not think of reference as a relation between a term and objects in the world. To be sure, denotation is a semantical relation in which terms stand to objects; but reference is not.

In the statement from Keynes's text quoted just above, Keynes tells us that 'a term is said to be distributed when reference is made to all the individuals denoted by it.' This *may* be read in such a way that it is the term that has reference, but it *need not* be so read. Keynes continues, and explains that when a term is undistributed '*information* is given with regard to a portion of the class denoted by the term' (italics added). It is, however, not a term that gives information – it merely denotes – *it is the proposition in which a term occurs that gives information*. Thus, when Keynes tells us that 'reference is made to all the individuals' it is the proposition that makes reference, rather than the term. That is, in Keynes's statement 'reference' is to be ascribed not to the word 'term' that comes earlier in the statement but to the word 'proposition' that occurs in the immediately preceding section title, 'The Distribution of Terms in a Proposition.' Thus, *Keynes's notion of 'reference' has to do with the truth conditions of propositions rather than the semantical relations of a term to its denotata*.

Keynes makes this clear when he explains the truth conditions of categorical propositions, and the distribution of terms, by means of the Euler diagrams. He first lays out the standard five relations of inclusion and exclusion between two classes *S* and *P* that are a priori possible; these five possibilities are mutually exclusive and jointly exhaustive. Keynes then states the truth conditions of categorical propositions by listing for each form those among the five possibilities under which the form is true: 'the force of the different propositional forms is to exclude one or more of (the five) possibilities' (*Studies and Exercises*, 117).

> Any information given with respect to two classes limits the possible relations between
> them to one or more of the following: [Keynes then gives the relations α, β, γ, δ, ε].
>   Such information may in all cases be expressed by means of the propositional forms
> A, E, I, O. (126–7)

Note how Keynes here speaks of 'information' being conveyed by the proposition. This confirms our reading of his statement on distribution.
    Keynes then explains how the four categorical forms have the following truth conditions:

> *SaP* ≡ All *S* are *P*: true on either α or β
>
> *SeP* ≡ No *S* are *P*: true on ε
>
> *SiP* ≡ Some *S* are *P*: true on α, β, γ, or δ
>
> *SoP* ≡ Some *S* are not *P*: true on γ, δ, or ε.

Thus, the semantics of the quantifier 'all' and the 'is' of class inclusion are such that any proposition insofar as it exemplifies the syntactical form 'All *S* are *P*' of A propositions is true if and only if the class denoted by the term substituted for *S* stands in either the relation α or the relation β to the class denoted by the term substituted for *P*. Again, the semantics of 'some,' 'not,' and 'is' are such that any proposition insofar as it exemplifies the syntactical form 'Some *S* are not *P*' of O propositions is true if and only if the *S* and *P* classes stand either in the relation γ or in the relation δ or in the relation ε. Similarly for the other two categorical forms.
    Keynes uses the Euler diagrams 'to illustrate the distribution of the predicate in a proposition.' 'In the case of each of the four fundamental propositions we may shade the part of the predicate concerning which knowledge is given us' (*Studies and Exercises*, 119). Note the word 'knowledge,' a clear synonym for 'information.' This again confirms our reading of the passage on distribution.
    The A proposition is an inclusion: the whole of *S* is included in *P* as the whole of *P* or as part of *P*. In the former case we have information about all of *P*, as in α, and in the latter case we have information about only part of *P*, as in ß:

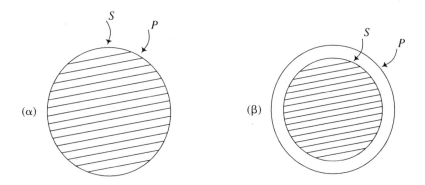

The E proposition is an exclusion: the whole of S is excluded from the whole of P. We thus have information about the whole of P, as in ε:

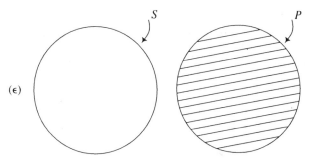

The I proposition is an inclusion: the whole of S is included in P as the whole of P (case α), or as a part of P (case β), or part of S is included in P as a part of P (case δ), or as the whole of P (case γ):

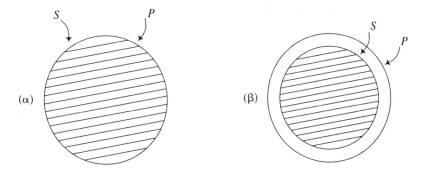

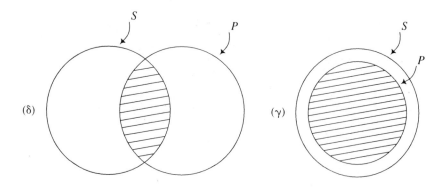

The O proposition is an exclusion: the whole of *S* is excluded from the whole of *P* (as in ε), or part of *S* is excluded from the whole of *P* (as in γ and δ):

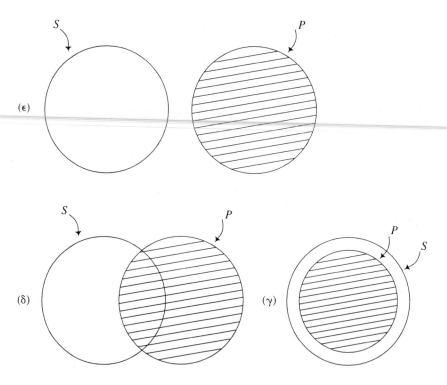

As Keynes now says: 'The result is that with A and I there are cases in which only part of P is shaded; whereas with E and O, the whole of P is in every case shaded; and it is made clear that negative propositions distribute, while affirmative propositions do not distribute, their predicates' (*Studies and Exercises*, 130). A similar exercise with subjects can establish that universal propositions do, while particular propositions do not, distribute their subjects.

The notions of 'inclusion' and 'exclusion' among classes are perfectly clear, as are the notions that the 'whole' or a 'part' of a class are included or not included in another class; the Euler diagrams provide a model that easily succeeds in making these notions clear. It follows that there is nothing problematic about Keynes's explanation of the doctrine of distribution. Geach's complaints are without substance, and stem from his failure to read Keynes with sufficient care to recognize that, when 'reference' is made in the case of a distributed term to all the individuals denoted by it, the reference is made by the proposition, not the term.

But, *if* the notions of 'whole,' 'part,' 'inclusion,' and 'exclusion' are not sufficiently clear, they can in fact be clarified further in two distinct, but equivalent, ways.

Before turning to these, however, we should look at two criticisms that Barker has made of Keynes's criterion of distribution. Barker objects to the claim that 'a term in a categorical sentence is distributed if and only if the sentence "refers to" all members of the class of things to which the term applies.'

[T]his explanation is obscure and misleading. The sentence 'All equilateral triangles are equiangular triangles' 'refers to' all equilateral triangles, and since necessarily these and only these are equiangular triangles the sentence would appear to 'refer to' all equiangular triangles also. Thus according to the old-fashioned account, it would seem that the predicate ought to count as distributed, however, and this illustrates one unsatisfactory aspect of the old-fashioned explanation of distribution. (*Elements of Logic*, 44)

Barker's argument is this. Suppose both

(1)   All equilateral triangles are equiangular triangles

and

(2)   All and only equilateral triangles are equiangular triangles

are true: then, since all of the subject is 'referred to' by (1), so must all the predicate also be referred to by (1); but in that case the predicate is distributed in (1), contrary to the traditional doctrine. However, as we say, 'refer to' is to be understood in terms of truth conditions. Now, the truth possibilities under which (1) is true are α and β

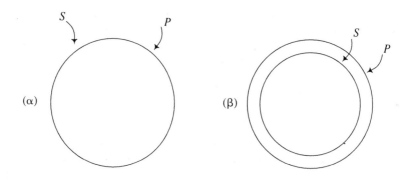

while (2) is true under α alone. It is certainly true that if (2) is true then P is distributed – after all, (2) entails

(3)   All equiangular triangles are equilateral triangles

and universal propositions distribute their subject terms. But from the fact that (2) is true, it does not follow that β is *not* among the truth possibilities for (1). The traditional doctrine counts P as undistributed in (1) because β is among its truth possibilities. Barker's conclusion follows only if one thinks the truth of (2) entails that β is not one of the truth *possibilities* for (1). This Barker attempts to secure by using an example (2) which is 'necessarily true.' Since (2) entails (1), the latter, too, will be necessarily true. If we let Δ represent the exclusive 'or,' then we have

(1) $\equiv (\alpha \; \Delta \; \beta)$

(2) $\equiv \alpha$.

If we let N represent the modality *necessary* and let L represent the modality *possible*, then with the modalities in place we have

(4) $N\,(\alpha \; \Delta \; \beta)$

(5) $N\,(\alpha)$

as Barker's premises. From these it of course follows that

(6) $N \sim (\beta)$

or, equivalently,

(7) $\sim L\,(\beta)$.

Thus, given that (1) and (2) are necessarily true, it is not possible that the situation represented by β obtains. From this Barker concludes that we have in (1) a proposition of the A form in which the predicate is distributed contrary to the traditional doctrine in which propositions of affirmative form do not distribute their predicates.

   However, Barker's conclusion that the predicate is distributed in (1) under the condition that (2) is necessarily true follows only if one can infer from the fact that β is not one of the *truth possibilities* of (1). But this does not follow. (1) exemplifies the A form

> All S are P.

The semantics of this form are such that β *is* among the truth possibilities of (1). It may be that one of the truth possibilities that (1) has by virtue of exemplifying the A form may for some other reason be impossible of fulfilment; but it does not follow that it is not among the truth possibilities of (1) insofar as the latter are determined by the semantics of the A form that (1) exemplifies. The semantics of the A form makes β one of the truth possibilities of (1), and Barker's argument does not change that semantics. Indeed, if we are even to state that argument, we must reckon β among the truth possibilities of (1). That, after all, is how one obtains the premise (4); if β is *not* among the truth possibilities of (1), then (1) = (2) and Barker's argument is

(4′) $N\,(\alpha)$

(5) $N\,(\alpha)$

from which the desired conclusion about β clearly does not follow. Hence, Barker's argument gets started only if it presupposes what it aims to refute.

   Perhaps Barker's error lies in supposing that, by virtue of the necessity of (2), (1) exemplifies the propositional form

(AA$_1$)   All S are S.[17]

The semantics of the quantifier 'all' and of the 'is' of class inclusion are such that any proposition insofar as it exemplifies the AA_1 form is true if and only if the relation α obtains; and it is evident that the AA_1 form guarantees that any proposition insofar as it exemplifies it will distribute both its subject and predicate terms. But of course, even though (2) is necessary, it does *not* follow that (1) exemplifies the AA_1 form *rather than* the A form. To suppose otherwise is akin to supposing that since

$$(p \lor q) \equiv [\sim(\sim p\ \&\ \sim q)]$$

is a necessary truth, any proposition exemplifying the logical form

$$p\ \&\ (p \lor q)$$

somehow does *not* exemplify that form but rather the form

$$p\ \&\ [\sim(\sim p\ \&\ \sim q)].$$

However, if Barker's example (1) will not secure his point, perhaps another example will, one in which it is explicit in the form of the proposition itself that subject and predicate are coextensive; for example,

(8)   All men are men.

This is of the A form, but distributes its predicate; so we have an A proposition that distributes its predicate, contrary to the traditional doctrine. But, while (8) exemplifies the A form, and also distributes its predicate, it does not do the latter by virtue of the former. Rather, it distributes its predicate by virtue of being of the form AA_1; insofar as (8) exemplifies the A form, β *is* among its truth possibilities.

The traditional doctrine of distribution as stated by Keynes holds that a proposition *insofar as it exemplifies the A form* does not distribute its predicate or, what is the same, does not exclude β from among its truth possibilities. One can't use examples such as (1) or (8) to show that this is wrong, any more than one can show that it is wrong to hold that a disjunction is true if and only if one or the other or both disjuncts is true because one can find disjunctions like

$$p \lor (q\ \&\ \sim q)$$

that are true if and only if the first disjunct is true.

It thus seems that Barker's first objection to the traditional doctrine of distribution is unsound.

Barker's second objection to Keynes's account of distribution 'is that it is unclear in its treatment of the predicate in the O form. To claim that the sentence 'some seamen are not prohibitionists' "refers to" all prohibitionists is to make an obscure and unsatisfactory claim' (*Elements of Logic*, 43). But this is simply to say that it is unclear what is meant when one holds that a proposition of the O form

> Some S are not P

asserts that *either the (non-empty) whole or a (non-empty) part of 'S' is excluded from the whole of 'P'*. I myself do not find this either obscure or unclear, especially in light of the explanation in terms of the Euler diagrams. If, however, it is obscure or unclear to one such as Barker, then one can do no less than attempt to clarify such crucial notions as 'whole,' 'part,' 'inclusion,' and 'exclusion.' I have already suggested that there are two, equivalent, ways of doing this. Let us turn to the first. As it turns out, it is precisely this that Barker proposes as a 'replacement' for Keynes's doctrine of distribution.

The *first* way to clarify the crucial notions consists in rephrasing each categorical proposition in an equivalent form which uses the notions of 'class' and 'subclass' (or 'portion of a class') and then, using these notions and the relations of identity and non-identity, unpacks the notions of 'whole,' 'part,' 'inclusion,' and 'exclusion':

A: All S are P

A':   For every subclass f of S there is a subclass g of P such that $f = g$

E: No S are P

E':   For every subclass f of S and every subclass g of P, $f \neq g$

I: Some S are P

I':   There is a subclass f of S and there is a subclass g of P such that $f = g$

O: Some S are not P

O':   There is a subclass f of S such that for every subclass g of P, $f \neq g$.

The unprimed and the primed versions of the categorical forms are logically equivalent. In order to see this clearly, one must have recourse to the symbolic techniques of modern logic. Consider the A form. This transcribes as

$$(x)(Sx \supset Px)$$

and A′ transcribes as

$$(f)[(f \subset S) \supset (\exists g)[(g \subset P) \& (f = g)]].$$

The other primed statements have similar translations. It is easy to prove that the members of each pair entail each other. The unprimed and the primed forms of the categorical propositions are thus logically equivalent.

The expansion of the ordinary categorical forms, the unprimed versions, into the logically equivalent primed versions, is, in fact, a traditional reading. In such an expansion we are considering the categorical forms in respect of their *comprehension*. Thus, Maritain explains the notion of the comprehension of a concept as follows: 'if it is true that a concept presents an *essence, nature,* or *quiddity* immediately to the mind, and that this essence is something real, then it must be said that the concept as such is essentially and originally characterized by its *comprehension*, that is to say by the sum of the constitutive notes of the nature that it presents to the mind' (*Formal Logic*, 23). Provided that we understand the 'constitutive notes' as subspecies of the concept, or, what is for these purposes much the same, subclasses of a class, then it is clear that the primed forms deal with the concepts S and P with regard to their comprehensions. Our primed expansions, then, are not at all foreign to traditional treatments of logic.

Now, with this reading, deriving from the tradition, it is easy to provide a clear meaning for the standard doctrine of the distribution of terms.

If T is a term in one of the categorical propositions, then reference is made to the whole of the class T just in case that T is attached to a universal quantifier, while reference is made to a portion only of T, that is, a proper subclass, just in case that T is attached to a particular quantifier. 'Inclusion' and 'exclusion' are explicated in terms of the relations of identity and non-identity among the classes and subclasses denoted by T and the other term T′ of the proposition. Distribution now falls out quite nicely: a term T is distributed just in case that T is attached to a universal quantifier in the primed versions A′ – O′ of the categorical forms.

Thus, what the primed formulae do is *provide an explicit syntactical criterion for which terms are distributed.* But, since the primed formulae are logically equivalent to the corresponding unprimed formulae, and *cannot differ from the latter in their truth possibilities, this syntactical criterion adds nothing that is not already implicit in the truth possibilities of the categorical propositions in standard form.* In short, the reformulation achieved by the primed formulae may add clarity, but, *far from its being shown that Keynes's doctrine of distribution is inadequate, what we see is why that doctrine is correct.*

Barker proposes to redefine the notion of distribution: 'A term S occurring as the subject of a categorical sentence is said to be distributed in that sentence if and only if the sentence, in virtue of its form, says something about *every kind of S*. Similarly, a term P occurring as the predicate is said to be distributed in that sentence if and only if the sentence, in virtue of its form, says something about *every kind of P' (Elements of Logic*, 43). He offers a 'more rigorous' formulation of the definition as follows: 'Suppose that T is a term which occurs as a subject or predicate in a categorical sentence s. Where T' is any other term, let s' be the sentence that is exactly like s except for containing the compound term T' and T, where s contains T. Now, T is said to be distributed in s if and only if, for every term T, s logically implies s'' (43). Katz and Martinich have objected to this criterion that, while one can indeed infer

(9a)   Some seamen are not rich prohibitionists

from

(9)   Some seamen are not prohibitionists

one cannot infer

(9b)   Some seaman are not sham prohibitionists

from (9), and so the predicate of *O* propositions cannot be distributed according to Barker's criterion but contrary to the traditional doctrine ('Distribution of Terms,' 117). But this is surely unjust. Barker restricts his *T* and *T'* to *terms*, where terms 'apply to ... individual things' (*Elements of Logic*, 37), many in the case of general terms and one only in the case of singular terms – a fairly traditional procedure, followed, for example, by Keynes (*Studies and Exercises*, 10, 53). The role of 'sham' is to negate; a sham prohibitionist is a *non*-prohibitionist, *not* a prohibitionist. Thus, the use of 'sham' is akin to that of 'non-' in 'non-*P*.' It is, in short, *syncategorematic*, and, as Keynes has said, 'using the word term in the sense in which it was defined ... it is clear that we ought not to speak of syncategorematic terms' (10).

The point to be noticed is that Barker's conjunctive terms T & T' denote *subclasses of T*. Thus, upon Barker's criterion, a term T is distributed in s just in case that what s says about T it also says about *every subclass* of T, or, as Barker puts it in his 'less rigorous' formulation, *every kind* of T. That is, T is distributed in s just in case that s can be reformulated as a statement that makes an assertion about every subclass of T. It is evident that the relevant reformulations are the primed formulae above. Indeed, the latter show why, precisely, it is that Barker's criterion works: it works because the formulae which make clear the terms to which it applies are logically equivalent to the standard categorical forms. But if we see why Barker's criterion works, we also see that it is unnecessary: it invokes nothing that is not already implicit in the truth possibilities of the categorical proposition in standard form. Far from replacing Keynes's criterion, Barker's is merely a reformulation of it.

Katz and Martinich propose a criterion apparently rather different from Barker's. In the end, however, as we shall see, it is, like Barker's, essentially Keynes's criterion, and therefore, despite appearances, is not after all radically different from Barker's.

The criterion is this ('Distribution of Terms,' 281): Let $C(T_1, T_2)$ be a categorical proposition. Then T_1 is distributed in $C(T_1, T_2)$ if and only if either

$$C(T_1, T_2)$$
$$T_1 a$$
$$\overline{}$$
$$\therefore (\exists x)(T_2 x \ \& \ x = a)$$

or

$$C(T_1, T_2)$$
$$T_1 a$$
$$\overline{}$$
$$\therefore (\exists x)(T_2 x \ \& \ x \neq a)$$

is valid. The general idea behind this criterion goes back well into medieval logic. Peter of Spain[18] described a term T in a proposition s as distributed if we can validly infer s' from s, where s' is obtained from s by replacing T, or T with its quantifier, by 'This T.' For example, from

Every man is an animal

we may deduce

This man is an animal

but not

Every man is this animal.

Hence, the subject is, but the predicate is not, distributed in an A proposition. Similarly, we may deduce from

No man is a horse

both

This man is not a horse

and

No man is this horse

and conclude that both subject and predicate are distributed in an E proposition. Peter of Spain did not use the doctrine of distribution to test syllogisms for validity. That idea came only later, perhaps with Pseudo-Scot.[19] Be that as it may, the point is that the medievals recognized that where a term T is distributed, one can deduce a conclusion about an *arbitrary member* of T, for example, '*this* man' in the case of 'Every man is an animal.' It is this idea about drawing a conclusion about an arbitrary member that the criterion of Katz and Martinich is designed to capture.

The rule (RD) prohibits inferring a conclusion where T is distributed from premises where it is not. It is easy enough to see why we must accept this rule if we accept the Katz and Martinich criterion. Suppose that T_1 is not distributed in

(i) $C_1(T_1, T_2)$

Then, from (i) and

(ii) T_1a

the conclusion

(iii) $(\exists x)(T_2x \,\&\, x = a)$

does not follow. Suppose, however, that T_1 is distributed in

(iv) $C_2(T_1, T_2)$

because (iii) follows from (iv) and (ii). If (iv) *were* to follow from (i) then we could deduce (iii) from (iv) and (ii) and therefore from (i) and (ii). But in that case, T_1 would, contrary to our original supposition, be distributed in (i). A violation of (RD) must therefore be invalid.

Nevertheless, to leave it at this, as Katz and Martinich do, is hardly satisfactory. We are told that '"All men are mortals and Socrates is a man" logically implies "For some mortal, Socrates is the same as that mortal." Hence the subject term is distributed' ('Distribution of Terms,' 281). But we are not told *why* the relation of logical implication holds as it is asserted to hold. We are told nothing about the logical form that determines the truth possibilities of categorical propositions so that we could see that the conclusion is indeed, as it is asserted to be, contained in the premises; that is, that every truth possibility for the premises is also a truth possibility for the conclusion. Nor are we told anything about the logical form that would enable us to see why the Katz–Martinich conclusions do *not* follow in the case of undistributed terms. Taking entailment to be an undefined term, as Katz and Martinich apparently do, is simply not helpful: one wants to know the logical feature that distributed terms have in the logical forms of categorical propositions that validates the Katz–Martinich inferences.

Katz and Martinich hint at what this form may be: 'Whether the statement does say or imply something about each particular depends on whether the term is distributed in the statement; if the term is distributed, it does, and if the term is undistributed, it does not' ('Distribution of Terms,' 281). If T occurs in s, then one can draw an inference from s about an *arbitrary* member a of T only if s makes an assertion about *all* of T. The hook-up with Barker's criterion is evident enough: one can draw an inference about an arbitrary member of T only if, as Barker says, the proposition s makes the same claim about every subclass of T that it makes of T. But the hook-up with Keynes's criterion is also evident enough: one can draw an inference about an arbitrary member of T only if s makes reference to *all* the individuals denoted by T; and one cannot draw an inference about an arbitrary member of T if s refers to them *only partially*. It thus appears that Katz and Martinich are not so far from either Keynes or Barker as appearances might suggest.

What we must do, of course, is attempt to make explicit those features of logical form that Katz and Martinich only hint at. This brings us to the *second* way to clarify the notions of 'whole,' 'part,' 'inclusion,' and 'exclusion' that we have, following Keynes, used to state the doctrine of distribution. This way consists in rephrasing each categorical proposition in an equivalent form which explicitly uses the relation of denotation of a term, and which unpacks 'whole,' 'part,' and so on, in terms of that relation and the notions of identity and non-identity among individuals. The equivalent forms are these:

A: All S are P

A^*:   For every x such that S denotes x there is a y such that P denotes y and $x = y$

E: No S are P

E^*:   For every x such that S denotes x and for every y such that P denotes y, $x \neq y$

I: Some S are P

I^*:   There is an x such that S denotes x and there is a y such that P denotes y and $x = y$

O: Some S are not P

O^*:   There is an x such that S denotes x and for every y such that P denotes y, $x \neq y$.

The starred and unstarred forms are logically equivalent. In order to see these clearly, one must, as before, have recourse to the symbolic techniques of modern logic. Consider again the A form. This transcribes as

$$(x)(Sx \supset Px)$$

while A^* transcribes as

$$(x)[Sx \supset (\exists y)(Py \ \& \ x = y)].[20]$$

The other starred statements have similar translations:

E^*: $(x)[Sx \supset (y)(Py \supset x \neq y)]$

I^*: $(\exists x)[Sx \ \& \ (\exists y)(Py \ \& \ x = y)]$

O^*: $(\exists x)[Sx \ \& \ (y)(Py \supset x \neq y)]$.

It is easy enough to construct formal proofs that the members of each starred/unstarred pair entail each other; but it is perhaps more revealing to think in terms of quantifier-free expansions, for when a transformation is made to the latter it becomes evident how the extra clauses in the starred formulae add nothing, that is, nothing that is non-tautological, to the unstarred forms.

The expansion of the ordinary categorical forms, the unstarred versions, into the logically equivalent starred versions, is in fact a traditional reading. In such an expansion we are considering the categorical forms in respect of their *extension*. Thus, Maritain explains the notion of the extension of a concept as follows: 'The extension of a concept includes ... the individuals ... in which it is realized' (*Formal Logic*, 27). What the starred versions do is show how the individuals in the extension of one concept are identical or are non-identical to the individuals in the extension of the other concept. Again, we have an understanding of the categorical forms that is familiar to the exponents of the traditional logic.

We use this traditional understanding of the categorical forms in terms of their extensions to make clear the notion of distribution. If T is a term in one of the categorical propositions, then, in Keynes's terms, 'reference is made to all the individuals denoted by it' just in case in the equivalent starred form it is attached to a universal quantifier; 'individuals denoted by it are only referred to partially' just in case it is attached to a particular quantifier. These quantifiers thus make explicit what it is for a proposition to refer to a whole or part of (the denotation of) T. As for the notions of 'inclusion' and 'exclusion,' these are explicated in terms of the relations of identity and non-identity among the individuals that are the denotata of T and the other term T' of the proposition.

In particular, then, a term T is distributed just in case that it is attached to a universal quantifier in the starred forms.

The starred forms thus provide a second syntactical criterion for which terms are distributed. But again, since the starred formulae do not differ from the unstarred formulae in their truth possibilities, this syntactical criterion adds nothing that is not already implicit in the truth possibilities of the categorical propositions in standard form. In other words, far from seeing that Keynes's doctrine is wrong, we see that the proposed alternative of Katz and Martinich, like that of Barker – to which it turns out to be logically equivalent – is in fact logically equivalent to that of Keynes. Once gain we have an 'alternative' to Keynes that in fact turns out to demonstrate that Keynes's doctrine is correct.

Among the inferences that are A-valid but not B-valid is the inversion of A:

All S are P

$\therefore$ Some non-S are not P.

This is not B-valid unless one adds the premise

There are non-P.

But that added premise is an A-tautology, since the latter excludes universal classes, guaranteeing for every class S that there are members in both S and non-S. This creates a problem for the rule (RD) of distribution. If we look at how the terms are distributed in this inversion, it turns out that P is *not* distributed in the premise 'All S are P,' but *is* distributed in the conclusion 'Some non-S are not P.' If we use the second of our syntactical criteria, then the premise is

SaP: $(x)(Sx \supset Px)$

$\equiv (x)[Sx \supset (\exists y)(Py \ \& \ x = y)]$.

The conclusion is

$nSoP$: $(\exists x)(\sim Sx \ \& \sim Py)$

$\equiv (\exists x)[\sim Sx \ \& \ (y)(Py \supset x \neq y)]$.

A term is distributed just in case that it is attached to a universal quantifier. In the premise, P is not so attached; in the conclusion it is. Hence, P is undistributed in the premise but distributed in the conclusion. (RD) is thus violated by an A-valid inference.

Geach[21] concludes that, since we here have an exception to the rule (RD), therefore the doctrine of distribution is worthless. Geach's criticisms are not quite to the point – some, in fact, are quite wrong – but he is right in directing our attention to this particular inference. He is, in fact, not the first to have pointed out that if we accept this as valid, then there is a conflict with the traditional doctrine of distribution.[22] The conflict is, however, revealing, and points towards an important feature of traditional logic.

Geach proposes the following interpretation of the categorical propositions:

A: $(S = \wedge \ \& \ P = \wedge) \vee (S = \vee \ \& \ P = \vee) \vee (S \neq \wedge \ \& \ P \neq \vee \ \& \ S \subset P)$

E: $(S = \wedge \ \& \ P = \wedge) \ \text{v} \ (S = \bigvee \ \& \ P = \bigvee) \ \text{v} \ (S \neq \wedge \ \& \ P \neq \bigvee \ \& \ S \cap P = \wedge)$

I: $(S = \wedge \ \& \ P \neq \wedge) \ \text{v} \ (S \neq \bigvee \ \& \ P = \bigvee) \ \text{v} \ (S \cap P \neq \wedge)$

O: $(S = \wedge \ \& \ P \neq \wedge) \ \text{v} \ (S \neq \bigvee \ \& \ P = \bigvee) \ \text{v} \sim (S \subset P).$

Upon this interpretation, all of the traditional square of opposition, including subalternation, is valid; so are all the traditional syllogisms, including those with weakened conclusions. All conversions and obversions permitted by the distribution rule (RD) are valid on this interpretation. The two basic forms 'All S are S' and 'Some S are S' are tautological on this interpretation. Finally, inversion is valid. It is assumed, as usual, that the universe is non-empty; but the terms S and P need not be non-empty and need not be non-universal.

Now, as we have said, on this interpretation of the categorical forms, the inversion of A is valid. Geach proceeds to argue this way: 'the question is whether the doctrine of distribution affords a formal test of validity. By constructing a system in which the usual relations of categoricals are maintained, but nevertheless an inference condemned by the doctrine of distribution is valid, I have shown that the doctrine of distribution is useless even as a mechanical test of validity.'[23] But let us see. A term T will be distributed in a disjunctive proposition just in case it is distributed in each disjunct; otherwise there will be a truth possibility in which we will have information about only part of T and it will be distributed in a conjunction just in case it is distributed in at least one conjunct. Consider the A proposition. In the first disjunct we have

$$P = \wedge \ \equiv \ (x)(\sim Px) \ \equiv \ (x)(y)(Py \supset x \neq y)$$

so P is distributed in that disjunct. In the second disjunct we have

$$P = \bigvee \ \equiv \ (x)(Px) \ \equiv \ (x)(y)(Py \supset x = y)$$

so P is distributed in that disjunct. In the third disjunct we have as a conjunct

$$P \neq \bigvee \ \equiv \ (\exists x)(\sim Px) \ \equiv \ (\exists x)(y)(Py \supset x \neq y)$$

so P is distributed in that disjunct also. Thus, in the A proposition on this 'unnatural' interpretation the predicate P is distributed. As for the conclusion 'Some non-S are not P' of the inversion, on this 'unnatural' interpretation we have in this first disjunct

$$P \neq \wedge \equiv (\exists x)(Px) \equiv (\exists x)(\exists y)(Py \,\&\, x = y)$$

so P is undistributed in that disjunct. Hence it is undistributed in the proposition. Thus, the inversion of A, on this 'unnatural' interpretation, yields a conclusion in which the predicate P is undistributed. Thus, Geach notwithstanding, even on this 'unnatural' interpretation of the categorical forms, the inversion of A is valid but does not constitute a violation of the distribution rule (RD). In short, Geach has not found a counter-example on which the doctrine of distribution fails.

But it does fail, as we have seen. Keynes himself proposed a defence of the traditional rules (*Studies and Exercises*, 139ff). This defence turns on adding an extra premise to the immediate inference of inversion. He points out that inversion of A is formally valid only if 'there are non-P' is added as a premise, that P is distributed in this added premise, and that we therefore do *not* have a case of a valid argument in which a term is distributed in the conclusion but not in the premises.

Keynes's point is correct. If we use the rules that we used above to obtain the starred versions of the categorical propositions to translate the premises needed to make the traditional inferences valid, then the statement that 'There are S,' that S is non-empty, that is, the statement

$$(\exists x)(Sx)$$

becomes

$$(\exists x)(\exists y)(Sy \,\&\, y = x)$$

and that statement 'There are non-P,' that is, the statement

$$(\exists x)(\sim Px)$$

becomes

$$(\exists x)(y)(Py \supset x \neq y).$$

The criterion of distribution is that a term T is distributed just in case it is attached to a universal quantifier. Thus, in a statement that a term is non-universal, the term is distributed, while in a statement that a term is non-empty, the term is undistributed. Thus, Keynes's claim is clearly correct that the rule (RD) of distribution is not violated by inversion of A, once one takes into account the added premise that is required for validity.

Keynes appeals, as we have seen, to the Euler diagrams to justify the rule (RD) of distribution. These presuppose that there are no empty terms. But they are consistent with there being universal terms. When Keynes justifies the rule (RD) in the case of inversion by adding the extra premise

There are non-*P*

he is in effect allowing that there can be universal terms and that inferences that depend upon their exclusion require an additional premise. Keynes is thus treating empty terms and universal terms very differently. The former do not, while the latter do, require elimination by an explicit premise. Keynes's treatment of traditional logic thus does not accept the notion of A-validity, which takes for granted that there are neither empty *nor universal* classes.

Keynes is of course free so to construe his notion of validity. At the same time, if we look at the traditional logic, it in fact rejects universal terms appearing in the subject or predicate places of categorical syllogisms. That means that it is working with the notion of A-validity, rather than Keynes's notion. It would seem, then, that there really is a problem with the rule of distribution (RD): given the notion of A-validity, one must take inversion as valid, and the rule (RD) reckons it as invalid.

The problem can be brought out by noting that the Euler diagrams fit Keynes's notion of validity but in fact do not fit A-validity. The Euler possibilities require that there be no empty classes. But they do not deal with all the possibilities that arise if there are no universal classes.[24] The propositions *SiP* and *nSinP* can in some cases both be true and in some cases both be false. Thus,

Some sailors are prohibitionists

Some non-sailors are non-prohibitionists

are both true; but, if we assume that everything that exists is either a substance or an accident and that nothing is both, then

Some substances are accidents

Some non-substances are non-accidents

are both false. The truth possibility of the first pair is represented by case δ, but the falsity possibility of the second pair is *not* adequately represented by ε:

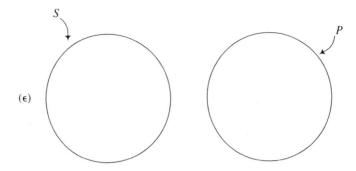

For, ε does not allow for the possibility that the classes S and P jointly exhaust the universe. We must therefore subdivide ε into ε' and ε":

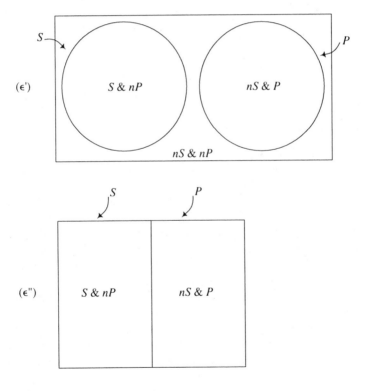

We need to subdivide δ into δ' and δ" for the same reason:

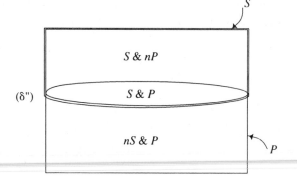

It is clear by now what has happened. The Euler diagrams as originally proposed did not allow for all the possibilities that arise *once one introduces negative terms*. The Euler diagrams set out the relationships that hold among positive terms, but once one adds negative terms, more possibilities open up. The Euler diagrams can be used to show which inferences are A-valid so long as there are no negative terms. But once negative terms are introduced we need to add more possibilities to those recorded in the original set of diagrams. Only then will we have a set of diagrams that can be used to test for A-validity.

Now, when Keynes explained the notion of distribution he used the Euler diagrams in the original version. This means that the doctrine of distribution as he explained it applies only so long as there are no negative terms. Once negative terms are introduced there is no guarantee that either the explanation nor the rule will be sound. But this is precisely what we do have in the case of inversion:

$SaP, \therefore nSoP$

We should therefore not be surprised that the rule of distribution (RD) breaks down in this case.

There are two things that might be done to remedy this problem that enters syllogistic and traditional logic when one introduces negative terms. One could modify the rules of distribution, as Miller has proposed.[25] Or one could stipulate, as Bird suggests,[26] that one would not judge of distribution until a proposition has been put into standard form containing no negative terms. This can always be done using the laws of obversion, inversion, and contraposition.

It is not our purpose here to decide which of these two alternatives is the most reasonable or convenient. The important point is simply the fact that the problems for the rule of distribution (RD) arise with the introduction of negative terms. The latter happened in the nineteenth century. Prior to that time, logics did not contain negative terms. If one looks at any standard logic text – for example, Aldrich,[27] Sanderson[28] or Burgersdijck[29] – there are no negative terms. Yet they contain the rule (RD) of distribution. In the context of these traditional logics, since they contain no negative terms, this rule is sound.

The interesting question is why these logics did not allow negative terms. As we shall see, they did so for reasons that are closely related to the reasons why the notion of A-validity was central to their conception of logic.

3) The Ontological Basis of Traditional Logic

In his *Formal Logic*, Jacques Maritain defends the traditional logic, including the traditional inferences on the square of opposition, conversion by limitation of A propositions, and the weakened moods of the syllogism which conclude a particular conclusion from universal premises. Here, for example, is how he explains conversion by limitation ('partial conversion'):

If we attach an existential sense to A and say:
    Every bat is a mammal (and bats exist)
we conclude correctly that: therefore some (existing) mammal is a bat.
   If, in abstracting from actual existence in A, we say:
    Every philosopher is a man (whether or not there are philosophers)
we conclude correctly that: therefore some man is a philosopher, on condition we understand ... that: therefore some man (as a possible creature) is a philosopher. (*Formal Logic*, 230).

At first glance the claim made here that partial conversion is valid seems simply to be the claim that there is always an extra existential premise:

All S are P

There are S

∴ Some P are S.

This cannot be the whole story, however, if only because Maritain draws a distinction between propositions taken existentially or concretely and propositions taken abstractly. But it is also clear that Maritain does not treat the existential claim simply as an extra premise. After all, he does segregate it when he presents his examples. This shows that in a way he assumes it as something that should really be granted. This course is appropriate for anyone who is defending A-validity. But *why* should it be taken for granted?

Maritain makes clear that not everything that goes on in logic, that is, logic as he understands it, is *merely* logic. Against Russell and the other 'logisticians' he argues that 'the existential or non-existential signification of a proposition depends upon its *matter*, and not upon its *form* alone. The fundamental error of the logisticians lies in their failure to distinguish between the form and the matter of propositions and in their belief that considerations bearing exclusively on the form suffice to explain the entire discourse' (*Formal Logic*, 226–7). This implies that there is something about the concepts themselves, the matter that appears in the categorical forms, that guarantees the 'existential signification' of these propositions. If, then, we can discover what this is we will understand why it is that A-validity is appropriate to the logic developed by Maritain.

What will become clear is that Maritain's logic, like that of A.C. Pegis, is rooted in the substance metaphysics that derives for Maritain more immediately from Aquinas and more distantly from Aristotle. In fact, this was the background for most logics until Locke's critique of syllogistic. If one looks at Aldrich, Sanderson, or Burgersdijck, to mention but three, one soon discovers that they take for granted a substance ontology. Indeed, if one looks at the Cartesian *Art of Thinking* (the 'Port Royal Logic') of Arnauld and Nicole, one again soon discovers that this, too, takes for granted a substance ontology, though it is, to be sure, a substance ontology that is somewhat different from that of Aquinas and Aristotle. Once we explore this ontological background that is taken for granted, several things become evident. It becomes evident, first, why these logicians hold that there are no empty classes, or, equivalently but in their terms, why there are no species that are not at least once exemplified. It becomes evident, second, why these logicians hold that there are no universal classes, or, equivalently but in their terms, why no transcendental appears as a term in a categorical proposition. And it becomes evident, third, why they do not allow negative terms to be among the things that can be substituted into the subject and predicate places of categorical propositions. Once these three points are recognized, it will have been established why these logicians take for granted that it is appropriate to use the notion of A-validity, and, moreover, since they do not have negative terms, it will also have been established why it is appropriate for them to use the standard rule of distribution for evaluating the validity of inferences.

We can approach the metaphysical or ontological background to traditional logic through the work of Maritain.

As we have seen, he distinguishes between propositions taken existentially or concretely and propositions taken abstractly. The former deal with actual things, the latter with possible things. This distinction is connected by Maritain with another distinction upon which he insists. This is the distinction between, to use his example, the concepts 'man' and 'humanity.' '[A] concept such as "man" ... presents to the mind a form *in the subject which it determines*, whereas a concept such as "humanity" ... presents a form *without the subject it determines*, or prescinding from this subject' (*Formal Logic*, 32–3). The first sort of concept is said to be 'concrete,' the latter sort 'abstract' (ibid.). The difference is that '[t]he concrete concept presents to the mind *what* such and such a thing is ... an abstract concept presents that *by which* the thing is such and such a thing' (33). The concept is the form or nature or essence of a thing. In fact, it is the form or nature or essence *qua* existing in the mind. Maritain distinguishes the mental concept and the objective concept (48). The proposition is primarily about the latter. The former 'we form in order to perceive the thing and to manifest it immaterially within us' (ibid.). This form, whether understood as the mental concept or as the objective concept, can in turn be taken in two ways, though to be sure it is still one and the same form. The concept understood concretely is the form or nature or essence as the form or nature or essence of an actual thing; the concept understood abstractly is the form as such, taken as the form which it is possible for an actual thing to have. Propositions formed using the concepts taken concretely are those that are understood existentially, while propositions formed using the concepts taken abstractly are those that are understood abstractly.

When we take a concept abstractly, and predicate of it those characteristics which constitute it, then we are looking at the *comprehension* of the term: 'if it is true that a concept presents an *essence, nature,* or *quiddity* immediately to the mind, and that this essence is something real, then it must be said that the concept as such is essentially and originally characterized by its *comprehension*, that is to say by the sum of the constitutive notes of the nature that it presents to the mind' (*Formal Logic*, 23). As we have seen, this reading of the categorical propositions can be given a sense in terms of ordinary formal logic. But for Maritain there is more to it than that. For him, as indeed, for others in the tradition of the substance metaphysics, propositions so understood are about forms or natures or essences – species – taken as such. They state relations among ideas, since these species are concepts in the mind. However, since these species are objective as well as in the mind, these propositions at the same time represent objective possibilities, possibilities for being. The relations among ideas are also relations among objective possibilities.

We can also take the concept concretely. In that case, the proposition is about the *extension* of the concept: 'The extension of a concept includes both the individuals and the objective concepts (universal but of less extension than itself) in which it is realized. For example, both the concept "man" and the concept "beast," are contained in the extension of the concept "animal," and consequently, "this man, that man, this horse, this butterfly, etc." are also and by that very fact contained in the extension of the concept' (*Formal Logic*, 27). Again, as we have seen, this reading of the categorical propositions can be given a sense in terms of ordinary formal logic. But also once again, there is more than a formal reading here; there is something rooted in the substance metaphysics.

In both sorts of proposition, the copula represents what Maritain, following the medieval tradition, refers to as 'identification.' This in a way appears in both the readings we gave above in terms of the formal logic of Russell and the 'logisticians'; it appears simply as the identity sign '=.' But, yet again, we are not surprised, there is more than a formal reading here; there is something rooted in the substance metaphysics.

Here is how Maritain presents the notion:

the copula does nothing but express the relation ... of the predicate to the subject; but what relation? The relation of identification of one with the other, the relation by which these two objects of thought, distinct as concepts ... are identified in the thing ... in *actual* or *possible real* or *ideal existence* ... The proposition 'a myriagon is a ten-thousand-sided polygon' is equivalent to: 'the object of thought myriagon *exists* (*in possible existence outside the mind*) with this essential determination: "ten-thousand-sided polygon." ' (*Formal Logic*, 52)

The example Maritain gives is an A proposition – SaP – taken abstractly. As he reads it, this proposition asserts that

$$SaP \equiv \text{the species } S \text{ is identical with either the species } P \text{ or a subspecies of } P.$$

Which is to say that it is true in either circumstances α or circumstances β as recorded in the possible cases of the Euler diagrams. Similar readings can be given to the other forms, taking negation to mean 'non-identity':

> SiP ≡ either the species S is identical with either the species P or a subspecies of P or there is a subspecies of S which is identical with either the species P or a subspecies of P
>
> ≡ SiP holds in α, β, γ, δ.

> SeP ≡ the species S is not identical with the species P nor with any subspecies of P
>
> ≡ SeP holds in ε.

> SoP ≡ either the species S is not identical with the species P or there is a subspecies of S which is neither identical to the species P nor to any subspecies of P
>
> ≡ SoP holds in γ, δ, ε.

The same characterization of the copula as identification holds for the concept whether it is taken concretely or abstractly, except that in the former case it is the concepts taken in extension that are identified, that is, the individuals within the extension of the subject are identified with the individuals within the extension of the predicate, in conformity to the quality of the proposition determined by the copula as it is positive or negative, and as determined by the quantity of the proposition as determined by the quantifier. In this case we would have:

> SaP ≡ for each individual within the species S there is some individual within the species P or within some subspecies of the species P with which it is identical
>
> ≡ SaP holds in α, β.

> SiP ≡ for each individual within the species S or within some subspecies of S there is some individual within the species P or within some subspecies of P with which it is identical
>
> ≡ SiP holds in α, β, γ, δ.

$SeP \equiv$ for each individual within the species S there is no individual within the species P with which it is identical

$\equiv SeP$ holds in ϵ.

$SoP \equiv$ for each individual within the species S or within some subspecies of S there is some individual within the species P or within some subspecies of P with which it is not identical

$\equiv SoP$ holds in γ, δ, ϵ.

These two readings of the categorical propositions correspond, of course, to the two readings which we gave to them above in our defence of the traditional doctrine of distribution. There we spoke of classes and subclasses, as is the wont of moderns, and here we speak of species and subspecies, more in tune with the traditional logic. But the point is the same. And that means, it is now clear, that there is no reason why the categorical propositions, understood as Maritain understands them, cannot be transcribed into the formal logic of Russell. Thus, for A propositions we have for the abstract reading

$SaP \equiv$ the species S is identical with either the species P or a subspecies of P

$$\equiv (f)[(f \subset S) \supset (\exists g)[(g \subset P) \& (f = g)]]$$

while we have for the concrete reading

$SaP \equiv$ for each individual within the species S there is some individual within the species P or within some subspecies of the species P with which it is identical

$$\equiv (x)[Sx \supset (\exists y)(Py \& x = y)].$$

There are similar transcriptions of the other categorical forms. The fact that we can render these propositions into the symbols of modern formal logic means that there is nothing in these propositions so understood that requires us to attribute to them such traditional relationships as appear on the square of opposition, immediate inferences such as conversion by limitation or the reduced moods of the syllogism. Or to put it another way, it seems, given these readings, that there is nothing special that requires us to use A-validity rather than B-validity.

Here is how Maritain presented the crucial inferences:

If we attach an existential sense to an A and say:

Every bat is a mammal (and bats exist)

we conclude correctly that: therefore some (existing) mammal is a bat.

The question is: what is the status of 'bats exist'? More generally, the various categorical propositions relate species *S* and *P*. It is assumed that each of these species has members. Why is it assumed that each species has members? Why is it assumed that each species is exemplified at least once? If we can establish why it is thought that this is somehow necessary, then we will have provided a justification for these philosophers – Maritain, Pegis, and the other defenders of traditional logic – taking for granted that the proper notion of validity is an A-validity.

Now, for these philosophers, the basic entities in the world are individuals. These entities are substances. When we look at how Maritain characterizes these basic entities in his *Introduction to Philosophy*,[30] we find the definition 'Substance is a thing or nature whose property is to exist by itself, or in virtue of itself (*per se*) and not in another thing' (*Introduction*, 224). Maritain tells us about substance that 'it is *a being existing by itself (per se)* or in virtue of itself, in virtue of its own nature, *ens per se existens*' (218). Substance is the subject of predication. As for the existence of a substance, we are told:

Existence pure and simple is undoubtedly Peter's primary or first existence. But it is not his sole mode of existence. He is sad to-day, yesterday he was cheerful; to-day he exists as *sad*, yesterday he existed as *cheerful*. He has lost the former existence and acquired the latter, but he has not therefore ceased to exist purely and simply. That is to say, he possesses a host of secondary qualifications in virtue of which he exists not only simply (*simplicitur*) but also under a particular aspect (*secundum quid*). (220)

The substance *is* in various ways – for example, Peter *is* sad. This is one of his modes of being, or modes of existence. Existence is thus tied to being in the sense given by the copula: to say that a substance exists in a certain way is to say it *is* in a certain way. Maritain indicates the nature of existence *simplicitur* when he points out that it is '*in virtue of*' the various specific qualifications that Peter 'exists ... simply.' Peter exists *simplicitur* in virtue of his being (among other things) sad. Thus, to be, or to exist, is to be in some way or other. To say that

This substance exists

is to say

There is some property or other that this substance possesses.

Now, to say that

This is coloured

is to say

There is a colour which this possesses.

The property of being coloured is a *determinable characteristic* relative to the various *determinate* colours, red, blue, and so on. Thus, to ascribe a determinable characteristic to a thing is to assert that there is some determinate characteristic which it has. But to say that a substance exists, on the account that we are considering, is to say that there is some property or other which the substance possesses. Hence, to ascribe existence to a substance is to ascribe a determinable characteristic to the substance, and in fact the most determinable of all characteristics, the characteristic of being or existence.

To say that a substance *a* exists is to say that there is a sentence of the form

a is *S*

which is true. Existence and true predication are thus inseparable concepts. Aristotle had already made this point with regard to the substance ontology when he said that 'is true' is the meaning of the verb 'to be' in the 'strictest' sense: 'being and non-being in the strictest sense are truth and falsity.'[31] Thus, a substance is made actual by virtue of having accidents or properties present in it; it *exists* or *has being* just to the extent that there are properties present in it, its *existence* or *being* is constituted by *what it is*, by *what is predicated of it*.

Maritain's teacher, Aquinas, takes up the substance ontology of Aristotle, and makes the same point as the Philosopher.[32] The individual, a substance having a certain form, – for example, this man – 'is said to be a suppositum, because he underlies [*supponitur*] whatever belongs to man and receives its predication' (*Summa*, FP, Q2, A3). This subject is made actual by virtue of its having accidents predicated of it; for the substance to exist is for the substance to have being, that is, for it to have certain accidents predicated of it. '[A] subject is compared to its accidents as potentiality to actuality; for a subject is in some sense made actual by its accidents' (FP, Q3, A6).

On the ontology which we are examining, then, being or existence is the broadest of all species, the supreme genus. But, for all that, it is not a species that appears as either the subject or predicate in a categorical proposition. It is a species but it is never S or P. It is, rather, the *copula*. This copula contains within itself, as determinates under the determinable, all the possible determinations of the subject. As Maritain puts it elsewhere, 'Being presents me with an infinite intelligible variety which is the diversification of something which I can nevertheless call by one and the same name.'[33] What the categorical proposition 'S is P' does in mentioning P is make more determinate what is said about the subject S when it is said simply (*simplicitur*, as Maritain would say) that 'S is.' The propositions 'All S is P' (*SaP*), 'Some S is not P' (*SoP*), and so on, make the same predication but distribute it differently, or, if you wish, indifferently, among the various S's (if any).

The subject of predication is always an 'S,' always a 'somewhat.' Thus, for example, 'we say that *Peter* is a *substance*,' but beyond this 'since his nature considered precisely as such (*what* he is, *that in virtue of which* he is what he is, that in virtue of which he is capable of existence pure and simple), like himself exists beneath the accidents, it also is entitled to the name of *substance*, and we can speak of *Peter's substance*' (Maritain, *Introduction to Philosophy*, 221). When we consider Peter, we are also considering Peter as a certain S, as having a certain nature. In that situation, of course, the concept or nature or form or essence is being taken objectively and also concretely. The nature or form or essence – the species, to speak generally – is the *possibility* of existence, it is that which makes Peter *capable* of existence. 'The nature or essence of the subject of action is therefore that in virtue of which it is capable of existence pure and simple (*simplicitur*); the nature of Peter considered as the subject of action is that in virtue of which I can say simply *Peter exists*' (220). Taken abstractly, the forms are objective possibilities of things, and the categorical propositions state relations among ideas or concepts, where, to recall, these forms are at once objective and also in the mind. It is through these possibilities that ordinary things, here taken to be substances, exist. 'A thing is said to exist by itself or in virtue of itself (*per se*), when it is brought into existence in virtue of itself, or of its own nature (by the causes on which it depends, if it is a created nature). In this sense Peter exists *per se*' (237).

To exist is also to be actual. It is to be 'in act,' as the traditional terminology goes. '*Act* may ... be defined as *being* in the strict sense of the term ...' (*Introduction*, 245). To be, to exist, and to be in act are, in fact, the same thing. Contrasted to 'act' is 'potentiality.' The forms as such, that is, taken abstractly, are

simple potentialities, *capabilities*. Taken concretely, they are in act, that is, they
are taken as the forms of substances that exist. For our purposes, the important
point is that every nature or form is the form of a substance that exists. Even the
form is simply considered abstractly; none the less it is the form of an existing
thing, something in act. '*Potentiality cannot exist in the pure state, apart from
any act*. This is evident. For, since existence is an act, potentiality can only exist
in beings which are in some other respect in act' (248). Maritain connects this
to another theme of the substance ontology, that potentialities are *active* potentialities, *active* tendencies. This is the 'dynamic character of being,' 'the fact that
I cannot posit this reality, grasped by my primary intuition of being as such,
without at the same time positing a certain tendency, a certain inclination'
(*Preface to Metaphysics*, 70). As Aquinas puts it, *omnia appetunt esse*: everything desires existence: 'all creatures naturally desire to be' (*Summa*, FP, Q104,
A1, Obj 3). Now, 'a thing is called natural because it is according to the inclination of nature' (FP, Q82, A1), and 'everything naturally aspires to existence
after its own manner' (FP, Q75, A6). Moreover, the dynamic structure of the
universe, its metaphysical structure, is such that 'a natural desire cannot be in
vain' (FP, Q75, A6). Thus, each form or nature or essence is an active potentiality and as such is an active tendency moving itself towards being. In other
words, it strives for existence. But in the metaphysical order of things, no such
tendency will go unfulfilled, since any such natural tendency, natural or real
potentiality, 'cannot be in vain.' However, to say that no such tendency is unfulfilled is to say that the species which, ontologically understood, that tendency
is, is actual. That is, it determines a class that has members.

On the substance ontology that underlies logic as traditionally presented it
thus turns out that all terms that appear in categorical propositions refer to species that as a matter of metaphysical necessity have members. *There are no
empty classes.* Moreover, no universal species is ever referred to by a term that
appears in the subject or predicate place of a proposition in categorical form. To
be sure, there is a universal species: the universal species is being as such. But
as we have seen, this is represented by the copula and not by one of the terms.
Thus, *there are no universal classes.* It follows immediately that *this ontological background to traditional logic legitimates the use of the notion of
A-validity.*

We can also see why there was no place for negative terms. A negative term
would represent *non-being*. Aristotle put the point succinctly: '"Not man" is not
a name, nor is there any correct name for it. It is neither a phrase nor a negation.
Let us call it an indefinite name' (*De interpretatione*, 16a30–1).[34] Such terms
were also called 'infinite' by the medieval logicians.[35] The medievals argued
that, in spite of the fact that these terms have the grammatical form of nouns,

they are in fact, from the viewpoint of the logician, very different. As Deborah Black has pointed out, this was a focus for the discussion of the logician's purposes in contrast to those of the grammarian.[36] Black cites Martin of Dacia who argued, as Black puts it, that '"Grammatically speaking" ... the infinite noun does signify substance with quality; but its indeterminacy prevents it from signifying any "nature" and thus it is excluded from the *ratio logica* of subjectibility and predicability.'[37] Black also cites Simon of Faversham, who puts the point in detail:

Note that the infinite noun and infinite verb are excluded from the consideration of the logician, because the noun and verb which the logician considers should be parts of enunciation; but infinite nouns are not parts of enunciation, because everything which can be a part of enunciation must signify some concept of the mind, for enunciation is principally for the sake of truth. But we cannot have truth except through that which expresses a determinate concept. For [the infinite noun and verb] are said indifferently of being and non-being, and therefore are neither verbs nor nouns for the logician, and thus are not his concern. However, they are not excluded from the grammarian's consideration, because they do possess those accidents of the noun and verb by means of which they can be composed with one another.[38]

The traditional logic does not of course eliminate negation. But negation is a form of the copula. Where the copula is 'is,' we join or identify concepts. Where the copula is 'is not,' we separate or distinguish concepts. This separation or distinguishing of concepts presupposes difference. And difference is, of course, real. But non-being is not real, it simply does not exist. There are, therefore, no concepts that refer to such 'things,' that is, no concepts that are negative. The occurrence of such terms in inferences is thus a figment of grammatical change rather than of logical form. They do have a place in any general account of inference, as Aristotle acknowledges in *De interpretatione*, chapter 10 (19b5ff), where various forms of opposition are discussed. In the same way, in the *Topics*, Aristotle discusses disjunctive terms such as 'expedient or honourable' (*Top.* 110b9),[39] even though such terms would not express, as Simon of Faversham put it, a 'determinate' concept. One can have truth-preserving relations, *rules for consistency*, while yet insisting that the *logic*, that is, the *logic of truth, the logic that perspicuously represents the ontological structure of the world*, does not contain such terms.

The traditional *logic*, then, *so far as it is tied as it is to the substance metaphysics*, thus has no place for negative terms. It follows that it can safely use the rule of distribution (RD) for evaluating the validity of inferences. Contrary to Geach and others, the presence of the rules of distribution constitute no weakness in traditional logic, they introduce no inconsistency. The point again is that *the ontological background to traditional logic legitimates the exclusion of negative terms from logic and thus justifies the usual appeal to the rules of distribution for evaluating validity.*

Now that this is clear we can also see why Russell and the nineteenth-century pioneers who preceded him could so easily opt for both B-validity and negative terms. *They rejected the substance metaphysics that underlies logic as traditionally understood.* Rejecting that metaphysics, they could see no reason for excluding empty terms or universal terms. And once these were admitted, they had no choice save to opt for B-validity. Moreover, for Russell and company, negation does not indicate, as it were, a hole in being; rather, it simply reverses the truth values of various propositions. The concept '*nS*' is nothing other than the predicate form '∼*Sx*' where the negation sign does nothing more than indicate that if an A proposition of the form '*Sx*' is true (false) then that of the form '∼*Sx*' is false (true). The ontological problem that arises for the substance metaphysics with the introduction of negative concepts does not arise once that ontology is rejected.

The point here is, of course, that the real disagreements between Russell and company, on the one hand, as defenders of the new logic, and, on the other hand, the defenders of traditional logic such as Maritain and Pegis, are ontological. Pegis was right to tell me, years ago, that the basis of his logic was a realistic metaphysics while the basis of Russell's logic was, if not nominalistic, then at least not realistic in the Thomist–Aristotelian sense. He was right to be impatient with me for not recognizing that this was a difference that went deeper than anything in formal logic. And he was right to resent the way the more 'modern' logicians represented his views and the views of others who, on the basis of their realism, defended the traditional logic. In effect, what Pegis was arguing was that the apparently purely formal logic of Aristotle's *Prior Analytics* was designed to be a logic that would fit the demonstrative logic of truth of the *Posterior Analytics*. To present the differences between the two camps, as Wittgenstein for one and Russell for another did, simply as differences about formal logic alone, as differences about the logic of the *Prior Analytics*, rather than as differences dependent upon the logic of truth of the *Posterior Analytics*, simply misses where the disagreement really is.[40] In fact, to the extent that Wittgenstein and Russell imply that the problem is, as it were, a sort of logical blindness on the part of traditional logicians, a blind clinging to outmoded tradition, they are less than fair to those who differ metaphysically. In fact, of course, one can argue that the blindness is on the part of the critics of traditional logic: these critics simply fail to recognize that the disagreement really is ontological. To be sure, they no doubt rejected that ontology. But that does not excuse their taking an ontological disagreement to be a blind clinging to tradition. It must be said, however, that simply to suggest, as Maritain does, that Russell and company miss the metaphysical point is also hardly adequate. What is required is an evaluation of the metaphysical arguments used to reject the substance ontology, not simply a reassertion of that ontology.

In fact, there is a very good argument against the 'realistic' metaphysics of Maritain, Pegis, and the rest. This argument derives from Locke, or, at least, it was placed on a firm basis by Locke. Given this argument, the case for Russell's logic against the traditional logic can certainly be made.

II: The Logic of Consistency

The translations into the formalism of Russell's logic are standard. Thus, SaP = All S are P is translated as

$$(x)(Sx \supset Px)$$

A general proposition, where S names or refers to some kind or species,

(g) $(x)(Sx)$

has standardly the following truth conditions: (g) is true if and only if all possible substitution instances of Sx are true, where a substitution instance of Sx is obtained by substituting for the variable x a constant term that names or refers to an individual thing. In particular, then, the proposition SaP under the translation is true if and only if all possible substitution instances of

$$Sx \supset Px$$

are true. And so, if a is an individual thing, then we obtain a substitution instance by substituting a for x to obtain

(si) $Sa \supset Pa.$

In order to determine whether this substitution instance is true we must in turn discover the truth values of the two atomic sentences

Sa

$Pa.$

The truth table for the horseshoe is this:

p	q	$p \supset q$
T	T	T
T	F	F
F	T	T
F	F	T

This allows all possible truth combinations; the truth values of p and q vary independently of each other. This means that the truth conditions of (si) are such that Sa takes on its truth value independently of the truth value that is taken on by Pa. But this presupposes that the kinds or species S and P are *independent* of each other. In terminology that has descended to us from the mediaevals, these concepts must be both *distinct* and *separable* – separable in the sense that if Sa is true then Pa can be either true or false. That is, there is nothing within the natures of the species or kinds S and P such that if S (or P) is (part of) the being of a then that determines that that being also includes, or also excludes, P (or S).

But now compare what Maritain has to say about verifying certain propositions.

From the point of view of matter we must distinguish between *propositions in necessary matter*, that is, those in which the Pr. is *essential* to the S ... and *propositions in contingent matter*, that is, those wherein the Pr. is *accidental* to the S ...

In the first case the proposition expresses an eternal truth and affirms only the relation (of identification) between the object of thought signified by the Pr. and the object of thought signified by the S ... Thus it does not require the actual existence of the subject in order to be true ... and has not necessarily and of itself an 'existential' sense.

In the second case the proposition expresses a contingent truth, and for this reason, that is, inasmuch as it does not express an eternal truth, it requires the (actual) existence of the subject in order to be true. It has an 'existential' sense ...; since the Pr. does not follow from the nature of the subject, it could not be verified except of a subject posited in existence. (*Formal Logic*, 227)

It is evident that the 'propositions in contingent matter' are in effect the propositions understood in the way in which we have just examined. It is the other class that is of present interest. These are the 'propositions in necessary matter.' These propositions are true by virtue of relations among the forms themselves. The forms or species are connected 'essentially' to each other. In a proposition of this sort in A form

All *S* are *P*

there is a *guarantee* that the proposition is true lying within the concepts or forms or species. These species or forms have natures such that if some individual is *S* then *S* guarantees that *P* will also be part of the being of that individual. *S* and *P* are in their natures such that being *S* and being *P* are inseparable. This guarantee of the truth of the proposition lying within the natures of the forms or species is a function of the matter of the proposition – it depends on the natures of the forms or species *S* and *P* – and is such as to render the proposition a necessary or eternal truth.

It follows that not all necessity is formal necessity. There are necessities in the matter as well as the form of propositions, and these necessities, too, determine the movement of discourse, of thought, and of reason. Form, of course, transmits truth. Form keeps discourse consistent; it guarantees that consistency. Its guarantee is one of necessity. But form does not yield any increase in truth: its connections are empty so far as matters of fact are concerned, tautological. Equally, form does not provide truth – that is, non-tautological truth – from which inferences may begin. But on the view of propositions and of necessity with which Maritain is working, there are non-tautological, substantive necessities. Reason grasps these necessities as well as the tautological necessities of the purely formal parts of discourse. Reason, considered as the capacity to grasp the necessities of things, has a broader task than simply keeping discourse consistent. The problem with Russell's view of logic is that it takes logic, and reason, to be simply the tautological logic of consistency, refusing to recognize that there is more to reason than this, refusing to recognize that reason has the further task of recognizing non-formal substantive necessities in the forms and species of things. To thus conceive reason and the discourse of reason as limited to the impoverished logic of consistency is the crucial error of Russell and the other logisticians, according to Maritain: 'the existential or non-existential signification of a proposition depends upon its *matter*, and not upon its *form* alone. The fundamental error of the logisticians lies in their failure to distinguish between the form and the matter of propositions and in their belief that considerations bearing exclusively on the form suffice to explain the entire discourse' (*Formal Logic*, 226–7).

One must also say, of course, that there is a case to be made for Russell's logic, that is, a case to be made that there are no substantive necessities of the sort that Maritain thinks are there, for the mind to grasp. This case was made by Locke and his followers, and came to be accepted by those concerned with the logic of science. It is empirical science which gives us knowledge of the world, not pure reason; and what empirical science is concerned with are contingent truths, truths known by sense experience. Within this context, reason, as the capacity to grasp the necessities of things, comes to be restricted to the logic of consistency.

Within a logic of consistency there is, of course, nothing wrong with negative terms. Nor is there anything wrong with empty and universal terms. When logicians began to break out of the strait-jacket of tradition in the nineteenth century, they very naturally introduced negative terms. Equally naturally, empty terms and universal terms were introduced. Not all parts of the set of rules from the traditional heritage of logic changed at once, however. It was only gradually recognized that the traditional rules were not everywhere compatible with the notions of negative terms and empty and universal species. Formal logic had to be revised. But, in the end, the needed changes were recognized, and B-validity eventually replaced an A-validity.

The logic of consistency is truth-functional. Bochenski has claimed that the origins of truth-functional logic go back to the Stoics.[41] There is a point to this, but in fact their concerns were, like those of Aristotle, more centrally focused on a logic of truth. They developed truth-functional arguments, but this was simply as a part of a broader logic of truth; more strongly, insofar as they allowed that there are truth-functional arguments and connectives, these are clearly second rate within the context of the logic of truth that they were concerned to develop.

The Stoics recognized as valid forms such as hypothetical syllogism, the validity of which is now reckoned to be in the domain of truth-functional propositional logic.[42] They recognized arguments like this:

If p then q

p
_____

$\therefore q$.

However, for Chrysippus, at least, these were *not* of the form

$p \supset q$

p
_____

$\therefore q$.

That is, for Stoics who followed Chrysippus, the 'if-then' of the premise of hypothetical syllogism did not have the truth-functional meaning assigned to it by formal logic. Rather, as W. Hay has pointed out,[43] the hypotheticals that the

Stoics allowed as premises for hypothetical syllogisms seem to have been conceived as necessary truths, non-formal substantive necessities. Philo,[44] and perhaps Zeno himself,[45] held that a hypothetical or conditional is true unless the antecedent is true and the consequent is false, that is, the truth conditions for the truth-functional '⊃.' Chrysippus, however, argued that a stronger notion of implication was needed, in which a conditional holds only if the contradictory of the consequent is incompatible with the antecedent.[46] The contradictory of the consequent and the antecedent must involve, as Cicero put it, a 'conflict' between themselves: *pugnant ... inter se*.[47] For a conditional to be valid, there must be, as Sextus said, a 'coherence' or 'connexion and consistency' (κοινωνίαν καὶ συνάρτησις) between the antecedent and the conditional.[48]

This was not simply a matter of the validity of arguments and conditionals. It is a matter of truth.[49] More specifically, it is a matter of the truth about the structure of the world that the Stoic wise man is able to grasp as he strives through knowledge of nature, including his own nature, to attain the good life.[50] Sextus makes the crucial connection in this way:[51]

Man does not differ in respect of uttered reason from the irrational animals (for crows and parrots and jays utter articulate sounds), but in respect of internal reason; nor <does he differ> in respect of merely simple impressions (for the animals, too, receive impressions), but in respect of the transitive and constructive impressions. Hence, since he has a conception of logical sequence, he immediately grasps also the notion of sign because of the sequence; for in fact the sign in itself is of this form – 'If this, then this.' Therefore the existence of sign follows from the nature and structure of Man.[52]

People and animals both have sense experiences; in this sense, they have the same presentations. Where humans differ from animals is in their ability to make inferences, and, in particular, inferences based on conditionals: if this, then this. It is the nature and constitution of humans, that is, of human reason, that they can grasp the connections recorded in conditional statements. This is connected to both the Stoic physics and the Stoic ethic. For Chrysippus the ultimate good is 'to live in accordance with one's experience of the things which come about by nature.'[53] This is equivalent to living in accordance with reason. Reason is, on the one hand, the active element in nature. Chrysippus, in his book *On Substance*, expounded the view that there are two principles of the universe: as Gould describes the view, 'one, the passive, is unqualified substance or matter; the other, the active, is god or reason, which is eternal and creates each thing throughout all matter.'[54] All things are interconnected by the necessity of reason – this is the Stoic doctrine of 'fate.' Cicero states the Stoic doctrine on this point in this way: 'If everything takes place with antecedent causes, all events take place in a closely knit web of natural interconnexion; if this is so, all things are caused by necessity ...'[55] The universe, guided by reason, proceeds deterministically along certain lines, consequences following regularly and necessarily from antecedent causes:[56] if this, then this. But reason is also, on the other hand, the capacity to grasp the reason of things, these reasons in things; and these reasons, these necessary connections, are what we find recorded in conditionals. The wise man is the person who can distinguish the genuine reasons of things, the necessary connections between antecedents and consequents – those connections, in other words, that are expressed in conditionals that are true.[57] With this knowledge the Stoic sage gains insight into the necessary structure of nature, and therefore into the duty and obligations of human beings.[58]

Thus, these Stoics, those who followed Chrysippus, as much as the Aristotelians, held that there are non-formal necessities grasped by the power of reason. For them, as for the Aristotelians, logic was not only a logic of consistency but also a logic of truth. The differences were metaphysical rather than about the demonstrative aims of logic. That is, the disagreements about the nature of logic or, what is the same, about the *organon* of philosophy, were about the metaphysical structure of the world, just as the disagreement between Maritain and Russell was metaphysical rather than purely logical. Of course, the metaphysical differences were different, but the point remains that in both cases what was crucial was a disagreement not simply about which forms are valid, a disagreement not simply about what to include in a logic of consistency, but rather about what sort of logic was needed to reflect the metaphysical or ontological structure of the world.

The relevant metaphysical differences seem to be these: For the Aristotelians there were many powers or souls, as many as there are different species of things. For the Stoics, in contrast, there is only one λόγος, the active, creative causal principle, also called God, that is the soul of the universe. For the Aristotelians, the forms of things constitute the terms that appear in categorical syllogisms. These forms come to be in the mind as concepts through a process by which they are abstracted from, separated from, the sensible appearances of things. Since for the Stoics there are no multiple forms of things, only one form or soul, there can be no separation of such a form. The Aristotelian sort of concepts simply cannot appear in inferences. The sort of demonstration that the Aristotelians put into syllogistic form to reveal the necessary structure of things is simply not possible within the Stoic framework. This, it seems to me, is what the celebrated nominalism of the Stoics amounts to. This is not to say, however, that the Stoics denied that there are objective necessities in things. To the contrary, these necessary connections are to be expressed in the conditionals that form the premises of demonstrative hypothetical syllogisms, as we have seen. As Kieffer has explained the metaphysical difference, for the Stoics 'knowledge consists of the necessary connectivity of events rather than of the necessary relation of genus and species.'[59] The necessary connections of the Stoic universe are not grasped by separating them from sense experience. They are, rather, found by reason *in* sense experience. Reason, first, grasps the necessary connections inseparably from our sense experience of things, and, second, grasps them as connections within the larger, all-encompassing pattern constituted by the single soul or λόγος of the universe.[60] The metaphysics/epistemology of demonstration is thus very different for the Stoics and the Aristotelians; and this is the real source of their differences concerning the nature of logic.

Some ancient philosophers construed conjunction in a way similar to the way in which Chrysippus insisted against Philo that the conditional be treated, that is, as being true on grounds of some necessary relation holding between its component propositions. The third of Chrysippus's indemonstrable argument forms is

$$\sim (p \;\&\; q)$$
$$p$$
$$\overline{}$$
$$\therefore \sim q$$

Galen, in his *Institutio logica*, argued that for a negated conjunction to be true there must also be a *conflict between p and q*. This is similar to the case of conditionals, where, for a conditional to be true, there must be a conflict between the antecedent p and denial $\sim q$ of the consequent – except that in the case of the conjunction it is a case of 'incomplete conflict' in the sense that both conjuncts cannot be true but both can be false, as in, for example, 'Dion is not both at Athens and on the Isthmus.'[61] Cicero, too, holds that if the conditional 'if p then $\sim q$' – for example,

If Fabius was born at the rising of the Dog Star, Fabius will not die at sea

– is true, then p is incompatible with q, and for this reason the conjunction

$$p \;\&\; q$$

is a conjunction of incompatibles – 'haec ... coniunctio est ex repugnantibus' – and asserted as an impossibility.[62] Here again we recognize the presence in logic of non-formal necessities, and the presence, therefore, of a logic which is not a mere logic of consistency but also a logic of truth.

Galen distinguished negated conjunctions the truth of which derives from the incomplete conflict of the parts from merely 'conjunctive' uses of 'and' such as 'Dion is walking and Theon is talking' which have neither consequence (as in true conditionals) nor conflict (as in negated conjunctions of the sort just described).[63] Here we have a truth-functional connective. Chrysippus himself suggested that the premise of his third indemonstrable should normally be construed in this truth-functional way. Aulus Gellius gives the required truth conditions for conjunction (*sumpeplegmenon*); they are the truth-functional truth conditions with which we are now familiar. As an example of a conjunction he cites 'P. Scipio, was a son of Paulus and twice consul, and he had a triumph and held the office of censor and L. Mummius was his colleague as censor.'[64] He then continues, 'in every conjunction if one [conjunct] is a falsehood then even if all the rest are true, the whole is said to be a falsehood. For if to all those truths which I have said about Scipio I were to add "and he conquered Hannibal in Africa" which is false, that whole which the things said conjoined are, on account of this one falsehood which would have been added, because they are said together, would not be true' (*Noctes*, xvi.8). These truth conditions differ from those proposed by Galen; unlike the truth conditions proposed by Galen, these require no non-formal necessities.

Galen argues in fact, that if we have a 'mere' conjunction, then it will be useless in demonstration. A 'mere' conjunction, a non-necessary connection, holds when we are talking about 'things that sometimes occur together and sometimes do not': 'all those facts that have neither necessary consequence nor conflict give material for the conjunctive proposition; e.g., "Dion is walking and Theon is talking" ...'[65] Here the negation is

It is not the case both that Dion is walking and Theon is talking.

This is of the form

$$\sim (p \ \& \ q).$$

If we are to use this in a demonstration in the form of the third Stoic indemonstrable, then we need to know that this is true. In order to know that it is true we have to know either that p is false or that q is false. But the indemonstrable form is this:

$$\sim (p \ \& \ q)$$
$$p$$
$$\overline{}$$
$$\therefore \sim q.$$

If we know the premise is true because we know '$\sim q$' is true, then we don't need the argument to demonstrate that fact to us; we know it already. At the same time, if we know the premise is true because we know '$\sim p$' is true, then we can't use the argument to demonstrate anything to us, since the minor premise will be false. Under the circumstances, then, the argument form in which the premise is a 'mere' conjunction is useless in demonstrations. As Galen puts it, where we have the argument form with a negated conjunction as a premise, and try to use it in a demonstration, then '[t]he additional assumption is, "But in fact Dion is walking," or "But Theon is talking," and the conclusion in the case of the first assumption "Therefore, Theon is not talking," in the second case, "Therefore Dion is not walking"; such material has been shown to be absolutely useless for demonstration' (*Institutio*, XIV. 7–8, 47).

This is parallel to C.I. Lewis's argument that the truth-functional 'if-then' of material implication, represented by the '$\supset$,' does not adequately represent the 'if-then' of ordinary language.[66] Lewis thought that the following argument forms are paradoxical:

$$q \qquad\qquad \sim p$$
$$\overline{} \qquad\qquad \overline{}$$
$$\therefore p \supset q \qquad\quad \therefore p \supset q$$

These inferences are not valid, Lewis suggested, for ordinary 'if-then,' though they are valid for the logician's sense of 'if-then' represented by the horseshoe. These forms are not valid for the ordinary 'if-then' because, in ordinary usage, for it to be true that

If p then q

what is required is not just that it *is* not the case that p is true and q is false, but more strongly that it *could* not be the case that p is true and q is false.

Lewis concluded that '$\supset$' does not adequately represent ordinary 'if-then,' and suggested that we add a new modal operator to logic, represented by N, where

Np

is read as

It is necessary that p.

The 'if-then' of ordinary discourse could then be represented, he proposed, by what he called 'strict implication,' which he defined as

$$p \mathbin{-\!3} q =_{\text{definition}} N\,(p \supset q).$$

A conditional, that is, a 'real' conditional and not merely a truth-functional conditional, can, in other words, be true only if there is some conflict between the antecedent and the negation of the consequent. This is parallel to Galen's argument.

There is an alternative conclusion, however. The argument form is useless in demonstration in case that the evidence for the truth of the truth-functional premise consists solely of a knowledge of the truth values of the component propositions. But if there are other sorts of evidence, then there is no reason why the form should not be used to establish that conclusions are true. One such sort of evidence would be the knowledge that the premise is necessarily true. This is in effect just what Lewis proposed. There is another sort, however. We might be able to say that '$\sim (p \,\&\, q)$' is true because we have *observed previous cases* of p and q and have concluded *inductively* that we *never as a matter of fact* have a case in which p and q are both true. This would allow us to have both a truth-functional conjunction in the premise of the form and also a form that is useful in providing grounds for the conclusion being true.

Chrysippus, in fact, took up this latter alternative.

Chrysippus proposes the use of conjunctive sentences with the now-standard truth-functional truth conditions in the premises of arguments of the form of his third indemonstrable. He makes this proposal in the context of a concern about astrological divination. He considers the conditional

If anyone is born under the Dog Star, then he will not be drowned in the sea

This conditional does not state a necessary connection; it can be construed not as statement of non-formal necessity but only as a statement the truth of which has been learned by experience. That means it must be taken to be a material or Philonian conditional, with the 'if-then' having the sense of the modern logician's '$\supset$.' Chrysippus recommends that it be expressed instead as a negated conjunction:

Not both: someone is born under the Dog Star and he will be drowned in the sea.

To do otherwise, to use the form of a conditional, as the Chaldeans do, would be to deceive themselves that they have, in fact, achieved knowledge of the reasons of things, the knowledge of the sage. Cicero, who reports this, goes on to note sarcastically, that this will require everyone or no one to so reformulate his or her conditionals:

O what amusing presumption! ... he tutors the Chaldeans as to the proper form in which to set out their observations! For I ask you, if the Chaldeans adopt the procedure of setting forth negations of indefinite conjunctions rather than indefinite sequences, why should it not be possible for ... geometricians and the other professions to do likewise? ... a geometrician will not speak as follows, 'The greatest circles on a sphere bisect each other,' but rather as follows, 'It is not the case both that there are certain circles on the surface of a sphere that are the greatest and that these circles do not bisect each other.' What is there that cannot be carried over in that sort of way from the form of a necessary consequence to that of a negation of conjoined statements?[67]

Cicero is not being entirely fair to Chrysippus here. The concern of the latter is genuine. The Stoic divided arguments into those that are (a) valid, that is, those in which the conclusion follows formally from the premises, from a subclass of these, those that are (b) valid and have true premises, and, further, these from a further subclass, those that are (c) demonstrative. The demonstrative arguments are those with necessary premises.[68] The aim of the dialectician is the production of demonstrative arguments. The ultimate premises, while present to the mind by virtue of the faculty of reason grasping these necessary truths, are not themselves the products of logic or dialectic narrowly conceived. As Bréhier once put it, 'the aim of the dialectician is not really the invention, the discovery of new theses; all his efforts bear upon the discussion of theses which present themselves naturally to the human mind.'[69] None the less, it is still true to say that it is reason that grasps the ultimate premises, and in that sense reason really is the faculty that discovers truth. As Chrysippus puts it in a fragment from his *On the Use of Reason*: 'It [the faculty of reason] must be used for the discovery of truths and for their organization ...'[70] But we cannot always obtain demonstrative truth. In the absence of insight into the necessary structure of the universe, we must rely upon empirically determined signs.

This is the basis of the art of divination. According to Cicero, it is necessary to distinguish two kinds of divination: 'one, which is allied with art; the other which is devoid of art.' 'Those diviners employ art, who, having learned the known by observation, seek the unknown by deduction.' The diviners without art are those 'who, unaided by reason or deduction or by signs which have been observed and recorded, forecast the future while under the influence of mental excitement, or of some free and unrestrained emotion.'[71] It is divination in the former sense which is of present concern. Divination in this sense is based on matter-of-fact regularities inferred inductively from long-continued and re-peated observation of signs. The result is not a knowledge of necessary truths – necessary causal connections. When we come to know these empirical general-izations, we do not really understand what goes on: such understanding comes only with insight into the necessary structure of the universe. But the empirical, inductive knowledge of signs is none the less useful. 'You ask ... why these things so happen, or by what rules they may be understood? I confess that I do not know, but that they do so fall out I assert that you yourself see.'[72] Because there are no necessary connections, the judgments are not certain, and, indeed, sometimes fail. But in this, the art of divination is no different from other non-demonstrative arts, such as medical science or military science.[73]

The art of divination, while not based directly on necessary connections, can confidently be relied upon because we also know that there are regular neces-sary connections in the universe, even if we do not know specifically what those connections are. Diogenianus quotes Chrysippus as saying that 'the prophecies of the diviners would not be true if all things were not encompassed by fate.'[74] Cicero states the Stoic position thus:

since ... all things happen by Fate, if there were a man whose soul could discern the links that join each cause with every other cause, then surely he would never be mistaken in any prediction he might make. For he who knows the causes of future events necessarily knows what every future event will be. But since such knowledge is possible only to a god, it is left to man to presage the future by means of signs which indicate what will follow them ... [Diviners] may not discern the causes themselves, yet they do discern the signs and tokens of those causes. The careful study and recollection of those signs, aided by the records of former times, has evolved that sort of divination, known as artificial ...[75]

Sambursky suggests that this reliance on induction in the context of a doc-trine of fate is a feature of Stoic philosophy that lies in the background of mod-ern science.[76] This, however, is misleading. In the first place, the doctrine of fate is a teleological doctrine, and not the mere assertion of the ubiquity of mat-ter-of-fact regularity that is the principle of determinism in modern science and the logic of the experimental method. In the second place, what is more impor-tant for present purposes, for modern science there is no method for which the method of induction is second best: the inductive method is *the* method of science. For the Stoics, in contrast, the inductive method is second best relative to the method of intuitive insight into the necessary structure of reality and the elaboration of this insight in demonstrative arguments. In regard to this latter point, the Stoics agreed with Aristotle against the new science of Galileo and Bacon.

We now see that when Chrysippus insisted that the Chaldeans and other diviners put their premises into truth-functional conjunctive form, the point was not merely logical. He was recognizing that their discipline was indeed restricted to sense experience and did not penetrate into the necessary structure of reality. Those judgments that did penetrate the necessary structure of reality should be expressed in the form of conditionals, and inferences based upon them should proceed via the first indemonstrable argument form (our hypothetical syllogism). In contrast, those judgments which were used to predict but were based only on induction should not be expressed in the form of conditionals, but rather as negated truth-functional conjunctions, and inferences based upon them should proceed via the third indemonstrable argument form. Chrysippus's point relies upon the distinction between the world of sense experience and of fallible induction, on the one hand, and the world beyond sense of the necessary structure of the universe and the infallible grasp of objective necessary connections. For the former world of sense experience, we must restrict ourselves to truth-functional connectives; in the other world of the necessary structure of reason, we must use non-formal necessities. Knowledge of the latter is true science, the goal of the Stoic sage; knowledge that restricts itself to the former is second best.

Bochenski's claim that one can find the beginnings of truth-functional logic within the logic of the Stoics is thus correct. None the less, it is misleading to say that this shows that the logical concerns of the Stoics were close to those of the modern logicians, such as Russell, who developed what we now call propositional logic. The concerns of the Stoics were to discover arguments with non-formally necessary premises, demonstrative arguments, and *merely* truth-functional logic was a second best of which we make use simply for want of something better. C. Normore is also wrong to declare that the truth-functional account of conjunction is *the* Stoic account and that it originates in Chrysippus.[77] None the less, he is correct to find that there is a truth-functional logic in the Stoics, as a subsidiary logic, a second-best logic.

Nor are Bochenski and Normore wrong to suspect that it is from this source that truth-functional logic entered the medieval world. Aristotle's is a logic of truth that hardly allows for a second-best logic. The Stoics such as Chrysippus did allow for such a logic, a truth-functional logic of opinion or mere belief, in contrast to the demonstrative logic of necessary connections. Thus, when the Stoics admitted truth-functional logic, they were prepared to admit a truth-functional 'and.' But there is little sense of such an 'and' in Aristotle. In fact, in Aristotle there is little sense that a conjunction '*p* and *q*' is even a statement that can be reckoned as either true or false, as opposed to a list of sentences each of which by itself is true or false but jointly are neither. Thus, he states that

> But if one name is given to two things which do not make up one thing there is not a single affirmation. Suppose, for example, that one gave the name 'cloak' to horse and man, 'a cloak is white' would not be a single affirmation. For to say this is no different from saying 'a horse and a man is white,' and this is no different from saying 'a horse is white and a man is white.' So if this last signifies more than one thing and is more than one affirmation, clearly the first also signifies either more than one thing or nothing (because no man is a horse).[78]

Boethius seems to have followed Aristotle in denying truth-valuable status to conjunctions, since he eliminates from his list of indemonstrable argument forms the third of Chrysippus's forms,[79] the one that contained the conjunction that Chrysippus insisted should be used for merely empirical predictions. This shows both the extent to which Boethius was not interested in the mere consistency of truth-functional logic and, at the same time, the extent to which he was not interested in merely empirical regularities.

It was via neither Boethius nor, equally unlikely, Aristotle, that the Stoic second-best logic entered medieval philosophy. But somewhere it did. Of course, it might well have been an independent discovery. Certainly, it would not be the first time something old but forgotten had simply to be re-invented when the need arose. In any case, as Normore has indicated, the route, if it exists, is obscure.[80] But however it came to be, it is true that such forms as hypothetical syllogism came to be treated in detail by later medieval logicians. For us, the important point is that they came to be treated in truth-functional terms, that is, in terms of a logic of consistency. They came, in other words, to be treated after the fashion of the second-best logic of the Stoics. By the time of Ockham, truth-functional connectives have become a commonplace. Normore cites Ockham's statement of the truth conditions for conjunction, which restates the Stoic truth conditions reported by Aulus Gellius. 'The copulative is that which is composed from several categorical sentences by means of this conjunction "and" or by means of some part equivalent to such a conjunction, just as in this copulative "Socrates runs and Plato disputes" ... Moreover for the truth of a copulative it is required that both parts be true and so if any conjoined part is false the copulative itself is false ...'[81] If one is thinking truth-functionally, simply in terms of a logic of consistency, then these truth conditions for conjunction are perfectly natural. It is immediately related to other connectives, such as (inclusive) disjunction. Thus, Ockham goes on to state that 'It should be known immediately that the contradictory opposite of a copulative is a disjunction composed from the contradictories of the parts of the copulative.'[82] He states this version of the De Morgan's law easily, as something well known to logicians. But if one is not thinking truth-functionally, if one is thinking instead in terms of demonstrations with premises whose necessity is non-formal, then one will, like Boethius, quite reasonably ignore 'mere' conjunction. The point is that around the middle of the twelfth century, roughly by the time of Abelard, the position stated by Ockham comes to be something of a commonplace, replacing the older tradition deriving from Boethius.[83]

It is an interesting issue to speculate: Why did this happen? Why did philosophers around this time come to take an interest in the logic of consistency?

Somewhere after the turn of the first millennium, after a long assault by pagans – the Norsemen – to the north, and infidels – the Saracens – to the south, and after an equally long retreat in the face of these assaults, western Europe, which also constituted western Christendom, regained its confidence in itself and began a counter-offensive. The now civilized Norsemen, the Normans, effected the reconquest of Sicily; by 1090 this had been completed. The *reconquista* of Spain was well under way; Toledo was taken in 1085. These efforts were soon to be supplemented by the Crusaders's recapture of the Holy Lands, and by crusades into the still pagan lands of the north-east. Accompanying these efforts was a corresponding intellectual flourishing equally aimed at extending Christianity. But to do the latter, there had to be an intellectual coming-to-grips with the problems in the faith. Many authorities were available – the Bible, the Fathers, the Saints, and so on. These gave to western Christendom a wealth of sources, but, unfortunately for that faith, these sources often contradicted one another. These conflicts were brought to the fore by Abelard in his *Sic et Non*[84] which was a collection of apparently contradictory pairs of propositions drawn from ostensibly reliable authorities.[85] It is the challenge of these apparent contradictions among accepted authorities that the intellectual leaders had to meet. They had to show that these authorities were *diversi, non adversi*.

They developed techniques for meeting this challenge. This was the well-known, and in time infamous, scholastic method. It required the definition and division of crucial terms, and the drawing of inferences to establish that there is in fact no conflict. A rule from one authority in apparent conflict with another is distinguished as a special case, and in its special nature a case requiring different treatment. And so on.

This method can be seen in the *Summa* of Aquinas for example. It was extended for use in areas other than the theological. Thus, the great jurist used the scholastic method in his discussion of Roman law. The great medical teacher Pietro d'Abano used it in his discussion of medicine. The authorities were different in these latter cases, of course. Bartolous had Justinian's code and the work of the glossators. These contained contradictions, and, like the theologians, he used the scholastic method to overcome these defects. Pietro d'Abano[86] looked back to medical authorities such as Hippocrates, Galen, and the Arabic Aristotelianism of Avicenna's *Canon*. Again, he saw his task as that of using the scholastic method to remove apparent contradictions: that is why his famous text has the title *Conciliator of Differences*.[87] As it has been put, '[t]he differences were real enough, being those between the physiological doctrines of Aristotle, the master of the philosophers, and those of Galen and his followers.' Authority was central to defining what constituted medical rationality. 'Adherence to a learned tradition expressed in a body of authoritative medical writings constituted a guarantee of the separate identity of rational inquiry about the human body by *medici* as an authentic intellectual enterprise, distinct from natural philosophy.' Hence, 'the elucidation, discussion, and, where possible, reconciliation of differences between the Aristotelian and the Galenic points of view on various physiological topics ... became and long remained a central preoccupation.'[88] Thus,

university medical education relied primarily on the study and exposition of authoritative texts. Hence, lectures on the texts and disputations, or formal academic debates, about problems of textual interpretation or the reconciliation of conflicting opinions, were as central to instruction in medicine as they were in other academic disciplines. But courses in theory (that is, philosophy of medicine and principles of physiology and pathology) and courses in *practica* (that is, the study of specifics of diagnosis and treatment) into which university curricula in medicine came to be divided were taught in this way, a method often termed 'scholastic.' In university lectures, medical books on essentially practical topics were treated to full scholastic analysis. Questions on which opinions differed were isolated, the views of authorities were listed and distinguished, objections to each were raised and solved in turn, and so on.[89]

This is not to say that there was no empirical investigation of matters of medical fact. In medieval universities, students attended anatomical demonstrations on the human cadaver. Public dissections were instituted at Bologna by 1316, and were made statutory at Montpelier in 1340.[90] However, all this was within the framework of the scholastic appeal to authority. As Siraisi notes, '[i]n reality, the goal of physiological or anatomical study was, in many instances, a better understanding of texts.' She then points out that this was entirely reasonable, given the cognitive aims of the age: 'In the context of the intellectual life of the thirteenth to the fifteenth centuries, that goal was, of course, entirely valid.'[91] And, of course, the philosophers also deployed the same method, trying in particular to sort out the philosophical system of Aristotle, and to reconcile it, as far as possible, with the authoritative texts of the Christian religion.

The scholastic method was one that was effective in achieving its end of reconciling authorities. The method has, in fact, been discovered several times in the history of humankind. Thus, for example, the Jewish rabbis confronted apparent inconsistencies in the Torah and other accepted religious texts, and in the tradition of Jewish law. The rabbis were confronted with the problem of reconciling these authorities. The method that they used was, in effect, the same method as that invented by the scholastic thinkers of Christian medieval Europe.[92]

The method did not aim to create or discover new knowledge; nor did those who used the method have any such end in view. To the contrary, they had already, ready to hand, all the knowledge they needed in the statements of the accepted authorities – whether the authority be religious, philosophical, legal, or medical, and, if religious, whether it be Christian or Jewish. It might indeed seem odd to us that medical lecturers, for example, would use the scholastic method to expound medical knowledge, rather than emphasizing observation of the natural history of diseases and the empirical discovery of causes. But these thinkers did not think in those ways: they thought rather in terms of accepted authorities. Mere practice was disdained. Not for nothing has it been called the 'age of authority.'

In the scholastic method, logic has an absolutely central role to play. Moreover, the relevant logic is the logic of consistency. Indeed, the whole problem is that of consistency. For, the problem that the method aims at resolving is that of the apparent *inconsistency* of different accepted authorities. The method aims to make them consistent. What is needed, then, is a set of tools that will help them maintain consistency in their arguments. This is, I would suggest, why it is that about the time of Abelard western European thinkers became interested in a logic of consistency.

These thinkers were also educators, and they inculcated their students with the scholastic method. This they did by setting them exercises in which consistency was the object of the exercise. These were exercises in which they were required to defend or dispute some set thesis. Logic became a sort of dialectic in the Aristotelian sense, for whom it denoted a kind of reasoning from opinions generally accepted, or, in their terminology, what is the same, probable premises.[93] Dialectic in this sense has as its purpose not the discovery of truth but the winning of debating matches, equipping its practitioners to attack and to defend both sides of any question proposed; and Aristotle's *Topics*, which deals with dialectic, is in fact a fascinating and subtle handbook for those who would like to become successful debaters. It was useful for lawyers and, where policy in civil society was decided in public assemblies, politicians – and for students aspiring to get ahead in an educational situation in which great emphasis was laid on success in disputations. For much of Europe there was as yet little need for either lawyers or orators. But there were students, and the skills they needed were those of dialectic. For the students and their educators, skill in the disputations sometimes in the course of time became an end in itself, something valued for its own sake, rather than for the extent to which it could help one contribute to locating truth among the sacred and valued authorities. Thinkers such as John of Salisbury could regret the attention paid in schools to purely formal exercises, at the expense of a study of the real logic that was supposed to open up to the human mind the truth of things, the necessary structure of reality. 'Dialectic ... is not great, if, as our contemporaries treat it, it remains forever engrossed in itself, walking 'round about and surveying itself, ransacking [over and over] its own depths and secrets: limiting itself to things that are of no use whatsoever in a domestic or military, commercial or religious, civil or ecclesiastical way, and that appropriate only in school.'[94]

In dialectic a central problem is that of locating premises to use in arguments in debate. This is the point of 'topics' or places: these places are places where premises may be located. Aristotle's *Topics* consists for the most part (six of the eight books) of strategies for finding arguments based on such logical topics as species or genus or any of the categories such as quality, relation, cause, and so on. The logic that was needed was a logic of consistency, and Aristotle restated the rules of the syllogism in this context, but now as a logic of consistency and not a logic of truth as in the *Posterior Analytics*. Included, too, were such argument forms as that of hypothetical syllogism, species of consistent argument forms that came down to the Middle Ages after elaboration by the Stoics. Topical rules for finding premises for these argument forms were provided to the Middle Ages by Boethius,[95] who proposed that the places were governed by certain maximal propositions which could be used to justify premises. Thus, for the topical place 'species' there is the maximal proposition 'of whatever the species is predicated so also the genus is predicated.' This proposition justifies various subsidiary propositions, called the 'topical differences.' Thus, the maximal proposition just stated justifies asserting the 'topical difference'

If Socrates is a man then Socrates is an animal.

This conditional in turn justifies the enthymematic inference

Socrates is a man; therefore Socrates is an animal.

This is a matter of hypothetical syllogisms; but Boethius also thinks in terms of categorical syllogisms.[96] Here he takes subject-predicate propositions of the *S–P* form, and uses the places to find arguments, that is, syllogisms, to prove that these are true. In other words, what he aims to locate in topical places are middle terms for the syllogisms. The 'topical differences' are then species under these various supreme genera.

The educators who took up this logic were often also philosophers. As such, they often found themselves in a peculiar position. On the one hand, they had their scholastic method, with its logic of consistency, for the purpose of studying and understanding Aristotle. On the other hand, that authority provided them with a logic of truth in the form of demonstrative syllogisms which, when their terms referred to the forms or natures of things, displayed the ontological structure of the universe. So there were two different logics with which they had to deal. This led to complicated, and often fruitful, interactions. Thus, for example, if one is concerned merely with consistency, then truth-functional connectives, whether conditionals or conjunctions, will not trouble one. Nor will inferences from false or impossible premises. It will not surprise one that inconsistent premises can validly imply any proposition whatsoever. But if one's concern is with a logic of truth, in which demonstration is the crucial notion, then one will be reluctant, as Chrysippus was, to accept merely truth-functional connectives. And, since good argument forms will include a component of non-formal necessity, one will be reluctant to accept the idea that any good argument could start from inconsistent premises. One will therefore be troubled by the discovery that inconsistent premises lead to any conclusion whatsoever. This will tend to move one away from a mere logic of consistency, a merely truth-functional logic, towards some stronger logic. This tension, this interest in truth-functional logic and also the pull away from it, can be found, for example, in Abelard.[97]

As for the logic of truth that the philosophers-educators found in Aristotle, they were in fact little interested in *using* that logic of truth. Theirs, to the contrary, was the method of authority. They did not need to apply a logic of truth to go beyond what they already knew. In particular, they had the authority of Aristotle for most issues in physics, and so on, supplemented where necessary (for example, on the question of the eternity of the world) by sacred texts. They no more thought of going about the process of exploring nature with this method than Pietro d'Abano thought that he might go outside the authority of Avicenna, Galen, and Hippocrates to explore empirically the development of diseases. On authority the medieval philosophers knew what the method of truth, the logic of truth, was, but they did not much have to use it.

In fact, there was, in due course, a gradual and mutual absorption of the doctrine of the syllogism, on the one hand, and the topical logic of Boethius, on the other. Syllogisms as useful arguments were thought to be dependent upon the topical logic. At the same time, however, since the syllogism provided a logic of truth, it also came to be held that the topically based inferences have to be necessary. So considered, topically based arguments came to be known as 'consequences.'[98] With Ockham and the terminist logicians who followed him in developing the doctrine of consequences, all topical arguments came to be treated as reducible to syllogisms.[99] Dialectic and syllogistic thus came in effect to be identified.

The humanists were subsequently to reject this merger of dialectic and syllogistic. They sought to retrieve the former from the latter.

For the schoolmen, however, as their educational methods developed, dialectic = syllogistic became the characteristic pattern of discourse in the disputational exercises central to those educational methods. Though these exercises were intended to prepare students to discover the truth by reconciling apparently contradictory authorities, it turned out that the dialectical method of the disputations enabled students to argue both sides of an issue, including the acceptability of various received authorities. This problem with dialectic had already been raised in the ancient world by Chrysippus, who said that the practice of arguing on opposing sides of an issue

is appropriate for men who advocate suspense of judgment with respect to all subjects ... but for the men who wish to produce knowledge in us, in accordance with which we shall live consistently, the contrary is appropriate, namely, to teach fundamental principles and to instruct beginners in the rudiments from start to finish. And in those cases in which it is opportune to mention the opposing arguments as well, it is fitting that they as in courts of justice destroy their plausibility.[100]

This makes the point that dialectics could provide a tool for the sceptical undermining of authorities. The new logic of Abelard, the new logic of consistency, appearing in the context of dialectic and the method of scholastics, could be, and was, attacked for calling into question accepted authorities. St Bernard of Clairvaux criticized the new logic on these grounds. He attacked the merely intellectual approach that was produced by the disputational method and the emphasis on logic: 'Virtues and vices are discussed with no trace of moral feelings, the sacraments of the Church with no evidence of faith, the mystery of the Holy Trinity with no spirit of humility or sobriety: all is presented in a distorted form, introduced in a way different from the one we learned and are used to.'[101] The dispassionate distance produced by the new logic of disputation was seen by Bernard, not without reason, as dangerous to faith itself: 'The faith of simple people is ridiculed, divine mysteries simplified, and the deepest matters become the subject of undignified wrangling. Human acumen presumes to everything and leaves nothing for faith. All that cannot be illuminated with the help of reason is regarded as trivial, something which is beneath one's dignity to believe in.'[102]

Bernard was not the only person ever to object in this way to these sorts of methods. The Jewish scholars of the late medieval period who elaborated the Talmud, using patterns already established in the Talmud, also faced the same sort of criticism. Their methods were called 'pilpul,' a term which denoted especially 'the use of subtle legal, conceptual, and casuistic differentiation.'[103] The Great Rabbi Loew of Prague objected that his contemporaries 'were more concerned with subtleties of their own judgment than with God's word, the Torah. Days are spent in useless matters, in finding pleasure in worthless complications. Would it not be better to become proficient in a regular handicraft where every refinement brings more positive result? Excessive subtlety on the other hand only leads to a confusion of truth with lies. This is what tears at my heart.'[104] Excelling in dialectics was what had become important, not sound knowledge of either the Torah or the Mishnah. The Great Rabbi stated that '[s]ince I began to think independently, my aim has been to find out what is clever and what foolish in the attitude of our generation to the Torah. I did not like their attitude.'[105]

Bernard's attitude was similar. As Gilson has paraphrased Bernard's words, Abelard was only one among a large group of dangerous teachers.

Some of them learn in order to know; others in order that it may be known that they know; and others again in order to sell their knowledge. To learn in order to know is scandalous curiosity ... – mere self indulgence of a mind that makes the play of its own activity its end. To learn for the sake of a reputation for learning is vanity. To learn for the sake of reputation for learning is cupidity, and, what is worse, simony, since it is to traffic in spiritual things ... The only proper thing to do is to make our choice between the sciences with a view to salvation ...[106]

The complaint of the saint is the complaint of the Great Rabbi. But the Great Rabbi was successful in his attack on logic chopping; the great defender of the monastic ideal was less so. Bernard did go on to succeed politically in getting some of Abelard's theological positions condemned at a provincial church council. But the scholastic method of using logic to reconcile conflicting authorities, and the method of disputational exercises, both made central by Abelard, were adopted, almost immediately, by everyone. The age was ready.

However, this same logic that was accepted in that age was in a later age to be attacked. In the early modern period, the logic of the scholastics was subjected to strong criticism – strong and successful: these later critics were to succeed where Bernard had failed. The new attack of the early modern period was directed in the first place towards the scholastic method, which was seen then, as it had been by Bernard and as it was later to be seen by the Great Rabbi Loew, as a method for winning disputes rather than as a method for discovering the truth. This is perhaps less than generous towards those who invented the method with the aim of reconciling apparently conflicting authorities. But in the early modern period, the method of authority itself was no longer acceptable – other methods of finding the truth were now in favour. The attack at this time was also directed at the method of syllogisms as a logic of truth. Syllogisms, it was now argued, were in fact not essential to laying out the ontological structure of the universe. They constituted no logic of truth but only a logic of consistency. The latter was in fact the logic of the disputational methods of the schoolmen. But if your concern is, as it was in the early modern period, with a new logic of truth, then you will not have much use for methods that do no more, at their best, than aim at reconciling authorities. Syllogisms, in short, will serve no good purpose.

These points are made by Descartes, who dismisses as being of little use in the search after truth 'that method of philosophizing which others have hitherto devised, nor those weapons of the schoolmen, probable syllogisms, which are just made for controversies. For these exercise the minds of the young, stimulating them with a certain rivalry; and it is much better that their minds should be informed with opinions of that sort – even though they are evidently uncertain, being controversial among the learned – than that they should be left entirely to their own devices.'[107] Furthermore, 'those chains with which dialecticians suppose they regulate human reason seem to me to be of little use here, though I do not deny that they are very useful for other purposes.'[108]

Locke makes much the same set of points. Syllogisms are useful for 'fencing,' as Locke puts it, but are not useful in the discovery of truth. They are a mere logic of consistency.

The rules of syllogism serve not to furnish the mind with those intermediate ideas that may show the connexion of remote ones. This way of reasoning discovers no new proofs, but is the art of marshalling and ranging the old ones we have already. The forty-seventh proposition of the first book of Euclid is very true; but the discovery of it, I think, not owing to any rules of common logic. A man knows first, and then he is able to prove syllogistically. So that syllogism comes after knowledge, and then a man has little or no need of it. But it is chiefly by the finding out those ideas that show the connexion of distant ones, that our stock of knowledge is increased, and that useful arts and sciences are advanced. Syllogism at best is but the art of fencing with the little knowledge we have, without making any addition to it. And if a man should employ his reason all this way, he will not do much otherwise than he, who having got some iron out of the bowels of the earth, should have it beaten up all into swords, and put it into his servants hands to fence with, and bang one another ... And I am apt to think, that he who shall employ all the force of his reason only in brandishing of syllogisms, will discover very little of that mass of knowledge, which lies yet concealed in the secret recesses of nature; and which, I am apt to think, native rustic reason (as it formerly has done) is likelier to open a way to, and add to the common stock of mankind, rather than any scholastic proceeding by the strict rules of mode and figure.[109]

These comments are just, even though they are no doubt rather annoying to those who defend formal logic. It is indeed the case that in formal logic there is a subject matter worth exploring. Aristotle did start that exploration. But Aristotle did more: he also proposed that syllogistic constituted a logic of truth as well as a logic of consistency. To this claim, the early moderns, both rationalists and empiricists are objecting. So to object is not to disparage the study for its own sake of the principles of a logic of consistency. In addition, for a variety of reasons, the schoolmen had given to syllogistic and, more generally, a logic of consistency, a primacy in both their method and their educational practice. This, too, was criticized by the early moderns: schoolwork should involve more than knowing how to 'fence.' To object to these sorts of things was perfectly reasonable, and those who like to do formal logic for its own sake should not take offence. Nor should they suppose that, because the remarks of Descartes and Locke do not touch the formal logic of consistency as something to be studied in its own right, therefore these thinkers have nothing reasonable to say about logic.[110] To the contrary, as I have been suggesting, their criticisms of logic as taught and used in the schools are very reasonable. Nor, finally, it should be noted, should it be thought that to reject the scholastic practice of disputation, as Descartes and Locke did, is to reject the idea of criticism.[111] There are more ways to conduct critical discourse than by the often stifling methods of the schoolmen.

The scholastic logic had earlier been attacked by the humanists of the Renaissance. The humanists had very different aims from the scholastics, and a very different method. The method that they invented was that of textual criticism. This was deployed both positively and negatively. Negatively, it was used to destroy the claims to authenticity of certain ancient texts. Thus, Lorenzo Valla, who was 'one of the first to investigate the authenticity of available sources by sound philosophical methods,'[112] was able to use the method to show that the so-called 'Donation of Constantine' of Rome to the Pope was in fact a late forgery. Again, where the earlier Renaissance figures had valued the Hermetic Corpus as a set of philosophical documents authoritative because they supposedly derived from ancient Egypt at the time of Moses, Isaac Casaubon was able to use the methods of textual criticism to show that they also were late forgeries. Positively, it was used to restore to a state as close to the original as possible various classical texts that had become corrupt through the ages. Humanists worked on establishing the text of Plato. Legal humanists worked to discover the earliest pre-Justinian Roman legal codes. Calvin produced a humanist's edition of Seneca. Erasmus tried his hand at the Bible. The challenge to St Jerome was too much for many. Be that as it may, the methods of textual criticism invented by the humanists were appropriate for the end for which they were designed, to wit, the recovery of the correct ancient text, or, at least, the text that is as close as is humanly possible to the ancient original.

The aim, however, was not merely the recovery of the ancient text. That text itself was valued because, as the version nearest to its source, it spoke most nearly with the voice of the ancients. The humanists, in other words, took the fact that a text derived from the ancient world as implying that the text is *true*: the ancient texts are valuable because they are *authoritative*. The method of textual criticism was combined with the method of authority for the humanists. They thus differed from the scholastics in the method that they used for discovering the authoritative statement; but they agreed with the scholastics that the appropriate method generally for attaining the truth is the method of authority. The disagreement was merely about *which* authority. As Sem Dresden has put it, for the humanists of the Renaissance, '[t]here is ... no question of free criticism having taken the place of authority. What was changing was the nature of authority; the previous age was now less important than the distant past, and according to the humanists there was or could be harmony and even similarity between contemporary and age-old views.'[113] For a humanist such as Lorenzo Valla, the standard for discourse was the Latin of the best authors of the classical age, not only for style but for the truth. Cicero, in particular, provided a standard for both style and truth, replacing in the latter case for many humanists the previously dominant Aristotle.

The reverence for ancient sources as the fonts of truth was ridiculed by both the rationalists and the empiricists in the early modern period. Thus, Locke held that 'there being no writings we have any great concernment to be very solicitous about the meaning of, but those that contain either truths we are required to believe, or laws we are to obey, and draw inconveniencies on us when we mistake or transgress, we may be less anxious about the sense of other authors; who writing but their own opinions, we are under no greater necessity to know them, than they to know ours.'[114] This echoes Descartes's view in the *Rules for the Direction of the Mind* that

We ought to read the writings of the ancients, for it is of great advantage to be able to make use of the labours of so many men. We should do so both in order to learn what truths have already been discovered and also to be informed about the points which remain to be worked out in the various disciplines. But at the same time there is a considerable danger that if we study these works too closely traces of their errors will infect us and cling to us against our will and despite our precautions. For, once writers have credulously and heedlessly taken up a position on some controversial question, they are generally inclined to employ the most subtle arguments in an attempt to get us to adopt their point of view. On the other hand, whenever they have the luck to discover something certain and evident, they always present it wrapped up in various obscurities, either because they fear that the simplicity of their argument may depreciate the importance of their finding, or because they begrudge us the plain truth.[115]

The method of textual criticism may be an efficient and effective method in establishing, as closely as possible, given the data, the earliest version of a text. But this method was practised among the humanists not merely for its own sake, but in order to locate the truth which they assumed was authoritatively present in the ancient texts. This ulterior purpose is what was attacked in the early modern period. These texts may indeed contain the truth about many things, but it is only because their authors have diligently used the logic of truth to discover that truth. But at the same time, they may have failed also to discover the truth, so that their texts contain error and falsehood. It is our task as searchers after truth ourselves to ourselves investigate the same concerns as the ancient texts, to separate in those texts what is true and what is false. The only thing that counts is the logic of truth, not the mere fact that a text has been handed down to us from Greece and Rome or, like the Bible, from the beginnings of our religious traditions.

   The practical concerns of the humanists were with the needs of the orator, whether a politician speaking in a civic assembly, or a lawyer speaking before a judge or perhaps a jury or a public speaker declaiming the virtues of some civic leader or tyrant. Humanism is rooted first of all in civic humanism.[116] This corresponded to a changing focus in the arts curriculum. Where formerly it had been a technical preparation in the linguistic tools needed for the practice of philosophy and theology, it became instead a general introduction to Greek and Latin language and literature for gentlemen and for those destined for professional careers.[117] Relative to these needs, the practice of Cicero was central. There was, in the first place, his style, which provided a standard of eloquence.

There was, in the second place, the tools that Cicero provided the orator. In particular, there was Cicero's logic. But for Valla in particular, there was, in the third place, the Academic scepticism of which Cicero was the defender.

The logic of the schoolmen, as we have seen, had two roots. One was the demonstrative syllogistic of Aristotle. The other was the topical logic of Aristotle's *Topics* and of Boethius, the latter deriving in part from Cicero. This topical logic was used by the schoolmen in their disputational exercises, but also professionally in the processes of reconciling conflicting authorities. In this logic, there was little need for *inventio* and none for probabilistic arguments. But these, to the contrary, were precisely what was needed by the orator.

The orator, of necessity, deals in matters in which there are dubitable things, trying to find which among alternatives is the most likely true. As Cicero points out in the *De oratore*, controversy in the law or politics requires the skills of the orator only where material is doubtful: 'those cases which are such that the law involved in them is beyond dispute, do not as a rule come to a hearing at all.'[118] If it is not doubtful, then the legal specialist can settle the matter, or the political arbitrator. There is no need to bring in persons – orators – to debate both sides of the issue to allow a jury or assembly to decide on which side the greatest probability lies. But, where there is doubt, persons are needed who can skilfully argue either, and therefore both, sides of the case (*in utramque partem*). '[S]o the orator may safely disregard all this region of unquestionable law, being as it certainly is by far the larger portion of the science; while, as for the law which is unsettled in the most learned circles, it is easy enough for him to find some authority in favour of whichever side he is supporting, and, having obtained a supply of thonged shafts from him, he himself will hurl these with all the might of the orator's arm.'[119]

But being able to argue both sides of a case is precisely the technique developed by the Academic philosophers in order to discover which side has the greatest probability. As Cicero puts it in the *Lucullus*:

The sole object of discussions [by Academic philosophers] is by arguing on both sides [*in utramque partem dicendo*] to draw out and give shape to some result that may be either true or the nearest possible approximation to the truth. Nor is there any difference between ourselves and those who think they have positive knowledge except that they have no doubt that their tenets are true, whereas we hold doctrines as probable which we can easily act upon but can scarcely advance as certain.[120]

We have seen that the Stoic Chrysippus criticized the position of those who argued both sides of the question as inherently sceptical, and antithetical to the truth. Chrysippus's point is well taken if indeed we can grasp with infallible certainty the necessary truths about things, the essential ontological structure of the universe. But if we restrict ourselves to the realm of opinion, to the world of ordinary experience, the world that we know by means of our senses, the world where there is no structure that can be infallibly grasped, then the Academic practice of arguing both sides to find where the balance of probabilities lies does make for a reasonable practice. The present point is that both the Academic sceptic and the orator deal with things that are dubitable. The Academic deals with what is most probable, most 'like the truth' ('verisimilitude'), because there are no deep necessities there to be known infallibly. The orator deals with individual cases where all the facts are not known, and all the law not settled. The similarity of the matter makes for similarity of method. The method of the orator is thus of a piece with that of the Academic sceptic.[121] As Cicero says, they share a common purpose: 'there is a close alliance between the orator and the philosophical system of which I am a follower [= Academic scepticism].'[122] For Valla this means that the orator is, in fact, for the most part closer to the requirements of philosophy than most philosophers, that is, the majority of philosophers who take for granted that they have infallible access to necessary truth: the orator is closer to 'sophos' than the 'philosophos.' 'Oratorem esse verum sapientum quantum in hominem cadit, hoc est plus quam philosophum, id est "sophum."'[123]

Since the Stoics were concerned with necessary premises infallibly known, they did not have a concern for the needs of the orator. In particular, Cicero pointed out, they did not deal with the problem of the topics, of *inventio*.

[I]n Deductive reasoning, they [the Stoic logicians] start with what they term self-evident propositions; from these the argument proceeds by rule; and finally the conclusion gives the inference valid in the particular case [true in the particular case: 'quid verum sit']. Again, how many different forms of Deduction they distinguish, and how widely these differ from sophistical syllogisms! ... Of the two sciences which between them cover the whole field of reasoning and of oratory, one the Science of Topics [*inventio*] and the other that of Logic [*iudicium*, rules of inference], the latter has been handled by both Stoics and Peripatetics, but the former, though excellently taught by the Peripatetics, has not been touched by the Stoics at all.[124]

In fact, according to Cicero, for the orator it is the science of topics which is most useful, and prior in the order of nature.[125] Cicero was to devote his *Topica* to this science ignored by the Stoics. The schoolmen did not ignore the topics, but the tendency was to assimilate these into the doctrine of consequences, making it part of the formal logic of consistency alongside syllogistic. However, if one is dealing with what is the most probable, it is useful to the orator to have available many argument forms that are not formally valid. These would include such forms as induction, example, enthymeme, and sorites. While these were often mentioned by the scholastic logicians in their texts, they were clearly of secondary interest. The interest of the schoolmen in consistency led to the enrichment of formal logic. But the same interest led to the impoverishment of the non-formally valid forms of inference that would be useful to the orator. The medievals did have an interest in disputational dialectic, but this dialectic was practised in the context of the scholastic method for reconciling conflicting authorities. This, too, contributed to the schoolmen ignoring the needs of the orator. Both these reasons for ignoring the needs of the orator derived from the same source, the humanists held, that prevented the Stoics from developing dialectic: the strain of dogmatism in their philosophy. A new logic could emerge only after the Academic philosophy of Cicero had removed the strain of dogmatism that tainted previous logics. This is why the Academic scepticism of Cicero was important to Valla: it provided the arugment that justified philosophically the new logic for orators.

If it was the philosophy of the Academic sceptic that cleared the way in Valla's thinking for the new logic, it is also true that it was directly in response to the needs of the orator that Valla undertook to publish his *Dialecticae disputatione* (1439). This work placed the topics and the problem of invention as central, and as equally central the non-formally valid argument forms that had been ignored or pushed to the side by the schoolmen. This humanistic logic was made possible by the Academic philosophy. At least, so it seemed to Valla. Unfortunately, this sort of philosophy, in this and his other works, led to a charge of heresy, and several rewritings were required.[126] Valla was, in fact, not part of the central philosophical movements of the age, which remained strongly Aristotelian and to a lesser extent Platonic.[127] If his reform of logic was to gain adherents it had to become acceptable to those who were in the mainstream, still committed to Aristotle.

Rudolph Agricola provided what was needed. Agricola's *De inventione dialectica* (*ca* 1479) was a logic that followed Valla's in making *inventio* central to logic. However, where Valla's book had been a propagandistic argument, aimed at persuading intellectual leaders about the reform of education, Agricola's was straightforwardly a textbook aimed at implementing for students the program that Valla was advocating.[128] The philosophical justification provided in Valla's book by Cicero's Academic scepticism is quietly set aside to be replaced by a pragmatic justification in terms of the needs of the new interests of students and the curriculum that was to serve those interests. The new logic could thus be accepted by those who remained in the still strong mainstream of Aristotelian philosophy. Agricola's text in turn was succeeded by the even more popular *Dialectica* of Ramus (French, 1555; Latin, 1556). In all these, a priority is given to invention (*inventio*) over inference (*iudicium*), and within inference syllogistic is downgraded relative to other forms of deductive inference (such as hypothetical syllogism) and probabilistic ratiocinative strategies.[129]

The doctrine of topics provides a way for securing premises that constitute the starting points of inferences, including the inferences of the formal logic of consistency, where the latter includes not only syllogistic but also such forms as hypothetical syllogism, disjunctive syllogism, simplification (conjunction), and so on. From the viewpoint of those interested in the new science, however, the doctrine of topics, or *inventio*, is not an appropriate starting point for inference. What is needed is a criterion of truth *prior* to the start of argument. The Ciceronian-humanist tactic of arguing both sides may well yield an evaluation of truth-likeness or verisimilitude in terms of probabilities, but can do this successfully only if its premises are truth-like, only if *their* probabilities are taken as given. The art of *inventio* does not enable one to do that. A logic of truth is still needed. So the rationalist and empiricist criticisms of scholastic dialectic apply equally well to the newer humanist treatments of dialectic.

However, if the rationalists and the empiricists were agreed that the logic of consistency required a logic of truth, and if they agreed that syllogistic could not provide that logic of truth, they strongly disagreed on the nature of that logic of truth. And this in turn shaped their very different responses to the traditional syllogistic. The traditional syllogistic, still defended by thinkers such as Pegis and Maritain, takes for granted that part of discourse is determined by non-formal necessities. The rationalists, while rejecting the notion that these necessities can be fitted into syllogistic form – rejecting, that is, the notion that the syllogism gives the ontological form of the world – do not reject the notion of non-formal necessities. In contrast, the empiricists do reject that notion. For the two, then, the inferential structures of discourse are very different. Their criticisms of traditional syllogistic, therefore, are also very different.

Recall Plato's brilliant metaphor of the line. Plato divides it into four parts; but for our purposes the major division between the upper portion and the lower portion is what is relevant. The upper portion represents knowledge, the lower portion opinion. Knowledge has as its object the necessary structure of being. Opinion has as its object the changing patterns of sense experience. Knowledge is infallible. Opinion is fallible. Knowledge is infallible because, on the side of the known object, it is knowledge of an unchanging unity, while, on the side of the knowing subject, there is a literal identity of the knower with the known. Opinion is fallible because the object known might change: its parts are all separable.

Philosophers apparently as diverse as Aristotle, Chrysippus, and Descartes all hold that there is knowledge in this sense – *scientia*, to use the traditional term – demonstrative science. Often we must for practical purposes rely upon opinion, as Chrysippus and Descartes both allowed. Logic, as a logic of consistency, holds at both levels. But at the level of knowledge the argument forms take as their premises certain non-formal necessities. These necessities reflect the essential ontological structures that account for the way the world of sense, the world of opinion, unfolds. These philosophers may well disagree, as Aristotle, Chrysippus, and Descartes did, about the nature of the non-formal necessities, the essential structures of the universe; but they all held that there was more to the necessary structure of the discourses of science in the sense of *scientia* than was provided by the logic of consistency.

These philosophers are to be contrasted to the empiricists, who were to subject to criticism the whole notion of *scientia*, knowledge in the sense of Plato's line, knowledge that is an infallible grasp of essential unities. The very idea of such essences, such objective necessary connections, was criticized as *meaningless*. Knowledge in the sense of *scientia* cannot occur: there is no object for it to be *of*. Thus, for the empiricists, knowledge can be nothing more than opinion. To be sure, we may want to distinguish, as no doubt we should, between those cases of opinion which are solidly founded and those cases of opinion which are not. The former we can reasonably refer to as 'knowledge,' the latter as 'opinion,' that is, *mere* opinion. Knowledge thus becomes, to use the fashionable phrase, *justified true belief*.

Of course, we cannot declare straight off that knowledge simply is justified true belief, as some recent philosophers have done.[130] That is simply to beg the question against the Platonists, the Aristotelians, the Stoics, and the rationalists. The earlier empiricists such as Locke and Hume did not in this way beg the question. They provided *arguments* to the effect that the discourse of their opponents about objective necessary connections, about essential truths, and so on, is empty, factually meaningless. Given these arguments, logic in the traditional sense, whether the syllogistic logic of Aristotle or the propositional logic of the Stoics, has no role to play other than that of a logic of consistency. Logic simply is truth-functional logic.

The argument, then, for this understanding of logic, is part of the case for empiricism. But these criticisms of traditional syllogistic and of rationalist logic are part of a broader case that Locke makes against much of the traditional way of doing and teaching philosophy. I refer, of course, to the criticisms of logic understood as dialectic and logic as used in the method of disputation. We have looked at the battery of criticisms that were directed at the traditional practices.

We must now turn to the criticisms that more specifically led to the conclusion that logic of the traditional sort can be nothing more than a logic of consistency. This is the argument – against the idea of Aristotle, of Chrysippus, of Descartes, of Maritain, and of Pegis – that there is something of the sort that was traditionally referred to as *scientia*.

But it is always useful, when we try to understand a position, to contrast it to others. What we shall do, then, in the next part of this study is to compare rationalist and empiricist critiques of syllogistic. This will involve looking at Descartes and Arnauld on the one hand, and at Locke on the other. As we shall see, only the latter reduces logic to a mere logic of consistency. To be clear on the logic that is being criticized by both we shall look at one of the more pellucid presentations in the seventeenth century of the syllogistic of the *Posterior Analytics*, that is, syllogistic not merely as dialectic but as the Aristotelian logic of truth. This is the logic of John Sergeant. John Locke not only criticized Aristotelian logics such as that of Sergeant; he also prepared written responses to Sergeant's criticisms of his (Locke's) position. In these responses, his critique of the doctrine of objective necessary connections is quite explicit. This argument, which restates the case made in the *Essay*, constitutes his attack on objective, non-formal necessities, and, at the same time, his case that ordinary logic is a mere logic of consistency. With this, Locke has made a place for a new logic of truth for empirical knowledge of matter-of-fact regularities, the logic of the experimental research of the new science.

III: Rationalist and Empiricist Critiques of Syllogistic[131]

The Art of Thinking[132] distinguishes between the *comprehension* of an idea and its *extension*. Pariente has argued that it is the former that is central to the logical theories of the Port Royal Logic,[133] while Buroker has argued that it is the latter.[134] In fact, the Port Royal Logic emphasizes both, so, on the face of it, neither should be seen as more important than the other. The distinction is important in traditional logic; we have seen it present in the logic of Maritain. What I shall argue is that, in fact, both are central to the logical doctrines of the Port Royalists, and, indeed, central to most of the standard doctrines of logic prior to Locke. It is central, too, as we shall see, to the very different critiques of syllogistic given by the empiricists on the one hand, and the rationalists – Descartes in particular – on the other.

1) Syllogistic

We are concerned with propositions in which there are two terms, a subject and a predicate. These terms are joined in a proposition by a copula. The result may be either affirmative or negative. If it is affirmative, then the two terms are joined; if it is negative, then the two terms are separated. Now, according to the Port Royalists, to join is to unite or *to identify*. Thus, we are told that

[i]t is certain that we can express a proposition to others only if we use two ideas, one for the subject and the other for the attribute, and another word which indicates the connection the mind conceives between them

This connection can best be expressed only by the same words we use for affirming, when we say that one thing is another thing.

From this it is clear that the nature of affirmation is to unite and identify, so to speak, the subject with the attribute, since this is what is signified by the word *is*. (*La Logique*, Pt II, ch. 17)[135]

We are also told that 'it is impossible for one thing to be joined and united to another without the other thing also being joined to it, and it is evident that if A is joined to B, B also is joined to A. Hence it is clearly impossible for two things to be conceived *as identified*, which is the most perfect of all unions, if this union is not reciprocal, that is, if we cannot make a mutual affirmation of these two united terms, in the way which they are united ...' (Pt II, ch. 18; italics added).[136] Hence, in the affirmative proposition that

S is P

we *identify* the subject and the predicate. However, we cannot have

$$S = P$$

because in general this will not be true for either the universal affirmative or 'A' proposition or the particular affirmative or 'I' proposition. More is needed, therefore, if we are to make sense of the claim that in affirmative judgments we identify the subject and the predicate.

In point of fact, this identification takes two forms, depending upon whether one is considering the *comprehension* of the idea or its *extension*. We have seen already, in fact, that the traditional doctrine of formal logic includes, on the one side, this distinction between considering propositions in terms of the comprehensions and their extensions. And we have seen, also, that it includes, on the other side, the notion that all such propositions can be given a sense in terms of the relation of identity. It will pay, however, to look specifically at the ways in which the Port Royalists treated these notions.

Let us consider comprehension first. Comprehension is explained by Arnauld and Nicole in the following fairly traditional way: 'The comprehension of an idea is the constituent parts which make up the idea, none of which can be removed without destroying the idea. For example, the idea of a triangle is made up of the ideas of extension, figure, three sides, three angles, and the equality of these three angles to two right angles, etc.' (*La Logique*, Pt I, ch. 6).[137] Notice that the genera or determinables extension and figure are parts of the determinate or specific idea of triangle. Or, in the terminology of the Cartesians, the *attribute* extension is contained in the *mode* triangle. In general, then, the genus (determinable, attribute) is contained in the species (determinate, mode).[138] Later on, the authors explain what this means in the context of judgments. Specifically, everything that is part of the comprehension of idea expressed by the predicate is contained as a part within the comprehension of the idea expressed by the subject.

[W]henever an idea is affirmed its entire comprehension is affirmed; for if one removes one of its essential attributes, it is no longer the same idea, rather, one destroys the idea itself, and entirely eliminates it; consequently, when an idea is affirmed, one affirms every idea which is included in its comprehension. For example, when I say 'A rectangle is a parallelogram,' I affirm of the rectangle everything that is contained in the idea of parallelogram; for, if there were some part of the idea of parallelogram which failed to be part of the idea of rectangle, then it would follow that the entire former idea, and not just a part of it, would not agree with the latter; and thus the word parallelogram, which signifies the idea as a whole, would have to be denied, rather than affirmed of, the rectangle. One sees that this principle of affirmation is the basis of all affirmative arguments. (Pt II, ch. 17)[139]

In general, there will be parts of the idea expressed by the subject term which are *not* part of the comprehension of the idea expressed by the predicate term. Thus, the genus of parallelogram requires a difference to be added to it, namely, right-angled, in order to obtain the species which is the subject of the judgment that 'a rectangle is a parallelogram.' There are more parts to the subject idea than there are in the predicate idea. Indeed, given that the predicate idea is a genus and the subject idea is a species, this is what we should expect.

Now consider the judgment that

All S are P.

We cannot have

$S = P$

but we can have

$$S = dP$$

where the d indicates a difference that is, as it were, added to the genus P in order to obtain the species S.

Now, as Pariente points out,[140] there is a similarity between this and the version of symbolic logic that was developed by George Boole.[141] We have seen above, in fact, that this way of putting things can be stated clearly within the formalism of ordinary modern formal logic. This reading of the categorical propositions corresponds to the abstract reading that we considered above, where we contrasted it to what we called the concrete reading, that is, the reading of the propositions as taken in extension. But in the present context, as Pariente has seen, Boole's notation has certain advantages over our more recent formalizations.

Boole let y stand for a class concept, and let v be 'a class indefinite in every respect but this, viz., that some of its members are' x (*Laws of Thought*, 67). Thus, Boole represents (67, 241) the A proposition

All y are x

by

$$y = vx.$$

Similarly, Boole represents (241) the I proposition

Some x are y

by

$$vx = vy.$$

The connection is clear in the case of these affirmative propositions. However, in the case of the negative propositions, the connection is not so clear. The E proposition

No x is y

and the O proposition

Some x is not y

are represented by Boole (*Laws*, 241) by

$$y = v(1 - x)$$

and

$$vy = v(1 - x),$$

respectively. There is nothing formally wrong with Boole's notation here. It is not especially helpful in understanding the Port Royalists, however. For, it does not connect up with the notion in the Port Royal Logic that negative propositions do *not* involve identity; rather, the negative copula represents *separation*.

We can do better if we adapt the notion just a bit. If vx represents *some* part of x, then we might equally let v^*x represent *all* parts of x. This would permit us to represent the universal negative or E proposition by

$$y \neq v^*x$$

and the particular negative or O proposition by

$$vy \neq v^* x.$$

But even this is not the best for our purposes. Recall that upon the traditional doctrine, the *species* is defined by the *genus* and the *specific difference*. For, example,

triangle is a figure bounded by three straight lines

or,

$$S = DP.$$

If we take the subject and predicate of the A proposition related as species and genus, then what it is asserting is that *there is a specific difference of the genus such that the species is identical to the genus so differenced*. That is, what the A proposition

All S is P

represents is, using more recent notation,

$$(\exists d)[\ S = dP]$$

This represents the traditional ideas somewhat more perspicuously than does the Boolean notation. Similarly, the E proposition

No S is P

will be represented by

$$(d)[\ S \neq dP]$$

while for the I and O propositions we obtain

Some S is $P \equiv (\exists d, d')[dS = d'P]$

Some S is not $P \equiv (\exists d, d')[ds \neq d'P]$

These representations are, as we suggested, those that deal with ideas in respect of their *comprehension*. What these propositions consider are the ideas themselves, and not the objects which the ideas are about, that is, their extensions. More specifically, they deal with *relations among ideas*, those *relations that involve species and genera*.

This account of syllogistic as involving relations among ideas or forms or essences of things is one that had wide acceptance. We might consider, for example, the exposition of logic given by Locke's Aristotelian critic, John Sergeant, in his defence of Aristotelian science, *Method to Science*.[142] Sergeant adopts the same position as the Port Royal logicians with regard to the copula, treating it as representing in judgment an *identification* of two ideas or forms or essences, or, to use Sergeant's phrase, 'notions.'

[t]he meaning of the word *is* which is the Copula, is this, that those Words are Fundamentally Connected in the same Thing and Identify'd with it Materially; however those Notions themselves be Formally different, provided they be not Incompossible ... As when we say *a Stone is Hard* the Truth of that Proposition consists in this, that the Nature of *hard* is found in that Thing or Suppositum call'd a *Stone*, and is in part Identify'd with it; however the Notions of *Stone* and *Hard* be Formally Distinct. Or, (which is the same) it is as much as to say, that that Thing which is Stone is the same thing that is Hard. (*Method*, 119)

Judgment does consist in bringing together notions in a relation in which they are compared, either to be identified or to be separated. In either case, however, the connection or lack thereof is an objective matter, a feature of the connections among the notions or essences themselves, and not just a subjective bringing together of the ideas.

There being ... a Real Relation between those Notions which are the Subject and Predicate, the latter being really in the understanding and That which is said of the Former, and the Former that of which 'tis said; and Relation being necessarily compleated and actually such, but the Act of a Comparing Power; it follows, that every Judgment is a Referring or Comparing one of those Notions to the other, and (by means of the Copula) of both of them to the same Stock of Being on which they are engrafted, or the same Ends; where they are Entitatively Connected (or the same Materially) before they are Seen or Judg'd to be so by our understanding. (121)

In the A proposition the subject is identified with the predicate, but, as with the Port Royalists, only with the predicate insofar as a suitable difference has been added to it: 'the Predicate has of it self a Large sense, taken alone and Abstractly; yet, when attributed to the Subject, it is restrain'd by ... only such a proportional part of its Notion as befits the Subject to receive' (127). We thus see that the Port Royal logicians and Sergeant shared much the same view of the nature of judgment.

It is judgments thus understood that enter into syllogisms. Syllogisms are necessary when propositions are not self-evident. Arnauld and Nicole put it this way: 'when consideration of two ideas alone does not suffice to enable one to judge whether to affirm or deny the one or the other, it is necessary to have recourse to a third idea ... which we call the *mean*' (*La Logique*, Pt I, ch. 1).[143] This third term, the middle, is compared in the syllogism with the two extremes, the minor term or subject of the conclusion, and the major term, or predicate of the conclusion.[144] Sergeant put the same point this way: 'a Syllogism can have but Three Terms in it; Two of which are given us in the Proposition to be Concluded; and the Third is that Middle Term, by finding which to be Identify'd with the other Two in the Premisses, we come to be assur'd, by virtue of the self-evident Proposition hinted above, that they are Identify'd in the Conclusion; or, which is the same, that the conclusion is True' (*Method*, 227). What makes a syllogism valid is the orderly progression of the notions from the more generic to the more specific: 'the Middle Term must be inferior in Notion to one of those Terms, and superior to the other. For, since ... Notions do arise orderly from the Inferiour to the Superiour ones; it follows, that that Notion is in the Middle between the other two which is Inferiour to one of those Notions and

Superiour to the other ... Wherefore the middle Term must, in the two Propositions which are the Premisses, be the Subject to one of the Terms, and the Predicate to the other' (ibid., 231).

2) *Demonstrative Syllogisms*

In order to discover the objections of the Cartesians to traditional syllogistic, we
shall have to look at the traditional doctrine of demonstration. This is perspicuously laid out by Sergeant.

As the Port Royalists indicate, a genuine demonstration requires two things,
'first, that the content be certain and indubitable; and second, that there be no
defect in the form of the argument' (*La Logique*, Pt IV, ch. 8).[145] Sergeant, too,
holds that first premises must be self-evident and indubitable. For, there is no
science without such self-evidence. Since propositions connect, or separate,
terms or ideas or notions, it follows that the self-evidence that is sought is the
self-evidence of a connection among ideas, or the self-evidence of an opposition among ideas: ''tis Connexion of Terms which I onely esteem as Proper to
advance Science. Where I find not such Connexion, and the Discourse
grounded on Self-evident Principles, or (which is the same) on the metaphysical Verity of the Subject, which engages the Nature of the Thing, I neither
expect Science can be gain'd, nor Method to Science Establish'd.'[146] Since
propositions are statements of identity, the self-evidence that accrues to them
derives from their form as statements of identity. 'The Self-Evidence belonging to First Principles consists in this, that the two Terms must be Formally
Identical. For, since ... the Terms in every Ordinary and Inferior Proposition,
nay, in every conclusion that is True, must be materially the same, and so the
Proposition it self materially Identical, it follows, that the Terms of the First
Principles, which ought to be more evident than They, as being Self-evident,
must be Formally Identical' (*Method*, 131–2). Science – that is, demonstrative
syllogism – thus consists of making explicit relations of identity between
notions. These notions or forms or essences are before the mind. In order to
see that a proposition formed from two of these notions is true, we must come
to recognize an identity between these notions. If the identity is not clearly
before the mind, then we must make it so by means of a syllogism. This syllogism will begin from premises which are formally and clearly identical. These
identities will make explicit the identity of the terms in the conclusion, an
identity which heretofore was only implicit, actually *there* in the notions or
essences, but unrecognized by us, that is, unrecognized by us until the demonstration made it clear by showing how the two extremes were each identical to
the middle.

Sergeant argues that to get on with the task of achieving knowledge that is scientific in the sense of *scientia*, that is, demonstrative knowledge, we must arrange our notions or essences in their true places in the species-genus hierarchy. '[I]t is one necessary and main Part of the Method to Science, to distinguish our Notions Clearly, and to keep them distinct Carefully,' and '[t]he best way to do this, is to rank all our Notions under distinct common Heads' (*Method*, 14). This injunction should not surprise one, given the position on demonstrative syllogisms. For, after all, the structure of the syllogism is essentially determined by the species-genus hierarchy. For Sergeant this implies that there is one supreme genus, that of existence or being: 'There is but one onely Notion that is perfectly Absolute, viz. that of Existence, and all the rest are in some manner or other, Respective ...' (15). The notions that fall under this supreme genus are formed by dividing the genus by the relevant specific differences. These differences include as a part the genus, as the Port Royalists also insisted. As Sergeant puts it, '[t]he Differences that divide each Common Head must be Intrinsical to it,' and these 'Intrinsical Differences can be no other but more and less of the Common Notion ...' (25). Thus, the specific difference entails the genus. The species will be identical with the specific difference, and the difference will entail the genus.

On this view, we begin with the proposition

$$S \text{ is } P$$

or, as we saw,

$$(\exists d)[\, S = dP\,].$$

This proposition does not express knowledge, let us suppose. That means that we do not see in terms of the notions S and P themselves why the identity holds. What this means, clearly, is that we have not located the difference that must *be there* if the proposition is to be true. What the syllogism

$$M \text{ is } P$$
$$S \text{ is } M$$
$$\overline{}$$
$$\therefore S \text{ is } P$$

does, provided that it is demonstrative, is supply the difference M that the conclusion asserts to be there, and which must be there if that conclusion is to be true. The syllogism expands the real definition

$$S = MP$$

where M is the specific difference. This thus entails, and makes evident, the proposition

$$S \text{ is } P = (\exists d)[\, S = dP\,]$$

about whose truth we were concerned. The entailment is, given our modern notation, simply that of existential generalization.

The specific difference M that forms the middle term of the syllogism will itself entail the genus P that it divides; that genus is part of the concept or notion which is the specific difference. The minor premise

$$S \text{ is } M$$

will thus express the identity

$$S = M$$

which will itself be the real definition

$$S = MP$$

with the genus located only implicitly in the notion M. The major premise

$$M \text{ is } P$$

will express the identity

$$M = MP$$

making explicit the genus that is located in, and entailed by, M. These identities will be, if the syllogism is genuinely demonstrative, all self-evident.

As Sergeant clearly recognizes, this depends upon the species-genus hierarchy, in which the lower species contain, and thereby entail, all the genera under which they fall. One obtains the ontological structure of reality and the logical basis for demonstrative syllogisms from this structure. The structure itself can be derived in the form of a Porphyryan tree by subdividing the supreme genus.

The various lower genera and species can all be defined in the usual genus-specific difference way. The exception is the highest genus which admits of no definition: 'the Notion of *is*, or Actual Being, is impossible to admit any Explication ...' (*Method*, 120). It is this notion which is the objective basis for truth in every true judgment: 'The Notion of *is* is the Determinate of its own Nature, and so most Fixt of it's self; and, therefore, most proper to fix the Judgment' (120).[147] But the notion appears in judgment not only as the copula but also in both the subject and the predicate which are species falling under, and therefore entailing, that genus: 'not only the Notion of the Copula, but of the Subject and Predicate too, is Existence' (134). It follows that '[t]his Proposition *Self-Existence is Self-existence* [or, equivalently, according to Sergeant, *what is, is* or *existence is existence*] is, of it self, most Supremely Self-Evident ...' (133). All other propositions in a body of knowledge that is scientific, that is, demonstrative, are identities that relate concepts that fall under the supreme genus of existence or (actual) being.

But, of course, the notions also yield the causal structure of reality. Thus,

all *Causality*, or the whole Course of Nature, is finally refunded into this Self-evident Principle, that *Things are such as they are*, that is, *are what they are*. For, since an Effect is a Participation of something that is in the Cause; and the Cause, as such, is that which imparts or communicates something it has to the Matter on which it works its Effect. Again, since the Effect *is such* as the Cause *is*, as to that which is imparted to it; and if the Cause be of *another* sort, the Effect still *varies* accordingly; there can be no doubt but that *Causality* is the Imprinting the *Existence* of that Essence or *Thing* which is the *Cause*, upon the *Matter*. Whence follows evidently, that the very Notion of *Natural Causality*, and the whole Efficacy of it, consists in the Causes *existing* (that is *being what it is*. (*Method*, 144–5)

The logical structure of demonstrative syllogism is thus not only a logical structure but also an epistemological structure and, further, an ontological and causal structure.

3) *The Cartesian Critique of Syllogistic*

Descartes offers a sound critique of the notion of a demonstrative syllogistic science in his *Rules for the Direction of the Mind*.[148] The traditional idea of a demonstrative science based on syllogistic is, he argues, of no help in the discovery of new truths.

To be sure, Descartes does not abandon the idea of science as *scientia*, the ideal of science as demonstrative from self-evident premises. Knowledge is, Descartes proposes, *scientia*, as it was for Aristotelians such as Sergeant. As Descartes states in his commentary to his Rule Two, 'All knowledge is certain and evident cognition' (*Rules*, 10), which is the basis for the rule itself: 'We should attend only to those objects of which our minds seem capable of having certain and indubitable cognition' (10). This is what we aim at when we attempt to solve our research problems.

Descartes formulates the notion of a research problem as involving something unknown which is what we are searching for, and something known that limits the range within which we must search for what we want to know. 'First, in every problem there must be something unknown; otherwise there would be no point in posing the problem. Secondly, this unknown something must be delineated in some way, otherwise there would be nothing to point us to one line of investigation as opposed to any other. Thirdly, the unknown something can be delineated only by way of something else which is already known' (*Rules*, 51). He then distinguishes two kinds of problems, the imperfect and the perfect. The latter requires nothing more for its solution than what is contained in the definition of the problem; the solution, that is, may be deduced from those givens. 'These conditions hold also for imperfect problems ... But if the problem is to be perfect, we want it to be determinate in every respect, so that we are not looking for anything beyond what can be deduced from the data' (51).

Descartes here distinguishes his own method from that of the 'dialecticians': 'when they expound the forms of the syllogisms, they presuppose that the terms or subject-matter of the syllogisms are known; similarly, we are making it a prerequisite here that the problem under investigation is perfectly understood. But we do not distinguish, as they do, a middle term and two extreme terms' (*Rules*, 51). To be sure, his method is like that of the dialecticians in proceeding in terms of self-evident inferences. As he puts it in his Rule Five, '[t]he whole method consists entirely in the ordering and arranging of the objects on which we must concentrate our mind's eye if we are to discover some truth. We shall be following this method exactly if we first reduce complicated and obscure propositions step by step to simpler ones, and then, starting with the intuition of the simplest ones of all, try to ascend through the same steps to a knowledge of all the rest' (20). In fact, these inferences are based on relations among ideas, those very relations of identity about which Sergeant and the Port Royalists speak.

This common idea is carried over from one subject to the other solely by means of a simple comparison, which enables us to state that the thing we are seeking is in this or that respect similar to, or identical with, or equal to, some given thing. Accordingly, in all reasoning it is only by means of comparison that we attain an exact knowledge of the truth. Consider, for example, the inference: all A is B, all B is C, therefore all A is C. In this case the thing sought and the thing given, A and C, are compared with respect to their both being B, etc. But, as we have frequently insisted, the syllogistic forms are of no help in grasping the truth of things. (57)

This Cartesian critique of syllogistic presupposes, however, an account of ideas which is essentially the same as that of Sergeant, that is, an account of knowledge in which the mind that knows has within it the essence or form or nature of the thing known.

Descartes takes perception to fall under the category of thought.[149] Thoughts in general and perceptions in particular are modifications of the mind; they are acts of the mind. But thoughts and perceptions are also about, or as Brentano taught us to say, they 'intend,' certain objects. They do this by virtue of *ideas* which inform them. An idea, then, is the form of an act of thought or perception; as Descartes puts it in his replies to the Third Set of Objections, 'by an idea I mean whatever is the form of a given perception.'[150] But an idea in this sense is not itself a content of thought that is apprehended by the mental act of which that idea is the form; we are *aware of* the idea but that idea is not the *object of perception*. Ideas, in other words, are not representative entities that intervene, as it were, between the act of the mind which is the perceiving and the object perceived.[151] For, Descartes distinguishes two senses of the terms 'idea.' In one sense it is a modification of the mind, and as such is something of which the mind is immediately aware. However, an act of thought or perception is also intrinsically connected to the object or body which it intends, by virtue of which connection the mind is aware of that object or body. Descartes draws the relevant distinction between two senses of the term 'idea' in the 'Preface' to the *Meditations*: 'in this word 'idea' there is here an equivocation. First it can be taken materially, as an operation of my intellect ... or it can be taken objectively for the body which is represented by this operation.'[152] This he elaborates in his reply to the Second Set of Objections:

II. *Idea.* I understand this term to mean the form of any given thought, immediate perception of which makes me aware of the thought.
III. *Objective reality of an idea.* By this I mean the being of the thing which is represented by an idea, in so far as this exists in the idea ... whatever we perceive as being in the objects of our ideas exists objectively in the ideas themselves.[153]

Descartes refers to actual being as 'formal' being or reality (*esse formale*): the formal being of a thing is constituted by its properties or modifications. This he distinguishes from the objective being of a thing, that 'mode of being, by which a thing exists objectively or is represented by a concept of it in the understanding.'[154] This is 'the way in which its [the intellect's] objects are normally there.'[155] When a thing has objective being in the mind, that thing is not a mode or property of the mind, that is, it is not 'formally' in the mind; otherwise, the mind would actually *be* the thing. None the less, though the thing that is objectively in the mind lacks formal reality as such, it is still not nothing[156] and is to be contrasted to ideas which are merely chimerical.[157] Thus, as Descartes put it to Caterus in the replies to the First Set of Objections, 'the idea of the sun is the sun itself existing in the intellect – not of course formally existing, as it does in the heavens, but objectively existing, i.e. in the way in which objects are normally in the intellect. Now this mode of being is of course much less perfect than that possessed by things which exist outside the intellect; but, as I did explain [in Meditation III], it is not therefore simply nothing.'[158] This makes clear that, for Descartes as for Sergeant, *the objective existence of things in the mind is constituted by the presence in the mind of the essence of the thing.*

Thus, the problem with syllogistic according to Descartes is not with the notion of inference, nor with the concept of judgment that is involved, nor with the thought that what one is after are relations among ideas or notions or essences. All these are shared by Descartes and the dialecticians. None the less, there is a gulf between Descartes and the defenders of syllogistic.

The defenders such as Sergeant argue that syllogism is *the* method to science. Descartes argues to the contrary that, in regard to the shared concept of science as *scientia*, the shared concept of science as involving a knowledge of objective necessary connections among essences, syllogism is of *no* use as a method to science, to knowledge of essential structural relations. The point is that *syllogistic is of no help in finding the relevant relations.* The syllogistic method may be of use in displaying knowledge that one has already obtained, but it is of no use in discovering the truth; it 'contributes nothing whatever to knowledge of the truth' (*Rules*, 36). What is crucial is the set of inferences

$$A = B, B = C, \quad \therefore A = C.$$

As Descartes puts it, we must 'think of all knowledge whatever – save knowledge obtained through simple and pure intuition of a single, solitary thing – as resulting from a comparison between two or more things. In fact the business of human reason consists almost entirely in preparing for this operation. For when the operation is straightforward and simple, we have no need of a technique to help us intuit the truth which the comparison yields; all we need is the light of nature' (57). Inferences of the sort just indicated are based on judgments of comparison. These judgments, when linked systematically, yield knowledge of connections among essences or notions or forms. It is the inferences that are essential to knowledge. What the defenders of syllogistic do – that is, the dialecticians – is take this inferential knowledge and *rearrange* it into syllogistic form. But that form does nothing to help us discover what the links are in the chain that leads to the solution of the research problem that we have posed to ourselves.

[O]n the basis of their method, dialecticians are unable to formulate a syllogism with a true conclusion unless they are already in possession of the substance of the conclusion, i.e. unless they have previous knowledge of the very truth deduced in the syllogism. It is obvious therefore that they themselves can learn nothing new from such forms of reasoning, and hence that ordinary dialectic is of no use whatever to those who wish to investigate the truth of things. Its sole advantage is that it sometimes enables us to explain to others arguments which are already known. It should therefore be transferred from philosophy to rhetoric. (36)

As the Port Royal Logic puts the point, 'we must discover the content of an argument before we can arrange its premises' (*La Logique*, Pt III, ch. 17).[159]

Given the Aristotelian notion of demonstration as laid out by, for example, Sergeant, and as implicit in the doctrine of judgment as involving identities (and non-identities) which is shared by the Aristotelians and the Cartesians, this criticism is wholly just.

However, having said this, it must also be emphasized that the basic notion that knowledge should be *scientia*, demonstrative from self-evident premises about the forms or essences or notions of things, is something that is shared by the Aristotelians and the Cartesians. For this view of knowledge, there is a deeper issue that we have not yet addressed. This issue was clearly faced by Descartes. It was ignored by his predecessors; it continues, alas, to be ignored by many others to this day.

4) *Relations of Ideas and Matters of Fact*

We have so far looked only at the comprehension of ideas, and considered truth as involving only relations among ideas. But, as Arnauld and Nicole insist, it is necessary to take account also of the extension of an idea; that is, in terms of our own terminology, we must also give categorical propositions their concrete readings.

The extension of an idea consists of 'the subjects which fall under the idea ... as the idea of a triangle in general has as its extension all the different species of triangle' (*La Logique*, Pt I, ch. 6).[160] There is a distinction, then, between a judgment considered as establishing a relation among ideas, and a judgment considered as establishing a relation among the subjects in the extension that it is about. Thus, when one says that

> All men are animals,

one asserts that 'everything which is a man is also an animal,' whereas if one says that

> Some men are just,

one is claiming 'only that the characteristic of being just is found in some men' (Pt II, ch. 17). Under this understanding of propositions, 'affirmation places the idea of the attribute in the subject, and the subject properly determines the extension of the attribute in an affirmative proposition; the identity which the proposition asserts concerns the attribute contracted into an extension equal to that of the subject and not in its full generality in case that it covers more than the subject ...' For example, 'it is true that lions are all animals, that is to say, that each lion falls under the idea of animal, but it is not true that they are all the animals' (ibid.).[161] Considered in terms of their extension, then, propositions mark an identity between the subjects in the extension of the subject with some at least of the subjects in the extension of the predicate term.

Considered in respect of the comprehension of ideas, the proposition

> S is P

asserts a relation between two ideas; it states that every part of the idea expressed by the predicate term is part of the idea expressed by the subject term. Let us represent this by

(*) $R(S, P)$.

Considered in respect of the extension of two ideas, the proposition asserts a relation among the individuals that are in the extensions of the ideas expressed by the subject and predicate terms; it states that

(@)   All S's are P's.

When propositions are considered in the former respect, then syllogisms formed by them are understood to be valid because they assert, as we have seen, identities between the various ideas expressed by the terms in them. When the propositions are considered in respect of the extensions of the terms, then syllogisms formed by them are understood to be valid because they assert relationships of class inclusion.

Now, one of the traditional doctrines is that the affirmative syllogisms of the first figure are rendered valid by the *dictum de omni*. The Port Royal Logic accepts this traditional doctrine, expressing the *dictum* in this way: 'What is given as holding of an idea taken universally is to be taken as also holding of all of which that idea is affirmed ... (*La Logique*, Pt III, ch. 5). Correspondingly the negative propositions are rendered valid by the *dictum de nullo*: 'What is denied of an idea taken universally is denied of all of which the idea is affirmed' (Pt III, ch. 5).

The question that arises is why it is believed that it is necessary to appeal to this *dictum de omni et nullo* in order to establish the validity of the syllogisms. (The *dictum* applies to syllogisms of the first figure only; but the syllogisms in the other two [or three] figures can be validated by proving them using the moods of the first figure.) On the one hand, it applies to the syllogisms insofar as the propositions they contain are understood in reference to their comprehension. In that case, the principle of validity is none other than the transitivity of identity (and the relations on the species-genus tree). Why do we need, in addition, the *dictum* to validate these syllogisms? On the other hand, the *dictum* applies to syllogisms insofar as the propositions they contain are understood in reference to their extension. In that case, the principle of validity is none other than the transitivity of class inclusion. Why do we need, in addition, the *dictum* to validate these syllogisms? It would seem that *the 'dictum de omni et nullo' plays no role in validating syllogisms that is not already implicit in their logical form*. Apparently it is, in other words, simply redundant. Why, then, the emphasis that is placed upon it? What, exactly, is its role conceived to be?

We can perhaps put ourselves in a better position to answer this question if we look at some more recent philosophers who have held a similar position.

Michael Tooley has defended the view that we should conceive of laws of nature as relations among universals.[162] The relations he has in mind are of the sort (*):

$R(S, P)$.

This relation is held to be stronger than matter-of-fact generalizations such as (@):

All S's are P's

though it is also held that the former logically implies or entails the latter:

(#) $R(S, P) \rightarrow$ All S's are P's.

The idea is that by introducing the stronger (*) as a way of representing laws of nature one will overcome the objections, based on inductive uncertainty, to a Humean account of laws that construes them as no more than matter-of-fact generalizations like (@). Tooley is here taking up a position that had earlier been advocated by C.D. Broad,[163] though Tooley apparently is unaware of the history of the view he advocates. Certainly, he fails to take account of the devastating criticism that Bergmann raised against Broad's earlier statement.[164] Bergmann's criticism of Broad was developed as a critique of Tooley in F. Wilson, *Laws and Other Worlds*.[165] James Brown seems equally unaware of the history and equally unaware of the criticism of Tooley. At any rate, in his *The Laboratory of the Mind*,[166] he repeats Tooley's thesis concerning the nature of laws as relations among universals, citing Tooley as an authority, but takes absolutely no account of the criticisms that have been made of the position. That, unfortunately, vitiates most of the argument of his book.

The point is as simple as it is obvious. *The defenders of the thesis that laws are relations among universals provide no clear account of entailment such that a statement of such a relationship entails a generalization about the particulars that exemplify those universals.* Thus, though Brown and Tooley both assert that the connection (#) is an entailment relation, they make no effort to indicate the logical form that provides the ground of the necessitation. The antecedent of (#) is a relation among universals, and therefore in some sense a necessary truth, at least in the sense of being timeless. However, on any standard account of entailment, there is no reason to expect a non-general relational fact about universals to entail a generalization about the particulars falling under those universals. Whatever else a law is, it is a regularity. The concern is whether there is anything more to laws. Tooley and Brown, insouciantly following Broad, argue that there is something more, to wit, the relation among universals. But if that is to make sense, then the relation among universals does indeed have to entail the matter-of-fact regularity, as they quite rightly see. But, alas! they provide no account of that entailment relationship. To assert that one exists is hardly to explain it. What precisely is the logical form that is the ground of the necessity alleged to hold between the antecedent and the consequent, between the relational fact about universals and the generalization about the particulars falling under those universals? In the absence of any reasonable account of logical form that would show the necessity of (#), the notion that we can construe laws as relations among universals is simply a non-starter.[167]

This indicates the central problem for those views that begin by construing the regularities that we observe in the world as reflecting structural relations among universals. This problem holds as much for the Port Royalists and Sergeant as it does for Broad, Tooley, and Brown. This is the fact that *there must be some guarantee that the relations among the universals are reflected in the regularities that hold among the particulars exemplifying those universals.* Tooley and Brown claim to find it in the supposed entailment relation (#), but until they explain their very special notion of entailment, they have not solved the problem. It is here, however, that we find the solution to the question that we asked above, why the Port Royalists spent time on the apparently redundant *dictum de omni et nullo*. Notice what the *dictum de omni* asserts: What is predicated of the subject is predicated of that of which the subject is predicated. This means that in

S is P

the predicate *P* is to be predicated of whatever the subject *S* is predicated. In particular, then, for every particular of which *S* is predicated, *P* will also, according to the *dictum*, be predicated. But in that case, if we have

S is P

then we also have

All S's are P's

Similarly, the *dictum de nullo* will guarantee that if *P* is denied of *S*

S is not P

then we will also have

No S's are P's.

But this is a way of ensuring that, given the relation among universals, the corresponding regularity will obtain among the particulars exemplifying the universals. Thus, *the 'dictum de omni et nullo' is a law that guarantees that structural relations among universals are reflected in the matter-of-fact regularities that hold among particulars exemplifying the universals.* Hence, taking the *dictum* to be in fact such a law, we see that the Port Royalists solved the problem that confronts Tooley and Brown as to how it happens that the order among particulars reflects the order among universals.

The *dictum* was understood by John Stuart Mill in just this way.

This maxim ... when considered as a principle of reasoning, appears suited to a system of metaphysics once indeed generally received, but which for the last two centuries has been considered as finally abandoned, though there have not been wanting in our own day attempts at its revival. So long as what are termed Universals were regarded as a peculiar kind of substances, having an objective existence distinct from the individual objects classed under them, the *dictum de omni* conveyed an important meaning, because it expressed the intercommunity of nature, which it was necessary on that theory that we should suppose to exist between those general substances and the particular substances which were subordinated to them. That everything predicable of the universal was predicable of the various individuals contained under it, was then no identical proposition, but a statement of what was conceived as a fundamental law of the universe. The assertion that the entire nature and properties of *substantia secunda* formed part of the nature and properties of each of the individual substances called by the same name – that the properties of Man, for example, were properties of all men – was a proposition of real significance when man did not mean all men, but something inherent in men, and vastly superior to them in dignity. Now, however, when it is known that a class, an universal, a genus or species, is not an entity *per se*, but neither more nor less than the individual substances themselves which are placed in the class, and that there is nothing real in the matter except those objects, a common name given to them, and common attributes indicated by the name; what, I should be glad to know, do we learn by being told, that whatever can be affirmed of a class may be affirmed of every object contained in the class? The class *is* nothing but the objects contained in it: and the *dictum de omni* merely amounts to the identical proposition, that whatever is true of certain objects is true of each of those objects. If all ratiocination were no more than the application of this maxim to particular cases, the syllogism would indeed be, what it has so often been declared to be, solemn trifling. The *dictum de omni* is on a par with another truth, which in its time was also reckoned of great importance, 'Whatever is, is.' To give any real meaning to the *dictum de omni*, we must consider it not as an axiom, but as a definition; we must look upon it as intended to explain, in a circuitous and paraphrastic manner, the meaning of the word *class*.[168]

We shall have some more to say about some of these points in due course. But right now another point needs emphasis.

This is the question about what it is that makes the *dictum* a metaphysical *truth*. What guarantees that it obtains? And how do we know it?

Among the recent philosophers who have defended the view that laws are to be construed as relations among universals, D.M. Armstrong alone has faced up to this problem, or, what amounts to the same, the problem why the structural relation (*) should somehow imply that the regularity (@) among particulars should also hold.[169] Armstrong considers the properties F and G to be causally related; this, the causal relation, he denotes by N. We therefore have

$N(F, G)$.

This N is the relation we have called R. Armstrong is concerned to understand how it is that this fact about universals can lead to there being a regularity

All F's are G's

among the particulars that exemplify F and G. He proposes that

> $N(F, G)$ is [itself] a universal, instantiated in the positive instances of the law ...

If we accept this view, he suggests, then 'it will be much easier to accept the primitive nature of N. It will be possible to see clearly that if N holds between F and G, then this involves a uniformity at the level of first-order particulars' (*What Is a Law of Nature?*, 88).

We have, therefore, certain facts about particulars exemplifying the (funny) universal $N(F, G)$. These facts Armstrong understands (91) to be perspicuously represented by something like

$(N(F, G))$ (a's being F, a's being G)

which we might equally well write as

(&) $(N(F, G))$ (Fa, Gb)

which makes clear that what we have is a *relation* holding between *states of affairs*. It is the 'causal relation,' but, because it holds between states of affairs rather than particulars, its status is rather more like that of a *connective* than it is like an ordinary first-order relation among ordinary particulars. This is the *first* feature of the causal relation that we should notice. *Second*, we should notice that this connective is not truth-functional; the sentences '*Fa*' and '*Gb*' that occur in (&) do not by themselves determine the truth-value of that sentence. In particular, *third*, the complex sentence (&) will be true only if the simple sentences appearing in it mention *kinds* that are causally related to each other. Finally, *fourth*, it is worth noticing that the nature of the causal relatedness will vary from case to case. It may be the simple kind involved in a regularity like

All F's are G's

or it may be a more involved connection such as

All F's that R an H are either G's or K's.

In the former case we will have

$N(F, G)$

while in the latter case we will have

$N[(F, R, H), (G, K)]$

or something like it. (Armstrong does not give details.) This will make N an odd relation indeed, a relation with a varying number of terms. Or perhaps there will be a variety of relations all sharing a common property, that of being 'causal necessitation relations.' How these themes are worked out by Armstrong or other defenders of the view we may leave to one side. (In general they aren't worked out.) It suffices to note only the requirement that the nature of the causal relation among the states of affairs will in fact have to vary from case to case depending upon the relevant law. These are the *four crucial features* of the nomological connective.[170]

These four crucial features give rise to *two philosophical problems*. The *first* of these is the fact that when one introduces a nomological connective such as that which appears in (&), one challenges the traditional analytic/synthetic distinction, which presupposes a language in which the only connectives are truth-functional. This problem amounts to saying in another way that there is a real question of the entailment relation that is supposed to hold between

$N(F, G)$

and

All F's are G's

or, what is the same, the notion of logical form that is supposed to make Brown and Tooley's (#) a necessary truth.

The *second* philosophical problem concerns how it is that we *know* that the non-truth-functional nomological connective actually holds between two facts or states of affairs.

In order to deal with the first of these problems, it will pay to look at the traditional doctrine of explanation a little more closely.

On the traditional view, we can (to take a simplified example) explain

Corsicus is dissolving

using the syllogism

Whenever Corsicus is put in water Corsicus dissolves
Corsicus is in water

∴ Corsicus is dissolving

provided that the premise is necessary. This is established as necessary through a *scientific syllogism*. Let Corsicus be *C* and let *A* be the pattern of sensible events of dissolving if in water. Then *C* is *A*. The scientific syllogism that we need will have this as its conclusion. The middle term that explains *C* being *A* is the capacity (essence, form, nature, notion, power) of *solubility*. Let *B* be this capacity. Then the explanatory syllogism is:

Whatever is soluble is such that whenever it is put in water it dissolves
(*B* is *A*)
(S′)   Corsicus is soluble (*C* is *B*)

∴ Corsicus is such that whenever it is put in water it dissolves (*C* is *A*).

Solubility is an active power, like the nutritive power, the activity of which accounts for the pattern of behaviour described in the conclusion. As the nutritive power is one of a set of powers, including the appetitive and the ratiocinative, which make up the soul or form or nature of human beings, so also solubility will be one of the set of powers that constitute the set that makes up the full nature or essence of Corsicus; for example, the nature of Corsicus may be sugar. Thus, the second premise of (S′) is necessary because the nature or essence is inseparable from the substance Corsicus. Provided that the major premise of (S′) is also necessary, then the conclusion will be necessary, and the syllogism will establish a necessary connection among the events in the pattern mentioned in the conclusion. The necessity of the law derives from the form or essence; thus, explanation of a pattern of behaviour is not obtained by fitting that pattern into a more complex pattern, as in the new science, but by *redescribing* the object whose behaviour is being explained.

Now, as Molière made clear, for the defenders of the new sciences, the major premise of (S') is indeed necessary, but only trivially so, since it is true by nominal definition. For that reason, for the critics of Aristotle, (S') is no more explanatory than

(S")

Whatever is a bachelor is an unmarried male
Callias is a bachelor

∴ Callias is an unmarried male.

(S") doesn't explain since the minor premise merely *restates* what the conclusion says. As for the major premise in (S"), that is in fact redundant; for the rule of language

'Bachelor' is short for 'unmarried male'

which licenses its assertion as a necessary truth equally licenses the rewriting of the minor as the conclusion and therefore also licenses the inference of the conclusion from the minor premise alone.

On the other hand, for Aristotle (S') *is* explanatory. The major premise cannot, therefore, be true by definition. It must be a substantive truth. Thus, for the empiricist, capacities are analysable in the sense that the term 'soluble' in the major premise of (S') can be defined by the right hand side using only (the principles of logic and) terms that refer to observable characteristics of things; while for Aristotle, since the major premise of (S') cannot be a definitional truth, capacities (dispositions, tendencies, powers) cannot be analysed into (patterns of) observable characteristics of things. *Aristotelian explanations are in terms of the unanalysable active dispositions and powers of things.*

If the major premise of (S') is not definitional, it is also not contingent; Aristotle holds, as we have seen, that the premises of a scientific syllogism are *necessary*. The empiricist divides propositions, as Hume does,[171] into two mutually exclusive and jointly exhaustive classes. There are, on the one hand, those propositions which state relations of ideas, and which are therefore necessary in the sense of being tautological, devoid of factual import, and, on the other hand, those propositions that which state matters of observable fact and which are therefore contingent. If the former propositions are, as they are often called, analytic and the latter synthetic, then for Aristotle, in contrast to the empiricist, there is a third category, that of propositions which are both synthetic and necessary. Kant's term, 'synthetic a priori,' is not quite appropriate for Aristotle, but the point is much the same. In any case, it will do for our purposes.

The connection with the views of Armstrong should be clear. Corsicus is involved in two states of affairs or facts, namely,

Corsicus is in water

and

Corsicus dissolves.

These are connected by the fact that

Corsicus is soluble

where this feature is no ordinary observable characteristic but the very nature or essence of Corsicus, a feature that Corsicus has necessarily. It provides the necessary connection between Corsicus's being in water and his dissolving; these two, observable, first-order properties, or, rather, the instantiations of these two, observable, first-order properties, are connected into a lawful pattern by this nature, form, essence, or notion. At the same time, this necessary connection entails the regularity

Corsicus is such that whenever it is put in water it dissolves.

This it does by way of the non-tautologically necessary or synthetic a priori truth that

(##)   Whatever is soluble is such that whenever it is put in water it dissolves

The connection is clear: *Armstrong's causal structural relations among universals are in fact the essences or forms or notions of things, in the Aristotelian sense.*

We now see why it is that he can reasonably hold that the structural connections among universals imply that the appropriate regularities hold among the particulars that instantiate the universals. While he does not say so explicitly, it is the solution implied by the Aristotelian structure of his position. The answer clearly is the Aristotelian pattern that the regularities among the particulars hold by virtue of the causal activity of those particulars. More fully, the regularity holds by virtue of the fact that it (the regularity) represents the exercise by the particular of its essential or natural powers to produce that pattern as the expression of its inner ontological structure in its (outer) being (its observable characteristics). *The causal activity of the particulars ensures that the patterns of observable characteristics of those particulars conform to the a priori structure of its essence.* In fact, even though Armstrong does not mention this Aristotelian solution to his problem, he does speak on occasion of how 'the regularity in things reflect[s] ... [a] power in the things themselves.' At least it is a power: 'Perhaps, as Berkeley thought, the regularities in things reflect no power in the things themselves, but only a particular determination of the will of God to have ordinary things ("ideas" for Berkeley) behave in a regular manner' (*What Is a Law of Nature?*, 105). In any case, just as for Armstrong the causal activity of particulars in this way ensures the conformity of matter-of-observable-fact regularities with the causal structural relations among universals, so in the same way, for the Aristotelians it is part of the idea of the essential causal structure of things that *the causal activity of particulars ensures that the 'dictum de omni et nullo' is true.*

Descartes, of course, as we saw in Study One, criticized the Aristotelian claim that ordinary things were active in the Aristotelian sense. On the Cartesian account of things, only God is active in this sense. Still, it is Her (or His) activity that ensures the conformity of matter-of-observable-fact regularities with the causal structural relations among universals, the essences of things. Thus, for the Cartesians as for the Aristotelians, it is part of the idea of the essential causal structure of things that *causal activity ensures that the 'dictum de omni et nullo' is true.* Or, at least, it does so provided that we can rely upon God to ensure that there is no powerful evil genius who can mislead us about the essential structure of the world – but of that issue, more in Study Seven.

If we leave aside this perhaps esoteric problem of the evil genius, then in one sense the resort to causal powers or activities in the Aristotelian sense, whether they are located in particular things as they are by Aristotle and Armstrong or located in God as they are by Descartes, does solve the problem that confronted Tooley and Brown. In another sense, it leaves things still problematic. Certainly, it fails to solve what we referred to above as the *first* of the philosophical problems confronting that position. For, it requires that we accept the proposition (#) or, what is the same, (##), as necessary in some sense that never gets characterized. We are left with the Aristotelian synthetic metaphysical necessities that are as puzzling today as they were when Molière ridiculed them. Neither Tooley nor Brown nor Armstrong can escape that criticism: it is as devastating today as it was when Molière first proposed it.

That leaves the *second* of the two philosophical problems. To this we now turn.

5) *Our Knowledge of Necessary Connections*

The problem of knowledge is the *second* of the two philosophical problems that confront the view that there is a primitive nomological connective, or, what we have now seen to be much of a piece, the Aristotelian doctrine that natures or essences or forms or notions of things provide necessary connections among the observable characteristics of things: how do we know when it obtains?

This is a question that not one of the moderns addresses. Brown introduces it as 'theoretical entity posited for theoretical reasons' (*Laboratory*, 83). Armstrong makes much the same point: 'The postulation of a connection between universals can provide an explanation of an observed regularity in a way that postulating a Humean uniformity cannot' (*What Is a Law of Nature?*, 104). It is totally different from the postulation of small parts that we cannot see, however – for example, the postulation of a speck of dust to explain why my watch won't work. In this latter sort of postulation, the entities postulated share certain features with the entities of which we are aware in ordinary experience. Although we may not experience the entities themselves, and, indeed, even though they may be too small or too far away for us ever to experience them, none the less, they are of *kinds* with which we are acquainted.[172] But this does not hold for the nomological connection: by hypothesis this is something totally different in kind from anything with which we are acquainted. This raises problems, surely, for anyone – problems that should not be brushed under the carpet by a blithe and unexamined use of the term 'postulation.'[173]

John Sergeant knew better: he, unlike his more recent successors, attempted to provide an account of knowledge which allowed that we are, in fact, aware of the essences, or, as he called them, the 'notions' in things.

For Sergeant, we begin with notions which are 'Naturally wrought in the Soul by the strokes of occurring Objects ...' For him, as for Aristotle, Descartes, and the Port Royal logicians, these notions are the essences of the things known *qua* existing in the mind: 'these Notions are the very Natures of the Thing, or the thing it self existing in us intellectually, and not a bare Idea of Similitude of it ...' (*Method*, 2). Sergeant objects to the Cartesian methodology that rejects appeal to the senses.[174] He holds that we must proceed by reasons and sense together (*Method*, c5), and emphasizes that sense alone is not sufficient for science, that is, *scientia*, (d5, d6). For *scientia*, experiments and induction are not enough, for one cannot get a universal from a group of particulars. So-called experimental philosophy 'is historical, and [a] narrative of particular observations; from which to deduce universal conclusions is against plain logick, and common sense' (d6). In order to move from sensible particulars to knowledge of universal conclusions one must grasp the natures or notions of the things with which we are in sensible contact. Thus, a regularity in sensible experience is *causal* and *scientific* just in case that it can be accounted for in terms of the natures or notions of things. The senses play a role in the *acquisition* of scientific knowledge but *not* in *evaluating* it. The role of the senses is simply to carry our notions to us (1–2). The 'Notions ... are the very Natures of the things in our understanding imprinted by outward objects' (b3; cf also page 2), and, contrary to a then-common position, Sergeant holds, as we saw, that the notion is 'the thing itself existing in us intellectually, and not a bare idea or similitude of it ...' (2). So all discourse, insofar as it aims to be a discourse grounded in truth, must be rooted in our notions, that is, grounded on these natures of things. Knowing is a matter of seeing the connections among notions (5). Understanding is a matter of finding middle terms in syllogisms, the terms of which are these notions, and these middle terms are, in effect, as we have seen, the real definitions of the things.[175] The premises of discourse are justified by reducing them to first principles (Sergeant, *Method*, 154–5; White, *Exclusion*, 5ff), of which, as we saw, the most basic (Sergeant, *Method*, 157–8; White, *Exclusion*, 6) is the self-evident (Sergeant, *Method*, 123; White, *Exclusion*, 7) principle that what is, is, or that self-existence is self-existence. Thus, principles and deductions, both based on notions, constitute the framework of knowledge around which any specific branch of human investigation can be arranged. 'For our notions being cleared, first principles established, the true form of a syllogism manifested, proper middle terms found, and the necessity of the consequences evidenced; all of those conclusions may be deduced with demonstrative evidence, which ly within our Ken, or which we can have occasion to enquire after' (Sergeant, *Method*, 12).

The fundamental starting point for any field of knowledge in which we aspire to attain such absolute certainty is the 'notion,' 'the thing itself existing in us intellectually.' Notions are grasped first in acts of 'simple apprehension': 'notion' and 'simple apprehension' are synonymous terms. He calls the immediate objects of mind 'notions' because 'they are the parts or elements of knowledge' which, when put together, constitute cognition, and 'they are called "simple apprehensions" to distinguish them from judgments which are compounded of more notions, and belong to the second operation of our understanding' (*Method*, 3). By means of notions 'we barely apprehend, that is, lay hold of, or take into us the thing, about which we afterwards judge or discourse' (ibid.). Correct judging consists of correctly joining notions – in the case of science, a matter of correctly seeing the *real necessary connections* among notions. Scientific understanding, in turn, is the elaboration of correct judgments of science into demonstrative syllogisms of the sort we have already discussed. Initial simple apprehension is succeeded by the mind in judgment as it were unpacking or relating notions according to their own internal structures, with discourse as a further elaboration of this insight into necessary connections. Objectively considered, truth is 'the conformity of our judgment to the nature of the thing,' while, formally considered, truth consists 'in the connexion of the terms of a syllogism,' but these amount to the same thing: since notions *are* the things, and since syllogisms are but notions elaborated according to their own internal necessary connections, it follows that, in a sense, things *are* syllogisms, or, at least, that the ontological structure of things is identical to the logical structure of thought in demonstrative knowledge. Thus, as for Aristotle, so for Sergeant: syllogistic is not merely a *logic* but also an *onto-logic* and an *epistemo-logic*. All this follows from Sergeant's having adopted the Aristotelian idea that 'actual knowing is identical with its object' (*De an.*, 430a20) and that 'in every case the mind which is actively thinking is the object which it thinks' (321b17–18).

All this works only if there is a mechanism which will guarantee the transmission of the notion of the thing to the mind that knows the things. Sergeant holds that the mechanism of sensation itself is atomistic (*Method*, 22), but this is the only point at which he differs in a major way from Aristotle. Sergeant's theory of arriving at truth is, in effect, that which was elaborated in the *Posterior Analytics*. The method consists in the orderly apprehension of the necessary connections among notions = the natures of things, and the role of sense experience is *not to judge* of scientific propositions but *merely to convey* to the mind the notions which it rationally intuits. These notions themselves constitute science, and they, not sense, constitute the judge of what is science. These intuited notions are unpacked by the rational faculties of the mind into judgments and syllogisms, at which point these notions constitute science as a body of self-evident and necessarily connected truths which by self-evidence and necessary connections *judge of themselves* and are not judged by anything (such as sense) external to them. Sergeant specifically draws the appropriate methodological conclusion: 'From what's said, 'tis deduced, that it is the one necessary and main part of the method to science to distinguish our notions clearly, and to keep them distinct carefully' (14). Knowledge, though dependent upon sense observation, goes beyond it, penetrating to the very natures of things and the necessary ontological structure of these natures. It is in terms of the causal activities of these natures or notions, activities that proceed in conformity to the logical structure of these notions, that Sergeant proposes that we can account for the way things appear in the world of our ordinary sense experience. In particular, this causal activity of the natures accounts for the possibility of our coming to know things, for our coming to have in our minds those very natures or notions of the things known.

Sergeant specifically connects 'notions' with God, asserting that an important aspect of notions is their 'metaphysical verity,' and that they 'partake of this [aspect] from the ideas in the divine understanding, from which they inerringly flow, and which are essentially unchangeable' (*Method*, 5). In this way, 'the God of truth' is the 'sole author of all the truth that is in us,' fixing *solidly* or unalterably the natures or essences of things upon which all science, that is, solid science or *scientia*, must be founded. Insofar as persons can penetrate to these spiritual natures they are also part of the higher order. Sergeant is quite clear on this. His theory of sense experience is elaborated in his *Solid Philosophy Asserted, against the Fancies of the Ideists, or, the Method to Science further Illustrated*,[176] where, as part of his atomism, he adopts the Cartesian theory of the pineal gland. But throughout he makes clear that it is his view that a major argument in support of his position on natures, notions, and our knowledge of things, is that if it were not this way then all science and all philosophy would be destroyed (*Solid Philosophy*, 29ff).

It all presupposes, on the one hand, the capacity of the mind to penetrate beyond the sensible appearances of things to their underlying notions or natures, and, on the other hand, the guarantee in those natures that their causal activities will produce in the mind natures which are in fact the natures of the things appearing to us sensibly, a guarantee that is ultimately rooted in the being and veracity of God. If this is granted, then of course Sergeant is correct in making his methodological focus these known natures or notions, rather than sensible knowledge, and adopting towards sense observation an attitude that is effectively one of contempt, and certainly contrary to the attitude of empirical science towards sense observation.

The contrast to empirical science is crucial. On the one hand, Sergeant's picture of science as *scientia* makes science equivalent to (demonstrative) syllogism. On the other hand, in order to achieve science one must grasp in an intellectual intuition the real definitions which function as ultimate premises, where this intuition is an intellectual act over and above the *mere* observation of sensible particulars, the function of the latter being no more than that of exciting or triggering the act of rational intuition. For empirical science, in contrast, we have, on the one hand, the position that explanation is not a matter of finding real definitions to function as middle terms of syllogisms, but of the subsumption of particulars under matter-of-fact causal laws; and, on the other hand, that in empirical science the observation of sensible particulars proceeds with the aim of confirming general hypotheses and eliminating false hypotheses. Thus, in respect of both aim and method, empirical science and the demonstrative Aristotelian science of Sergeant are at odds. And, specifically, their attitudes towards sensible particulars are totally contrary. For the Aristotelian science, sensible observation merely triggers but in no way controls scientific knowledge; for empirical science, sensible observation provides logical control over the hypotheses rejected and accepted as confirmed.

Sergeant's claims about the proper 'method to science,' the proper method to achieve 'solid philosophy,' depends upon his claim that behind sensible corporeal objects are the spiritual, true natures of things, the logical structure of which is the cause of the being of things and knowledge of which constitutes our scientific understanding of those things. It was Locke who challenged all this. *Locke's world, unlike that of Sergeant – and Aristotle and Descartes – is all at one level, the level of sense.* And with the elimination of the higher, or, perhaps, deeper, level, the need to introduce rational intuition is gone. No longer are there two ways of knowing – sense and intuition: all knowing now begins and ends in sense. For Sergeant, human reason is one's spiritual aspect: it is one's point of contact with God. For Sergeant, human beings are semi-divine, split between a heavenly, rational part and a sensible, corporeal part. Locke attacks this idea of human being, insisting that humans are thoroughly of *this* world. The rational powers that supposedly lift humans out of this world are otherwise to be accounted for. Reason is as natural as sensation. There is nothing about science and knowing that marks a break in human being between a sensible level and some higher or deeper level. Any claims that reason may do more than begin and end in sensible knowledge is dismissed. Locke, the first of the great modern philosophers to dismiss the inhuman view of human being (that we are partly not of this world in which we live and breathe), was the first to adopt the truly humane Enlightenment view of human being as human and no more than human.

It is Locke's position that '[i]f we can find out how far the understanding can extend its view, how far it has faculties to attain certainty, and in what cases it can only judge and guess, we may learn to content ourselves with what is attainable by us in this state.'[177] His purpose is 'to enquire into the original, certainty and extent of human knowledge, together with the grounds, and degrees of belief, opinion and assent' (*Essay*, I.i.2); the method he proposed to use, we all know, was the 'historical, plain method' (I.i.2) (where 'historical' meant experimental or observational, that is, the method of empirical science). The result was an attack on the metaphysics that from Plato and Aristotle had supported the view of human being as a sort of being divided from itself, a being with two ways of knowing. The 'real essences,' that is, natures or notions of things, that were the core of that metaphysics Locke declared unknowable. Since the method of empirical science relies not on rational intuition for knowing particulars but on sense observation alone, Locke declares 'real essences' not a proper object of science. For Aristotle and for Sergeant, and, in many respects for Descartes also, method moves from sense to intuition and concentrates on the latter, valuing the former only as a trigger for the latter. Method, for such a philosopher, is not unitary. For Locke, in contrast, method *is* unitary. He succeeds in making it unitary by distinguishing empirical science from metaphysics: the latter is a 'science' of essences, or notions, or natures, or what-have-you, lying, as Aristotle, Descartes, and Sergeant would have it, beyond or above or behind the world of sense. For Locke that world is simply unavailable.

This sort of point needs to be made against those who, like Sergeant, lay claim to a knowledge of a world beyond that known by sense experience. In his marginal notes to Sergeant's *Solid Philosophy*,[178] Locke repeatedly makes points of just this kind. Where Sergeant asserts that the fundamental error of the methodology of the ideists like Locke is that they determine the clearness of their ideas from 'the fresh, fair, and lively appearances they make to the Fancy' (*Solid Philosophy*, 371), whereas 'only the definition, by explicating the true essence of a thing, shews us directly the true spiritual notion of it' (372) (which notions can't be wrong because they owe their truth to God); and, 'if I, mistaking or not mistaking, have such a meaning [notion] in my mind ... that meaning is truly in us' (374), Locke comments:

Where are those definitions that explicate the true essence of things? And (excepting mathematical) how many of them has J.S.? He would oblige the world by a list of them if it were of noe more but those things he has talked of in his books and pretends to know. (*Solid Philosophy*, 372; Note 84, Yolton, 'Locke's Unpublished,' 545)

He that has a meaning to any word has it no doubt while he has it. But he that varies the meaning of his terms or knows not precisely what he means by them (as nothing more ordinary) fils his discourse with obscure and confused ideas. (*Solid Philosophy*, 374; Note 85, Yolton, 'Locke's Unpublished,' 545)

The charge here is that what the methodology claims it can deliver is never produced.

At times, Sergeant does try to deliver. For example, Locke's argument that we do not know how or why parts of bodies cohere is taken by Sergeant to mean that Locke claims not to know what extension is in its real nature. Sergeant argues that if Locke 'means that we can give no physical reason for it, or such a one as fetch'd from the qualities or operations of bodies, I grant it; for all those qualities and operations are subsequent to the notion of extension and grounded on it; but, if he thinks there cannot be a far better and clearer reason given from the supreme science, metaphysics, I deny it' (*Solid Philosophy*, 247–8). He then proceeds to give his own, metaphysical, explanation: 'If then divisibility be the essence of quantity [Sergeant has argued for this earlier] and divisibility signifies unity of the potential parts of quantity; and continuity (as making those parts formally *indivisas in re*) be evidently the unity proper to those parts; it follows, the quantity being the common affection of the body, does formally, and as necessarily, make its whole subject, that is all its parts continued or coherent, as duality does make a stone and a tree formally two' (*Solid Philosophy*, 278).

But this is no answer. As Locke comments, this account makes bodies cohere 'because they do cohere' (*Solid Philosophy*, 249; Note 44, Yolton, 'Locke's Unpublished,' 541). 'The sum of which argument is that we make the word *body* stand for an idea of solid parts united together or cohering. Therefore we know what makes those parts cohere' (*Solid Philosophy*, 248; Note 42, Yolton, 'Locke's Unpublished,' 541). To be sure, such propositions are indeed necessary, but only because they are trivial: of course this triviality prevents their being explanatory. Locke's criticism of Sergeant is just the same as the more general criticism that Molière made of the traditional Aristotelian doctrine of causes and powers.

Sergeant thought Locke's problem was that he sought empirical causes when what were needed were essences or notions of things, metaphysical causes of regularities. What Locke construes as a trivial proposition resulting from unpacking simple ideas that have as a matter of convention been included in a more complex idea, Sergeant construes as a necessary but *non*-trivial proposition. Persons who, like Locke, restrict themselves to the world of sense experience and ignore the 'transnatural or ... *altissimus causae* which only metaphysics give us' (*Solid Philosophy*, 117), are handicapped in the search for truth: 'Let any one ask a naturalist [empiricist] why Rotundity does not formally make a thing round, and you will see what a plunge he will be put to, not finding in all nature a proper reason for it' (ibid., 249). But for Locke, 'Transnatural causes in natural philosophy are not natural causes ...' (*Solid Philosophy*, 248; Note 43, Yolton, 'Locke's Unpublished,' 541). The metaphysical explanations Sergeant presents add nothing to what the 'historical, plain method' reveals of the world as we experience it. The *Essay* itself makes this point systematically. One can find in experience none of the entities, neither notions nor natures nor necessary connections, that the Aristotelian or rationalist asserts are there, known, familiar as objects in everyday experience, and centrally ingredient in the apprehensions we have of things (*Essay*, III.vi.24–7). Locke argues, on the basis of the 'historical, plain method,' that neither simple apprehension nor judgment nor discourse provide any grounds for the claim that we know the real essences or notions of material objects. Our ideas of substances (which in the Aristotelian tradition consist of parts necessarily tied together in a real definition and which are therefore in that tradition called 'simple apprehensions') are disclosed by the 'historical, plain method' to be not a group of ideas tied, of their necessity, into a unity, but no more than a collection of independent ideas of sense: 'our specifick ideas of substances are nothing but a collection of a certain number of simple ideas, considered as united in one thing. These *ideas* of substances, though they are commonly called simple apprehensions, and the names of them simple terms; yet in effect, are complex and compounded ... of various properties, which all terminate in sensible simple *ideas*, all united in one common subject' (II.xxiii.14).

Upon the Aristotelian view, notion or nature or essence provides the ontological ground of the unity; but, given the view that this very notion is in the mind when we know the substance, it is also this notion that must provide the epistemological ground for our knowing that certain sense properties are regularly connected in experience and others are not. If the Aristotelians were correct, then according to Locke, these simple notions must always be present in our judging. Yet the 'historical, plain method' shows that *they are in fact never there when we judge*, and that our ideas of substances are complexes of simple ideas *in no way* bound together by ontological or logical necessity.

Locke returns to the same point later, in Book III of the *Essay* (438–71); and again in Book IV. His discussion in Book IV is particularly revelatory of his attitude towards explanations of the sort Sergeant offered of coherence, and, more generally, of his attitude towards necessary connections that the Aristotelians believed bind simple ideas together into resulting real definitions. It is evident, Locke says, that we do not know the necessary connections required for an Aristotelian understanding of why parts of things cohere (IV.iii.26). But even if we knew why the parts cohere, we still would not know everything necessary for a grasp of the *notion* of the thing in Sergeant's sense. For the notion must account for all the causal activities of the substance of which it is the notion, insofar as these activities are not merely occasional. Now, the *regular* activities of external substances include the production of the ideas of the secondary qualities, that is, the production of simple ideas such as red, sweet, and so on. For these activities to be knowable scientifically, in Sergeant's Aristotelian sense, regularities revealed by sense about such activities must be demonstrable by syllogisms grounded in notions. But for that to be possible, there must be necessary connections between red, sweet, and so on and the notions or natures of the substances that cause these qualities to appear. These necessary connections must be both ontological, in the entities themselves, and epistemological, giving us, when in the mind, scientific knowledge of those entities. But, Locke argues, we grasp no such connections:

'Tis evident that the bulk, figure, and motion of several bodies about us, produce in us several sensations, as of colours, sounds, tastes, smells, pleasure and pain, etc. These mechanical affections of bodies, having no affinity at all with those ideas, they produce in us, (there being no conceivable connexion between any impulse of any sort of body, and any perception of a colour, or smell, which we find in our minds) we can have no distinct knowledge of such operations beyond our experience; and can reason no otherwise about them, than as effects produced by the appointment of an infinitely wise agent, which perfectly surpasses our comprehensions ... (IV.iii.28; see also IV.vi.10)

Locke's appeal is to the empiricist's principle of acquaintance (PA).[179] This is not to say that Locke was systematic in his deployment of this principle: he did not give up a substantialist account of mind largely incompatible with the principle;[180] and his nominalism and nominalistic account of relations was also very much incompatible with the principle.[181] None the less, we can see in Locke the first systematic use of PA. He did at least deploy the principle sufficiently systematically and sufficiently deeply to remove forever from Aristotelianism and rationalism whatever plausibility they had previously had.

We should also note, and indeed emphasize, that when Locke appeals to PA to deny the existence of necessary connections he is excluding not only Sergeant's 'notions' but also any sort of Aristotelianism that holds that there are non-tautologically necessary or synthetic a priori necessary connections of the sort

(##)   Whatever is soluble is such that whenever it is put in water it dissolves

or, what amounts to the same, the principle

(#) $R(S, P) \rightarrow$ All S's are P's

of our latter-day Aristotelians Tooley, Brown, and Armstrong.

The whole notion of a necessary connection is based in large part on a simple confusion, Locke, with some plausibility, goes on to suggest. Natural-kind terms appear in the purported demonstrative syllogisms of the Aristotelian. On the view of the latter, these kinds are explanatory through their logical/ontological/causal structure. But these objective necessary structures do not exist, given the appeal to PA. They are, to the contrary, to be understood as simply nominally defined. Indeed, whatever the illusions of the Aristotelian may be, this is all that they *are* or *could be*. The Aristotelian illusion of a real connection arises from treating some property as once part of the nominal definition and once not part of it. Treated the first way, we have a necessary connection but one that is trivial. Treated in the second way, we have a non-trivial and substantive connection, but one that is only contingent, not necessary. The Aristotelian illusion arises from blurring these two things together.

When 'All gold is yellow' appears in a demonstrative syllogism, in Sergeant's view, 'gold' refers to a universal notion, present in every sample of gold. The proposition itself expresses a necessary connection between this notion and that of yellow, which was contained in it as part of its real definition; and this proposition, though expressing a necessary connection, also is expressive of a substantive and explanatory connection. For Locke, Sergeant's notions, real essences, are not found in the world given in sense, and there is no other way for them to be given to us : the world is all on one level. The notions are therefore quite strictly unknowable. Syllogisms and the propositions in them can *never* be based on notions. 'Gold' cannot refer to a notion, or any universal, and must, instead, be a concept *the meaning of which is wholly explicable in terms of sense experience* – which can do *nothing more than* describe in terms of their sensible qualities the members of a class of sensible particulars. The proposition 'All gold is yellow' must be taken as making an assertion about sensible particulars, not about some notion that lies 'behind' and somehow 'explains' sensible appearances. As Locke says, 'in truth, the matter rightly considered, the immediate object of all our reasoning and knowledge, is nothing but particulars' (*Essay*, IV.xvii.8).

The concept of gold is the arbitrary creation of the human mind, designed to permit it to refer easily to all the members of a class of sensible particulars: it is an arbitrary creation in the sense that 'gold' is used to refer to particulars having sensible properties P_1, P_2, ..., and so on, where what is included in the list P_1, P_2, ..., is a matter of *decision*. The only idea of gold that an examination of our experience by the 'historical, plain method' reveals to us is not a 'real essence,' not one of Sergeant's 'notions,' but only a *nominal essence*: 'The real essence of ... any ... sort of substances, it is evident we know not; and therefore are ... undetermined in our nominal essences, which we make ourselves ...' (*Essay*, III.vi.27). As for P_1, P_2, ..., these are, in the last analysis, concepts referring to unanalysable, or simple properties of things: that is, in Locke's terms, certain simple ideas given in sense experience: 'our specific ideas of substances are nothing else but a collection of a certain number of simple ideas, considered as united in one thing. These ideas of substances, though they are commonly simple apprehensions, and the names of them simple terms; yet in effect are complex and compounded' (II.xxiii.14).

Among these simple ideas that go to make up our complex ideas of the nominal essences of substances will be many ideas (for example, that of yellow) of secondary qualities, and among these the mind can discover none of the necessary connections that should be there if the neo-Aristotelian and the rationalists are correct. In this sense, these simple ideas are logically, ontologically, and epistemologically *independent* of each other – 'Our senses, conversant about particular sensible objects, do convey into the mind several distinct perceptions of things' (*Essay*, II.i.3) – or, in the language of the time, distinguishable and therefore separable – just as for the Aristotelians and the rationalists these are not separable, neither logically nor ontologically nor epistemologically independent of each other. Now, in the proposition 'All gold is yellow' the concept of gold will be a complex idea, formed out of certain simple ideas which are logically independent of each other, while yellow will be itself a simple idea. Either the idea of yellow will be among the ideas in the definition of gold, or it will not be. If it is, then the proposition is trivial, as Molière argued:

Alike trifling it is, to predicate any other part of the definition of the term defined, or to affirm any one of the simple ideas of a complex one of the name of the whole complex idea; as, 'All gold is fusible.' For fusibility being one of the simple ideas that goes to the making up the complex one the sound gold stands for, what can it be but playing with sounds, to affirm that of the name gold, which is comprehended in its received signification? It would be thought little better than ridiculous to affirm gravely as a truth of moment, that 'gold is yellow'; and I see not how it is any jot more material to say, 'it is fusible,' unless that quality be left out of the complex idea, of which the sound gold is the mark in ordinary speech. (IV.viii.5)

A proposition that is trivial is, of course, certain, and known a priori, but equally, because it is not substantive, it cannot be explanatory. Again as Locke puts it,

'Every man is an animal, or living body,' is as certain a proposition as can be; but no more conducing to the knowledge of things, than to say, 'a palfry is an ambling horse, or a neighing ambling animal,' both being only about the signification of words, and make me know but this: That body, sense, and motion, or power of sensation and moving, are three of those ideas that I always comprehend and signify by the word man; and where they are not to be found together, the name man belongs not to that thing: And so of the other, that body, sense, and a certain way of going, with a certain kind of voice, are some of those ideas which I always comprehend, and signify by the word palfry; and when they are not to be found together, the name palfry belongs not to that thing. (IV.viii.6)

Thus, on the one hand, if the idea of yellow is not part of the complex idea of gold, then the proposition that 'All gold is yellow' is substantive and not trivial, while, on the other hand, in the absence of the required necessary connections, neither can it be a proposition knowable 'scientifically,' in Sergeant's sense, by means of a demonstrative syllogism. Nor can it be known with certainty, but only with the uncertainty that attaches, in the absence of any observable necessary connection, to every inference from an observed sample of the population to a general matter-of-fact proposition concerning a regularity in the population.

Locke's critique of the traditional notion of demonstrative science is thus much more radical than that of Descartes. In fact, the criticism of the latter is remarkably slim. After all, it is not a critique of the notion of demonstrative science itself, but only a modest criticism of the notion that syllogistic can be useful in the discovery of new necessary connections of the sort that one must come to know if one is to achieve the ideal of *scientia*. Locke, in contrast, challenges not just the notion that syllogistic can yield demonstrative science, but much more deeply the very idea that there are necessary connections of the sort required if any demonstrative science is to be possible, whether it be the demonstrative science of Descartes or that of Aristotle and Sergeant – or that of Tooley, Brown, and Armstrong who have much more recently argued that knowledge of laws consists in our grasping necessary connections among universals. For these latter philosophers, as for Sergeant and Descartes, there remains what we above referred to as the 'second philosophical problem' : how are these connections known? This question applies as much to the more recent defenders of objective necessities as it does to those whom Locke had in mind, philosophers such as Sergeant and Descartes. Locke provides the clear answer: we are not presented with such connections, and they are therefore, by an appeal to the principle of acquaintance, denied a place in one's ontology. It refutes his contemporaries: it continues to refute those who would revive that antiquated position.

6) *Method Made Empirical: (a) The Logic of Consistency*

Under the traditional account of science as *scientia*, demonstrative from necessary premises, syllogisms played a central role in displaying the (onto-)logical structure of reality. Given Locke's critique, based on the empiricist's principle of acquaintance, of the notion of such necessary connections, the role of logic had to be rethought. Locke makes the major point in redefining the role and status of logic in the search after truth.

Syllogistic has to do with inference. Now, for Locke, as for everyone, 'To infer is nothing but, by virtue of one proposition laid down as true, to draw in another as true, i.e. to see or suppose such a connexion of the two ideas of the inferred proposition ...' (*Essay*, IV.xvii.4). Syllogism consists of those 'forms of argumentation, wherein the conclusion may be shown to be rightly inferred' (IV.xvii.4). Knowledge of this is useful because it is useful to maintain consistency in our reasonings. Thus, its invention by Aristotle 'did great service against those who were not ashamed to deny any thing' (ibid.). It can 'help us in convincing men of their errours and mistakes' (IV.xvii.6). It is not, however, the grand instrument of truth that Aristotle and, following him, Sergeant claimed it to be. It is of no help in finding the middle terms that are necessary if we are to have inferences.

> The rules of syllogism serve not to furnish the mind with those intermediate ideas that may show the connexion of remote ones. This way of reasoning discovers no new proofs, but is the art of marshalling and ranging the old ones we have already. The forty-seventh proposition of the first book of Euclid is very true; but the discovery of it, I think, not owing to any rules of common logic. A man knows first, and then he is able to prove syllogistically. So that syllogism comes after knowledge, and then a man has little or no need of it. (IV.xvii.6)

Locke makes the same point with regard to the maxims such as 'whatever is, is' about which Sergeant makes so much, claiming them to be the foundation of human knowledge. They establish consistency, but they do not generate new truth: 'they sometimes serve in argumentation to stop a wrangler's mouth, by showing the absurdity of what he saith, and by exposing him to the shame of contradicting what all the world knows, and he himself cannot but own to be true. But it is one thing to show a man that he is in an error; and another to put him in possession of truth ...' (IV.xvii.11). Reasoning proceeds by means of distinct ideas; its very possibility presupposes that we perceive this distinction. Moreover, it is a distinction in ideas with regard to what they are about – their objective reality, to speak with the Cartesians. The distinction is thus a distinction *in things*. The perception of such distinctions among ideas insofar as they are of things is the starting point of knowledge. But such knowledge of distinctions does not establish connections among them, that is, connections of the sort that must be there if we are to obtain new knowledge, or even to make inferences. Indeed, as we saw, Locke's appeal to PA establishes that our ordinary ideas of things are in fact given to us as *distinct*, and given to us as *not* involving connections – objective necessary connections – to other things. '[T]he immediate perception of the agreement or disagreement of identity being founded in the mind's having distinct ideas, this affords us as many self-evident propositions, as we have distinct ideas. Every one that has any knowledge at all, has as the foundation of it, various and distinct ideas: And it is the first act of the mind (without which it can never be capable of any knowledge) to know every one of its ideas by itself, and distinguish it from others' (IV.vii.4). Thus, maxims such as 'whatever is, is' not only do not form the basis of knowledge but, in addition, are not even necessary to establish such propositions as 'red is red' or 'blue is not sweet':

It is not therefore alone to these two general propositions, 'whatsoever is, is'; and 'it is impossible for the same thing to be, and not to be'; that this sort of self-evidence belongs by any peculiar right ... These two general maxims, amounting to no more in short but this, that the same is the same, and the same is not different, are truths known in more particular instances, as well as in those general maxims; and known also in particular instances, before these general maxims are ever thought on, and draw all their force from the discernment of the mind employed about particular ideas ...' (IV.vii.4)

Indeed, so far from being a help in the pursuit of truth, syllogistic and tautological maxims can lead us away from the truth. As Locke says, as well as stating truths, and helping us attain consistency, it is also the case that 'general maxims will serve to confirm us in mistakes; and in such a way of use of words, which is most common, will serve to prove contradictions' (IV.vii.12). Consistency can preserve falsehood as well as truth, if it be falsehood from which our – consistent – inferences begin. Locke gives the example of Cartesian reasoning by means of Sergeant's maxim that 'whatever is, is' to establish that there is no vacuum:

the idea to which he annexes the name body, being bare extension, his knowledge, that space cannot be without body, is certain. For he knows his own idea of extension clearly and distinctly, and knows that it is what it is, and not another idea, though it be called by these three names, extension, body, space. Which three words, standing for one and the same idea, may no doubt, with the same evidence and certainty, be affirmed one of another, as each of itself; and it is as certain, that whilst I use them all to stand for one and the same idea, this predication is as true and identical in its signification, that 'space is body,' as this predication is true and identical, that 'body is body,' both in signification and sound. (ibid.)

Far from being the mark of science, syllogistic has been reduced to the criterion for valid inference, whether that inference be from true premises or false premises.

The traditional logic has thus radically changed its role in the search after truth. Formerly it was the touchstone of truth: there could be no science – that is, *scientia* – unless that knowledge could be exhibited in the form of a syllogism. With Locke, however, the role of logic is to secure consistency in one's thought, where now thought is limited to the empirical, and the standards of *scientia* have been abandoned. Or, rather, that is the role of formal logic. But that is no longer the whole story about logic. Where once there had been syllogistic both as a logic of consistency and as a standard of scientific truth, there is now syllogistic, or, more generally, formal logic, as the standard of consistency, and something more, a new *logic of truth*, to use John Stuart Mill's apt phrase. When Locke complained about the way traditional logic was taught using Burgersdijck's or some other traditional logic text, he was making the point that what was needed beyond this was tutoring in the new logic of truth of the science of Bacon, Boyle, and Newton.[182] In studying only formal logic, 'we learn not to Live, but to Dispute.'[183] It is to this new logic of truth upon which Locke was insisting that we must now turn.

7) *Method Made Empirical: (b) The Logic of Truth*

Locke is scathing in his comments on those who propose that maxims such as 'whatever is, is' are the clue to the advancement of knowledge. 'They are,' he tells us, 'not of use to help men forward in the advancement of sciences, or new discoveries of yet unknown truths.'

> Mr. Newton, in his never enough to be admired book, has demonstrated several propositions, which are so many new truths, before unknown to the world, and are farther advances in mathematical knowledge: But, for the discovery of these, it was not the general maxims, 'what is, is,' or, 'the whole is bigger than a part,' or the like; that helped him. These were not the clues that led him into the discovery of the truth and certainty of those propositions. Nor was it by them that he got the knowledge of those demonstrations; but by finding out intermediate ideas that showed the agreement or disagreement of the ideas, as expressed in the propositions he demonstrated. This is the greatest exercise and improvement of human understanding in the enlarging of knowledge, and advancing the sciences; wherein they are far enough from receiving any help from the contemplation of these, or the like magnified maxims. (*Essay*, IV.vii.3)

But what is the logic of truth, the methodological rules that are now needed to supplement the logic of consistency and to replace the mistaken rules of syllogistic demonstrative science?

We can begin to see what Locke has in mind by noting that he is, of course, prepared to grant that there is something non-arbitrary about the grouping of simple ideas into our complex ideas of substances. Our ideas of substances are all only nominal definitions, not the substantive real definitions of Sergeant, Aristotle, and company. Since they are nominal, they are framed by the mind; they are the mind's creation, not forms or essences imposed upon it by the object known. However, although these complex ideas are of the mind's own making, the mind does not proceed in an arbitrary manner. There is method to its conventions. Thus, 'all the ideas we have of particular distinct sorts of substances, are nothing but several combinations of simple ideas, coexisting in such, though unknown, cause of their union, as makes the whole subsist of itself' (*Essay*, II.xxiii.6). Certain simple ideas of sense, or, rather, the sensible properties from which these are derived, are *regularly coexistent*. That is, matter-of-fact regularities do obtain in world. Knowledge of these is uncertain in the sense that knowledge of all matter-of-(empirical-)fact regularities is uncertain. None the less, experience testifies to these regularities. It is a convenience if we form or define a complex idea of a substance out of those which we have 'usually observed, or fancied to exist together' (II.xxiii.6). It is, in other words, *a convenience in the use of thought and in the use of language to state facts if we represent the cluster of contingently coexisting properties by a single complex idea to which we refer with a single name.*

Men, observing certain qualities always join'd and existing together, therein copied nature; and of ideas so united, made their complex ones of substances. For though men may make what complex ideas they please, and give what names to them they will; yet if they will be understood, when they speak of things really existing, they must in some degree, conform their ideas to the things they would speak of: or else men's language will be like that of Babel; and every man's words, being intelligible only to himself, would no longer serve to conversation, and the ordinary affairs of life, if the ideas they stand form be not some way answering the common appearances and agreements of substances, as they really exist. (III.vi.28)

Locke's criterion or test for definitions is *empiricist* and *pragmatic*.[184] The criterion adopted by Sergeant and company refers to an entity beyond the world of sense experience. Does the definition conform to the *real definition* of the essence or notion of the thing? Locke's rejection of Aristotelian ontology leads him to a radically new concept of causal inference. For Sergeant and company, it is the real definition, the real essence, that is crucial; for Locke, definitions are irrelevant as judgments of lawfulness.

Suppose that we observe that a group of sensible properties are constantly conjoined: will they continue to be so? For the Aristotelian the conjunction will continue to hold provided that it is part of the real essence or notion of the thing.

If sense experience conveys this notion to us and we intuit it in a simple apprehension, we thereby acquire knowledge of necessary constant conjunction: it is a law. Since Locke denies the epistemological significance of real essences, he cannot appeal to them for judgment of lawfulness. None the less he still must provide for judgments of lawfulness – whether a regularity observed to hold in a sample will continue to hold in the population. Since observation of the sample is the only basis for the judgment, Locke recognizes that such judgments cannot be certain, that they cannot attain to the standard claimed to be possible by the defenders of science as *scientia*: 'We may take notice, that general certainty is never to be found but in our ideas. Whenever we go to seek it elsewhere in experiment, or observations without us, our knowledge goes not beyond particulars. 'Tis the contemplation of our own abstract ideas that alone is able to afford us general knowledge' (*Essay*, IV.vi.16). This incapacity to achieve absolute certainty is a fact about the limits of human knowledge. It follows, as Locke says, that 'natural philosophy is not capable of being made a science' (IV. xii.10), that is, science in the sense of *scientia*. To be sure, although absolute certainty is not possible, there are goals that we can attain, more modest goals. 'Experiments and historical observations we may have, from which we may draw advantages of ease and health, and thereby increase our stock of conveniences for this life: but beyond this, I fear our talents reach not, nor are our faculties, as I guess, able to advance' (ibid.). And because it *is* a *fact*, we ought not, at least ought not if we are reasonable persons, to attempt to overcome it.

If, by this enquiry into the nature of the understanding, I can discover the powers thereof; how far they reach; to what things they are in any degree proportionate; and where they fail us: I suppose it may be of use to prevail with the busy mind of man, to be more cautious in meddling with things exceeding its comprehension; to stop when it is at the utmost extent of its tether; and to sit down in a quiet ignorance of those things which, upon examination, are found to be beyond the reach of our capacities. We should not then perhaps be so forward, out of an affectation of an universal knowledge, to raise questions, and perplex ourselves and others with disputes about things, to which our understandings are not suited; and of which we cannot frame in our minds any clear or distinct perceptions, or whereof (as it has perhaps too often happened) we have not any notions at all. If we can find out how far the understanding can extend its view, how far it has faculties to attain certainty, and in what cases it can only judge and guess; we may learn to content ourselves with what is attainable by us in this state. (I.i.4)

We must, as it were, lower our standards, that is, our *cognitive standards*, in the light of our knowledge of the ontological structure of the world and our rational capacities. Knowledge in the traditional sense is not possible concerning matter-of-fact regularities; we must therefore settle for judgment 'which is the putting ideas together, or separating them from one another in the mind, when their certain agreement or disagreement is not perceived, but presumed to be so' (*Essay*, IV.xiv.4.653), allowing that standard of probability 'to supply the defect of our knowledge' (IV.xv.4), where probability 'is the admitting or receiving any proposition for true, upon arguments or proofs that are found to persuade us to receive it as true, without certain knowledge that it is so' (IV.xv.3). We must, in other words, not attempt to achieve the impossible goal of the certainty of *scientia* but instead settle for the attainable goal of probability. We must adapt our cognitive ends to cognitive means that we know to be available.

At the same time, however, we must adopt rules that do, in fact, so far as we can tell, lead to attain truth with the degree of certainty that is possible. 'We must ... if we will proceed, as reason advises, adapt our methods of enquiry to the nature of the ideas we examine, and the truth we search after' (*Essay*, IV.xii.7). Thus, although there are limits to human knowledge, limits that prevent us from attaining the ideal of demonstrative *scientia*, this does not mean that we cannot develop *within these limits* rules to help us judge of truth so far as we can ascertain it, that is, matter-of-empirical-fact truth, with at best a moral and not an absolute certainty. In particular, the 'grounds' of probability are, first, 'The conformity of any thing with our own knowledge, observation, and experience' and, secondly, 'The testimony of others' (IV.xvi.4). A proposition is more or less probable 'as the conformity of our knowledge, as the certainty of observations, as the frequency and constancy of experience, and the number and credibility of testimonies, do more or less agree' (IV.vxi.6). The highest degree of probability so attained raises our affirmations and denials 'near to certainty' (IV.xvi.6). 'Where any particular thing, consonant to the constant observation of our selves and others, in the like case, comes attested by the concurrent reports of all that mention it, we must receive easily, and build as firmly upon it, as if it were certain knowledge; and we reason and act thereupon with as little doubt, as if it were perfect demonstration' (IV.xvi.6). With rules of this sort we can develop strategies for attaining the truth, insofar as we can really attain it, given the ontological structure of the world we know and the faculties that we have.

To be sure, Locke does not develop in any great detail the rules for attaining the truth. He does not elaborate, beyond these few hints, the nature of the strategies we can reasonably adopt to attain the truth or, at least, most probably attain it, and still less does he elaborate a set of reasons for preferring one strategy to another. But he does enough to give us the general idea of how we should seek rules for attaining truth – rules, based on experience, for attaining, at least probably, the truth concerning matters of empirical fact.

These rules will be the test of empirical knowledge and will constitute the logic of truth that is the necessary supplement to the logic of consistency in the Lockean or empiricist account of our knowledge of matters of empirical fact, the only sort of fact that, given PA, we can accept as knowable. In terms of these rules we judge whether properties observed to be constantly conjoined will continue to be so. Lawfulness in nature is no longer to be judged by whether observed regularities can be accounted for in terms of notions or real essences. The whole apparatus of self-evident first principles and demonstrative syllogisms is no longer relevant to evaluating matter-of-fact regularities. The only test remaining is observation and rules that in practice lead one to truth about matters of fact. Sense experience alone controls claims to scientific knowledge. Experiment and hypothesis testing is done in terms of experience. It is Locke's claim that this makes possible an adequate intellectual understanding of the possibility of a new, probabilistic science by the empirical method.

'Truth' is 'nothing but *the joining or separating of signs, as the things signified by them, do agree or disagree one with another*' (*Essay*, IV.v.2; italics added). Since 'joining or separating of signs here meant is what by an other name, we call proposition,' it follows that '[t]ruth properly belongs only to propositions' (ibid.). Such an account of truth is on the surface not much different from that of Sergeant or the Cartesians and the Port Royal logicians. They were, however, confronted by a problem which does not confront Locke. This is the problem that we saw to be present and which the earlier philosophers tried to solve by invoking the *dictum de omni et nullo*.

For the Cartesians, the Port Royalists, and Sergeant, knowledge is primarily a matter of grasping the logical structure of ideas. But then we have the question why this a priori structure should be reflected in the relations among the extensions of the ideas in the propositions, among the particulars that fall under the concepts. The *dictum de omni et nullo* is the claim that IN FACT the ontological structure of ideas determines the relations among the particulars falling under the ideas. But what guarantees that the *dictum* is true? The answer of Aristotle and Sergeant – and of their latter-day followers such as Armstrong – is that it is the *causal activity* of the particulars, structured by those very forms, that guarantees the truth of the *dictum*. Locke's appeal to PA, however, effectively eliminates this attempt to hold that we have reason to believe the *dictum* to be true. What is important about the appeal to PA, however, is that it eliminates the need for the *dictum* itself.

The *dictum* is introduced in order to show the *ontological gap* between ideas or notions or real essences, on the one hand, and, on the other, the particulars falling under those essences or forms. It is needed in order to ensure that the structure among the essences or forms is reproduced, as it were, in the regularities among the particulars. Time, or, at least, the temporal world thus becomes the moving image of eternity, to use Plato's pregnant metaphor. However, Locke's appeal to PA *eliminates the ontological structure of forms or ideas, and, since this no longer exists, or, at least, is dismissed, there is no longer any need to ensure that it is reproduced among the particulars!*

For Locke, to be sure, there is a structure among ideas. They are joined or separated in propositions, and truth consists in these ideas being joined or separated as the extensions of the ideas are joined or separated. The point is that, when a proposition is true, the structure of the proposition *reflects the relations among the extensions; it is not supposed that the relations among ideas somehow guarantee the relations among the extensions.* The relations among extensions determine the truth of the proposition; the structure of the proposition does not somehow have to guarantee the relations among the extensions. The problem that the *dictum* was intended to address has thus disappeared. As we saw Mill put the point,

> when it is known that a class, an universal, a genus or species, is not an entity *per se*, but neither more nor less than the individual substances themselves which are placed in the class, and that there is nothing real in the matter except those objects, a common name given to them, and common attributes indicated by the name; what, I should be glad to know, do we learn by being told, that whatever can be affirmed of a class may be affirmed of every object contained in the class? The class *is* nothing but the objects contained in it: and the *dictum de omni* merely amounts to the identical proposition, that whatever is true of certain objects is true of each of those objects. (*System of Logic*, II.II.2)

One final point must be made. For Aristotle, Sergeant and Armstrong, the logical structure of ideas or forms guarantees their own truth. They do this by virtue of their objective existence in things, forming the causal activities of the latter, and guaranteeing that those things reflect the structure of the ideas that inform them. In the case of Locke, or, more generally, the empiricist, there is a rather different sort of guarantee that the structure of our ideas reflects the objective relations among things falling under those ideas.

For Locke and the empiricist, our ideas are derived from things. This is partly a matter of the automatic operation of natural mechanisms of the mind. It is also partly a matter of deliberate structuring of one's own ideas by one's own reflective mental activity. Many of our complex ideas of substances are formed automatically. Others are formed quite deliberately, and those that are formed automatically come to be reformed as a result of deliberate reflective activity. In general, on the basis of observation and experience we join and separate our ideas. *What guarantees the truth of our ideas, insofar as it can be guaranteed, is our deliberate effort to conform our thought to the rules of the logic of truth.* This is not the sort of grand metaphysical guarantee of truth that we find in Aristotle and Sergeant, and in Tooley, Brown, and Armstrong. To the contrary, the Lockean guarantee of truth is not only fallible but depends upon some contingent and perhaps accidental properties of the human mind, some apparently rather trivial features of the mind. But for all that, it is the best that we can do. More strongly, it is in fact *better* than the grand metaphysical schemes of the anti-empiricist philosophers, the defenders of objective necessities. For what Locke has shown, with his appeal to the empiricist's PA, is that the objective necessities the anti-empiricists suppose to be there are in fact *illusions*, fancies of those who want to prove that their minds have some superhuman and non-natural capacities that enable them somehow to penetrate deeper into the structure of reality than can their senses. It is thus Locke, not Sergeant, who in the end is the defender of solid philosophy. *It is Locke's achievement to have defended solid philosophy against the fancies of the metaphysicians.*

Study Three
Berkeley's Metaphysics and Ramist Logic

'What is truth?' asked Pilate and did not wait for an answer. Aristotle would have told him that '[t]o say of what is that it is not, or of what is not that it is, is false, while to say of what is that it is, and of what is not that it is not, is true ...' (*Metaphysics*, 1011b 27–9).[1] But Aquinas[2] tells us that

a house is said to be true that expresses the likeness of the form in the architect's mind; and words are said to be true so far as they are the signs of truth in the intellect. In the same way natural things are said to be true in so far as they express the likeness of the species that are in the divine mind. For a stone is called true, which possesses the nature proper to a stone, according to the preconception in the divine intellect. Thus, then, truth resides primarily in the intellect, and secondarily in things according as they are related to the intellect as their principle. (*Summa*, FP, Q16, A1)

Saying of what is that it is, is a rather complicated thing, in other words, in spite of its apparent simplicity. For, what is depends upon one's ontology. According to Aquinas, the truth of a thing is a standard to which the thing bears a likeness. This standard is the form or species of the thing in the divine intellect. The person who knows abstracts from the thing judged about the species or standard of its truth, and compares the object to this standard. 'When ... [the mind] judges that a thing corresponds to the form which it apprehends about that thing, then first it knows and expresses truth. This it does by composing and dividing: for in every proposition it either applies to, or removes from the thing signified by the subject, some form signified by the predicate ...' (ibid., A2). When the mind is thus judging the thing, the form of the thing known is in the mind. '[S]ince everything is true according as it has the form proper to its nature, the intellect, in so far as it is knowing, must be true, so far as it has the likeness of the thing known, this being its form, as knowing. For this reason truth is defined by the conformity of intellect and thing; and hence to know this conformity is to know truth' (ibid.). As for the individual of which forms are predicated, this, for example, this man, 'is said to be a suppositum, because he underlies [supponitur] whatever belongs to man and receives its predication' (TP, Q2, A3). This subject is made actual by virtue of its having accidents predicated of it. '[A] subject is compared to its accidents as potentiality to actuality; for a subject is in some sense made actual by its accidents' (FP, Q3, A6). The substance is, however, an individual thing that is distinct from those accidents: 'Although the universal and particular exist in every genus, nevertheless, in a certain special way, the individual belongs to the genus of substance. For substance is individualized by itself; whereas the accidents are individualized by the subject, which is the substance; since this particular whiteness is called "this," because it exists in this particular subject' (FP, Q29, A1). The unity of the individual derives from the substance. The collection of properties or accidents which characterize the individual are unified into the set of characteristics of a particular individual precisely because each of them inheres in the substance. The unity of any ordinary thing, such as a die, derives from the substance in which all the properties of the thing inhere.

As for Pilate's question, 'What is truth?', in order to answer that we must lay out our ontology. In the case of the traditional Aristotelian ontology that we have been looking at, a substance is true – true to its own self, if you wish – provided that the accidents which make the substance actual are those required by the species or form of the substance.

This doctrine of truth depends upon each substance having a certain form or nature or essence, or, what is the same, from the side of mind, an intelligible species. In the seventeenth century a different ontology developed, one that rejected the intelligible species of Aristotle and Aquinas. It retained, however, the doctrine of substance. As we saw in Study One, this entailed a rejection of the traditional doctrine in which a substance was said to be true to the extent that it conformed to the specific form or nature the activity of which determined its being. But the retention of the doctrine of substance still tied truth to the substance ontology. To speak the truth is to say of what is that it is. And to say of what is that it is, is to predicate a property of a substance.

Now, what this means is that the relation of predication occupies a special place in any account of logic and language, and is contrasted to other 'ordinary,' if you wish, relations. Logic and grammar as accounts of language deal with propositions. These are either true or false, as Aristotle said. When we say of what is that it is, when we say, for example, that a is S

then what we are saying, on the traditional view, is that the *substance*, referred to by the subject term *a*, has the property referred to by the predicate term *S*. If *a* is in fact *S*, then we are saying of what is that it is, or, in other words, we are saying what is true. And if *a* is after all not *S*, then we are saying of what is not that it is, or, in other words, we are saying what is false. On the view that substances are the basic entities in one's ontology, as Aquinas, following Aristotle, indicates, then all other, 'ordinary,' relations are among the properties that are predicates of substances. This view is standard enough to constitute the pattern laid out in the common logic texts of the age.

Let us look, to take one example, at the *Monitio logica* of Franco Burgersdijck, a widely used text in the early modern period. Burgersdijck was professor of philosophy at Leyden, and his *Monitio logica* was republished in his native Holland several times, as well as being published in England, including a translation of 1697.[3] In spite of its condemnation by Locke,[4] it continued to be used.[5] Among other things, it discusses relations, presenting a view that was uncontroversial in the early modern period, putting them in a context of an ontology of substances.

Burgersdijck, in fact, discusses relations in two contexts. One is in the context of the categories or predicaments and the other is in the context of the topics.[6]

For the medievals, the entities about which one could talk were divided into the categories. These categories or predicaments are those of substance, quality, quantity, relation, action, and so on. Burgersdijck gives the traditional list, which, of course, like almost everything else in logic, both medieval and modern, derives from Aristotle. A sentence or proposition makes a predication of an entity in a category when it attributes a certain property to it. A predication was legitimate, or, as we would say, 'well formed,' no matter the category to which the subject of predication belonged, provided that the property predicated of that subject fell within one of the predicables: relative to the subject, it had to be a genus, species, difference, property, or accident. The list of predicables defines the various sorts of judgment that can be made, and which can enter into discourse and into the reasoning that appears in syllogisms.

In Burgersdijck the discussion of the predicables follows immediately upon the discussion of the predicaments or categories. This discussion is immediately followed by a discussion of the various topics, that is, the places from which arguments may be drawn for use as premises in disputations and arguments. These places include those of property and accident, whole and part, cause and effect (which includes matter and form, efficient cause, and end), subject and adjunct, similarity (what Burgersdijck calls 'convenience'), difference ('diversity or distinction'), opposition, and order. In fact, as the text proposes, there is no real separation of the predicables from the topics, so that the various predicables themselves become topical places, or, what in effect amounts to the same, the topics come to define different sorts of legitimate predication. Once this is recognized, we see that Burgersdijck has in effect extended, whether he knew it or not, the list of legitimate forms of predication. Bringing predicates together as cause and effect, as in 'fire causes heat,' for example, becomes one form of predication, alongside the predications allowed by the predicables, for example, that of a genus, as in 'man is an animal.' For Burgersdijck, then, the list of legitimate forms of judgments has been extended beyond that of the medievals: the forms of legitimate judgment now include not only that defined by the relation of copulation, but also those defined by such relations as cause and effect, whole and part, similarity or resemblance, and difference.

Here we find a difference between Burgersdijck's treatment of relations and that deriving from Aristotle. For the latter, relations are one of the categories; they are among the things predicated of substances. But in Burgersdijck, relations are discussed in two ways, once in the context of the list of categories, and once in the list of topics. Relations are at once in a special category of entity and are also ways of relating such entities to each other in judgments.

Burgersdijck describes the category of relation in this way – which of course goes all the way back to the first chapter of Aristotle's *Categories*: 'Those things are said to be related, which in Respect of what they are, are said to be others, that is, of others or in any other Manner or Respect are referred to another' (*Monitio*, 19). According to Burgersdijck every relation involves two entities: that from which the relation originates and that in which it terminates. 'In every Relation,' he tells us, 'are required Subject and Term ... That is called the Subject to which the Relation is attributed; or that which is referred to some other thing ... That, the Term to which the Subject is referred' (21). The subject is the relate and the term the correlate. Relational predications thus presuppose qualities in both the subject and the term. The quality may be either a property, that is, a quality present universally in subjects of the relevant sort, or an accident, that is, a property that is not present universally. In the latter case, Burgersdijck speaks of the relation's having a foundation. 'Some Relations are supposed, supposing the Subject and Term: Others besides these do require a Foundation ... And a Foundation is that by whose Means the Relation acrews to the Subject' (22). He gives the following examples:

When an Egg is said to be like an Egg, the Similitude between these two Eggs arises in each as soon as they begin to exist; nor is there any thing required towards their

Relation, besides the Existence of two Eggs. But the Relation of Servant does not presently arise in the Subject so soon as the Term exists; but it behoves that something else also do intercede upon which this Relation is founded: For a Servant is therefore the Servant of one, because by him he has been either saved or purchased, &c. (22)

The distinction between property and accident was to become increasingly unimportant; so, for the sake of convenience, let us speak of those qualities in a thing by which the relation 'acrews to the subject' as the *foundation*.

Now, in the basic instance the subjects and terms of relations are substances. The substances are related to each other by virtue of the foundations which are present in them. As the examples make clear, the qualities which are the foundations of the relation are themselves *non-relational*. Moreover, each of the substances which are the subject and the term is, as Burgersdijck puts it, following the ancient formula, 'a Being subsisting of it self, and subject to Accidents' (*Monitio*, 8). Unlike accidents, substances do not depend for their existence on something else: 'To subsist by it self is nothing else but not to be in any thing as a Subject ...' (ibid.). Now, an accident is 'a being inherent in a Substance'; it 'cannot exist without a Subject'; nor can it 'pass from one Subject to another' (10). Thus, an accident cannot exist apart from the substance in which it is present. So substances can exist apart from other substances. If a substance ceases to exist, so do its accidents; but since substances subsist by themselves, a substance can exist unchanged, that is, remain the same in its being, even if other substances were to cease to exist. Since it is not predicated of other things, the non-existence of other things does not affect its existence or being.

What is distinctive of relation, then, is not that it introduces an entity over and above substances and the qualities that are present in them as properties and accidents. What is important about relation is that, by virtue of the qualities in them, one substance is *referred* to another. In a *relational judgment*, that is, one of the judgmental forms given in the list of topical relations, one substance becomes the subject of the judgments, another the term. Here, the substance that is the subject becomes so through a non-relational quality that has been picked out as the foundation of the judgment in that substance; and the substance that is the term becomes so through a non-relational quality that has been picked out as the foundation of the judgment in that substance. Thus, substances become subjects and terms, and qualities become foundations, through the fact that they appear in a relational judgment. *Relations do not constitute an ontological union among substances; such union as occurs is a union in the judgment.*

We find this view of relations repeated in Locke, but in the context of a systematic account of the operations of the mind.[7] Thus he tells us that

The nature therefore of relation, consists in the referring, or comparing two things, one to another; from which comparison, one or both comes to be denominated. And if either of those things be removed, or cease to be, the relation ceases, and the denomination consequent to it, though the other receive in it self no alteration at all, v.g. Cajus, whom I consider to say as a father, ceases to be so to morrow, only by the death of his son, without any alteration made in himself. (*Essay*, II.xxv.5)

A little later Locke adds that 'there can be no relation, but betwixt two things, considered as two things. There must always be in relation two ideas, or things, either in themselves really separate, or considered as distinct, and then a ground or occasion for their comparison' (*Essay*, II.xxv.6). Here we have the same doctrine of relations that one finds in Burgersdijck: relational judgments are *about* facts that are essentially non-relational, facts in which a non-relational quality is present in a substance, while the bringing together of these two into a *unity* of the subject and term is provided by the mind's referring the one to the other in a *mental act of comparison*.

This view can be represented as follows.[8] Consider the relational statement that

($@$) a is R to b.

On the Lockean account, such a statement has a twofold analysis. On the one hand, there are the *objective truth conditions*, the objective facts concerning a and b which determine whether the statement is (objectively) true or false. On the other hand, there is the *subjective mental state* that the *use* of the relational statement *expresses*. As for the former, the objective facts represented by ($@$) are *non-relational*:

(#) a is r_1 and b is r_2.

The non-relational properties r_1 and r_2 are the (objective) *foundations* of the relation. Both

(+) a is r_1

and

(++) b is r_2

will be true. What must be noted is that a and b are *independent* of each other in the sense that even if one ceases to exist, this will not affect the being of the other. Thus, for example, if *a* were to cease to exist, the relational fact (#) would no longer obtain. Moreover, (+) would also cease to obtain. None the less, the being of *b* would be unaffected. For, (++) would still be true: the non-existence of *a* will imply that (+) could no longer be true, but this does not affect the predications that would continue to be made of *b*, including the predication represented by (++). As for the subjective state that the use of (@) *expresses*, this is a *judgment of comparison*. The two substances, *a* and *b*, are entities capable of subsisting by themselves on this account of relations. On the usual account, with which we are more familiar, deriving from Russell, this is not so. On this latter account, (@) will be represented by a primitive relational predicate:

(*) $R(a, b)$.

This in turn means that we will have both

(&)   *a* is (R to *b*)

and

(&&)   *b* is (R-ed by *a*).

However, if *a* ceases to exist, then not only will (*) cease to be true, and not only will (&) cease to be true, but in addition (&&) will also cease to be true. In other words, upon Russell's ontology of relations, if one of the relata in a relational fact were to cease to exist, the being of the other relatum would change. Upon Russell's account, then, the two relata are *not* independent of each other as they are upon the traditional account for relations defended alike by Burgersdijck and Locke.

Locke locates this traditional account of relations which he adopts in the context of his more general account of thought and language. Human beings are sociable animals, and language is the medium by which they communicate. Articulate sounds express our ideas and enable us to communicate those ideas to others (*Essay*, III.i.1, 2). Various complex ideas, such as the mixed modes, consist of complex ideas whose unity is not a matter of objective connection but of the mind putting them together. 'Nobody can doubt but that these ideas of mixed modes are made by a voluntary collection of ideas put together in the mind, independent from any original patterns in nature, who will but reflect that this sort of complex ideas may be made, abstracted, and have names given them, and so a species be constituted, before any one individual of that species ever existed' (II.v.5). The examples that he gives are the relations of sacrilege and adultery; he could equally have used brothers or fathers. There are three aspects to the creation of these ideas, and their representation in language, according to Locke.

[W]e must consider wherein this making of these complex ideas consists; and that is not in the making any new idea, but putting together those which the mind had before. Wherein the mind does these three things: First, it chooses a certain number: Secondly, it gives them connexion, and makes them into one idea: Thirdly, it ties them together by a name. If we examine how the mind proceeds in these, and what liberty it takes in them, we shall easily observe how these essences of the species of mixed modes are the workmanship of the mind; and consequently, that the species themselves are of men's making. (III.v.4)

Indeed, it is the habit formed by the word that provides the cement as it were that keeps the parts of the idea together over time in a single, unchanging complex (III.v.10). In any case, the point is that the *unity* of the complex is a unity that derives from an act of the mind. The word that we use not only expresses the idea but expresses, too, the *act of unifying* that binds the simple ideas together in the mixed mode or other abstract idea.

Communication thus consists in grasping not only the simple ideas that make up our complex ideas, but also in grasping the act of the mind that unifies these ideas into wholes. The relational acts in my mind must correspond to the relational acts in your mind, if our communication is to be successful.

There is one exception, one case where the unity of the complex derives from an objective ontological basis, and not simply from the act of mind that unifies the ideas into a whole. That, of course, is the unity that derives from a substance.

Whatever ideas we have, the agreement we find they have with others, will still be knowledge. If those ideas be abstract, it will be general knowledge. But, to make it real concerning substances, the ideas must be taken from the real existence of things. Whatever simple ideas have been found to co-exist in any substance, these we may with confidence join together again, and so make abstract ideas of substances. For whatever have once had an union in nature, may be united again. (*Essay*, IV.iv.12)

The Lockean account of language and relations is thus very much of a piece with that found in Burgersdijck. What we find in Berkeley, however, is a radical change: *substance disappears*.[9]

As to what philosophers say of subject and mode, that seems very groundless and unintelligible. For instance, in this proposition 'a die is hard, extended, and square,' they will have it that the word *die* denotes a subject or substance, distinct from the hardness, extension, and figure which are predicated of it, and in which they exist. This I cannot comprehend: to me a die seems to be nothing distinct from those things which are termed its modes or accidents. And, to say a die is hard, extended, and square is not to attribute those qualities to a subject distinct from and supporting them, but only an explication of the meaning of the word *die*. (*Principles*, sec. 49)

We have here a new ontology, in which predication no longer represents the relation between a property and a substance but rather the relation between a property and a whole of which it is a part. To say of what is that it is, is to attribute a property to a whole of which it is indeed a part; while to say of what is not that it is, is to attribute a property to a whole of which it is not a part.

It is the origins of this radical new ontology that I wish to explore in this essay. In fact, of course, there are several things that are in the background to this amazing innovation in the history of philosophy. What I want to argue is that we can see in the logic of Petrus Ramus *one* of the things that we can reasonably suppose to have contributed to showing Berkeley the way to this breakthrough.

The immediate background is, of course, Locke. Locke retains a substance ontology for ordinary things. Berkeley exorcises substances in this context. What leads him to see that he can do this? What leads him to see that one need not give predication a special place in accounting for the structure of the world? What I want to suggest is that there is in the logic of Ramus a view of logic and of language, a view of discourse, that carefully avoids tying logic and language to an ontology of substances.

Berkeley himself, of course, places his new account of the nature of predication within the context of a view of the world in which properties are the basic characters or words of a natural language, the language of God.

Hence it is evident, that those things which, under the notion of a cause co-operating or concurring to the production of effects, are altogether inexplicable, and run us into great absurdities, may be very naturally explained, and have a proper and obvious use assigned them, when they are considered only as marks or signs for our information. And it is the searching after, and endeavouring to understand those signs (this language, if I may so call it) instituted by the author of nature, that ought to be the employment of the natural philosopher, and not the pretending to explain things by corporeal causes; which doctrine seems to have too much estranged the minds of men from that active

principle, that supreme and wise spirit, 'in whom we live, move, and have our being.'
(*Principles*, sec. 66)

Winkler has argued that Berkeley 'transforms the natural world from a system of bodies with powers and operations into a system of inert signs – a *text* – with no existence apart from the spirits who transmit and receive it.'[10] As he puts it a little later, 'experience is a text, authored by God in the language of ideas for the sake of our well-being.'[11] There is an important sense in which Winkler's characterization of Berkeley's world as a *text* is true and important. Berkeley's metaphysics did indeed involve a radical reconceptualization of the universe. But Winkler, none the less, has not got it exactly right either.

Already the new science had given up much of Aristotle's metaphysics. In particular, it had given up the unanalysable powers which provided the explanations of ordinary events. Thus, Robert Boyle, in his *A Free Enquiry into the Vulgarly Receiv'd Notion of Nature*,[12] attacked specifically the appeal to natures or forms as something that at once was no more than an appeal to ignorance – the sceptical arguments made acceptance of any such metaphysical entities unreasonable – and at the same time something that interfered with the progress of experimental science. The 'Nature ... is so dark and odd a thing, that 'tis hard to know what to make of it, it being scarce, if at all, intelligibly propos'd, by them that lay the most weight upon it' (*Free Enquiry*, 129). At the same time, Boyle 'observe[s] divers Phaenomena, which do not agree with the Notion or Representation of Nature ...' (136). Boyle cites the phenomenon of a vacuum as contrary to many of the things that people have said about Nature as a causal and explanatory force. That is, when one appeals to such an entity, one not only settles for explanations which are bad because they are obscure but settles for explanations which are in fact wrong, misdescribing the phenomena in question. He also cites (145ff) the alleged explanations of motions in terms of the qualities gravity and levity. Bodies of the former sort, that is, with that sort of 'Innate Appetite' (147), move in straight lines towards the centre of the Earth, while bodies of the other sort move in straight lines away from the centre and towards the heavens. Boyle points out how this doctrine makes very little sense with regard to the motion of a pendulum (148–9). Boyle also argues in detail (sec. VIII, 347ff) that various propositions which are supposed to be established in regard to the notion of 'nature,' – such as 'nature does nothing in vain' – explain either too much or too little, and, in fact, in general can be understood, when taken as scientific and referring to patterns of behaviour of objects in the world of sense experience, as making assertions compatible with the mechanical philosophy.[13]

Or, to take another example, E. Halley argues against those who would explain gravity in terms of 'a certain *Sympathetical attraction* between the *Earth* and its *Parts*, whereby they have, as it were, a desire to be united ...' This, he says, 'is so far from explaining the *Modus*, that it is little more, than to tell us in other terms, that *heavy bodies descend*, because they *descend*.'[14] In this respect Winkler is correct: the world had lost its active dispositions. The world had become inert.

However, if the world had lost active forms, and all these features of the traditional Aristotelian metaphysics, it continued to be a world of substances. The new science had not (yet) attacked the substance metaphysics. Thus, while Locke provided the metaphysical/epistemological framework in which the empirical methodology of the new science was put to work, he none the less retains substances as entities that we must hypothesize, though we know not what they are in themselves. They are simply those things, 'we know not what,' which underlie and support qualities: 'our idea of substance is ... obscure ... It is but a supposed I know not what, to support those ideas we call accidents' (*Essay*, II.xxiii.25). Although the idea is obscure, we can use it to think about substances as the support of qualities; and to judge that the latter exist is to judge that they are supported by a substance. This much of the traditional doctrine of substance remains.

However, if there are substances, what we in fact perceive are collections of qualities.

The mind being, as I have declared, furnished with a great number of the simple ideas, conveyed in by the senses, as they are found in exterior things, or by reflection on its own operations, takes notice also, that a certain number of these simple ideas go constantly together; which being presumed to belong to one thing, and words being suited to common apprehensions, and made use of for quick dispatch, are called, so united in one subject, by one name: Which, by inadvertency, we are apt afterward to talk of, and consider as one simple idea, which indeed is a complication of many ideas together; because, as I have said, not imagining how these simple ideas can subsist by themselves, we accustom ourselves to suppose some substratum wherein they do subsist, and from which they do result, which therefore we call substance. (*Essay*, II.xxiii.25)

Accidents may well be predicated of an 'I know not what,' but, so far as perception is concerned, what is given to us is simply a collection.

For Locke, events are in principle connected by the relation of causality. But, in fact, we do not know this relation. '[W]hatever change is observed, the mind must collect a power somewhere able to make that change, as well as a possibility in the thing itself to receive it. But yet, if we will consider it attentively, bodies, by our senses, do not afford us so clear and distinct an idea of active power, as we have from reflection on the operations of our minds' (*Essay*, II.xxi.4). For most qualities of things, there simply are no forms that could provide a connection any stronger than matter-of-fact regularity: the necessary connections are not given to us in ordinary experience; we therefore have no idea of such connections; and in the absence of such an idea, such necessary connections are simply inconceivable.

> The reason why the one ['primary qualities'] are ordinarily taken for real qualities, and the other only for bare powers, seems to be, because the ideas we have of distinct colours, sounds, &c. containing nothing at all in them of bulk, figure, or motion, we are not apt to think them the effects of these primary qualities, which appear not, to our senses, to operate in their production; and with which they have not any apparent congruity, or conceivable connexion. (II.viii.25)

Or, as he puts it elsewhere, the complex idea we have of any substance 'cannot be the real essence ... for then the properties we discover in that body would depend on that complex idea, and be deducible from it, and their necessary connexion with it be known' (II.xxxi.6).

There are no perceivable necessary connections among the properties that things are presented as having in the world of ordinary experience. Words that philosophers use that purport to refer to real essences are in fact meaningless, sounds without referents. '[W]hen I am told that some thing besides the figure, size, and posture of the solid parts of that body, is its essence, some thing called substantial form, of that I confess, I have no idea at all, but only of the sound form, which is far enough from an idea of its real essence, or constitution' (*Essay*, II.xxxi.6). In the absence of real essences, Aristotelian forms, or necessary connections, we are reduced to regularity. But this is, after all, all that we need for our practical purposes in our ordinary life in our ordinary world: we seem able to get on in life just fine with regularity alone.

> Tell a country gentlewoman that the wind is south-west, and the weather lowering, and like to rain, and she will easily understand it is not safe for her to go abroad thin clad, in such a day, after a fever: She clearly sees the probable connexion of all these, viz. south-west wind, and clouds, rain, wetting, taking cold, relapse, and danger of death, without tying them together in those artificial and cumbersome fetters of several syllogisms, that clog and hinder the mind, which proceeds from one part to another quicker and clearer without them; and the probability which she easily perceives in things thus in their native state would be quite lost, if this argument were managed learnedly, and proposed in mode and figure. (IV.xvii.4)

So, for the most part, what we have are not causal relations but merely constant conjunctions. In these constant conjunctions, some qualities are in effect *signs* for other qualities.

For Locke there are the metaphysical relations on the one hand – the relations of substance to quality, and of causation – and, on the other hand, there is what we perceive – collections and constant conjunctions.

What Berkeley did was reject the doctrine that the world of ordinary things is a world of inert substances. The world of ordinary things is, rather, a world of qualities tied together by various relations. Ordinary things are qualities tied together by the part-whole relation, and predication is a matter of a quality's being predicated of the whole of which it is a part. The whole, so far as it is perceived, is a conjunction of qualities. As he put it in his Commonplace Book,[15] with specific reference to the relations of space and extension which were supposed by the Cartesians to constitute the essences of things, uniting them into wholes: 'We think by the meer act of vision we perceive distance from us, yet we do not, also that we perceive solids yet we do not, also [planes], yet we do not. Why may I not add? we think we see extension by meer vision, yet we do not' (*Philosophical Commentaries*, No. 215). Ordinary things are tied together by the relation of causality. But these causal relations are conjunctions. As Locke argued, so did Berkeley: we are not acquainted with necessary connections.

All our ideas, sensations, or the things which we perceive, by whatsoever names they may be distinguished, are visibly inactive; there is nothing of power or agency included in them. So that one idea or object of thought cannot produce, or make any alteration in another. To be satisfied of the truth of this, there is nothing else requisite but a bare observation of our ideas. For since they and every part of them exist only in the mind, it follows that there is nothing in them but what is perceived. But whoever shall attend to his ideas, whether of sense or reflection, will not perceive in them any power or activity; there is therefore no such thing contained in them. (*Principles*, sec. 25)

So far as concerns the world of ordinary things, causation is constant conjunction:

There are certain general laws that run through the whole chain of natural effects: these are learned by the observation and study of nature, and are by men applied as well to the framing artificial things for the use and ornament of life, as to the explaining the various phenomena: which explication consists only in showing the conformity any particular phenomenon hath to the general laws of nature, or which is the same thing, in discovering the uniformity there is in the production of natural effects; as will be evident to whoever shall attend to the several instances, wherein philosophers pretend to account for appearances. (*Principles*, sec. 62)

For Locke, words are signs of ideas, which are in turn the representatives of entities. For Berkeley, in contrast, words are directly the signs of entities. These entities in turn are the signs of other entities. The connection between words and ideas, as in Locke, or – as in Berkeley – between words and entities, is conventional; the connection is not intrinsic but rather is established by human artifice. For Berkeley the same holds for the way in which entities are signs of other entities. There is no intrinsic connection: there are simply conjunctions. The connection is, however, established by artifice: the artifice of the Great Artificer, God. The structure of the world is established by the activity of spirits:

We perceive a continual succession of ideas, some are anew excited, others are changed or totally disappear. There is therefore some cause of these ideas whereon they depend, and which produces and changes them. That this cause cannot be any quality or idea or combination of ideas, is clear from the preceding section. It must therefore be a substance; but it has been shown that there is no corporeal or material substance: it remains therefore that the cause of ideas is an incorporeal active substance or spirit. (*Principles*, sec. 26)

What holds for causal relations also holds for the relations that bind qualities into ordinary things, and, indeed, all other relations. Relations as such are not perceived by sense. As Berkeley put it, 'we know and have a notion of relations between things or ideas, which relations are distinct from the ideas or things related, inasmuch as the latter may be perceived by us without our perceiving the former' (sec. 89). Relations are not perceived by sense because they involve activity: 'all relations including an act of the mind, we cannot so properly be said to have an idea, but rather a notion of the relations or habitudes between things' (sec. 142).

The structures of the world, including the structures of qualities into ordinary things, are thus a matter of the relating activities of active substances.[16] Qualities are the words of the language of God, and the conjunctions that we perceive and learn are the syntax of God's language. This syntax is, of course, the relational structure of the world. Thus, the structure of the world is provided by the activity of the deity, or, sometimes, by the activity of lesser spirits.

In Burgersdijck's logic, there is a double account of relations: one of relations considered objectively, and one of relations considered as forms of judgment. The former takes them to be predicated of substances. The latter takes them to be forms of mental activity. What Berkeley does is eliminate substances. The role of substance in accounting for the unity of things is replaced by the part-whole relation. Now, this relation, that of part to whole, is one of the relational forms of judgment, alongside causation, and so on, that Burgersdijck considers among the topics where arguments are to be found. Thus, in Berkeley, predication is assimilated to the other relations, and loses the special status that it had in Aristotle, Aquinas, Descartes, Locke, Burgersdijck, and the rest. In Berkeley, predication reflects an act of the mind that has no objective basis in things, or, at least, no more objective basis than any other relation, such as the relation of causality; predication is, rather, an act of the mind that unifies sense qualities into wholes. Except, of course, there is in a way an objective basis for this relation, as for the other structural relations among the sense contents of the world, such as the relation of causation: the objective basis is the structuring activities of God.

Berkeley in this way, as it were, *reanimates* the world. Where Descartes and Locke, and, in general, the new science, had removed activity from the world, Berkeley restored it. Berkeley's world as a linguistic text became, once again, as it was for Aristotle, a world that is filled with and moved by activity, the primitive unanalysable activity of substances. But if Berkeley succeeded in refilling the world with activity, he did so only by giving up the substance analysis of ordinary things. In this respect he broke away from the tradition in a way in which neither Descartes nor Locke nor the scientists of the Royal Society were able to do.

One understands the language of God, the natural language of sense contents, provided that the acts in one's own mind mirror the acts by which God structures his language. I get it right, for example, if the acts by which I join up ideas into causal sequences correspond to the way in which God structures the regular connections among sense contents. And I get it right if the acts of unification by which I join the simple sense contents that I receive into things correspond to the acts by which God unifies those ideas.

There are no doubt many strands to the story of how Berkeley was able to achieve this break. What I hope to do, as I have said, is trace what I think is one of these strands. It is an important strand, however, because it led to later developments of attempts to account for structure.

The strand that I want to examine lies in a logical tradition different from that of Burgersdijck. In this other logical tradition, the role of substance is *not* central. I refer to the account of logic developed by Petrus Ramus, Pierre de la Ramée.

Ramus developed certain ideas that had already been presented by Rudolphus Agricola.[17] This was the reorientation of the traditional doctrine of topical places. In the earlier tradition of logic, one that would be more familiar to us from our own texts of formal logic, one proceeded from concepts through judgments to syllogisms. The discussion of concepts included the discussion of the categories and the predicables, as well as the different sorts of supposition of terms. The discussion then passed on to judgments or propositions, which included such things as affirmative and negative, universal and particular, and what we call the immediate inferences. The discussion then turned to reasoning, which emphasized the categorical syllogisms of Aristotle, but included hypothetical and disjunctive syllogisms, together with sorites, induction, example, and enthymeme, taken as special cases of syllogism. Scientific method as we know it, that is, the method of empirical science, had no place here, of course. What followed the discussion of the traditional forms of argument was their use, on the one hand, in science in the sense of *scientia*, that is, demonstrative science where the premises are self-evident, and the use of the forms of argument, on the other hand, in ordinary arguments with only probable premises. It was in the latter context that the doctrine of topical places was located: the places were storehouses for different sorts of premises that could be used in the sort of discourse that was relevant in legal and political contexts.

This organization of the traditional logic can be found, for example, in Robert Sanderson's *Logicae artis compendium*.[18] Sanderson had been appointed Reader in logic at Lincoln College, Oxford, in 1608. His was also Regius Professor of Theology prior to the deposition of Charles I. A royalist, he was removed from the chair during the interregnum, only to be reinstated after the restoration. His career ended with a brief episcopate at Lincoln. Sanderson first published his logic in 1618. It is an excellent text, and, not surprisingly, proved very popular, subsequently going through many editions.

We would recognize most of the questions discussed by Sanderson. What we do not find in our own logics is the orientation towards the standard substance metaphysics deriving from Aristotle. In our own logic texts, we would not find a discussion of the categories or predicaments. Nor would we find an orientation of the reference of terms to the members of the first of these categories, substance, which Sanderson defines in the usual way: 'Substantia est Ens per se subsistens' (*Logicae*, 17). Nor would we find listed the other categories, such as quality, about which Sanderson asserts that '[q]ualitas est forma accidentalis, a qua Substantia denominatur Qualis: ut Albedo. Quale est, quod a Qualitate denominative dicitur: ut Album' (21). Nor, finally, would we find these things discussed in a context where it has been laid down that the legitimate forms of predication for the entities in any category are defined by the list of predicables.

Then there are relations. We would, of course, find these discussed in our own logics, but not in the way that Sanderson does, namely, in conformity with the Aristotelian metaphysics in which substances and properties are the only entities. Relations are properties in the other categories that are referred one to another: 'Veteres Logici Relata definienbant esse, ea, quae altgerius esse dicuntur, aut alio quopian modo ad aliud referentur: ut Scientia et Scibilie' (*Logicae*, 24). Sanderson classifies relations in much the same way that Burgersdijck does.

Where Sanderson differs from Burgersdijck is where he locates the topics. Sanderson locates them in the context of probable, or non-demonstrative, syllogisms. They are places where we store premises for arguments for use in legal and political oratory and for sermons. If we wish to advance an argument in the context of such oratory, then we turn to our places and search for argument that will fulfil our purposes. For Sanderson, the topics have nothing to do with demonstrative syllogisms.

Burgersdijck locates the topics much earlier in his discussion, treating them as, in effect, forms of relational judgment, as we have seen. Burgersdijck retains the distinction between, on the one hand, the topics, where arguments are to be found, and, on the other, the categories and predicables, which define the various forms of judgment. But at the same time he blurs the distinction, as Sanderson does not: for Burgersdijck, the categories and the predicables, while logically different from the topical places, are also, like the topics, *places*, that is, places where arguments are to be located.

Burgersdijck is, in effect, trying to have it both ways. He wants to defend a logic which is, like that of Sanderson, Aristotelian in the strong sense of reflecting the Aristotelian ontology of substances as independent entities, and accidents as non-relational entities dependent upon substances. But he also wants a logic that will be practical in its orientation, helping students not only to evaluate propositions for well-formedness and arguments for validity but to find arguments that they can use in their own discourses. For the latter, the difference between topical places, on the one hand, and the categories and predicables, on the other, tends to disappear, as indeed we have seen happen in Burgersdijck. It is the logic of Ramus that is pulling the resisting Burgersdijck in this direction, away from the Aristotelian logic of Sanderson.

The logic of Ramus in one way or another incorporates much of the traditional doctrine. But this material is covered only in the context of a logic that is oriented towards its practical uses in the discourse of politics and law. Ramus takes up Agricola's distinction between invention and judgment; but, where Agricola dealt only with the subject of invention, Ramus goes on to write a logic that incorporates both aspects.[19] The result is a logic that must have been like a breath of fresh air. Not only does the logic have a clear appeal to practical purposes, compared to a logic such as that of Sanderson. In this it has the sort of appeal that informal logic has in our own day. But it is also remarkably well written, with examples that have a wide literary appeal. Sanderson's book is reasonably well written, but like most logic texts in our own day is rather dull reading. If there are any examples, they are pedestrian. Burgersdijck's is even worse: it is not even well written. Ramus's *Dialectique* is well organized for its purposes, well written, and filled with lively and interesting literary examples that provide useful samples for the poet and orator. For our purposes, however, the important point is that it separates logic from its former role as a language that reflects perspicuously the ontological structure of the world as determined by the substance metaphysics deriving from Aristotle. Ramus aims to give to students a presentation of logic that will be useful for the purposes of legal and political oratory and for sermons. For these purposes, the doctrine of substances need play little role. Thus, when Ramus provides examples of cause, and, in particular, final cause, he cites as explanatory of certain events, first, the end of marriage, and second, the end of taking up arms against Caesar in civil war. In neither case does the subject of the discourse seem to be a substance (*Dialectique*, 64–5). This orator's logic allows for judgments in which we have connections among things that are not substances. Thus, in Ramus's logic we are no longer committed to the view that, when we say of what is that it is, what we are doing is attributing a property to a substance. Ramus, in other words, provides a logic in which discourse is no longer tied to an Aristotelian account of the world. The point of the logic is not to fit an ontology that is no longer substantialist; it is to provide a logic that will be useful to orators. But in producing the latter sort of logic, the result is an account of logic and language which has little place for the traditional doctrine of substance. So the result is, in effect, an account of logic and language that will fit a world that, like Berkeley's, argues against construing ordinary things as substances.

We know that Hume did read Berkeley. Did Berkeley read Ramus? This I do not know. I know of no reference in Berkeley to Ramus. Ramus's logic was widely known, however, as is clear from the history given in Walter J. Ong's *Ramus and Talon Inventory*.[20] Logic as a two-part art was first clearly laid out in the *Dialectique* published in Paris in 1555. This was followed by several other editions, including, of course the Latin edition.[21] The first English-language edition of Ramus was *The Logicke of ... P[eter] Ramus*, translated by Rolland M'Kilwien, published in London in 1574.[22] There was also J. Piscator's commentary on Ramus, *In P. Rami dialecticam animadversiones* (London: H. Bynneman 1581). William Temple made his name as a Ramist with his *P. Rami dialecticae libri duo* (Cambridge: Thomas Thomas 1584). Ong notes that 'there seem to be more English translations of Ramus's *Dialectica* than translations in any other language.'[23] These include editions of 1656, 1658, 1685, and

1699. There were also Latin editions published in England. Thus, there was John Seton's popular *Dialectica* (annotated by Peter Carter, London 1611). There was published at Cambridge in 1672 a *P. Rami dialecticae libro duo* with commentary by Guilielmi Amesii. George Downame's scholastic elaboration of Ramus's logic was published at Cambridge in 1699. We know that Ramus's logic was popular as a text in several dissenting academies in the later seventeenth century.[24] But this is not yet to connect Berkeley with Ramus. We know that Oliver Goldsmith attended Trinity College, Dublin, and complained about the 'cold logic of Burgersdiscius' and 'the dreary subtleties of Smiglecius.' Burgersdijck we know about; his logic was indeed influenced by Ramus, though it still retained the crucial Aristotelian notions of substance, categories, and predicables. The *Logica* of Martin Smiglecki (1638), was essentially Aristotelian, being distinguished only by its mode of presentation, which was in the form of disputations. So Smiglecki and Burgersdijck were used to teach logic at Trinity College, Dublin, at about the time Berkeley was a student.[25] Provost Narcissus Marsh published his *Institutiones logicae* (Dublin: S. Helsham 1681), and this, referred to as the 'Provost's logic,' was well known at Trinity. It is not, however, a Ramist text. It gives the older division of logic into terms, propositions, and arguments. In this section, there is a discussion of the predicables followed by the usual Aristotelian categories, beginning with substance, which is defined as 'Ens per se Existens' (*Institutiones*, 39). In the context of propositions, Marsh includes a section on supposition. The section on argumentation includes a section on demonstration. This is restricted to *scientia*: 'certa et evidens rei cognitio' (205). It is in this context that Marsh discusses causes. He then turns to dialectical syllogisms, and here he includes the traditional material on the topics. Decidedly not Ramist. At the time of Berkeley, then, there seems to be little official use of Ramus, and no clearly Ramist text, despite the fact, to which the publication record testifies, of the continued popularity of Ramist texts in English. There is, however, one text that must be mentioned in this context. This is the already-mentioned annotated edition of Ramus's *Dialectica* published in 1584 by William Temple. Temple was a tutor in logic at King's College, Cambridge, and made a name for himself as a defender of Ramus against the latter's critics. He responded to criticisms of Ramus by Oxford's Everard Digby (Temple's reply, 1580) and Johannes Piscator (from Strasbourg). The latter (1581) was appended to a second reply to Digby, and was thought to be of sufficient merit to be reprinted at Frankfurt in 1584. Meanwhile, in 1582 Temple had written a second reply to Piscator. The point of this discussion of Temple is simply that this logician became the fourth provost of Trinity College, Dublin, in 1609 and retained the position until his death in 1627. He was an able administrator and left a deep imprint on the college. It is hard to believe that the work of such a vigorous and able defender of Ramus would not have been remembered in the college of which he had been an impressive provost. Other provosts were also Ramist, including the seventh, William Chappell.[26] It is more than likely, then, that Ramist views were well known at Trinity College, Dublin, when Berkeley was a student there. None of this is proof that Berkeley ever read or heard of Ramus while at Trinity, but it leads me to suspect that he did. However, there is one more fact that is relevant.[27] Berkeley was a student at Kilkenny College before going up to Trinity, and there is considerable circumstantial evidence that the headmaster at the time, Edward Hinton, was a Ramist. Again, this is no proof that Berkeley had ever encountered Ramus's views on logic and language. But it is hard to conceive that it is not so: the circumstantial evidence from Kilkenny College and Trinity College, Dublin, seems strongly in its favour.

What, however, is in Ramus's logic?

The *Dialectique* begins by defining logic as 'art de bien disputer' (61). Its practical orientation is thus established immediately. But Ramus intends this in full generality. Disputes are resolved by discovering the truth. Logic or dialectic is thus the means to discover and display the truth; it is the root of all knowledge: 'devons-nous apprendre la Dialectique pour bien disputer à cause qu'elle nous déclare la vérité ...' (61). Logic does this whether we are considering matters that are scientific, or matters that are contingent, or, what is the same, matters of opinion. Dialectic or logic is 'l'art de cognoistre, c'est-à-dire Dialectique ou Logique, est une et mesme doctrine pur apercevoir toutes choses' (62).

Ramus, following Agricola, divides logic into two parts. The parts of dialectic, or logic, according to Ramus, are invention and judgment. Of these he says that 'La première déclaire les parties séparées dont toute sentence est composée. La deuxiesme monstre les manières et espèces de les disposer ...' (*Dialectique*, 63).

The first part is described by Ramus as 'la doctrine des lieux Topiques' (*Dialectique*, 63), but insofar as it gives the parts of sentences that are to be admitted into the discourse of reason, of logic, it also covers what would be, for the traditional logic the various categories. Indeed, Ramus makes this connection himself (ibid.). Inartificial arguments derive their worth from authority; the places for inartificial arguments are such things as the law and testimony (96ff). But it is the places for artificial arguments that interests us. These are arguments that logic alone helps us construct. What we begin with is a concept. We then wish to discover concepts that can be related to it in propositions which will subsequently be used in syllogisms to bring out in discourse the truth of things. The concept from which we begin and the concept discovered in the topical place will be united to form a proposition. The list of topics, therefore, gives a list of acceptable sentence forms, the list, in other words, of 'well-formed' sentences.

Although these places occupy in logic the role of Aristotelian categories, together with the predicables, it turns out that neither the traditional categories nor the predicables appear in Ramus's list. What he gives instead are four quite different species of categories (which are themselves subdivided as the discussion proceeds). These are 'Causes et Effectz, Subjectz et Adjoinctz, Opposez, Comparez' (64).

For our purposes we should note, first, that the list contains only relations. These relations are the forms of acceptable sentences. These relations may be either necessary or contingent; if the former, then we have demonstrative science, and if the latter then we have probabilities.

Ramus indicates that his opposition of subject and adjoint is the same as that of Aristotle who 'en plusieurs lieux oppose le subjet et l'accident' (*Dialectique*, 74), though he is also careful to distinguish essential attributes of the subject, where the connection is necessary, and mere accidents, where the connection is contingent (74). We should therefore note, second, that, where the Aristotelian logic gives a special place to the relation of predication, contrasting it to other relations, Ramus instead includes the relation of subject and adjoint *parallel* to other relations, including the relation of cause to effect. In this respect, Ramist logic separates itself from all other logics in divorcing itself from the substance metaphysics. At the same time, it provides the form that a logic might reasonably take if predication is to be a relation alongside other relations such as cause and effect, similarity and difference.

Finally, we should note, third, that the apparatus of species and genus also loses its special place in logic and is relocated in the topical place 'comparez' (*Dialectique*, 80). Ramus first discusses 'comparison de quantité' (80), and under this head refers to the relations of equal (80), more (82), and less (83). He then turns to 'comparison de qualité' (85), and the allocation of things to species and genera is made a matter of the relations of similarity (85) and dissimilarity (87). In this respect, Ramist logic separates itself from most logics in divorcing itself from a metaphysics in which the relations of species and genus provide the necessary connections among the terms of scientific syllogisms. The Cartesians, especially in the Port Royal Logic, were to make a similar break with the traditional Aristotelian reliance upon species and genus, and in this they were followed by Locke. This leads to similarities between the accounts of inference that we find in Ramus, on the one hand, and the Cartesians and Locke on the other. At the same time, however, it must be emphasized that both the Cartesians and Locke retained the core of the substance metaphysics: the basic entities are substances. This means that the accounts of logic found in the Cartesians and Locke do not yet make the radical break with the tradition that marks both Ramus's logic and Berkeley's metaphysics.

To see this latter point more clearly, we must turn to Ramus's discussion of argumentation or inference. This occurs in the second part of logic. This second part is called 'judgment,' and is contrasted to invention. Judgment 'monstre les voyes et moyens de bien juger par certain régles de disposition ...' (*Dialectique*, 115). Judgment in turn has three subdivisions: 'Enonciation, Syllogisme, et Méthod' (115). The traditional logics had three parts: term, proposition, and syllogism. 'Enonciation' covers roughly the same material as was covered in the traditional logics under the heading of proposition, while 'syllogisme' covers the inferences in the traditional way, omitting, however, much of the material of mood and figure and all of the traditional rules. Ramus's view seemed to be that one would pick up the relevant norms more quickly by studying effective examples than by internalizing a set of formal rules whose use depends upon a rather artificial categorization in terms of mood and figure. This disdain for the traditional apparatus was to find itself repeated in the Cartesians and in Locke.

'Enonciations' or propositions are divided into simple or complex. The latter include disjunctive, conjunctive, and conditional propositions. As examples of simple propositions, Ramus gives (*Dialectique*, 115)

> Le feu brule.

> Le feu est chault.

> Le feu n'est eau.

The first derives from the topical place *cause and effect*, the second from *subject and adjoint*, and the third from *opposition*.

Thus, if we have, for example, *burning* as the object of our interest, then we can locate an argument to be joined to it in a proposition in the topical place *cause*. This argument is *fire*, and we form the proposition

> fire burns

Similarly, if we have *fire* as the object of interest, then we can locate an argument with which to join it in the place *adjoint*; this argument is *hot*, and yields the proposition

> fire is hot.

The connection may be necessary or contingent. Ramus wrote a peroration concerning basic judgments or first principles prior to his discussion of syllo-gisms, and in this he has pointed out (*Dialectique*, 123), that *scientific* proposi-tions are self-evident. In this case the connection between the terms of the proposition will be necessary. Judgments which are contingent and not neces-sary are called 'opinion' (124). In this case, the connection between the terms of the judgment will be only *probable*. In general, the places where probable argu-ments are to be located are those of inartificial arguments, that is, law or testi-mony divine or human. In the case of necessary judgments, the places to be searched are those of artificial arguments, that is, those of 'Causes et Effectz, Subjectz et Adjoinctz, Opposez, Comparez.' The justification for asserting the proposition is, in the case of inartificial arguments, the authority, and, in the case of artificial arguments, self-evidence.

It should be noted, however, that one can continue to use the loci or places of artificial arguments even if not all connections are necessary or self-evident. It can be held both that causal connections are contingent, and that terms are con-nected to each other by these relations. The authority will be that of *experience*, and the tie that will lead the mind from one term to another located in the topical place will be not a necessary connection but a psychological habit created by the experience which is the authority that testifies to the truth of the connection.

Ramus next passes on to syllogism: 'Syllogisme est disposition par laquelle la question disposée avecques l'argument est nécessairement conclue ...' (*Dia-lectique*, 125). We need a syllogism when we have an 'énonciation doubteuse' and 'pour la preuve d'icelle, est besoing de quelque moyen et tiers ...' (ibid.).

This notion that syllogisms involve the interpolation of third terms between extremes in order to justify the linking of those extremes is a commonplace. The extremes in the syllogism

> All *M* is *P*
> All *S* is *M*
> ───────
> All *S* is *P*

are the subject and predicate of the conclusion. For the Aristotelians, in a scien-tific syllogism, these are, respectively, the species and genus, and the middle that links them is the specific difference. In other words, for the Aristotelians, the syllogism exhibits the real definition of the subject. Since understanding a thing involves the grasp of its real definition, to grasp a scientific syllogism and to understand a thing amount to the same: syllogism provides insight into the ontological structure of reality. Once logic is released from the ontology of

forms understood in species-genus terms, inference can be based on other relations than these traditional connections. In particular, for example, as Ramus
points out (*Dialectique*, 129), Aristotle never treats an argument such as

> Octave est hériter de César
> Je suis Octave
>
> ———————————————
>
> Je suis donques hériter de César

as a syllogism – science does not deal with singulars. It was called instead an
'exposition' (129). But for Ramus, this form is one among the several forms of
syllogism. No longer bound by the ontology of forms, relations other than those
of genera and species can yield valid arguments.

Ramus's point was later taken up by other philosophers. Thus, Descartes distinguishes his own method from that of the 'dialecticians':[28] 'when they
expound the forms of the syllogisms, they presuppose that the terms or subject-
matter of the syllogisms are known; similarly, we are making it a prerequisite
here that the problem under investigation is perfectly understood. But we do not
distinguish, as they do, a middle term and two extreme terms' (*Rules*, 51). To be
sure, his method is like that of the dialecticians in proceeding in terms of self-
evident inferences. As he puts it in his Rule Five: 'The whole method consists
entirely in the ordering and arranging of the objects on which we must concentrate our mind's eye if we are to discover some truth. We shall be following this
method exactly if we first reduce complicated and obscure propositions step by
step to simpler ones, and then, starting with the intuition of the simplest ones of
all, try to ascend through the same steps to a knowledge of all the rest' (20). In
fact, these inferences are based on relations among ideas.

This common idea is carried over from one subject to the other solely by means of a simple comparison, which enables us to state that the thing we are seeking is in this or that
respect similar to, or identical with, or equal to, some given thing. Accordingly, in all
reasoning it is only by means of comparison that we attain an exact knowledge of the
truth. Consider, for example, the inference: all A is B, all B is C, therefore all A is C. In
this case the thing sought and the thing given, A and C, are compared with respect to
their both being B, and so on. But, as we have frequently insisted, the syllogistic forms
are of no help in grasping the truth of things. (57)

There are relations other than those of species and difference that validate
inferences, including the relations that justify

$$A = B, B = C, \quad \therefore \ A = C.$$

As Descartes puts it, inference rests on a comparison of our ideas, and the discovery therein of relations that connect those ideas. We must 'think of all knowledge whatever – save knowledge obtained through simple and pure intuition of a single, solitary thing – as resulting from a comparison between two or more things. In fact the business of human reason consists almost entirely in preparing for this operation. For when the operation is straightforward and simple, we have no need of a technique to help us intuit the truth which the comparison yields; all we need is the light of nature' (*Rules*, 57). Inferences of the sort just indicated are based on judgments of comparison. These judgments, when linked systematically, yield knowledge of connections among essences or notions or forms. It is the inferences that are essential to knowledge; the claim that knowledge is not obtained until the argument has been put in strict syllogistic form is simply not true. What the defenders of syllogistic do, that is, the dialecticians, is take the inferential knowledge and *rearrange* it into syllogistic form. But that form does nothing to help us to discover what the links are in the chain that leads to the solution of the research problem that we have posed to ourselves.

[O]n the basis of their method, dialecticians are unable to formulate a syllogism with a true conclusion unless they are already in possession of the substance of the conclusion, i.e. unless they have previous knowledge of the very truth deduced in the syllogism. It is obvious therefore that they themselves can learn nothing new from such forms of reasoning, and hence that ordinary dialectic is of no use whatever to those who wish to investigate the truth of things. Its sole advantage is that it sometimes enables us to explain to others arguments which are already known. It should therefore be transferred from philosophy to rhetoric. (36)

As the Port Royal Logic puts the point,[29] 'we must discover the content of an argument before we can arrange its premises' (*La Logique*, Pt III, ch 17).[30]

Locke makes much the same point. The 'original way of knowledge' is 'by the visible agreement of ideas' (*Essay*, IV.xvii.4). 'To infer is nothing but, by virtue of one proposition laid down as true, to draw in another as true, i.e. to see or suppose such a connexion of the two ideas of the inferred proposition ...' (ibid.). Syllogism presupposes that we have grasped the inferential structure, the connections among our ideas, rather than itself being essential to such knowledge. As we have already noted, any country gentlewoman

clearly sees the probable connexion of all [her ideas regarding the weather] ... without tying them together in those artificial and cumbersome fetters of several syllogisms, that clog and hinder the mind, which proceeds from one part to another quicker and clearer without them; and the probability which she easily perceives in things thus in their native state would be quite lost, if this argument were managed learnedly, and proposed in mode and figure. For it very often confounds the connexion: And, I think, every one will perceive in mathematical demonstrations, that the knowledge gained thereby comes shortest and clearest without syllogism. (ibid.)

Ramus is thus followed by Descartes and Locke in giving up the idea that the only relations that can legitimate inferences, syllogisms if you wish, are those of species and genera. However, while Descartes and Locke abandon the insistence that only species-genera relations give scientific inferences, their notions of logic and language are still bound up with a substance metaphysics. It is precisely this that is *also* given up by Ramus: connections of qualities to the things of which they are predicated, that is, the relations of adjoints to subjects, do not hold a special place but are simply relations among relations alongside those of, say, causation.

Walter J. Ong has been highly critical of Ramus's logic.[31] We must distinguish, according to Ong, *reasoning*, which is 'the drawing of consequences from one or more propositions' (*Ramus*, 73), from *understanding*. Reasoning is where formal logic has its home. Formal logic constructs its rules with regard to the extensions of terms, but even this abstraction from content is not a total abstraction from matter: its movement reflects, and therefore presupposes, an ontological structure in the reality that it aims to illuminate. 'Logic is a study of the reflection of [the] material world – the world with which man is directly confronted – in the structures of the mind' (74). Thus, 'the presence of discursive reason in the human intellectual apparatus is due to the material component in man's cognitional make-up and in the make-up of the reality he is faced with' (73–4). Reason brings to reality the light of the intellect, but its role is, as it were, the transmission of light from one part of reality to another. It cannot do this task of transmitting light unless it has ready for use some already-illuminated piece of reality. It is the *understanding* that provides this initial light. Any knowledge which we have by way of logic, then, presupposes knowledge by way of understanding. The upshot of logical inference is understanding, but that is an understanding that logic can achieve only by taking for granted that understanding has already been achieved. And so, as Ong points out, 'Aquinas takes a rather dim view of discursive reason or ratiocination as compared to sheer understanding, which in its pure state would be intuitive' (74).[32]

With this it is unlikely that anyone would disagree: the inferences of logic transmit truth, they do not justify their premises as true. There must therefore be something, call it the 'understanding,' if you will, by which we come to know the truth of our premises. Ramus, in particular, would not disagree. Inference is located within the part of logic called *judgment*, the *second* part of logic. The premises are, in contrast, discovered by reference to the topical places, which occur in the first part of logic, the part called *invention*. Ong, however, criticizes Ramus and Ramists for their 'insistence that it is reasoning or ratiocination, not understanding, which differentiates men from animals, and on the insistence ... that God prescribes "method," which is conceived of as a kind of protracted ratiocinative process ...' (*Ramus*, 74). According to Ong, then, Ramus must omit from logic whatever it is that is required if we are to recognize the role of understanding in logic. Since the understanding yields premises, that which Ramus omits when he ignores the understanding must be omitted from the list of topics. Whatever Ramus puts into his list of topics or categories, it does not include a list of those things that must be there if we are to recognize correctly the role of understanding in grasping the ontological structure of reality. But what is missing? As we have seen, it is precisely the traditional Aristotelian categories and predicables, precisely those things that make the traditional logic of Sanderson or Burgersdijck or Aquinas fit so neatly a world the ontology of which is given by the metaphysics of substances and intelligible forms.

Ong's criticism of Ramist logic, then, amounts to a statement that this logic is inappropriate to a world whose ontology is that of the Aristotelian metaphysics of forms and substances. However, this statement is a criticism only if the ontology of the world is indeed of that sort. Since Ong does not defend his statement, his running stream of contemptuous dismissals of Ramus is little more than a begging of the ontological question. His complaint about Ramus and the 'decay of dialogue' is really a complaint with regard to the decay of the Aristotelian metaphysics. But he offers no defence of that metaphysics.

In fact, in many respects, the criticism that Ong makes of Ramus applies also to Descartes, Arnauld, and Locke, all of whom insist that, while syllogism is the interpolation of middle terms, the logical structure of the relations does not have to be that of real definitions, in terms of species and genus, and, moreover, for Locke, not even a necessary connection.

None the less, there remains for Descartes, Arnauld, and Locke, the substance metaphysics. Thus, while the Port Royal Logic rejects the apparatus of predicables and real definitions, and most of the traditional categories, it none the less includes a categorical scheme based on the Cartesian metaphysics, and which therefore presupposes an ontology of substances and properties. That is, it presupposes an ontology in which the relation of predication occupies a special place, different from that of all other relations.

It is in precisely this respect that Ramus develops a view of language which is similar indeed to that of Berkeley. It is a view of language and logic in which predication is like the relation of causation rather than occupying a special place. Given that we take causation to be a relation which is, on the one side, regularity, but, on the other side, a connection established by the activity of the mind, then we should say the same thing about subjects and adjoints. There will be, on the one side, a collection of properties including the adjoint, and on the other side, a connection of these into a unified whole established by the activity of the mind. The structure of contents into things will parallel the structure of things into causal sequences. But this, of course, is precisely what Berkeley argues.

The question that I raised was what enabled Berkeley to make the important breakthrough that permitted philosophers for the first time to conceive of things as collections of qualities rather than as substances with properties. There are probably many strands that met to account for this radical change in the way of conceptualizing things in one's ontology. But what I am suggesting is that Ramus's logic is one such strand. Unlike the traditional logic, Ramus's logic was not wedded to the ontology of substances. At the same time, he clearly put predication as a relation among relations, and permitted philosophers for the first time to think of predication in ways that differed from the tradition. In particular, it could be assimilated to the ways in which one thought about causation, or fatherhood, or sacrilege. Berkeley could therefore take Locke's account of the latter relations and apply it to the relation that unified sense contents into things. The result is the new account of predication, the new ontology of things.

In any case, there remains the structure of Ramus's logic. I have argued that this structure could be one of the strands that enabled Berkeley to come to his rejection of the old substance metaphysics. I have little doubt that the influence of Ramus was so pervasive, even where he was rejected, as in Burgersdijck, that it was in fact a contributing factor for Berkeley's great achievement in ending the hegemony of the old Aristotelian ontology that insisted that ordinary things are substances: thanks to Berkeley, but, behind him, thanks to Ramus, things could now be conceived as bundles of properties.

Study Four

Empiricist Inductive Methodology: Hobbes and Hume

It is usually stated that Hobbes's methodology of science is purely deductivist and, moreover, an extreme conventionalism in which the premises of the demonstrations of the deductive science are all true by definition. Thus, A.E. Taylor[1] states that '[o]nly ... the truly deductive type of reasoning is rigidly certain and yields perfectly determinate conclusions' (*Hobbes*, 34–5), while 'the ultimate first principles of deductive science are all ... *definitions*, that is, statements of the meaning of *names*' (36). 'Everything in science, therefore, turns upon the original definitions; science is merely the correct deduction of the consequences implied in the giving of *names*' (35). Hence there is no role for experiment in science:

[Hobbes] held that the Royal Society was proceeding on altogether false lines in attempting to advance physical science by direct experiment rather than by reasoning deductively from preassumed general theories. (35)

[There was] no place for 'Baconian induction' in his ... conception of scientific method. Bacon's zeal for experiment ... is entirely alien to the essentially deductive and systematic spirit of the Hobbesian philosophy. (6)

Another who argues for this interpretation of Hobbes on method is Copleston,[2] who speaks of Hobbes's 'monolithic idea of science' in which there is 'a progressive development from first principles in a deductive manner' (*History*, 29). Watkins[3] similarly holds that Hobbes was 'unimpressed by the inductivist philosophy of Bacon' (*Hobbes's System*, 31). For Hobbes, in contrast, the appropriate method is that of *demonstration*; the demonstrative method consists of 'laying down first principles and proceeding deductively therefrom...' (44). As for the first principles, these 'first truths were arbitrarily made by those that first of all imposed names upon things' (106). The deductive methodology is more or less endorsed by Watkins as akin to the hypothetical-deductive methodology defended by Popper; but he suggests that the (non-Popperian) radical conventionalism is inadequate; the appropriate premises in any hypothetical-deductive explanation are empirical truths, not statements true by definition or *ex vi terminorum*.

Clearly, there must be something to this reading of Hobbes if it is endorsed by thinkers as diverse as Taylor, Copleston, and Watkins. None the less, the reading is, I think, distinctly inadequate. Most have concluded that Hobbes learned nothing from Bacon when he worked as the latter's secretary. What I propose to show is that there is in fact a good deal of the inductive method in Hobbes. In particular, I hope to show that Hobbes is struggling to articulate the empiricist inductive methodology that Hume was later to articulate and defend. I hope to show, moreover, that the methodology that Hobbes defends includes something akin to Bacon's method of eliminative induction. We do not in fact know what Hobbes might have learned from Bacon. But if my argument is correct, then the usual conclusion that, given his (alleged) deductivism, he must have learned nothing, will have been shown to be unfounded. To the contrary, the presence of something like the method of eliminative induction shows that, when Hobbes worked as Bacon's secretary, he might well have become infected by the Lord Chancellor's inductivism.

This is not to say that there is nothing to be said for the traditional account of Hobbes's methodology of science that we find in Taylor, Copleston, and Watkins. To the contrary, Hobbes does hold that laws are in some sense necessary, and tries to account for that necessity by appeal to a doctrine about definitions. The problem of the necessity of laws is one thing. Hobbes's solution is another.

As we saw in Study One, Galileo refocused the intellectual concerns of those who seek the truth of things on matter-of-fact regularities. But this is not yet to reconstruct the notion of cause or of causal law. If one is going to make the move to identifying causation with regularity, then one has to be able to account for the necessity of causes. To put the point another way, one has to be able to account for the distinction between *post hoc* and *propter hoc*, that is, between those regularities which are 'mere' regularities and those that are causal or necessary. Hobbes argues that we should identify cause and regularity, rejecting the Aristotelian notion that one can account for the distinction between causal laws and mere regularities in terms of active natures or powers or forms. He must then himself present some account of the necessity of causal laws. This he does in terms of a doctrine of definition. Unfortunately, it will not do. In this, his critics such as Taylor, Copleston, and Watkins are correct. The problem that Hobbes was trying solve remained, however. As we shall argue, it remained for Hume to present a solution to this problem of accounting for the necessity of causes in a way that was compatible with empiricism.

I: Hobbes's Baconian Induction

Hobbes[4] gives us 'an exact notion of that which we call cause' as follows: 'a cause is the sum or aggregate of all such accidents, both in the agents and the patient, as concur to the producing of the effect propounded; all which existing together, it cannot be understood but that the effect existeth with them; or that it can possibly exist if any one of them be absent' (*Elements*, 77). Causal relations thus involve regularities: 'it cannot be understood but that the effect existeth with' the cause. The cause is the set of conditions that are sufficient for the effect: the cause is the set of conditions that 'concur to the producing of the effect.' Moreover, the cause is also necessary: the cause is absent, if one or more of its parts are absent, and if the cause is absent, then so is the effect, for 'it cannot be understood ... that [the effect] can possibly exist if [the cause] be absent.' Statements of causation are therefore general statements of necessary and sufficient conditions.

How are we to discover these causal relations? Hobbes continues:

we must examine singly every accident that accompanies or precedes the effect, as far forth as it seems to conduce in any manner to the production of the same, and see whether the propounded effect may be conceived to exist, without the existence of any of these accidents; and by this means separate such accidents, as do not concur, from such as concur to produce the said effect; which being done, we are to put together the concurring accidents, and consider whether we can possibly conceive, that when these are all present, the effect propounded will not follow; and if it be evident that the effect will follow, then that aggregate of accidents is the entire cause, other wise not; but we must still search out and put together other accidents. (*Elements*, 77)

If the effect can exist without a certain accident, then the latter is not (part of) the cause. Thus, the cause is common to all cases where the effect is present. *This is the method of agreement*. Again, we must consider cases where the effect is absent. The cause is then the set of accidents wherein those cases where the effect is present differ from those cases where the cause is absent. *This is the method of difference*. The method of agreement yields necessary conditions, while the method of difference yields sufficient conditions. Hobbes applies these jointly. That is, what he describes is in effect the joint method of agreement and difference. This joint method yields necessary and sufficient conditions – that is, precisely those conditions that Hobbes has described as constituting the cause of an effect.

The methods of agreement and difference and the joint method are, of course, the eliminative methods of experimental science that Bacon had been the first to describe. *Hobbes is thus arguing that the methods appropriate to the discovery of causes are the eliminative methods of Bacon.*

There is little doubt that Bacon was the first to describe these methods of experimental science.[5] Hobbes was secretary to Bacon during the latter's retreat in disgrace from the world when he was devoting most of his time to experimentation at his residence at Gothambury. We shall never know whether it was here that Hobbes learned the logic of experiment. But, as we have just seen, learn it he did. And it is not unlikely that he learned it at Gothambury. Certainly, Hobbes's clear knowledge of the methods and his connection with Bacon make doubtful indeed Taylor's claim that 'the influence of Bacon ... has left no trace on Hobbes's own matured thought' (*Hobbes*, 6). Moreover, it makes clear that Hobbes's method is, contrary to Watkins, *not* simply hypothetical-deductive: Baconian induction, too, is present.

II: Hobbes's Inductive Principles

Philosophy aims to know the causes of things, either things by means of their causes or causes by means of their effects. As Hobbes puts it: 'Philosophy is the knowledge we acquire by true ratiocination, of appearances, or apparent effects, from the knowledge we have of some possible production or generation of the same; and of such production, as has been or may be, from the knowledge we have of effects' (*Elements*, 68). What we find in the world are bodies qualified by certain accidents. Bodies, in fact, are singular things. These singular things are compounded of accidents; the former are distinguished by their unique combinations of the latter (68). These accidents taken as common to several singular things are referred to as universals (68–9). What we aim to know in the first place are the causes of universals. Then we can infer from these the causes of singular things. For 'the causes of singular things are compounded of the causes of universal or simple things' (68). In order to find these causes, it is necessary to consider not singular things but the universals which are the parts of these things. '[S]eeing universal things are contained in the nature of singular things, the knowledge of them is to be acquired by reason, that is, by resolution ... [B]y resolving continually, we may come to know what those things are, whose causes being first known severally, and afterwards compounded, bring us to the knowledge of singular things. I conclude, therefore, that the method of attaining to the universal knowledge of things, is purely analytical' (68–9). The Baconian methods are therefore to be understood as part of the *analytic* method.

As for universal things, 'they have all but one universal cause, which is motion' (*Elements*, 69). This makes clear that Hobbes holds that all universal things – and therefore all singular things – have causes, and, moreover, that these causes are all of a certain kind, namely, motions of bodies. In order to discover the cause of a particular universal thing, we can limit our search to the variety of possible motions. From among this range, the Baconian methods will locate the cause.

Now, the Baconian methods of elimination will yield a true statement of cause only if two conditions are fulfilled: *first*, that *there is* a cause there to be discovered, and, *second*, that this cause is one member from among a certain *limited variety*. If the first condition is not fulfilled, then the elimination of all but one possible case will not guarantee that that case is a cause. And if there is no limit to the number of possibilities that must be eliminated, then no application of the eliminative mechanisms, however systematic and extended it might be, will succeed in reducing the possible cases to one. These two conditions have usually been called the *principle of determinism* and the *principle of limited variety*, respectively. Unless these principles hold, the eliminative mechanisms are logically incapable of yielding as a conclusion that such and such a property ('universal thing') is the cause of some other property.[6] Hobbes, we have just seen, holds that these two principles obtain.

It is clear, then, that Hobbes's account of what he calls the 'analytic method,' or, what we have seen to be the same, the methods of experimental science, is quite in order.

III: Hobbes's Account of Reason

Bacon's methods, however, are the methods of *empirical* science. As such they can yield no necessity – unless, that is, the first principles themselves are somehow necessary. Hobbes does, of course, hold that these principles are somehow necessary. Specifically, he holds precisely this with respect to the principles of determinism and limited variety, as he understands them, those principles which are required if the experimental methods are to yield conclusions concerning causes which are certain. As he puts it with regard to his version of those principles, 'the causes of universal things ... are manifest of themselves, or (as they say commonly) known to nature; so that they need no method at all; for they have all but one universal cause, which is motion' (*Elements*, 69).

However, it is notorious that such knowledge of first principles is held by Hobbes to be based on definitions. These principles are necessary precisely because they are simply true by definition, *ex vi terminorum*. 'By the knowledge ... of universals, and of their causes ... we have in the first place their definitions, (which are nothing but the explication of our simple conceptions)' (*Elements*, 70). This in turn leads into the standard objection to Hobbes's account of method, that, since definitions are arbitrary, so are all first principles;

this implies that there is no objective structure to the world to distinguish those parts of discourse which are true from those which are false; and this is then taken to be a *reductio ad absurdum* of Hobbes's position. And it indeed would be if it were an accurate reading of Hobbes's thought. However, one is left wondering how so competent a philosopher as Hobbes could have fallen so quickly into such an absurdity. It cries out for an alternative interpretation. Indeed, so does the presence of the Baconian method that we have already noted. If Hobbes does what we have seen that he does, namely, insist that one proceed by the analytical method, that is, the method of Bacon, appealing to the observed facts of the case, how could he also conclude that all principles are arbitrary, true by definition? Again, it cries out for an alternative interpretation.

Such a reinterpretation is possible only if we recognize that Hobbes is proposing to substitute an account of reason that is very different from the traditional one. This becomes clear if we recall what he wrote about reason in his *Objections* to Descartes's *Meditations*.

Now, what shall we say if it turns out that reasoning is simply the joining together and linking of names or labels by means of the verb 'is'? It would follow that the inferences in our reasoning tell us nothing at all about the nature of things, but merely tell us about the labels applied to them; that is, all we can infer is whether or not we are combining the names of things in accordance with the arbitrary conventions which we have laid down in respect of their meaning. If this is so, as may well be the case, reasoning will depend on names, names will depend on the imagination, and imagination will depend (as I believe it does) merely on the motions of our bodily organs; and so the mind will be nothing more than motion occurring in various parts of an organic body.[7]

For Hobbes, reasoning is a matter of words. As he later elaborated this account, words are names: 'Words so connected as they become signs of our thoughts, are called SPEECH, of which every part is a *name*' (*Elements*, 15). Names serve two functions, that of being marks and that of being signs: 'seeing (as is said) both marks and signs are necessary for the acquiring of philosophy, (marks by which we may remember our own thoughts, and signs by which we may make our thoughts known to others), names do both these offices; but they serve for marks before they be used as signs' (15). Words taken to be marks are associated with thoughts in the sense of images. These images resemble each other, and in particular resemble the past images or phantasms from which they derive. '[M]arks [are] ... sensible things taken at pleasure, that, by means of them, such thoughts may be recalled to our mind as are like those thoughts for which we took them' (14). At the same time, they can function as signs by calling to the minds of others similar or resembling ideas or images. 'A NAME is a word taken at pleasure to serve for a mark, which may raise in our mind a thought like to some thought we have before, and which being pronounced to others, may be to them a sign of which thought the speaker had, or had not before in his mind' (16). Names can have meaning without naming any actual thing; Hobbes gives as examples the names 'nothing' and 'less than nothing.' But where they do name, they name bodies. But these bodies are not to be considered to have essences or real natures of the sort which the Aristotelians and rationalists suppose things to have. '[T]hat the sound of this word *stone* should be the sign of a stone, cannot be understood in any sense but this, that he that hears it collects that he that pronounces it thinks of a stone. And, therefore, that disputation, whether names signify the matter or form, or something compounded of both, and other like subtleties of the *metaphysics*, is kept up by erring men, and such as understand not the words they dispute about' (17; Hobbes's italics).

Descartes rejected Hobbes's account of reasoning. To Hobbes's objection he replied that

I did explain the difference between imagination and a purely mental conception in this very example, where I listed the features of the wax which we imagine and those which we conceive by using the mind alone. And I also explained elsewhere how one and the same thing, say a pentagon, is understood in one way and imagined in another. As for the linking together that occurs when we reason, this is not a linking of names but of the things that are signified by the names, and I am surprised that the opposite view should occur to anyone.[8]

What he is doing is simply insisting that there are ideas other than images, and that these ideas are signified by our words and provide the ground for our reasoning about things. Now, traditionally, for Aristotelians, ideas are the essences or forms or natures of things *qua* existing in the mind. But the same position is defended by such rationalists as Descartes.[9] Thus, Descartes refers to actual being as 'formal' being or reality (*esse formale*): the formal being of a thing is constituted by its properties or modifications. This he distinguishes from the objective being of a thing, that 'mode of being, by which a thing exists objectively or is represented by a concept of it in the understanding.'[10] This is 'the way in which its [the intellect's] objects are normally there.'[11] When a thing has objective being in the mind that thing is not a mode or property of the mind, that is, it is not 'formally' in the mind; otherwise, the mind would actually *be* the thing. None the less, though the thing that is objectively in the mind lacks formal reality as such, it is still not nothing,[12] and is to be contrasted to ideas which are merely chimerical.[13] Thus, as Descartes put it to Caterus in the replies to the First Set of Objections, 'the idea of the sun is the sun itself existing in the intellect – not of course formally existing, as it does in the heavens, but objectively existing, i.e. in the way in which objects are normally in the intellect. Now this mode of being is of course much less perfect than that possessed by things which exist outside the intellect; but, as I did explain [in Meditation III], it is not therefore simply nothing.'[14]

One must, of course, distinguish the thing from the thing *qua* formally existing. It is the thing *qua* formally existing that is *there* in the world, existing as the entity with which other real objects, including ourselves, interact. Thus, when Descartes distinguishes the thing (say, the sun), from the sun *qua* formally existing, the thing that is said to be in the mind, that is, the sun insofar as it does not formally exist, must be the essence of the thing. In other words, what is present to the mind when the thing is objectively in it is not this real object but its essence. For Descartes, the objective existence of things in the mind is constituted by the presence in the mind of the essence of the thing. But once we recognize this we also recognize that the Cartesian account of perception is structurally that of Aristotle, in which the mind knows the object by virtue of having the form or essence of that object literally in it. To be sure, Descartes rejects the mechanisms by which Aristotle explained the presence of the form or essence in the mind. For Aristotle, the form was transported from the object known to the knowing mind through a process of abstraction. Using the wax example of Meditation II, Descartes argued that no such process could provide us with the essence of any thing. Such essences are, he argued, following Plato, innate. This distinction between innatism and abstractionism is, however, a difference that is, as it were, within the ring. The important point is that both adopted a position concerning things, which involved distinguishing things from their natures or forms or essences, and, depending upon that, a position concerning the nature of ideas, in which the essences of things are in the mind of the knower.

Descartes thus retained in its central points the Aristotelian notion that science was, *ideally, scientia* in the Aristotelian sense, that is, in the sense that at its best science consists of *demonstrations proceeding from self-evident propositions concerning the natures or essences of things*. For both, the eternal or necessary truths are connections among the essences of things, connections that hold independently of the actual existence of things. In fact, it is precisely this latter that is the ground of the necessity of these truths.

It is precisely this concept of reason that Hobbes is rejecting, whether it be the Aristotelian version or that of Descartes. There are no essences or forms or natures of things that could in any way ground the necessity of the truth of the basic premises of science. Thus, for Hobbes there is *no objective ground* for the necessity of causal propositions.

But knowledge, by the traditional definition, is *scientia*: it is a matter of demonstration from necessary truths. Wherein, then, can the necessity of the first principles be located, given the Hobbesian framework? The answer that Hobbes gives is clear in his Objection to Descartes: *the necessity does not lie in the things; it lies rather in our thought about those things.*

But to say this is not yet to give an account of such necessity. However, Hobbes does go on to attempt to provide an explanation of why we attribute necessity to causal propositions and why we treat science as a matter of demonstration from premises understood to be – somehow – necessary. The model that he believes he has available to account for the necessity of our causal judgments and of the first principles by which we know causal relations is that of definition – not, of course, the real definition of the Aristotelians nor the necessary connections of the rationalists, but simple *nominal definition*.

IV: Hobbes's Supposed Nominalism

Before elaborating this point, there is a possible misreading of Hobbes that must be excluded. For, it has determined some of the more radical readings of Hobbes's account of science.

I refer to the charge that Hobbes is a radical nominalist who makes every truth a matter of convention. This is the reading, for example, of Taylor. Hobbes holds that the only names that denote realities are the names of individual bodies. Among these names are those that are *proper* to one thing and those that are *common*. Upon Hobbes's account of things, there are no forms or natures or essences for these terms to denote. '[W]hen *a living creature, a stone, a spirit*, or any other thing, is said to be *universal*, it is not to be understood, that any man, stone, &c. was or can be universal, but only that these words *a living creature, a stone*, &c. are *universal names*, that is, names common to many things; and the conceptions answering them in our mind, are the images and phantasms of several living creatures, or other things' (*Elements*, 20). But, as we have seen, 'a name is a word taken at pleasure to serve for' a mark or sign of our ideas (16). It would seem, therefore, that all truth is arbitrary or conventional. Indeed, Hobbes says almost this when he remarks that 'the first truths were arbitrarily made by those that first of all imposed names upon things, or received them from the imposition of others. For it is true (for example) that *man is a living creature*, but it is for this reason, that it pleased men to impose both names on the same thing' (36).

The result is an apparently silly position. For if is literally true that all truth is conventional, then all propositions about individual bodies are either necessary or contradictory. If *a* by convention denotes one individual and *F* is by convention a common name that denotes the same individual, then the sentence

   Fa

is true by virtue of its meaning, that is, the meanings of the terms that it contains. If, similarly, it is decided that *a* is not among the bodies commonly denoted by *G*, then

   Ga

is false by virtue of the meanings of the terms that it contains. But to say that the former is true *ex vi terminorum* while the latter is false *ex vi terminorum* is to say that the former is necessarily true while the latter is necessarily false. If it is further held that propositions that are self-contradictory are empty, or say nothing, then all false propositions will be empty or say nothing. If one further understands this to mean that, strictly speaking, one succeeds in making an assertion only if one asserts a proposition that says something, then one never says anything when one says anything false. Indeed, since all is a matter of convention, a matter of what the *speaker* decides, it becomes questionable whether it is even possible to err with regard to the truth. At most one would have violated the conventions of the language, which would make it a matter of misspeaking rather than one of genuine error.

Such a position seems to have been held in antiquity by Antisthenes,[15] of whom it is reported by Aristotle that he held the paradoxical belief that 'contradiction is impossible' (*Topics*, 104b20),[16] and that he 'foolishly claimed that nothing could be described except by its own formula, one formula to one thing; from which it followed that there could be no contradiction, and almost that there could be no error' (*Metaphysics*, 51024b30). It is likely that this view was the object of Plato's criticism in the *Sophist*.[17] In any case, it is certainly true that Plato, in the *Theaetetus*[18] proposed a solution that could avoid the paradoxes created by Antisthenes' view. Specifically, Plato argued that one must draw a distinction between the object to which the subject term of a proposition applies and that by virtue of which the predicate term applies to that thing. One must distinguish, in other words, between a thing and its properties. Predicate terms will have, as it were, a double reference. On the one hand, they apply to the same things to which subject terms apply. On the other hand, they apply to the properties of things. Thus, the predicate term *F* will apply to the property *F*, and then apply to all those things which have this property. John Stuart Mill was later to draw the relevant distinction in terms of 'connotation' and 'denotation.'[19] A name which applies properly to a thing, that is, to only one thing, is said to *denote* that thing. A name which applies commonly to several things is said to *denote* those things, but is also said to *connote* a certain property. Then, it applies *correctly* to a thing if that thing in fact exemplifies the property that it connotes. In this way one can consistently hold both that common names name or denote individual things and also that falsehood does not amount to contradiction and that it is after all possible to err and to assert what is false.

Hobbes does not speak of *connotation*. But he says much the same. The imposition of common names is not wholly arbitrary. Rather, it depends upon the resemblances of things, that is, the properties or accidents which they exemplify. '*Positive* [names] are such as we impose for the likeness, equality, or identity of the things we consider; *negative*, for the diversity, unlikeness, or inequality of the same. Examples of the former are, *a man, a philosopher*; for a *man* denotes any one of a multitude of men, and a *philosopher*, any one of many philosophers, by reason of their similitude; also, *Socrates* is a positive name, because it signifies always one and the same man' (*Elements*, 18). Propositions are formed from subjects and predicates joined by the copula. The subject and predicate are both names, and these

names raise in our mind the thought of one and the same thing; but the copulation makes us think of the cause for which those names were imposed on that thing. As, for example, when we say *a body is moveable*, though we conceive the same thing to be designed by both those names, yet our mind rests not there, but searches farther what it is *to be a body*, or *to be moveable*, that is, wherein consists the difference betwixt those and other things, for which these are so called, others are not so called. (31)

The *cause* for our saying that the body is moveable is 'that it *is moved* or the *motion* of the same' (32). These reasons justifying the copulation, making the assertion true, are the similarities and dissimilarities in things. These similarities and dissimilarities are objective, not merely the result of some arbitrary imposition of names upon things. For, Hobbes holds, there is a ground in things for these similarities and dissimilarities. Specifically, the ground in things for these objective similarities and dissimilarities is the *accidents* in things. The things in which accidents inhere are all conceivable apart from those accidents (save for the one essential accident of extension), though we should not, Hobbes warns us, conceive of their separability as a matter of abstracting these parts and making them into independent existents.

[The] causes of names are the same with the causes of our conceptions, namely, some power of action, or affection of the thing conceived, which some call the manner by which any thing works upon the senses, but by most men they are called *accidents*; I say accidents, not in that sense in which accident is opposed to necessary; but so, as being neither the things themselves, nor parts thereof, do nevertheless accompany the things in such manner, that (saving extension) they may all perish, and be destroyed, but can never be abstracted. (32–3)

Truth and falsity are therefore not merely a matter of arbitrary choice, and Hobbes's cannot, therefore, be construed as some sort of radical nominalist on the model of Antisthenes.

To be sure, it is arbitrary that we have chosen to apply *this* predicate rather than *that* to things by virtue of some objective similarity. For example, it is arbitrary that we have chosen 'red' rather than 'der' to name commonly all those things that are similar to each other in respect of a certain colour. Given his position on the objective basis of similarities and dissimilarities in the accidents of things, it is no doubt just this that Hobbes had in mind when he stated that 'the first truths were arbitrarily made by those that first of all imposed names upon things, or received them from the imposition of others. For it is true (for example) that *man is a living creature*, but it is for this reason, that it pleased men to impose both names on the same thing' (*Elements*, 36). Contrary to Taylor, then, this passage cannot be used to ascribe to Hobbes a radical conventionalism.

We may conclude that, whatever Hobbes has in mind concerning his account of first principles as definitions, it is not something that follows simply from a radical conventionalism along the lines traditionally ascribed to Antisthenes. For Hobbes holds no such radical view: like Aristotle and most subsequent philosophers, he agrees with Plato's solution to the problem posed by Antisthenes, that we must distinguish between a thing and the properties of that thing, between a thing and the objective similarities and dissimilarities between that thing and other things.

V: Hobbes's Account of Causal Necessity

The traditional account of science as *scientia* is that this proceeds deductively from self-evident premises. For the Aristotelians there were many self-evident truths to function as premises in scientific demonstrations. For the new science the range was rather more limited. It was, in fact, restricted to that paradigm of demonstrative science the geometry of Euclid. Galileo had advanced geometry as the tool of science. Descartes pursued the same Galilean themes, but, unlike Galileo, put those themes in a rigidly metaphysical context. The axioms of geometry were the essential truths of body. These axioms were the necessary connections to be found in the essence or nature of body, that essence which was the objective idea of body as innately present in our minds. Hobbes, as is well known, was immensely impressed by the work of Galileo, and took up the same themes as Descartes. But he strongly objected to the metaphysical context upon which Descartes insisted: there are no essences or forms or natures of things, and in this sense no metaphysical necessity.

Yet it is science and therefore necessary. What, then, is the source of this necessity? This is the question we asked previously. We have seen one answer – that it derives from the conventionality of all truth – that will not do. We must find another.

If Hobbes was committed to the Galilean ideal of a mathematical science, he was also committed to the traditional logic, including the account of demonstration that is part of the traditional logic.

On the traditional view,[20] we can (to take a simplified example) explain

> Corsicus is dissolving

using the syllogism

(S)   Whenever Corsicus is put in water Corsicus dissolves
         Corsicus is in water
         _____

         ∴ Corsicus is dissolving

provided that the premise is necessary. This is established as necessary through a *scientific syllogism*. Let Corsicus be C and let A be the pattern of sensible events of dissolving it in water. Then C is A. The scientific syllogism that we need will have this as its conclusion. The middle term that explains C's being A is the capacity (power, nature, essence, form, soul) of *solubility*. Let B be this capacity. Then the explanatory syllogism is:

(S′)   Whatever is soluble is such that whenever it is put in water it dissolves
         (B is A)
         Corsicus is soluble (C is B)
         _____

         ∴ Corsicus is such that whenever it is put in water it dissolves (C is A).

Solubility is an active power, like the nutritive power, the activity of which accounts for the pattern of behaviour described in the conclusion. As the nutritive power is one of a set of powers, including the appetitive and the ratiocinative, which make up the soul or form or nature of human beings, so also solubility will be one of the set of powers that constitute the set that makes up the full nature or essence of Corsicus – for example, the nature of Corsicus may be sugar. Thus, the second premise of (S') is necessary because the nature or essence is inseparable from the substance Corsicus. Provided that the major premise of (S') is also necessary, then the conclusion will be necessary, and the syllogism will establish a necessary connection among the events in the pattern mentioned in the conclusion. The necessity of the law derives from the form or essence; thus, explanation of a pattern of behaviour is not obtained by fitting that pattern into a more complex pattern, as in the new science, but by *redescribing* the object whose behaviour is being explained.

Now, as Molière made clear, for the defenders of the new sciences, the major premise of (S') is indeed necessary, but only trivially so, since it is true by nominal definition. For that reason, for the critics of Aristotle, (S') is no more explanatory than

> Whatever is a bachelor is an unmarried male
> (S")   Callias is a bachelor
> _____
>
> ∴ Callias is an unmarried male

(S") doesn't explain since the minor premise merely *restates* what the conclusion says. As for the major premise in (S"), that is in fact redundant; for the rule of language

> 'Bachelor' is short for 'unmarried male'

which licenses its assertion as a necessary truth equally licenses the rewriting of the minor as the conclusion and therefore also licenses the inference of the conclusion from the minor premise alone.

On the other hand, for Aristotle, (S') *is* explanatory. The major premise cannot, therefore, be true by definition. It must be a substantive truth. Thus, for the empiricist capacities are analysable in the sense that the term 'soluble' in the major premise of (S') can be defined by the right hand side using only (the principles of logic and) terms that refer to observable characteristics of things; while for Aristotle, since the major premise of (S') cannot be a definitional truth, capacities (dispositions, tendencies, powers) cannot be analysed into (patterns of) observable characteristics of things. *Aristotelian explanations are in terms of the unanalysable active dispositions and powers of things.*

If the major premise of (S') is not definitional, it is also not contingent; Aristotle holds, as we have seen, that the premises of a scientific syllogism are *necessary*. The empiricist divides propositions, as Hume does,[21] into two mutually exclusive and jointly exhaustive classes. There are, on the one hand, those propositions which state relations of ideas, and which are therefore necessary in the sense of being tautological, devoid of factual import, and, on the other hand, those propositions which state matters of observable fact and which are therefore contingent. If the former propositions are, as they are often called, analytic, and the latter are synthetic, then for Aristotle, in contrast to the empiricist, there is a third category, that of propositions which are both synthetic and necessary.

The tradition tried to fit the inferences of Euclid into this syllogistic framework. To be sure, as is well known, there are many inferences in Euclidean geometry that cannot be captured within the limited logical apparatus of syllogistic. Be that as it may, it was the traditional view that the inference patterns of Euclid did indeed fit those of syllogistic. It is this view that Hobbes accepts.

What he rejects is the traditional account of unanalysable powers. None the less, he still accepts the traditional doctrine of *scientia* in which science consists of demonstrative syllogisms having as their premises propositions which are necessary truths. These necessary truths, traditionally, express the real definitions of things. What Hobbes does is reject the doctrine of unanalysable dispositions but retain the idea that they are none the less *necessary* – except that he now construes them as *nominal definitions*. *Hobbes takes over the traditional model of science as demonstrations the premises of which are necessary propositions, save that upon his account the necessity derives not from the ontological structure of things but from the linguistic conventions which make the premises true by definition.*

This, however, does not seem to leave us in any better a position in interpreting Hobbes than did Taylor's account of the necessity of propositions in terms of a radical conventionalism. After all, if the premises of demonstrative science are all true by virtue of certain nominal definitions, then in what way have we escaped attributing to him a conventionalism that makes all causal propositions true by definition? And if causal propositions are all true by definition, then why do we have to employ the Baconian methods of elimination in order to discover their truth?

VI: Hobbes on the Logical Structure of Science

Hobbes does address this problem. He tells us that

the reason why I say that the cause and generation of such things, as have any cause or generation, ought to enter into their definitions, is this. The end of science is the demonstration of the causes and generations of things; which if they be not in the definitions, they cannot be found in the conclusion of the first syllogism, that is made from those definitions; and if they be not in the first conclusion, they will not be found in any further conclusion deduced from that; and, therefore, by proceeding in this manner, we shall never come to science; which is against the scope and intention of demonstration. (*Elements*, 82–3)

Three things are clear here. First, it is clear, as we have argued, that on Hobbes's view, science aims to be demonstrative and, if it is to be demonstrative, must begin from definitions. Second, it is clear that Hobbes holds that the definitions which he seeks are definitions of things, that is, of things insofar as they are effects. Third, these definitions of the things insofar as they are effects must include the causes of those effects.

The traditional doctrine was, of course, that, in giving a real definition of a substance, one was giving a definition of the essence of the thing. In giving the real definition of being human one was giving the real definition of the essence of Socrates. One thereby defines the thing or substance in which the effect occurs. Hobbes's point is much the same, except that he does not distinguish the thing and its essence: there is no essence or form or nature. What one defines, then, upon Hobbes's scheme is the thing itself.

Now, as noted earlier, Hobbes's discussion of cause makes clear that we have a cause only if we have a general proposition. We can correctly say that

> *A* is the cause of *B*

only if we can affirm that

> Whenever *A* then *B*

or, in the more traditional phraseology,

(@)   (All) *A* is *B*.

The Baconian methods of elimination are designed to discover the truth of such propositions. Such a proposition cannot be necessary. For a proposition is necessary only if it is true by definition. For, as Hobbes also says, 'in every *necessary* proposition, the predicate is either equivalent to the subject ... or part of an equivalent name ... But in a *contingent* proposition this cannot be' (*Elements*, 38). However, even if it is true that *universal* causal propositions, as Hobbes calls them, are not necessary, there is for all that no reason why the proposition that

(@')   Corsicus is such that whenever it is *A* then it is *B*

cannot be constituted as necessary.

Now, according to Hobbes, we achieve knowledge of what a thing is by *resolution* or *analysis*. Thus, he tells us that 'if any man propound to himself the conception of *gold*, he may, by resolving, come to the ideas of *solid*, *visible*, *heavy*, (that is, tending to the centre of the earth, or downwards) and many other more universal than gold itself ...' (*Elements*, 69). Since a singular thing is distinguished from other things by its accidents, our knowledge of singular things similarly proceeds by resolution or analysis. Now, it is true that

> something is heavy if and only if whenever it is unsupported then it move downwards.

This, in fact, is true by definition. There are similar definitions of the other powers that enter into the definition of *gold* itself. Indeed, the concept of *gold* is just the conjunction of a set of such defined powers. Thus, the concept of a sort of singular thing and, since singular things are distinguished by their accidents, the concept of any singular thing itself, will be a conjunction of defined powers. But those powers are *patterns* of behaviour: they relate *universal* things to their causes. These statements of the causes of universal things that enter into the definitions of singular things are contingent. But, though they are contingent, *that they describe the behaviour of a certain singular thing will be true by definition*. Thus, although (@) as a statement of the cause of a universal thing is contingent, the statement (@') as the statement of causation applying to a singular thing will be necessary.

To see this, consider a simple example. Since to understand a singular thing is to resolve it into the set of powers that define it, we can suppose Corsicus to be defined by two powers. If, for example, we have as true by definition the two disposition statements

> *D* if and only if whenever *A* then *B*

D' if and only if whenever A' then B'

we may then further define

D* if and only if D and D'.

This is what Hobbes refers to as a 'compounded name,' wherein a simple name 'by joining another name to it, is made less universal, and signifies that more conceptions than one were in the mind, for which that latter name was added' (*Elements*, 24). Now, if we also take D* as defining the essence of form of Corsicus, then we have the syllogism

> Whatever is D* is such that whenever it is A then it is B
>
> (S")   Corsicus is D*
>
> ———————————————————
>
> ∴ Corsicus is such that whenever it is A then it is B.

The concept D* functions as a middle term. The major premise is true by definition of this concept. Moreover, this concept is (part of) the concept defining Corsicus. So the minor premise is also necessary. That makes the conclusion necessary. We therefore have a demonstrative syllogism that establishes that (@') is necessary. But, to repeat, this does not mean that the statement (@) concerning the cause of the universal thing A is not necessary.

The syllogism here is demonstrative in the sense of proceeding from necessary premises. It is synthetic in the sense that it presupposes that we have synthesized our knowledge of the causes of universal things into a concept of a singular thing. Analysis is needed to discover the causes of universal things, that is, the parts of singular things. But in order to demonstrate the necessity of the causation of singular things, our method must be 'compositive' (*Elements*, 71), putting together these statements of causation or power into the 'compounded names' of things. It is the compounded names or definitions of singular things that enable us to construct a demonstrative science based on the synthetic methods of syllogistic.

We thus obtain the following picture of science according to Hobbes: *Bacon's methods of eliminative induction are used to establish the truth of the contingent statements of causation with regard to universal things, and these statements of causation or power then enter into the definitions of the concepts of singular things. These definitions of things form the ultimate premises of demonstrative science, which proceeds synthetically from them by means of syllogisms. This science establishes the necessity of causal statements about singular things.*

VII: Hume's Account of Causal Necessity

Hume gives two definitions of 'cause' (T 172).[22] There has been controversy about it: How can one give two different definitions of the same notion without contradiction?[23] J.A. Robinson claims that the only significant Humean definition of 'cause' is the first. He suggests ('Hume's Two Definitions,' 138–9) that this is the *real* definition and that the second is *not* really a definition but is a psychological comment on the relation so defined. Now, there is a point to this, since it *is* clear that the second definition introduces an element of psychology derived from and explained by the associationist theory. On the other hand, it does fly in the face of Hume's explicitly saying that he is giving two definitions. More seriously, however, if one construes Hume as giving only the first definition as *the* definition of 'cause,' then one is attributing to him what A.C. Ewing has called the 'regularity' theory of causality, which holds that there is nothing more to causal assertions than assertions of *de facto* regular connection.[24] Thus, Fogelin,[25] Penelhum,[26] and Robinson,[27] to mention only a few, all attribute to Hume an analysis of causation as, in Robinson's terms, 'nothing more than an instance of a general uniformity of concomitance between two classes of particular occurrences.'[28] But this theory may be criticized, as it is by Ewing, as being unable to distinguish causal associations from those that occur by chance, in other words, for being unable to distinguish laws from accidental generalities. The difficulty is that Hume never held this 'regularity' theory.[29] The *Treatise* fully recognizes the distinction Ewing's 'regularity' theory denies between *post hoc* and *propter hoc*. As it points out, 'an object may be contiguous and prior to another, without being consider'd as its cause. There is a NECESSARY CONNEXION to be taken into consideration; and that relation is of much greater importance, than any of the other two above-mention'd' (T 77). And most of the rest of Book I, Part III is devoted in one way or another to an investigation of the nature of this necessary connection. Robinson thus commits Hume to a theory Hume explicitly denies. It is the point of Hume's second definition, I suggest, to introduce the element of necessary connection that a 'mere regularity' theory fails to capture. What must be done is to make clear how the two definitions are related to each other.

Fundamental to understanding Hume's account of causation is the distinction that can be drawn with respect to our beliefs between the propositional attitude and the propositional content. If one person believes that Toronto is west of Hamilton and a second disbelieves it, the propositional content of both beliefs is the same, namely, the proposition that Toronto is west of Hamilton. But the propositional attitudes in the two beliefs are different. In the one case, the attitude is that of believing or asserting; in the other, it is that of disbelieving or denying. The propositional content is true or false. Its truth-value depends only on the facts it is about. In particular it does not depend upon what propositional attitude one has with respect to that content. This independence is important, since it is precisely this independence that enables truth and falsity to function as *standards* justifying the attitude one has. The attitude of believing or asserting with respect to a proposition is *objectively justified* if and only if that proposition is true. And the attitude of disbelieving or denying is objectively justified if and only if the proposition is false. Clearly, if truth-value depended on attitude, the former could not provide an objective standard for evaluating the latter.

The first definition of 'cause' is this: '*an object precedent and contiguous to another, and where all objects resembling the former are plac'd in a like relation of priority and contiguity to those objects, that resemble the latter*' (T 172).[30] This is 'cause' defined as a philosophical relation (T 94, 170). Upon this definition, one event causes a second where the two are subsumable under a matter-of-fact generality. Particular instances of causation are to be understood in terms of *de facto* regularities. This means that any proposition stating a causal regularity has the logical form $(x)(fx \supset gx)$, where the descriptive predicates are logically independent of each other. This renders in the language of Russell the Humean point that cause and effect are logically independent of each other (T 139).[31]

The rationale for this first definition of cause derives from Hume's critique of the Aristotelian and Cartesian claim that there are *objective* necessary connections. He argues that if we take the idea of some cause and the idea of its effect, then there is no contradiction in supposing the former to exist and the latter not: 'the actual separation of these objects is so far possible, that it implies no contradiction nor absurdity; and is therefore incapable of being refuted by any reasoning from mere ideas; without which 'tis impossible to demonstrate the necessity of a cause' (T 80). There is no contradiction in separating the ideas because these ideas derive from perceptions and in perception there is no necessary connection that is presented to us: 'as all our ideas are deriv'd from impressions, or some precedent *perceptions*, 'tis impossible we can have any idea of power and efficacy, unless some instances can be produc'd, wherein this power *is perceiv'd* to exert itself ... [But] these instances can never be discover'd in any body ...' (T 160).

Hume's appeal here is to the empiricist principle of acquaintance (PA). Among the entities that are given to us in our ordinary sensory experience, we discover no connections such that one of these entities cannot exist unless another of these entities exists. In identifying the property that characterizes the cause we do not have to refer as a matter of necessity to the property that characterizes the cause, one to which it is necessarily tied; the properties are presented as logically self-contained rather than as necessarily tied to one another. We are, in other words, acquainted with no such entity as a necessary connection in ordinary experience, and there is therefore no necessary connection among the ideas that we use to describe whatever is given to us in experience.

This argument from acquaintance derives from Locke. Locke considers the regular activities of external substances. These include the production of the ideas of the secondary qualities, the simple ideas of red, sweet, and so on. For these activities to be explained as those who defend necessary connections require, there must be necessary connections between red, sweet, and so on, and the natures or real essences of the substances that cause these qualities to appear. These necessary connections must be both ontological, in the entities themselves, and epistemological, giving us, when in the mind, scientific knowledge of those entities. But, Locke argues, we grasp no such connections.

'Tis evident that the bulk, figure, and motion of several bodies about us, produce in us several sensations, as of colours, sounds, tastes, smells, pleasure and pain, etc. These mechanical affections of bodies, having no affinity at all with those ideas, they produce in us, (there being no conceivable connexion between any impulse of any sort of body, and any perception of a colour, or smell, which we find in our minds) we can have no distinct knowledge of such operations beyond our experience; and can reason no otherwise about them, than as effects produced by the appointment of an infinitely wise agent, which perfectly surpasses our comprehension. (*Essay*, IV.iii.28)[32]

Locke's appeal to a principle of acquaintance (PA) is clear. Properties are presented as logically self-contained; there is nothing about them as presented that requires us when we are identifying them to refer as a matter of necessity to other properties, those to which they are necessarily tied. We are, in other words, not presented with necessary connections, and we can therefore, by PA, not admit them into our ontology.

The Cartesians and Aristotelians admit the same point: objective necessary connections are not given in sense experience. For the substance tradition, certainty of inferences from samples to populations is achieved through a grasp of the essences or natures of things. Causal activity of a thing in accordance with its essence guarantees that always in similar circumstances it would again behave in similar ways. Hume, of course, argues vigorously against this view (T 157ff), adopting arguments from Malebranche (T 158). Malebranche argued that there are no objective necessary connections other than those of God's causal activity:[33] 'les causes *naturelles* ne sont point de véritable causes ... Il n'y a donc que Dieu qui soit véritable cause, et qui ait véritablement le puissance de mouvoir les corps' (*Recherche*, II: 200, 203). Hume refers to this 'Cartesian' doctrine (T 159), and argues that it is untenable (T 160). The Cartesians 'have had recourse to a supreme spirit or deity, whom they consider as the only active being in the universe, and as the immediate cause of every alteration in matter' (T 160). These Cartesians appeal to innate ideas to provide themselves with an idea of cause, but if we reject that, as Hume does, following Locke, then there is no source for our ideas other than sense experience, acquaintance. In that case, the Cartesian hypothesis of a deity cannot do the job required of it.

[T]he principle of innate ideas being allow'd to be false, it follows, that the supposition of a deity can serve us in no stead, in accounting for that idea of agency, which we search for in vain in all the objects, which are presented to our senses, or which we are internally conscious of in our own minds. For if every idea be deriv'd from an impression, the idea of a deity proceeds from the same origin; and if no impression, either of sensation or reflection, implies any force or efficacy, 'tis equally impossible to discover or even imagine any such active principle in the deity. Since these philosophers, therefore, have concluded that matter cannot be endow'd with any efficacious principle, because 'tis impossible to discover in it such a principle; the same course of reasoning shou'd determine them to exclude it from the supreme being. (T 160)

If, as these philosophers hold, there are no objective necessary connections among bodies because we have no impression of such a connection, then neither do we have any idea of it – ideas being derived from impressions – and therefore we cannot have an idea of God that includes within it the idea of causal power or activity.[34] 'We never therefore have any idea of power' (T 161).

Thus, since we are not acquainted with any objective necessary connections, they are not to be admitted to one's ontology. Causation, then, considered objectively, involves nothing more than regularity. It is this *objective aspect* of causation that is captured in Hume's first definition.

It follows that, so far as propositional content is concerned, the assertion of a causal regularity involves no difference in logical form from the assertion of an accidental generality. Both are of the form $(x)(fx \supset gx)$, from which it follows that *objectively there is no logical difference between causal regularities and accidental generalities*. Objectively, then, there is no difference of a logical sort between *post hoc* and *propter hoc*.

None the less, such a distinction is to be drawn, as we have seen Hume recognize: 'An object may be contiguous and prior to another, without being consider'd as its cause. There is a NECESSARY CONNEXION to be taken into consideration; and that relation is of much greater importance, than any of the other two above-mention'd' (T I.III.ii, 77). Hume therefore raises the question, *'What is our idea of necessity, when we say that two objects are necessarily connected together'* (T 155). What he finds is that when two sorts are causally connected, upon the appearance of an object of the one sort, 'the mind is *determin'd* by custom to consider its usual attendant' and that it is this *'determination,* which affords [him] the idea of necessity' (T 156). This yields the second definition of 'cause' as *'an object precedent and contiguous to another, and so united with it in the imagination, that the idea of the one determines the mind to form the idea of the other, and the impression of the one to form a more lively idea of the other'* (T 172). This is 'cause' defined as a natural relation (T 97, 170).[35] The idea of necessary connection, which is an ingredient in the idea of cause, is therefore the propensity of the mind to make inferences in the case of causal connections which is absent in the case of accidental generalities (T 167).

If a propositional content of the form $(x)(fx \supset gx)$ is such that as *a matter of psychological fact* it is used to support assertions of subjunctive conditionals ('the idea of the one determines the mind to form the idea of the other') and to make predictions ('the impression of the one [determines the mind] to form a more lively idea of the other'), *then* the assertion of that proposition is the assertion of a causal generality. A generality is lawlike just in case it is in fact used to predict and to support subjunctive conditionals; otherwise it is a statement of 'mere regularity.'

The connection between lawlike generalities and subjunctive conditionals has often been remarked upon.[36] What Hume does is use this connection to *define* lawlikeness: a generality is causal or not, depending upon its psychological context. Lawlikeness is a matter of the propositional attitude that obtains with respect to the generality. The generality is lawlike if and only if it is not merely believed or asserted but is asserted in a certain more specific way, namely, with a preparedness to take risks with it, that is, to predict and to assert subjunctive conditionals. We may call this the law-assertion attitude.

Lawlikeness thus becomes a subjective matter. It is not objective, a feature of the logical form of causal propositions. The thesis in ethics that value is a matter not of propositional content but rather of the psychological attitude is often called 'emotivism.' Hume's account of causation may therefore perhaps not unreasonably be characterized as an emotivist account of causation. Indeed, I think it is possible to push this analogy with various ethical theories – Hume's in particular, of course – some good distance. But one should be careful not to call Hume's ethical views 'emotivist' *simpliciter.*[37] In particular, one must account for our correcting and adjusting our moral judgments, a point he emphasizes many times.[38] It is a point he also emphasizes in aesthetics: in his essay 'Of the Standard of Taste,' Hume discussed how we correct aesthetic judgments, giving a very detailed analysis of the process. This is, of course, the basic objection to the thesis of Blanshard[39] and Coleridge,[40] among many others, that, on associationist principles, the mind cannot be guided by certain ideals and regulate its own activities in terms of those standards. In particular, I would propose that Hume similarly holds, in a way compatible with his associationism, that the mind, guided by the ideal of truth, by the passion of curiosity, can actively correct its causal judgments.[41]

But now let us return to the point where Hume's theory of causal judgments is similar to emotivism – that causality, like value, is a matter not of propositional content but of psychological attitude. As we noted, given this similarity, Hume's account of causation might be called an emotivist account. Once this is noted, one also recognizes similar objections that can be raised against both the thesis about moral judgments and the thesis about causal judgments. As with the emotivist theory in ethics, the immediate question about the emotivist theory of causation is this: Does not emotivism, when it denies the existence of an objective standard, entail that adopting the attitude is something that admits of no justification? Or, in other words, does not Hume's account reduce causal reasoning to irrationalism?[42]

Now, the reply to this, in the case of Hume's account of causal inference, just as in the case of ethics, is to challenge the presupposition of the objector's question and ask what one might *reasonably* mean by 'justification' in this context. In ethics, if the emotivist's arguments that objectivism is false are accepted, then it is no serious objection to his account that he leaves no room for an objective justification of value judgments. Of course he has left no room: he has just finished arguing that such justification is not possible. And if it is not possible, it is not reasonable to insist upon it. Whatever justification amounts to in ethics, the one thing that cannot *reasonably* be demanded is objective justification. And, in the case of Hume's account of causation, the same sort of response is called for. *Given* Hume's argument that, objectively considered, all causal assertions are assertions of constant conjunction,[43] *then* it is not reasonable to demand an *objective* justification for the adoption of the law-assertion attitude towards some generalities rather than others.[44]

However, from the fact that no objective justification is possible, it does not follow that all adoptions of the law-assertion attitude are equally justified; though the demand for objective justification is unreasonable, not all law-assertive attitudes are equally justified. Though the demand for objective justification is unreasonable, not all law-assertive attitudes are therefore equally reasonable. Hume clearly recognizes this point and explicitly draws our attention to cases where the law-assertion attitude holds but where it is also not justified. His discussion of credulity (T 112), of the often adverse effects of education (T 116), of the role of imagination (T 123), of unphilosophical probability (to which a whole chapter is devoted) (T 132ff), all make evident that the *Treatise* draws a distinction between these cases where the attitude is unjustifiably held and cases where it is justifiably held, where its adoption is in accordance with the '[r]ules by which to judge of causes and effects,' which appear in their own section with that very title (T I.III.xv).

An assertive attitude is objectively justified if and only if the proposition in question is true. But this holds equally for laws and for 'mere regularities'; provided that the latter are true, they may justifiably be asserted. So truth, while *sufficient* to justify the attitude of assertion or mere assertion, is only *necessary* to justify the attitude of law-assertion. Thus, if a generality is false, one is objectively unjustified in holding towards it the law-assertion attitude. This will be so even if the other necessary conditions of justification (whatever they may be) are all fulfilled. And it will be so even if we have *all possible reason* to believe that that necessary condition of truth is fulfilled.

A generality is a statement about a total population. Normally, all one ever observes is a sample. Between sample and population there is a logical gap. This gap is such that properties regularly associated or constantly conjoined in the observed sample may not be constantly conjoined in the population. Hume argues for the existence of this logical gap (T I.III.iii), which he at one point (T I.III.xii, 139) expresses as the principles that '*there is nothing in any object, consider'd in itself, which can afford us a reason for drawing a conclusion beyond it*' and '*that even after the observation of the frequent or constant conjunction of objects, we have no reason to draw any inference concerning any object beyond those of which we have had experience.*' Here, of course, what is meant is reason drawn from those objects 'consider'd in themselves' rather than 'not any sort of reason'; for, after all, Hume does go on to give us the 'rules by which to judge of causes,' where he sketches the conditions under which one *can* reasonably infer from a sample to a population. That properties are constantly conjoined in a sample is a necessary condition for their being constantly conjoined in the population. A necessary condition for justifiably making a law-assertion is that the generality asserted be true. Given the *logical* gap between sample and population, it is *not possible* to know whether this necessary condition is fulfilled simply by observing that the regularity obtains in the sample. The *best* we can do is to know that a necessary part of this necessary condition obtains, namely, that the regularity holds in the sample. That is the *best* we can do, short of omniscience. And since it is the best we *can* have, we must make do with it. *If we observe that a regularity holds in a sample, we thereby have every objective reason it is (at that point)*[45] *possible to have to justify our believing that the regularity holds in the population.*[46] Subjectively, the only and best objective evidence a regularity obtains overall is that we have observed it to obtain among the facts we already know. So, subjectively, we may be justified in asserting a generality when objectively the assertion is not justified.[47] Still, if we have done the best we can do, if we assert only when we are subjectively justified, then we cannot be blamed for not having done more, even where we are objectively unjustified. Fallibility is not a vice.[48]

We have still not distinguished causal from 'mere' regularities. The remarks just made, drawing attention to our fallibility, apply equally to both sorts of regularity. An observed constant conjunction is a necessary condition for our being subjectively justified in adopting the law-assertion attitude towards a generality, but it is not a sufficient subjective condition. It is sufficient only if the acquisition of this evidence has proceeded in accordance with the 'rules by which to judge of causes.' Hume notes carefully that there are many situations in which the mind is confronted with a set of contrary hypotheses, each of which initially fits the data, and fulfils the necessary condition for being subjectively justified (T I.III.xii, 131–5; xiii, 154). Since these are contraries, the law-assertive attitude towards any particular one cannot be justified initially. But now, suppose more data are sought, in accordance with the 'rules by which to judge of causes.' Suppose further that these data ultimately eliminate all but one of the contraries. In that case, the adoption of the law-assertive attitude towards that hypothesis will be (subjectively) justified, that is, justified to the extent that it is reasonable to seek such justification. On the other hand, other principles of selection are possible.

Hume cites the principle that we choose as worthy of law-assertion the hypotheses we want to be true, quoting Cardinal de Retz on the principle that the wish is the father of the belief, '*that there are many things, in which the world wishes to be deceiv'd*' (T 153). Where the world wishes to be deceived it can avoid trying to gather together the *evidence* relevant to reasonably deciding among the possible hypotheses (T 152–3). In the section 'Of unphilosophical probability' (T I.III.xiii), a number of such unreasonable principles are mentioned. Those who desire a *reasonable decision* among contrary hypotheses must go out and *actively collect* additional observational evidence that will permit a decision to be made. The data that are given are often acquired only with great difficulty. As Hume puts it, directly after stating the 'rules by which to judge of causes':

There is no phaenomenon in nature, but what is compounded and modify'd by so many different circumstances, that in order to arrive at the decisive point, we must carefully separate whatever is superfluous, and enquire by new experiments, if every particular circumstance of the first experiment was essential to it. These new experiments are liable to a discussion of the same kind; so that the utmost constancy is requir'd to make us persevere in our enquiry, and the utmost sagacity to choose the right way among so many that present themselves. (T 175)

One makes a reasonable decision among alternative possible hypotheses when one has *actively* sought out such data as would permit one logically to make such a decision, eliminating or falsifying hypotheses until exactly one is rendered subjectively worthy of law-assertion. *The choice among hypotheses may be made* EITHER by collecting relevant observational evidence according to the 'rules by which to judge of causes' OR by some other principle. If on some basis other than such observational evidence, then the resulting law-assertion is unjustified.

VIII: Correcting Hobbes

We can see what Hobbes is trying to do. Bacon was the first to describe the logic of the experimental method. These were the methods that were used by such scientists as Harvey. Others besides Bacon were prepared to defend the use of these methods. As we argued in Study One, Descartes, in the *Discourse on Method*, for example, defended their use. We now see that Hobbes was another defender of these methods. Hobbes and Descartes both accepted that, at their best, causal judgments have to be necessary. For Descartes, one achieved such necessity if the causal judgments could be derived from necessary truths about the essences of things. Hobbes accepted this point. To this extent, commentators such as Taylor, Copleston, and Watkins are correct. But, as these commentators have not noticed, Hobbes also advanced a much more empiricist account of both things in the world and our knowledge of such things than that of Descartes. Specifically, for Hobbes, unlike Descartes, there are no essences, nor, therefore, any objective necessities in things that our reason could grasp. But that means that the Hobbesian concept of reason is very different from that of Descartes. To be sure, they both use the language of ideas, but where, for Descartes, the ideas that our reason grasps are the essences – the reasons – of things, for Hobbes, in contrast, ideas are images as, on the one hand, representative of resemblance classes of images and appearances of things and as, on the other hand, associated with certain words that stand as marks and signs of those images. Since there are no objective necessities to be reflected in our ideas, those ideas or images in and of themselves stand in no logical relations to one another. One obtains logical connections among ideas only when one establishes those conventions which yield nominal definitions of things. The only necessity, therefore, that is possible for Hobbes is this definitional necessity. This is, however, but a weak simulacrum of the ontologically grounded necessity of Descartes and the rationalists. In particular, it simply disguises the fact that what is crucial for any causal explanation are the *contingent causal relations* that enter into the definitions of things. These turn out to be *merely* contingent, and, in fact, the distinction between causation and accidental generality disappears.

The disguise did indeed have some success. It seems to have played some role at least in leading most commentators to overlook the fact that for Hobbes there are certain basic causal propositions which are, on his own view, contingent, and which are to be established by the Baconian eliminative methods of experimental science. None the less, we now see that a clear view of Hobbes's position allows that these empirical and non-demonstrative methods have a central place in his account of science, and that they cannot be eliminated in favour of the idea of a demonstrative science proceeding from premises that are simply true by definition, *ex vi terminorum*, however central to science Hobbes took such demonstrations to be.

In any case, as the commentators usually make clear, the conventionality of nominal definitions hardly provides a secure ground wherein to attempt to locate causal necessity. It is likely for this reason that Hobbes's approach to causal judgments proved unattractive to his empiricist successors. Hume adopted much the same view of ideas, and argued in detail the Hobbesian claim that there are no objective essences of things nor, therefore, any objective causal necessities.[49] Both thus agreed, for much the same sort of reason, that basic causal judgments are matter-of-fact generalities. Hume also defended what we have seen to be the Hobbesian view, derived from Bacon, that such judgments are established by the methods of eliminative induction, or, as Hume called them, the 'rules by which to judge of causes and effects' (T I.III.xv). But Hume gave up the whole idea that science should somehow have demonstration as its goal. Locke had developed the empiricist ontology and epistemology far enough that by Hume's day it was no longer implausible to hold that our capacity to come to make reasonable causal judgments has an empirical and contingent basis rather than the a priori demonstrations of traditional *scientia*.[50] This meant that Hume did not feel constrained, as Hobbes apparently felt constrained, to locate such necessity as causation has in the notion of a demonstrative science. And so, according to Hume, what distinguishes causal propositions from others is indeed their necessity. But it is neither the ontological necessity of the Aristotelians and the rationalists, nor the conventional necessity of Hobbes. It is rather the simple psychological necessity of being moved to make such inferences as a matter of habit induced, in the first instance, by the observation of constant conjunctions.[51] This Humean alteration to the basic empiricist position of Hobbes proved in the long run a more congenial account of the (non-objective) necessity of causal judgments.[52]

Study Five

'Rules by Which to Judge of Causes' before Hume

Hume's 'rules by which to judge of causes and effects' (*Treatise*, I.II.xv)[1] are divided into two parts by his fourth rule. The first three state the regularity account of causation. This is followed by a rule stating the principle of 'same cause / same effect.' The remaining rules state the rules of eliminative induction, in a way that is both accurate and a model of concision. With regard to both parts, Hume makes an original contribution. If we locate this contribution within the context of what had gone before we can begin to grasp both Hume's genius and the central place that he must be seen to occupy in the development of the methodology of empirical science.

To get a hold on Hume's contribution, it is necessary to begin with the contrast with what went before. Specifically, we must begin with the older philosophical tradition, deriving from Aristotle, about the nature of causation.

Thus, we find that Aquinas states the following rule for judging of causes: 'if the cause be removed, the effect is removed' (*Summa*, FP, Q77, A3, Obj 3).[2] This sounds very much like some things that Hume himself states. For, Hume provides, among others, the following rule for judging of causes: 'The difference in the effects of the two resembling objects must proceed from that particular, in which they differ' (T 174). Hume's point is apparently much the same as Aquinas's. Hume's is one of his 'rules by which to judge of causes and effects.' This would make it seem that these rules have a long history prior to Hume, and that the contribution of the latter is not so significant after all. But in fact that is not so. Apparent similarities should not be allowed to mask profound differences.

Hume puts this rule – it is the sixth of his 'rules by which to judge of causes and effects' – in a context. Two points are relevant to the difference between what he is about and what Aquinas is about. In the first place, in his third rule Hume has stated the principle that '[t]here must be a constant union betwixt the cause and the effect.' About this point he has argued that ''[t]is chiefly this quality, that constitutes the relation,' that is, the relation of cause to effect (T 173). In the second place, Hume argues that his sixth rule is 'founded' upon another. This further reason is the principle stated in his fourth rule, to the effect that '[t]he same cause always produces the same effect, and the same effect never arises but from the same cause' (T 173). 'For as like causes always produce like effects, when in any instance we find our expectation to be disappointed, we must conclude that this irregularity proceeds from some difference in the causes' (T 174). As for the 'same cause / same effect' rule, '[t]his principle we derive from experience, and is the source of most of our philosophical reasonings' (T 173). The subsidiary rule is, it is clear, that method of eliminative induction to which John Stuart Mill gave the name 'method of difference.' What this rule of inference leads to is a conclusion that a certain general truth holds, a truth to the effect that one condition is a sufficient condition for another. Since the conclusion is a general truth, the criterion of causality established by the third rule is met. But the rule known as the method of difference yields such a conclusion, Hume is clear, only if we accept antecedently the principle of 'same cause / same effect,' that is, what later thinkers were to refer to as the principles of determinism and of limited variety.[3]

Aquinas's rule, while it sounds like that of Hume, is not designed to yield statements of matter-of-fact regularity as conclusions. Aquinas simply does not place it in the context of principles of determinism and limited variety that are necessary if such inferences are to be made. In fact, Aquinas's rule has very little to do with matter-of-fact regularities. For, as we have seen in Study One, it is Aquinas's view that causal patterns can have exceptions. He puts it this way:

There is a difference between universal and particular causes. A thing can escape the order of a particular cause; but not the order of a universal cause. For nothing escapes the order of a particular cause, except through the intervention and hindrance of some other particular cause; as, for instance, wood may be prevented from burning, by the action of water. Since then, all particular causes are included under the universal cause, it could not be that any effect should take place outside the range of that universal cause. So far then as an effect escapes the order of a particular cause, it is said to be casual or fortuitous in respect to that cause; but if we regard the universal cause, outside whose range no effect can happen, it is said to be foreseen. Thus, for instance, the meeting of two servants, although to them it appears a chance circumstance, has been fully foreseen by their master, who has purposely sent to meet at the one place, in such a way that the one knows not about the other. (*Summa*, FP, Q77, A3, Obj 3)

This indicates that for Aquinas, the notion of cause is very different from that located by Hume in matter-of-fact regularity. Particular causes involve the exercise of powers that bring about certain patterns of events. But these patterns have exceptions. Such an exception occurs when a more powerful cause interferes with and hinders the activity of the particular cause in question. There is only one case where exceptions are not possible. This is the case of the most powerful cause, namely, God or the universal cause. It is this most powerful cause that creates all subsidiary patterns. Precisely because it is most powerful, there are no exceptions to the order that it creates. But, to repeat, for all lesser causes, those which are not infinitely powerful, there can always be exceptions brought about by external forces or powers.

It is important to recognize that for Aquinas, a cause is a *power* or *force*. This is, of course, denied by the regularity theory. The notions of a causal pattern and of exceptions to these patterns is understood in terms of the exercise of these powers, either by the object moving or by the external object that moves it. 'One thing,' Aquinas tells us, 'causes another in two ways: first, by reason of itself; secondly, accidentally. By reason of itself, one thing is the cause of another, if it produces its effect by reason of the power of its nature or form, the result being that the effect is directly intended by the cause ... Accidentally, one thing is the cause of another if it causes it by removing an obstacle: thus it is stated in Phys. viii ..., that "by displacing a pillar a man moves accidentally the stone resting thereon"' (*Summa*, FP, Q14, A1). When the obstacle is removed, the object that moves *moves itself*, that is, it exercises its power to move itself according to its nature or form. Such motion is natural, where natural motion is contrasted to *violent* motion, and, indeed, violent rest, where an object is held in a motion or in a state of rest that is contrary to its nature, or natural inclination. As Aquinas says, 'violent movement is that which is contrary to nature' (FS, Q6, A4, Obj 3). There is motion that proceeds from an 'interior principle' and that which proceeds from an 'exterior principle': the latter is violent motion. '[W]hat is compelled or violent is from an exterior principle ... [V]iolence ... is ... contrary to the nature of a natural inclination or movement. For a stone may have an upward movement from violence, but that this violent movement be from its natural inclination is impossible. In like manner a man may be dragged by force: but it is contrary to the very notion of violence, that he be dragged of his own will' (FS, Q6, A4).

We have, then, motion which is natural, that is, motion which is caused by a natural inclination of the object moving. When an object so moves, it is moved by its own nature or form; the motion results from the exercise of its natural power. In contrast, there is violent motion, where an object is moved by some external object. Such motion is contrary to the nature of the object moved.

Thus, as we discussed in detail in Study One, there are two patterns of explanation. The one pattern is that of natural motions, which are explained by the exercise of an internal natural inclination or power. The other pattern is that of violent motions, which are explained by an external object exercising a greater power to move a second object contrary to the latter's natural inclinations. Natural motions constitute certain patterns, but these patterns are not exceptionless. The exceptions that occur are violent motions. This is in contrast to the Humean idea of explanation where explanation is of one sort, subsumption under exceptionless matter-of-fact regularity.

Crucial to these patterns is the notion of *form* or *nature*. This is a metaphysical entity. It is not given to us in sensible experience. It is, rather, known by a rational intuition. The form or nature is an intelligible rather than a sensible entity. For Aquinas, this rational intuition is arrived at through a process of abstraction. The rational intuition is the product of this process of abstraction. When the intuition occurs, the form or nature of the thing known is literally in the mind of the knower. The act of abstraction takes the form from the thing known, or, at least, takes the form from the sensible appearances of the thing known, and places it in the mind of the knower. As an intelligible entity, the form or nature is at once in the thing of which it is the nature or form, but also the idea of that thing in the mind, when that thing is known. '[B]y ideas are understood the forms of things, existing apart from the things themselves. Now the form of anything existing apart from the thing itself can be for one of two ends: either to be the type of that of which it is called the form, or to be the principle of the knowledge of that thing, inasmuch as the forms of things knowable are said to be in him who knows them' (*Summa*, FP, Q15, A1). Furthermore, the nature or form of the thing is also the idea of the thing in the mind of God. 'It is necessary to suppose ideas in the divine mind.' '[T]here must exist in the divine mind a form to the likeness of which the world was made. And in this the notion of an idea consists' (ibid.).

In the science of Aquinas, the forms or natures or essences of things constitute a set of necessary connections among observable events. For this tradition, deriving from Aristotle, the covering law pattern of explanation of the new science is not enough. Rather, as others have also pointed out, it is the underlying nature or essence which both explains the observable events and accounts for why the *de facto* regularity of sense is a *law*, that is, why it is a regularity that holds of *necessity*.[4]

Consider once again our (somewhat simplified) example of Corsicus and solubility. Let C, that is, Corsicus, be some object or, rather, substance, and suppose that Corsicus is such that whenever it is put in water it dissolves. Let A be this pattern of sensible events. Then C is A. The middle term that explains C

being *A* is the capacity (power, nature, essence, form, soul) of *solubility*. Let *B* be this capacity. Then the explanatory syllogism is:

Whatever is soluble is such that whenever it is put in water it dissolves
(*B* is *A*)

(S')   Corsicus is soluble (*C* is *B*)

∴ Corsicus is such that whenever it is put in water it dissolves (*C* is *A*).

Solubility is an active power, like the nutritive power, the activity of which accounts for the pattern of behaviour described in the conclusion. As the nutritive power is one of a set of powers, including the appetitive and the ratiocinative, which make up the soul or form or nature of human beings, so also solubility will be one of the set of powers that constitute the set that makes up the full nature or essence of Corsicus (for example, the nature of Corsicus may be sugar). Thus, the second premise of (S') is necessary because the nature or essence is inseparable from the substance Corsicus. Provided that the major premise of (S') is also necessary, then the conclusion will be necessary, and the syllogism will establish a necessary connection among the events in the pattern mentioned in the conclusion. The necessity of the law derives from the form or essence; thus, explanation of a pattern of behaviour is not obtained by fitting that pattern into a more complex pattern, as in the new science, but by *redescribing* the object whose behaviour is being explained.

Now, as we have noted and as Molière made clear, for the defenders of the new science, the major premise of (S') is indeed necessary, but only trivially so, since it is true by nominal definition. For that reason, for the critics of Aristotle (S') is not more explanatory than

Whatever is a bachelor is an unmarried male

(S'')   Callias is a bachelor

∴ Callias is an unmarried male.

(S'') doesn't explain since the minor premise merely *restates* what the conclusion says. As for the major premise in (S''), that is, in fact, redundant; for the rule of language

'Bachelor' is short for 'unmarried male'

which licenses its assertion as a necessary truth equally licenses the rewriting of the minor premise as the conclusion and therefore also licenses the inference of the conclusion from the minor premise alone.

For the older tradition, however, as we saw in Study One, the major premise of (S') is both necessary and substantive. That means that it cannot be true simply *ex vi terminorum*, nor that the disposition can be defined as Molière suggests. Rather, the disposition, taken to be the form or nature or essence of the thing Corsicus, must be a primitive and unanalysable power. In order to recognize the pattern that whenever Corsicus is in water he dissolves, we must have a rational intuition of this form. For Aquinas, of course, as for Aristotle, this intuition is the product of an act of abstraction. For the Platonistic variant on this tradition, in contrast, its defenders held that there were indeed such forms yielding necessary connections among observable events, but one's knowledge of these forms is argued to be innate.

The behaviour of a particular object is understood in terms of its nature or form. The forms or natures or essences themselves have an internal logical or ontological structure. This structure is the object of study of syllogistic science. The logical structure is given in terms of genus-species relation: it is the same as the forms of certain other things, and differs from other forms. The similarities are given in terms of the genus of which it is a part; the differences are given in terms of the specific difference that serves to distinguish it from other species within the genus. Thus, to understand a form or nature we must grasp this logical or ontological structure. We need, that is, to grasp the definition of the species – its *real* definition – in terms of its genus and its specific difference. This real definition can then be expressed in a syllogism: the species is the subject of the conclusion, the genus is the predicate of the conclusion, while the difference is the middle term. These logical relations are, of course, eternal and necessary – the definition reveals the *ontological structure* of the form in the mind of God; and so, as we noted in detail in Study One, the search for understanding is at once the search for real definitions and the search for middle terms of demonstrative syllogisms – indeed, these two amount to the same, or, at least, to merely different phases in the one activity of reason. Syllogism is thus a way of coming to know, while at the same time it is the logical, or ontological, structure of things. Syllogism is not just logic, but also onto-logic and epistemo-logic.

It was Galileo who shifted the interest of science from natures or forms to general regularities. His concern was for projectile motion. For Aquinas, 'the movement of heavy bodies towards the center' (*Summa*, FS, Q54, A2, RObj 2) is their natural motion. Knowing this is a matter of intuiting the nature of objects that gravitate. The motion of projectiles is contrary to this natural motion. This motion is therefore violent, and must be explained in terms of external forces and, in particular, in terms of the entity that set the projectile in motion. Aquinas therefore says 'the necessity whereby an arrow is moved so as to fly towards a certain point is an impression from the archer, and not from the arrow' and tells us that 'the violent necessity in the movement of the arrow shows the action of the archer' (FP, Q103, A1, RObj 3). The end of its movement derives not from its own nature but from the nature of the external object that is causing the violent motion: 'an arrow tends to a determinate end through being moved by the archer who directs his action to the end' (FS, Q1, A2). The object of Aquinas's science, so far as it concerns projectile motion, is a matter of locating more precisely the external force that accounts for this violent motion. As is well known, this in fact proved a difficult task. For, once the projectile leaves the hand of the archer, or whoever or whatever sets it in motion, there seems to be no external force capable of maintaining that violent motion. And if there is no such force, then it is as if the obstacle to natural motion was removed: the violent motion should cease and the projectile should immediately return to a natural motion towards its natural end. But the projectile does not cease its apparently violent motion. Where, then, is the external force that maintains that violent motion? We need not go into attempts to solve this problem. Suffice it to say that it is a problem only so long as we maintain both that there are forms or natures of things in terms of which we must explain the motions of objects, and that the forms of heavy objects determine that their natural motion is straight downwards towards the centre of the earth, or, what was for many the same, the centre of the universe. What Galileo did was reject the idea that the main task of science was to search for these forms that determine natural motion and violent motion, and in particular that the main task for the physicist with regard to projectiles was to discover the external force accounting for the violent motion. As Stillman Drake has argued, Galileo gives up this old task of science (*scientia*) and replaces it with a new task.[5]

In Galileo's dialogue on *Two New Sciences*,[6] Salviati represents Galileo. Sagredo provides wise and intelligent commentary. It is the latter who introduces 'the question agitated among philosophers as to the possible cause of acceleration of the natural motion of heavy bodies.' He then proposes an answer in terms of an external force impressed upon the object by the substance that initiated the non-natural or violent motion that is contrary to its natural motion as a heavy (or gravitating) object straight downwards (*Two New Sciences*, 158). The traditional philosopher, Simplicio, immediately raises some objections to this account in terms of active forces. As Stillman Drake points out in his note to this text (158n), this reasoning of Sagredo is precisely that which Galileo himself had used in his earlier thinking concerning free fall and projectile motion. But Galileo's spokesman, Salviati, instead of replying to the points raised by Simplicio, now suggests a very different move. He suggests that this old problem simply be abandoned, as a search after fantasies, and that they search instead for a simple description of the patterns in which these objects move. Recall the argument:

The present does not seem to me to be an opportune time to enter into the investigation of the cause of the acceleration of natural motion, concerning which various philosophers have produced various opinions ... Such fantasies ... would have to be examined and resolved, with little gain. For the present, it suffices ... to investigate and demonstrate some attributes of a motion so accelerated (whatever be the cause of its acceleration) that the momenta of its speed go increasing, after its departure from the rest, in that simple ratio with which the continuation of time increases, which is that same as to say that in equal times, equal additions of speed are made. And if it shall be found that the events that then shall have been demonstrated are verified in the motion of naturally falling and accelerated bodies, we may deem that the definition assumed includes that motion of heavy things, and that it is true that their acceleration goes increasing as the time and the duration of the motion increases. (158–9)

What we here see is Galileo changing the question that the physicist was to ask: instead of forces grounded in the metaphysical natures of things, he proposed instead to discover the *exceptionless patterns* or *regularities* of motion. Giving up the quest for natures and natural forces, Galileo was remarkably successful in discovering regularities in the motions of things. Not only did he discover the precise form of the motion of objects in, as one says anachronistically, 'free fall,' but he also discovered the regularities that describe the motions of all projectiles. The latter, as he discovered, always moved along curves having the form of parabolas.

In this context, mathematics did not attempt to describe a deeper underlying metaphysical essence as the syllogistic described the deeper ontological structure of the essences of things; rather, mathematics was for Galileo merely a tool to describe the observable motions of things, a tool to record the regularities in a convenient language.

As we argued in Study One, following Stillman Drake in his *Galileo: Pioneer Scientist*,[7] Galileo was able to make his great breakthrough with regard to the motions of objects precisely because he gave up the search after metaphysical forces of the sort that Aquinas thought were there. The very notion of what science was up to had changed.[8] But, it is equally clear, not the notion of cause. To insist that science search after regularities *rather than* causes, is not yet to replace the notion of cause as it had come down in the philosophical tradition with the notion of cause as regularity. It is not yet to replace the Aristotelian notion of cause that we find in Aquinas by the notion of cause as regularity that found its first clear statement in Hume. In order for this to happen, the whole notion of Aristotelian causes as active powers, forms, natures, and essences had to be subjected to criticism and eliminated. Only upon that elimination could the discovery of causes in the sense of regularities be proposed as the goal of science.

Now, as Popkin has argued,[9] the early modern period was characterized by a widespread development of sceptical arguments and positions. These derived from Montaigne, of course, and found their greatest development perhaps in Bayle. There were many motives for this development, not least of which was the attempt to argue for tolerance on the basis of the fact that no one really knew anything for certain. Many of those who were active in pursuing the new science of Galileo were moved by just these motives. But at the same time, these same arguments were an attack on the Aristotelian forms or natures or essences that were the central explanatory entities of the older science that was defended by the scholastics such as Aquinas.

The older tradition deriving from Aristotle and Aquinas continued to be defended as an account of the structure of the world, and therefore as an account of what, ideally, causal science should aim to discover. In particular, the Platonistic variant on the older tradition continued to be defended by Descartes.

Descartes argues that things or substances have essences, and that these essences are forms in the mind of God. Further, these essences are also present in our own minds. To this extent – to this very great extent – the Cartesian ontology and epistemology is of a piece with that of Aristotle and Aquinas.

Descartes takes perception to fall under the category of thought.[10] Thoughts in general and perceptions in particular are modifications of the mind; they are acts of the mind. But thoughts and perceptions are also about, or 'intend,' certain objects. They do this by virtue of *ideas* which inform them. An idea, then, is the form of an act of thought or perception; as Descartes puts it in his replies to the Third Set of Objections, 'by an idea I mean whatever is the form of a given perception.'[11] But an idea in this sense is not itself a content of thought that is apprehended by the mental act of which that idea is the form; we are *aware of* the idea but that idea is not the *object of perception*. Ideas, in other words, are not representative entities that intervene, as it were, between the act of the mind which is the perceiving and the object perceived. For, as we know, Descartes distinguishes two senses of the terms 'idea.' In one sense it is a modification of the mind, and as such is something of which the mind is immediately aware. However, an act of thought or perception is also intrinsically connected to the object or body which it intends, by virtue of which connection the mind is aware of that object or body. Descartes draws the relevant distinction between two senses of the term 'idea' in the 'Preface' to the *Meditations*: 'in this word 'idea' there is here an equivocation. First it can be taken materially, as an operation of my intellect ... or it can be taken objectively for the body which is represented by this operation.'[12]

Descartes refers to actual being as 'formal' being or reality (*esse formale*): the formal being of a thing is constituted by its properties or modifications. This he distinguishes from the objective being of a thing, that 'mode of being, by which a thing exists objectively or is represented by a concept of it in the understanding.'[13] This is 'the way in which its [the intellect's] objects are normally there.'[14] When a thing has objective being in the mind, that thing is not a mode or property of the mind, that is, it is not 'formally' in the mind; otherwise the mind would actually *be* the thing. None the less, though the thing that is objectively in the mind lacks, as such, formal reality, it is still not nothing, and is to be contrasted to ideas which are merely chimerical. In short, *the objective existence of things in the mind is constituted by the presence in the mind of the essence of the thing.* In other words, to make the point once more, *Descartes's metaphysics is in its basic structure essentially the same as that of Aristotle.* Where they differ is in the relatively minor point that for Aristotle forms come to be in the mind through a process of abstraction, whereas for Descartes those forms are innately present in the mind. Descartes, in other words, is defending the Platonistic variant of Aristotle's metaphysics.

This variant of the traditional ontology deriving from Aristotle was also defended in England at the time of Galileo by the Cambridge Platonists, such as Ralph Cudworth and Henry More. Thus, Cudworth, in his *True Intellectual System* (1678),[15] tells us of the 'essences of things, that have no generation nor corruption,' and are the 'intelligible natures, species, and ideas, which are the standing and immutable objects of science.' These are ideas or forms in the mind of God: 'If therefore there be eternal intelligibles or ideas, and eternal truths, and necessary existence do belong to them; then must there be an eternal mind necessarily existing, since these truths and intelligible essences of things cannot possibly be anywhere but in a mind' (*True Intellectual System*, 2:736). Cudworth argues that we need to appeal to active powers or souls to explain the motions of things. Thus, 'in the bodies of animals, the true and proper cause of motion, or the determination thereof at least, is not the matter itself organized, but the soul either as cogitative, or plastickly self-active, vitally united thereunto, and naturally ruling over it' (2:669). Indeed, this is true of all matter: 'since it appears plainly, that matter or body cannot move itself, either the motion of all bodies must have no manner of cause; or else must there of necessity be some other substance besides body, such as is self-active and hylarchical, or hath a natural power of ruling over matter' (ibid.). Atheists deny that there is anything beyond matter and what is knowable by sense. There is much 'mundane regularity,' but the atheist can ascribe no cause to it; for him it is all inexplicable: 'they either make the mere absence and want of a cause to be a cause, fortune and chance being nothing else but the absence or want of an intending cause; or else do they make their own ignorance of cause, and *they know not how*, to be a cause ...' (ibid.). '[N]o Atheists [are] able to assign a true cause of motion, the knowledge of whereof plainly leadeth to a God' (ibid.).

It was this system that came increasingly to be subjected to sceptical attacks. Thus, to give one example, there was Joseph Glanvill, who, it can be argued, in his *Vanity of Dogmatizing*[16] and *Scepsis scientifica*,[17] presented a viewpoint on knowledge that was in certain respects a precursor of that of Hume.[18] Glanvill is, interestingly, prepared to accept something like the traditional framework in order to account for our knowledge of mathematics and theology. In his *Vanity of Dogmatizing* he tells us that mathematical truths are

superstructured on principles that cannot fail us, except our faculties do constantly abuse us. Our religious foundations are fastened at the pillars of the intellectual world, and the grand Articles of our Belief as demonstrable as Geometry. Nor will ever either the sub-tile attempts of the resolved Atheist; or the passionate Hurricanoes of the phrentick Enthusiast, any more be able to prevail against the reason our Faith is built on, than the blusting windes to blow out the Sun. And for Mathematical Sciences, he that doubts their certainty hath need of a dose of Hellebore. (*Vanity*, 209)

Glanvill's is therefore no empiricist metaphysics or theory of knowledge. None the less, he argues that all our knowledge, save possibly mathematical and theological knowledge, derives from our senses: 'The knowledge we have comes from our Senses, and the Dogmatist can go no higher for the original of his certainty' (218). Because our senses deceive us at times, because our imaginations mislead us, and because of other natural infirmities of humankind, all our knowledge is fallible. The 'Peripatetick Philosophy,' in particular, is subjected to criticism. We are simply unable to know 'the hidden things of Nature,' contrary to what the Aristotelians claim; we lack the capacity to 'see the first springs and wheels that set the rest agoing' (*Scepsis scientifica*, I:15). In particular, 'the causality it self is insensible' (*Vanity*, 190); we do not, for example, observe the causality between fire and heat. Since, on the one hand, necessary causal connections are not perceived, while, on the other hand, all our knowledge derives from our senses, it follows that 'we cannot conclude anything to be the cause of another; but from its continual accompanying it' (189–90). But in that case, there is a gap between evidence and judgment; we lack certainty: as Glanvill puts it, 'to argue from a concomitancy to a causality, is not infallibly conclusive' (190). As for the possibility of demonstrating causal propositions, Glanvill argues that we know no premises sufficient for any such demonstration; the evidence that he offers – it is hardly conclusive – is the fact that no one has produced demonstrations that are in fact capable of commanding general assent. 'Our demonstrations are levyed upon Principles of our own, not universal Nature' (193). Glanvill concludes with the argument that for our human problems a probable, rather than certain, knowledge of science suffices, and that a moderate scepticism is more appropriate than any dogmatism. If we give up dogmatizing we will, in fact, be able to improve our knowledge of nature, and therefore in turn our own well-being, as well as adhering to the one true religion (224–50).

There are features here that point forward to Locke and to Hume. But at the same time, the scepticism is limited by Glanvill's equally clear commitment to at least certain aspects of the traditional view. Interestingly enough, somewhat similar arguments can be found in the Cartesians. These philosophers agreed that there are substances and that these substances have essences. But they argued, contrary to the traditional view found in Aquinas, that there was but one essence needed to explain material bodies. This essence of material things was extension. This essence is also an idea in the mind of God, and is what our own mind grasps when it attempts to understand things. Insofar as this position argues that in order to understand things we must grasp their essences and not merely the observable patterns of behaviour, they agree with the older tradition of Aristotle and Aquinas against the position of Galileo. There is another disagreement with the tradition that has to be noted, however. The essences for the Cartesians are *not* powers; they are not active forces. Rather, there is no active power or force in created things. The only power is God. It is His activity alone that provides necessary connections among events and things. The Cartesians agree, then, with critics of the traditional view such as Glanvill that we are not presented with the active powers of things. Indeed, not only are we not presented with them, but such powers do not exist: to repeat, God alone truly exercises power.

This Cartesian doctrine holds that what seem to be causal patterns among observable events are not really causal, but only the *occasions* for God to exercise His power. This occasionalism, this denial that ordinary things have any causal power, feeds into the sceptical case against real connections, and Hume was to exploit this opening towards the regularity view of causation, as we shall see. None the less, it is important to recognize the difference between Hume and the Cartesian occasionalists. The latter, unlike the former, still hold that there are necessary connections effected by the exercise of real power: it is just that this power is exercised by God alone. Thus, as Malebranche puts it,

God created the world ... and He moves all things, and thus produces all the effects that we see happening, because He also wills certain laws according to which motion is communicated upon the collision of bodies; and because these laws are efficacious, they act, whereas bodies cannot act. There are therefore no forces, powers, or true causes in the material, sensible world; and it is not necessary to admit the existence of forms, faculties, and real qualities for producing effects that bodies do not produce and for sharing with God the force and power essential to Him.[19]

In fact, 'not only are bodies incapable of being the true causes of whatever exists: the most noble minds are in a similar state of impotence. They can know nothing unless God enlightens them. They can sense nothing unless God modifies them. They are incapable of willing anything unless God moves them toward good in general ...' (*Search after Truth*, 449). Moreover, 'it appears to me quite certain that the will of minds is incapable of moving the smallest body in the world; for it is clear that there is no necessary connection between our will to move our arms, for example, and the movement of our arms' (ibid.). Malebranche goes on to indicate, however, that '[i]t is true that they are moved when we will it, and that thus we are the natural cause of the movement of our arms' (ibid.). However, these natural causes 'connecting' mind and body are like the natural causes between different states of bodies, as when one body collides with another, or between different states of mind, as when we proceed methodically to come to know certain things, or when we strive for the best. They are all alike in being *merely* natural, and '*natural* causes are not true causes; they are only *occasional* causes that act only through the force and efficacy of the will of God ...' (ibid.).

This last makes the point clearly: while for ordinary purposes regularity and cause tend to merge in the doctrine of occasionalism, none the less, mere regularity does not exhaust the notion of cause as it does for Hume. For the Cartesians, unlike Hume, there is more to the notion of cause than regularity. Indeed, for the Cartesians, that more is the notion of necessary connection and power, just as it was for Aquinas and Aristotle. It is just that for the former, unlike the latter thinkers, all power lies in God and none in ordinary things. But for both the Cartesians and the Aristotelians, there is more to causation than regularity; there is also the notion of power and of necessary connection. If we must resort to regularity as our criterion of causation, then what we are here resorting to is clearly *second best*.

Arguments of the sort that Glanvill offered did find their place among the arguments defending the new science of the Royal Society. Indeed, we can find them located within the context of stronger arguments directed against any form of appeal to abstract forms, anything beyond mere observable regularity. In this respect, the sceptical arguments used to defend the new science were directed not only at the Aristotelianism of Aquinas and the Cambridge Platonists but also at the occasionalism of the Cartesians, insofar as the latter insisted that explanation had in the end to refer to essences which transcend the world of ordinary things given in sense experience. These arguments were sufficiently widespread and sufficiently accepted to find their place in Thomas Sprat's semi-official *History of the Royal Society* (1667).[20] Indeed, this was not just a history but an apology and defence.[21] As Sprat himself says in the 'Advertisement to the Reader,' the 'Objections and Cavils against it [the Royal Society], did make it necessary for me to write of it, not altogether in the way of a plain History, but sometimes of an Apology.' The apology proceeded by means of sceptical attacks on the tradition of forms or necessary connections and by means of a defence of a Baconian method for the search after truth. According to Sprat, the experimenter of the Royal Society 'labours the plain and undigested Objects of his Senses, without considering them as they are joyn'd into common Notions' (*History*, 334), that is, forms or natures of things. The scholastic philosophy made no progress in the knowledge of the world 'Because it rely'd on general Terms, which had not much foundation in Nature; and also because they took no other course, but that of disputing' (16–17). This argument against essences or forms applies equally to the Cambridge Platonists and the Cartesians as it does to the scholastics such as Aquinas. The members of the Royal Society adopt a modest scepticism appropriate to their aim of discovering matter-of-fact truth: 'They have been cautious, to shun the overweening dogmatizing on causes on the one hand: and not to fall into a speculative scepticism on the other ...' (101). Human judgment is fallible: 'the thing it self is of that nature; that it is impossible to place the minds of men beyond all condition of erring about it' (101). None the less, within these limits, progress can be made. The task is to examine the facts until they make clear themselves which pattern it is that they genuinely exemplify. The 'critical, and reiterated scrutiny of those things, which are the plain objects of their eyes; must needs put out of all reasonable dispute, the reality of those operations, which the Society shall positively determine to have succeeded' (99). The idea here is, as it was for Bacon, to refute in order better to arrive at the truth: 'To the Royal Society it will be at any time almost as acceptable, to be confuted, as to discover: seeing, by this means, they will accomplish their main Design ...' (100). Truth will emerge from the discovery of error: 'the tracing of a false Cause, doth very often so much conduce; that, in the progress, the right has been discover'd by it.' 'It is not to be question'd, but many inventions of great moment, have been brought forth by Authors, who began upon suppositions, which afterwards they found to be untrue' (108). It was Bacon who led the revival of knowledge; in his 'Books there are every where scattered the best arguments, that can be produc'd for the defence of Experimental Philosophy; and the best directions, that are needful to promote it' (35). He is, however, not perfect: 'His Rules were admirable: yet his History not so faithful, as might have been wish'd in many places, he seems rather to take all that comes, then to choose; and to heap, rather, then to register' (36).

The scepticism about forms or notions or natures as providing explanatory necessary connections was pursued vigorously by defenders of the new science. They had clear objects to attack. Thus, Bartholomäus Keckermann's *Systema physicum* went through a number of editions early in the seventeenth century, and was used in particular as a text at Cambridge.[22] Keckermann tells his students that '[n]ature is the principle or cause of motion or quiet in a natural body.'[23] He is here following Aristotle, who argued with respect to things which exist 'by nature' that 'each of them has *within itself* a principle of motion and of stationariness (in respect of place, or of growth or decrease, or by alteration),'[24] later adding that 'nature is a source or cause of being moved and of being at rest in that which it belongs primarily ...'[25] As Ross points out, Aristotle 'habitually identifies nature as power of movement with nature as form.'[26] This makes clear the teleology of explanations in terms of natures: 'The form ...' Aristotle says, 'is "nature" rather than matter; for a thing is more properly said to be what it is when it has attained to fulfillment than when it exists potentially.'[27] It is in Aristotle that the later commentators and elaborators could find appeals to such 'explanatory principles' as 'nature does nothing in vain.'[28] It is doctrines of this sort, whether those of Aristotle or of Keckermann, that are the object of sceptical attack in, to take one important example, Robert Boyle's *A Free Enquiry into the Vulgarly Receiv'd Notion of Nature* (1686).[29] In this polemic, Boyle attacked specifically the appeal to natures or forms as something that at once was no more than an appeal to ignorance – the sceptical arguments made acceptance of any such metaphysical entities unreasonable – and at the same time something that interfered with the progress of experimental science. The 'Nature ... is so dark and odd a thing, that 'tis hard to know what to make of it, it being scarce, if at all, intelligibly propos'd, by them that lay the most weight upon it' (*Free Enquiry*, 129). At the same time, Boyle 'observe[s] divers Phaenomena, which do not agree with the Notion or Representation of Nature' (136). Boyle cites the phenomenon of a vacuum as contrary to many of the things that people have said about nature as a causal and explanatory force. That is, when one appeals to such an entity, one not only settles for explanations which are bad because they are obscure but settles for explanations which are in fact wrong, misdescribing the phenomena in question. He also cites (145ff) the alleged explanations of motions in terms of the qualities gravity and levity.

Bodies of the former sort, that is, with that sort of 'Innate Appetite' (147), move in straight lines towards the centre of the Earth, while bodies of the other sort move in straight lines away from the centre and towards the heavens. Boyle points out how this doctrine makes very little sense with regard to the motion of a pendulum (148–9). Boyle also argues in detail (sect. VIII, 347ff) that various propositions which are supposed to be established in regard to the notion of 'nature,' as in 'nature does nothing in vain,' explain either too much or too little, and, in fact, in general can be understood, when taken as scientific and referring to patterns of behaviour of objects in the world of sense experience, as making assertions compatible with the mechanical philosophy.[30]

The various arguments came to find their place in a metaphysics and epistemology that ensured that the notion of cause as regularity that was central to the new science of Galileo was the only notion of cause that could reasonably be defended. This was the metaphysics and epistemology as systematically developed by Locke.[31] We do know that a central motive in Locke was the development of a metaphysics and epistemology that could be used to defend political tolerance in areas of politics and religion. To do this he used sceptical arguments against the older tradition, and used them far more systematically than did Glanvill and Sprat. But at the same time, he also developed a positive alternative to the older metaphysics and epistemology: an empiricist metaphysics and epistemology as opposed to the rationalism of Aristotle and Aquinas – and, for that matter, Descartes.

For such a one as Glanvill, the forms that provided necessary connections among events were there, in the mind of God at least, and no doubt also in things. It was just that we have difficulty in coming to know them. For Locke it is otherwise: for most qualities of things, there simply are no forms that could provide a connection any stronger than matter-of-fact regularity: the necessary connections are not given to us in ordinary experience; we therefore have no idea of such connections; and in the absence of such an idea, such necessary connections are simply inconceivable.

The reason why the one ['primary qualities'] are ordinarily taken for real qualities, and the other only for bare powers, seems to be, because the ideas we have of distinct colours, sounds, &c. containing nothing at all in them of bulk, figure, or motion, we are not apt to think them the effects of these primary qualities, which appear not, to our senses, to operate in their production; and with which they have not any apparent congruity, or conceivable connexion. (*Essay*, II.viii.25)

Or, as he puts it elsewhere, the complex idea we have of any substance 'cannot be the real essence ... for then the properties we discover in that body would

depend on that complex idea, and be deducible from it, and their necessary connexion with it be known' (II.xxxi.6). There are no perceivable necessary connections among the properties that things are presented as having in the world
of ordinary experience. Words that philosophers use that purport to refer to real
essences are in fact meaningless, sounds without referents. '[W]hen I am told
that some thing besides the figure, size, and posture of the solid parts of that
body, is its essence, some thing called substantial form, of that I confess, I have
no idea at all, but only of the sound form, which is far enough from an idea of
its real essence, or constitution' (ibid.). In the absence of real essences, Aristotelian forms, or necessary connections, we are reduced to regularity. But this is,
after all, all that we need for our practical purposes in our ordinary life in our
ordinary world: we seem able to get on in life just fine with simple regularity.

Tell a country gentlewoman that the wind is south-west, and the weather lowering, and
like to rain, and she will easily understand it is not safe for her to go abroad thin clad, in
such a day, after a fever: She clearly sees the probable connexion of all these, viz.
south-west wind, and clouds, rain, wetting, taking cold, relapse, and danger of death,
without tying them together in those artificial and cumbersome fetters of several syllogisms, that clog and hinder the mind, which proceeds from one part to another quicker
and clearer without them; and the probability which she easily perceives in things thus in
their native state would be quite lost, if this argument were managed learnedly, and proposed in mode and figure. (IV.xvii.4)

To aspire to more than mere regularity where we know that the world is such
that it cannot be obtained is something that is simply unreasonable, since it is
doomed to failure.

When we know our own strength, we shall the better know what to undertake with hopes
of success: And when we have well surveyed the powers of our own minds, and made
some estimate what we may expect from them, we shall not be inclined either to sit still,
and not set our thoughts on work at all, in despair of knowing any thing; nor on the other
side, question every thing, and disclaim all knowledge, because some things are not to be
understood. (I.i.6)

None the less, Locke still argues that, although there are many aspects of the
world of which we are aware in ordinary experience in which there are no necessary connections, we have the *idea* of a necessary connection, or, as he says,
of a real essence. To be sure, we do not know these real essences – 'The ... ignorance as I have of the real essence of this particular substance [gold], I have also
of the real essence of all other natural ones; of which essences, I confess, I have

no distinct ideas at all; and I am apt to suppose others, when they examine their own knowledge, will find in themselves, in this one point, the same sort of ignorance' (*Essay*, II.xxxi.6). In this respect, he is more of a sceptic about essences than are the Cartesians. For, both Locke and the Cartesians deny that events in the world of ordinary experience are linked by necessary connections and powers. Yet the Cartesians hold that we do know the real essences of these things, that is, of material objects. This Locke denies.

However, Locke does not deny that we can form the idea – an abstract idea – of that *kind* of entity. We can have no idea of a specific real essence, save in one specific case, but from this one specific case we can abstract the genus, and thereby obtain the generic idea. This enables us to *think* about other instances, even if those other instances are never presented to us.

We can form the abstract idea of a real essence and thereby think about real essences because we are presented with one instance of such a connection in our experience. This is the connection between a volition and its immediate upshot. '[W]hatever change is observed, the mind must collect a power some-where able to make that change, as well as a possibility in the thing itself to receive it. But yet, if we will consider it attentively, bodies, by our senses, do not afford us so clear and distinct an idea of active power, as we have from reflection on the operations of our minds' (*Essay*, II.xxi.4). In this alone among all possible cases we are presented with an instance of a real connection, a real exercise of power, a real essential connection. So from this alone we can form the abstract idea of a real necessary connection and active power. We therefore have the idea of a real essence, though we are, for by far the greatest part, not presented with such entities in ordinary experience. With this abstract idea to hand, we can think about such entities, and can affirm that every thing has a real essence, even though it is never presented to us in experience. This means that for Locke there is always a standard of knowledge and of causation that goes beyond mere regularity, condemning those cases in which knowledge is indeed of mere regularity to be as it were second best, and not *genuinely causal*.

It was Hume who broke this standard. He argued that, as everyone agreed, we have no impression of an objective necessary connection: nothing like that is given in ordinary experience. We do not even obtain it from the experience of the workings of our own mind, he argues against Locke, agreeing with Male-branche.

Some have asserted that we feel an energy or power in our own mind; and that, having in this manner acquired the idea of power, we transfer that quality to matter, where we are not able immediately to discover it. The motions of our body, and the thoughts and sentiments of our mind (say they) obey the will; nor do we seek any further to acquire a just notion of force or power. But to convince us how fallacious this reasoning is, we need only consider, that the will being here considered as a cause has no more a discoverable connexion with its effects than any material cause has with its proper effect. So far from perceiving the connexion betwixt an act of volition and a motion of the body, it is allowed that no effect is more inexplicable from the powers and essence of thought and matter. Nor is the empire of the will over our mind more intelligible. The effect is there distinguishable and separable from the cause, and could be foreseen without the experience of their constant conjunction. We have command over our mind to a certain degree, but beyond that lose all empire over it: and it is evidently impossible to fix any precise bounds to our authority, where we consult not experience. In short, the actions of the mind are, in this respect, the same with those of matter. We perceive only their constant conjunction; nor can we ever reason beyond it. No internal impression has an apparent energy, more than external objects have. Since, therefore, matter is confessed by philosophers to operate by an unknown force, we should in vain hope to attain an idea of force by consulting our own minds. (T 632; appendix to I.III.xiv)

We therefore have no impression of a necessary connection. Nor, accepting Locke's argument against innate ideas, do we have any other idea of such a necessary connection. Since all our ideas derive their meaning by virtue of their connection to sense experience, to our impressions of sense and of inner awareness, it follows that we have no idea of an objective necessary connection.

Since these philosophers [Cartesians, such as Malebranche] ... have concluded that matter cannot be endowed with any efficacious principle, because it is impossible to discover in it such a principle, the same course of reasoning should determine them to exclude it from the Supreme Being. Or, if they esteem that opinion absurd and impious, as it really is, I shall tell them how they may avoid it; and that is, by concluding from the very first, that they have no adequate idea of power or efficacy in any object; since neither in body nor spirit, neither in superior nor inferior natures are they able to discover one single instance of it. (T 160)

Insofar as there is any necessity involved in causation, it lies on the side of the mind: we judge those regularities to be causal where we are inclined to infer the effect from the cause or conversely. 'The necessary connexion betwixt causes and effects is the foundation of our inference from one to the other. The foundation of our inference is the transition arising from the accustomed union. These are, therefore, the same' (T 165).

It follows that the notion of cause as an objective necessary connection, as something more than 'mere' regularity, is meaningless – a word without any legitimate idea.

Thus, upon the whole, we may infer, that when we talk of any being, whether of a superior or inferior nature, as endowed with a power or force, proportioned to any effect; when we speak of a necessary connexion betwixt objects, and suppose that this connexion depends upon an efficacy or energy, with which any of these objects are endowed; in all these expressions, so applied, we have really no distinct meaning, and make use only of common words, without any clear and determinate ideas. (T 162)

Or, rather, it is more likely that the word is more misapplied than meaningless.

I can only reply to all these arguments, that the case is here much the same, as if a blind man should pretend to find a great many absurdities in the supposition, that the colour of scarlet is not the same with the sound of a trumpet, nor light the same with solidity. If we have really no idea of a power or efficacy in any object, or of any real connexion betwixt causes and effects, it will be to little purpose to prove that an efficacy is necessary in all operations. We do not understand our own meaning in talking so, but ignorantly confound ideas which are entirely distinct from each other. I am, indeed, ready to allow, that there may be several qualities, both in material and immaterial objects, with which we are utterly unacquainted; and if we please to call these power or efficacy, it will be of little consequence to the world. But when, instead of meaning these unknown qualities, we make the terms of power and efficacy signify something, of which we have a clear idea, and which is incompatible with those objects to which we apply it, obscurity and error begin then to take place, and we are led astray by a false philosophy. This is the case when we transfer the determination of the thought to external objects, and suppose any real intelligible connexion betwixt them; that being a quality which can only belong to the mind that considers them. (T 162)

The idea of cause as an objective necessary connection is thus either meaningless – no idea attaches to it – or contradictory – ascribing to the objective world a feature that is purely subjective, the mental habit or necessity of inference.

It follows that there is nothing more to causation than regularity. Hume makes the Lockean point that this is what we should rest satisfied with.

For nothing is more certain, than that despair has almost the same effect upon us with enjoyment, and that we are no sooner acquainted with the impossibility of satisfying any desire, than the desire itself vanishes. When we see, that we have arrived at the utmost extent of human reason, we sit down contented; though we be perfectly satisfied in the main of our ignorance, and perceive that we can give no reason for our most general and most refined principles, beside our experience of their reality; which is the reason of the mere vulgar, and what it required no study at first to have discovered for the most particular and most extraordinary phenomenon. (T xvii)

We should not aspire to the standards of necessary connection that are there in the notion of causation that we found in Aquinas. The new science, when it focuses on matter-of-fact regularity, is therefore on philosophically sound ground. Indeed, the alternative of Aquinas is either meaningless or contradictory. It follows that the sorts of rules that Aquinas proposed have nothing to do with causation correctly understood. What we need instead is a set of rules that provide standards for judging of causes in the sense of *regularity*.

The old logic of science was syllogistic. This was the onto-logic and epistemo-logic that was supposed to give us understanding of the natures or forms or essences of things. Given that there are no such entities, it follows that the old logic of syllogistic is no longer an onto-logic and an epistemo-logic. No longer does the form of the syllogism equal the ontological structure of being, and no longer does the movement of thought reflect this ontological order in its own epistemological order. Logic becomes simply the logic of consistency. '[S]yllogism serves our reason ... to show the connexion of the proofs in any one instance, and no more' (*Essay*, IV.xvii.4). And the truth which it keeps consistent is, in the case of our knowledge of things, simple matter-of-fact truth, truth which, because there is no necessary connection, is contingent and probable.

As demonstration is the showing the agreement or disagreement of two ideas, by the intervention of one or more proofs, which have a constant, immutable, and visible connexion one with another; so probability is nothing but the appearance of such an agreement or disagreement, by the intervention of proofs, whose connexion is not constant and immutable, or at least is not perceived to be so, but is, or appears for the most part to be so, and is enough to induce the mind to judge the proposition to be true or false, rather than the contrary. (IV.xv.1)

The evidence for such propositions consists in experience and testimony. Since neither extends as far as the propositions that we assert, that is, since what we assert that when we assert that causal propositions are assertions of regularity about a population, and since all that we ever know either by experience or by testimony is a sample, it follows that there must in all such assertions be an element of risk: none is more than probable. 'Probability then, being to supply the defect of our knowledge, and to guide us where that fails, is always conversant about propositions, whereof we have no certainty, but only some inducements to receive them for true' (IV.xv.4). The evidential grounds that justify our judgments of probability are these:

First, the conformity of any thing with our own knowledge, observation, and experience.

Secondly, the testimony of others, vouching their observation and experience. (ibid.)

Locke further qualifies this statement of relevant evidence. '[A]s the conformity of our knowledge, as the certainty of observations, as the frequency and constancy of experience, and the number and credibility of testimonies, do more or less agree or disagree with it, so is any proposition in itself more or less probable' (IV.xv.6).

If we no longer can use syllogism to discover the truth of things, then we need a new guide to truth. Syllogism is now merely the logic of consistency and we need a set of rules conformity to which will lead us to the truth, insofar as we can get hold of it. This logic of consistency must, as we saw in Study Two, be supplemented by a logic of truth, to use again the felicitous phrase of John Stuart Mill.[32] This logic of truth is a set of rules for asserting propositions about matter-of-fact regularities, which is what causation has now become and which is the subject matter of the new science. Locke, we now see, has recognized the need for this new logic of truth, this new set of rules for affirming statements of causal regularity, or, rather, of causation, since regularity *just is* what causation now amounts to. We see that Locke has recognized the basic inductive nature of this new logic of truth, as well as certain elementary features of how different sorts of empirical evidence justify different degrees of certainty.

But Locke's *Essay* was a work in metaphysics and epistemology. What was needed was a systematic exposition linking this new logic of truth with the old logic, now become a logic of consistency. This need was recognized shortly after Locke. In fact, the latter's *Essay* provided, I would suggest, just that sort of metaphysical and epistemological justification of the project of the new science, of Galileo's project, that was needed in order to make possible this very new sort of textbook on the 'art of thinking.'

Prior to Locke in England, there had been a variety of logic textbooks in use. Ramus had had considerable impact,[33] and even texts which were not strictly Ramist, such as that of Franco Burgersdijck,[34] showed that influence. Burgersdijck presented the traditional material of logic – definition, immediate inference, syllogistic, as well as a few arguments from what we now call sentential logic, such as hypothetical syllogism – but he also gave a central place to the doctrine of topics or commonplaces. The latter was intended to provide a set of locations in which one could discover premises for arguments to be used in disputations and oratory. But how were those places to be filled? For that, one needed a logic of truth, and nothing of the sort was provided. Essentially, it remained a logic of consistency. It is no wonder that Locke rejected the logic of Burgersdijck as useless for science.

There was, about the time of Bacon, a reaction in England against the topical and Ramist logics.[35] R. Sanderson's *Logicae artis compendium*[36] is typical of this reaction, against the popular logics, and back towards a purer logic. The whole apparatus of both the scholastic terminology and the doctrine of commonplaces is eradicated. The result is a clear and uncomplicated discussion of the components of what we now call 'traditional logic.' It remains, however, merely a logic of consistency. Sanderson towards the end briefly introduces the idea of method,[37] where he distinguishes invention and doctrine. Another example of the same sort of logic is John Wallis's *Institutio logicae* (1687).[38] Wallis was a leader in the new science, and no mean scientist and mathematician himself. But his logic lectures were of the age, and remained a logic of consistency. Like Sanderson, Wallis divides the subject matter of logic into concept, proposition, and syllogism. At the end of the discussion of the latter, he mentions imperfect induction (*Institutio*, 168–9), and indicates it is not conclusive; that it yields probability rather than certitude; and that it is open to refutation by negative instances. He then (169) touches very briefly on the method of investigation, or experimental philosophy, which aims at knowledge of causes from a knowledge of effects: this proceeds from particular observations and examinations to universal conclusions (cf 188–9). But this is incidental: Wallis's logic, like that of Sanderson, added little to the idea of logic as a logic of consistency.

However, we should also note that Wallis, again like Sanderson, effectively eliminates the whole doctrine of commonplaces from a central, if not dominant, position. There is, to be sure, a reference to the doctrine (*Institutio*, 183–7), but it, like the reference to the experimental method, is brief and incidental. Wallis and Sanderson thus separate logic from the methodology of the orator, and this makes possible the relocation of logic in the context of a methodology of empirical science. However, though both Sanderson and Wallis nod in the direction of a methodology of empirical science, it is at best a nod. The possibility is never actualized in either. Their remarks on invention, investigation, or the discovery of truth are in fact so brief and formulaic as to permit their being interpreted in a way that would make them incompatible with any such methodology. In fact, their logics were capable of being taken up by those – such as John Sergeant – who continued to defend the older scholastic doctrines of Aquinas, that is, the Aristotelian metaphysics of forms or essences.[39]

It was not until after Locke had developed the metaphysics and epistemology of empiricism that the framework was available in which one could locate traditional logic as a logic of consistency in the context of a logic of truth appropriate to empirical science. It was Isaac Watts who recognized the need for a new logic of truth in his textbook *Logick* (second edition, 1726).[40] Watts was, of course, a dissenting minister – he was well known as a writer of hymns, and in that capacity becomes the only logician to have a place in Johnson's *Lives of the Poets* – and teacher in a dissenting academy. He was therefore out of the mainstream of university texts, but being so situated outside that tradition enabled him to incorporate the latest up-to-date material.

Watts accepts the traditional division of the subject of *Logick* into (a) perception, conception, or apprehension, (b) judgment, and (c) argumentation or reasoning. The last is essentially syllogistic. The arguments with which the latter is concerned take judgments as their premises. What is crucial is how Watts treats judgment. His approach is in fact radically different from that of earlier writers like Wallis. For Watts, unlike Wallis, the treatment of judgment is no longer a matter merely of contraries, and contradictories, and such like. This is not simply a logic of consistency. To be sure, little of the traditional material is missed by Watts. (He even discusses the doctrine of commonplaces, though only to dismiss it as useful only to men of less than 'moderate sagacity' (*Logick*, 307).) It is what is added that is of importance.

Part 2 of the *Logick* deals with judgment. Two chapters state 'Rules for judging a right,' indicating general rules, and various more specific rules; for example, chapter V, section I, gives rules for judging of sense; section 5, rules for judging of human testimony; and section 7 'Principles and Rules of judging, concerning Things past, present and to come, by the mere use of reason.' The general rules of chapter IV embody the fallibilistic attitude of science – the attitude accepted by Locke when he advised us to reject real essences and to rest content with formulating, so far as we are at any time able, accurate nominal essences of things. Thus, Watts tells us that we should examine all our judgments afresh (*Logick*, 231), and that we should search for evidence, whether for the agreement or disagreement of ideas (240).

In the section containing the rules for judging of things past, present, and to come (*Logick*, 275ff), he states the principle of the uniformity of nature in his first rule, and in the second and fourth rules draws attention to the distinction that John Stuart Mill was later to refer to as that between unconditioned and conditioned antecedent, and indicates what this distinction implies with respect to different permissible inferences from causal propositions of the two kinds. None of this is taken very far in the *Logick*.

But what is interesting about this description by Watts, on how both reasonably and systematically to obtain premises by methods which are the methods of empirical science, is that it should occur at all. A start has been made. For the first time an empirical logic of truth is being self-consciously developed as being necessarily presupposed by any mere logic of consistency. The empirical method of science had become a living reality: an intellectual understanding of it had, for the first time, been attained. It was made possible by Locke's campaign against the old world view of essences, forms, and natures and in favour of the new science.

None the less, it is only a start. It is a long way from Watts's very simple rules to a reasonably detailed logic of truth. What, precisely, we want to know, are the rules of the *experimental* method? *What, precisely, are the rules for discovering by experiment matter-of-fact causal regularities?*

In fact, the relevant logic of truth had been laid out by Bacon.[41] This was the logic of eliminative induction.[42] Bacon had clearly recognized that what was important to the new science was matter-of-fact regularity. According to Bacon in his *New Organon*,[43] what science considers is natures, that is, properties of things, and its aim is to discover 'the Form of a Nature assign'd,' that is, the feature such that, when it is present in an object, it brings about in that object the presence of the nature in which we are interested, or, as Bacon puts it, the 'real Difference naturizing Nature' (*New Organon*, II.i). This knowledge is knowledge of causes; in Bacon's terms, 'true Knowledge is Knowledge of Causes' (II.ii). This knowledge of forms will be better than a mere knowledge of the particular thing which has the relevant form, that is, as Bacon under-stands it, the material and efficient cause of the nature being in that thing. For, 'he who understands forms, will perceive the Unity of Nature in the most dis-similar Cases' (II.iii). This knowledge must be certain in the sense that it 'will not deceive him in the result nor fail him in the trial' (II.iv), that is, it must be as certain as possible and without exceptions: the form 'accompanies it [the nature] every where' (II.iv). The knowledge must thus be of an *exceptionless general law* or *regularity*: 'the Form of any Nature, is such, that where it is, the given Nature must infallibly be ...' (II.iv).

Specifically, the ideal is knowledge of necessary and sufficient conditions: 'The Form ... is perpetually present, when that Nature is present; ascertains it universally, and accompanies it every where. Again, this Form is such, that when removed, the given Nature infallibly vanishes: Therefore the Form is perpetually wanting, when that Nature is wanting; and thus confirms its Presence, or Absence; and goes and comes with that Nature alone' (*New Organon*, II.iv).

Bacon carefully rejected the 'abstractions' of the schools, the search after metaphysical essences of things.

Nor do we judge it material to the Fortunes of Mankind, what abstract Opinions any one entertains of the Nature and Principles of Things ...

We are not solicitous of such useless Things, as depend upon Opinion; but, on the contrary, resolve to try whether we can lay any firmer Foundations of the human Power and Greatness; and enlarge the Bounds thereof. (*New Organon*, I.cxvi)

The problem is that the mind can construct many hypotheses. It is not our capacity to invent hypotheses that is important but our other capacity to curb it. '[T]he Understanding left to its self, and its own spontaneous Motion, is unequal to the Work [that of discovering 'what are the Works and Laws of Nature'], and unfit to enter upon the raising of Axioms; unless it be first regulated, strengthened, and guarded' (II.x). What is needed, Bacon saw clearly, is not a rule for inventing hypotheses to fit what data we have, but a rule for eliminating those which are incorrect. What is crucial is, as Popper put it, falsification. Bacon was able to formulate clearly what we have come to call the methods of agreement and difference, those methods that yield necessary conditions and sufficient conditions, respectively, and, when used jointly, necessary and sufficient conditions.

Knowledge of these regularities cannot be based on induction by simple enumeration, which is puerile and a 'childish Thing' (*New Organon*, I.cv). Such inferences are liable to be upset by 'contradictory Instances' (I.cv). This method is highly inefficient as a means for achieving the cognitive goals of the new science. One needs instead a method that will, as it were, take advantage of the negative instances. The correct method should rather 'by proper Rejection and Exclusion ... conclude upon Affirmatives, after the due Number of Negatives are thrown out' (I.cv). Knowledge satisfying the cognitive goals of the new science can more efficiently be arrived at by this method, what has come to be known as the method of induction by elimination. Bacon was the first to articulate this method clearly; as he said, this attempt to found knowledge upon negatives and elimination is 'a Thing never yet done, nor attempted' (I.cv).

These methods aim to discover necessary conditions, sufficient conditions, and necessary and sufficient conditions. Bacon himself was concerned only with the more limited, yet more cognitively desirable, case of necessary and sufficient conditions. He lays out the eliminative methods clearly, telling us that in induction (by elimination) there are two steps. First, 'all the known Instances agreeing in the same Nature, tho' in the most dissimilar Subjects, are to be brought together, and placed before the Understanding' (*New Organon*, II.x). Second,

those Instances are to be brought before the Understanding, which have not the Nature assigned; because the Form ... ought no less to be wanting, where the given Nature is wanting; than to be present, where that is present: but as it would be endless to pursue these Instances throughout; Negatives are to be subjoined to the Affirmatives; and the want of the given Nature, to be considered only in such Subjects as are nearest related to those wherein it resides and appears. (II.xii)

These tables provide the data for solving the problem of finding 'upon a particular and general View of all the Instances, such a Nature ... as may be continually present, or absent, and always increase and decrease, with that Nature; and ... limit the more common Nature'; this problem is solved with the affirmation of a law through the method that 'can only first proceed by Negatives, and lastly, after a perfect Exclusion, end in Affirmatives' (II.xv).

Bacon clearly recognizes the need for principles of determinism and limited variety. Thus, he states that 'in nature nothing really exists besides individual bodies, performing pure individual acts according to a fixed law' (*New Organon*, II.ii); that is, the individual or particular events in nature all occur according to 'fixed law.' Thus, we know that *there are* laws, there to be discovered. He also assumes a principle of limited variety. Again, it is clear that there can be what Bacon calls a 'perfect exhaustion' that ends in a definite affirmative only if the variety to be exhausted is indeed limited.

Bacon held the traditional positions up to ridicule. He speaks of the 'pernicious and inveterate Custom of dwelling in abstract Notions' (*New Organon*, II.iv), and of the error of 'attributing the Essences of things to forms' (II.ii). What he is here condemning has earlier been made perfectly clear:

The Understanding is, by reason of its own Nature, carried on to Abstraction; and fancies those things to be constant, which are wavering: but it is better to dissect Nature, than to abstract her ...

... [B]ut for the Aristotelian forms, they are Idols, or Figments of the Mind ...
(I.li)

This condemnation of the traditional doctrine of cause that was held by Aquinas and the other scholastics was followed, as we have seen, by the Royal Society. But like the latter, Bacon, too, did not provide a metaphysical and epistemological alternative. That was Locke's achievement, to be made more consistent by Hume.

Even so, though Bacon's defence of the new science was imperfect to the extent that he did not argue for the empiricist metaphysics and epistemology, there none the less was ready to hand in Bacon's works a set of principles that constituted the needed logic of truth that would provide the supplement necessary to the logic of consistency.

By the end of the eighteenth century, the Scottish thinkers who followed Hume and Reid had clearly incorporated Bacon's rules into their thinking about truth.[44] They explicitly adopted the Humean point that what empirical research was after was matter-of-fact regularities. Dugald Stewart put the point this way quite clearly in his *Elements of the Philosophy of the Human Mind*: Hume's view of causation is true, and it is moreover the commonly accepted view.

It seems now to be pretty generally agreed among philosophers, that there is no instance in which we are able to perceive a necessary connexion between two successive events; or to comprehend in which manner the one proceeds from the other, as its cause. From experience, indeed, we learn, that there are many events, which are constantly conjoined, so that one invariably follows from the other: but it is possible, for any thing we know to the contrary, that this connexion, though a constant one, as far as our observation has reached, may not be a necessary connexion; nay, it is possible, that there may be no necessary connexions among any of the phenomena we see: and if there are any such connexions existing, we may rest assured that we shall never be able to discover them. (*Elements*, 71)

Stewart is here stating the views of a number of Scottish thinkers, including the physicists Playfair, Robison, and Leslie, and the physician Cullen. It should be noted, however, that these philosophers did qualify Hume's views of causation just enough to allow them to infer the existence of God, but beyond that they accepted his account of causation.[45]

In was in their articles in the *Encyclopaedia Britannica* that these thinkers explicitly adopted the Baconian method of eliminative induction as the logic of truth that was needed to supplement the logic of consistency. It was expounded at length by Robison in his *Encyclopaedia Britannica* article on 'Philosophy,' and later by John Playfair, in his 'Dissertation Second,' published in the supplemental volumes to the fifth edition of the *Encyclopaedia*.[46] Playfair emphasizes, as did Bacon, that the aim of science is to find the 'form' of a given 'quality,' where 'The form of any quality in body is something convertible with that quality; that is, where it exists, the quality is present, and where the quality is present, the form must be so likewise' ('Dissertation,' 41). The method consists in listing the properties that are present when the quality of interest is present, and also the 'negative instances' where the properties are absent when the quality is present. Using these comparisons, one will by the method of comparison and exclusion discover the form of the quality (ibid.). Playfair is explicit on the fact that limiting variety creates a range of hypotheses among which experiment is to decide which is true by eliminating the false. He expresses the requirement for limited variety in this way: 'In order to inquire into the *form* or cause of any thing by induction, having brought together the facts, we are to begin with considering what things are thereby excluded from the number of possible forms. This exclusion is the first part of the process of induction: it confines the field of hypothesis, and brings the true explanation within narrower limits' (42).

The Scottish defenders of science correctly saw that these Baconian rules constituted the logic of truth that was needed to supplement traditional logic and syllogistic as the logic of consistency. In the first edition of the *Encyclopae-dia*, the article on 'Logic' dealt almost wholly with the traditional concept, judgment, and reasoning in the sense of syllogistic, with some reference to demonstration.[47] The article is anonymous, but the editors refer in their intro-ductory remarks to William Duncan's *The Elements of Logic*.[48] This was a well-known text, first published in 1748 in volume one of Robert Dodsley's *The Pre-ceptor*.[49] Duncan does note that experience is the foundation of natural knowl-edge, and that in order to improve such knowledge we use 'the Method of Trial and Experiment' (*Preceptor*, 152). Duncan then goes on in Book IV to deal with method, and chapter 3 concerns 'the method of science' (312ff). He again makes the point that experience is the foundation of natural knowledge (325) ; and argues that such knowledge is 'founded upon Induction and Experiments made with partciular Objects' (329). This is not very far into a logic of truth; it is rather less than we find in Watts. But the *Encylopaedia* article fails to go even this far. By the time of the third edition of the *Encyclopaedia*, the article on 'Logic,'[50] like the *Encyclopaedia* itself, was much enlarged; it is almost itself an elementary textbook. In this entry, the Scots methodologists provided a stan-dard account of the traditional logic, placed in the context of an account of ideas and judgment similar to that of Locke. But the article also indicated that syllo-gistic as a logic of consistency required to be supplemented by a logic of truth. To do this, the article on 'Logic' refers (Part II, ch. 5, 212–14) to the logic of Edward Tatham,[51] *The Chart and Scale of Truth*,[52] but on this topic the article provides, like Tatham's book (which is extensively quoted), not so much a text-book presentation as a programmatic statement that strongly urges the next gen-eration of textbook writers to include Bacon's rules in their texts.

Another treatment of logic at the turn of the century can be found in Thomas Belsham's exposition of psychology, *Elements of the Philosophy of the Mind, and of Moral Philosophy. To Which Is Prefixed a Compendium of Logic*[53] (1801), which contains a brief exposition of logic as a preface. But this presen-tation is once again very traditional, hardly going as far even as Watts. Belsham does include a remark upon evidence in the context of judgment, thus acknowl-edging that an art of inventing truth precedes a mere logic of consistency. Yet all that he says is that '[o]bservations of the senses are the foundations of natu-ral knowledge, and of what is called Experimental philosophy' (*Elements of the Philosophy*, xxxii), about what Duncan said and rather less than Watts had pro-vided. Later in his exposition, Belsham categorizes induction as in the standard logics as a special case of syllogism, blurring the incomplete inductions of the empirical sciences with the formally valid complete inductions of the logicians. Given this analogy, incomplete induction is treated as nothing more than induc-tion by simple enumeration (lxxxi) – something which Bacon of course had long ago dismissed as 'childish.' It must be supplemented, Belsham tells his readers, by arguments from analogy based on the principle that 'the same circumstances will invariably produce the same effects' (lxxxii–lxxxiii). Belsham does not see that these considerations amount to remarks on evidence for the truth of judgments additional to those that he had earlier made about observation. And further, he fails to see, as Hume had clearly seen, that this 'same cause / same effect' principle is the foundation for the Baconian 'rules by which to judge of causes and effects.' James Mill, in his review of Belsham,[54] rightly complained about this absence of any logic of truth, particularly in the context of an exposition of the science of psychology, where it was important that the inferences be placed on a methodologically secure basis. The logic is, Mill tells us, quite correctly, 'a short view of the vulgar old system.' 'It is no more than an account of the syllogistic method of reasoning, with an appendage, which since the time of Mr. Locke has been generally prefixed to it, an abridgement of his doctrine of ideas, all copied chiefly from Dr. Watts' (Review, 2). Mill justly comments that '[i]t was not to be expected that an author of this cast should produce any of the improvements, which logic, as still taught, stands so much in need of.' However, Mill goes on, 'an author, who at this time of day undertakes to deliver a system of logic, should certainly know that the syllogistic art is a very small part of that important subject.' He then points out that this logic of consistency must be supplemented by what Bacon called the *ars inveniendi*, a logic of truth. The latter 'Lord Bacon pronounced to be entirely wanting in his time, and exerted all his abilities to supply, producing his glorious doctrine of induction. And it is truly astonishing that none of the authors who since his time have produced systems of logic, have thought of delivering fully so much as what he has left us on that subject, not to speak of perfecting what he left uncompleted' (3). Belsham in particular 'is so perfectly unacquainted with the nature of Lord Bacon's Induction that he evidently confounds it (see his account of Induction) with the old induction of the schools ...' (3).

Mill's remarks make it clear that the standard treatments of logic, though they made some pretence at being an aid to science, did not yet incorporate a logic of truth; for the most part they had not progressed as far as Watts, let alone gone beyond him.

What makes this perhaps odd is that in fact there was available one source for a clear statement of Bacon's rules, a statement that is in fact far more concise and to the point than Bacon's own statement, and more clear in outline than either the exposition of Herschell in his *Preliminary Discourse*[55] or of John Stuart Mill in his *System of Logic*. This was the statement in Hume's *Treatise* to which we have already referred. Hume's rule five is the method of agreement and rule six is the method of difference. Rule four is the required statement of the principles of determinism and limited variety. At least, given what Hume says about the logic of the inferences involved, that is how rule four must be understood. The Scottish methodologists undoubtedly read Bacon, and in part derived their account of science directly from that source. At the same time, however, it is entirely reasonable to suppose that they also got their knowledge of the rules of eliminative induction from the same source from which they got their views on causation, namely, Hume. One wonders why those who were developing texts in logic could not utilize the same source.

Indeed, perhaps there were other sources to which they might have turned for the same information. The existence of such a source or sources would not be surprising. For, after all, Hume was a young man when he wrote the *Treatise*. The concise statement of the rules of eliminative induction is a major achievement, and one can ask from what sources Hume himself derived his statement of the 'rules by which to judge of causes and effects.'

Where did Hume get these rules from? One thing at least is clear, and that is that he did not find them incorporated in any available textbook of logic and scientific method. Logic texts then in use did not incorporate rules for experimental science. They are not to be found in even the best texts, such as the Port Royal Logic.[56] The University of Glasgow was very much in the forefront in incorporating the new empirical philosophy into its curriculum, but the logic taught there was traditional. Gershom Carmichael, at the end of the seventeenth century used the Port Royal Logic. His student, and Hume's teacher, Francis Hutcheson, taught logic for many years at Glasgow, and, while he locates the discussion within the context of the new way of ideas, the logic itself deals with the traditional issues of the predicables, immediate inference, and syllogism. Induction occurs only as a special case of the latter, and method is the method of exposition in discourse.[57] The best text that was available was that of Watts, and that hardly incorporated Bacon's *ars inveniendi*, the logic of truth of empirical science. But that merely dismisses one possible answer to the question; it is not to provide the answer that we want. So we must ask again, where did Hume get these rules from?

Now, there were other sets of rules that were, as it were, 'in the air' prior to Hume's setting out to write down his list of methodological principles. One important such set of rules was that of Newton. These rules were sufficiently regarded as standard to be incorporated into Ephraim Chambers's *Cyclopaedia*,[58] in the article 'Philosophizing,' under the subheading 'Rules of Philosophizing.' This *Cyclopaedia* was an immensely popular work, and many would have met Newton's rules here. This group would likely include Hume. But there are other places where he could have encountered them

Michael Barfoot[59] has uncovered an early catalogue of 'The Physiological

Library' assembled by Robert Steuart, professor of natural philosophy at Edin-burgh, as a class library for the use of his students. This catalogue was printed in 1725 and includes a list of 149 'benefactors' – or 150 if we include Steuart himself. Among the latter are the students of natural philosophy. This list was extended in December by a further 57 students, and in this latter list is found the name 'David Hume.' This establishes that Hume attended Steuart's class during the 1724–5 session. As Barfoot points out, this is important for understanding Hume's background knowledge of science. Barfoot looks at the contents of the library, as given by this catalogue, and at Steuart's lectures and infers that Hume was, in fact, well acquainted with the central points of the new science of Boyle and Newton. Hume would also have attened classes in which various experiments in hydrostatics and pneumatics would have been performed. He would in this way have acquired a basic knowledge of the theories of the new science and of the practice of the experimental method.

Clearly, Hume might have found among the books in this class library indications of the rules that are proper for carrying out experimental research. Thus, among the books in Steuart's collection is Watts's *Logick*,[60] from which we can infer that Hume might well have been acquainted with it. We have seen, however, that, in spite of its virtues as a text in logic which went some way towards incorporating the new 'logic of truth,' it is unlikely to be a source for Hume's statement of the rules of eliminative induction. The library also contained several works by Descartes, including the *Meditations*, the *Principles*, and the *Discourse on Method*.[61] We should note, too, that it also contained Glanvill's *Scepsis scientifica*.[62]

Several of the books that were in the library include references to Newton's rules for philosophizing. Thus, among the books was Benjamin Worster, *A Compendious and Methodical Account of the Principles of Natural Philosophy*,[63] containing (3) a statement of Newton's rules. Another volume was W.J. 'sGravesande, *Mathematical Elements of Physics*,[64] which refers to Newton's rules in the Preface (15, unpaginated), and then states them fully in the main body of the text (4). He argues that '[t]he Properties of Body cannot be discovered *a priore*' (Preface, 11, unpaginated), and defends the rule (Preface, 21–2, unpaginated) that induction is the proper method for science. He notes that 'food that nourished me yesterday will do so today,' and several other similar propositions such as that hemlock is a poison or that bricks, having been solid up till now, will remain so. 'All these Reasonings,' he tells his readers, 'are founded upon Analogy: And it is not to be doubted but we are put under the Necessity of Reasoning by Analogy, by the Creator of all Things. This therefore is a proper Foundation of Reasoning' (ibid.). There are other Newtonian texts that were in the library such as J.T. Desaguliers, *Lectures of Experimental Philosophy*,[65] and W. Whiston, *Praelectiones physico-mathematicae*,[66] but these do not contain any statement of rules for experimental philosophy. In any case, just as Newton's science was 'in the air,' so were his rules, and there are many places where Hume could have run into them.[67] But could they have been the source of his rules for judging of causes and effects?

The first of Newton's 'Rules of Reasoning in Philosophy'[68] asserts that 'We are to admit no more causes of natural things than such as are both true and sufficient to explain their appearances.' What counts is the *capacity to explain and predict*. There is no requirement, for example, that the alternatives to the axioms be inconceivable, as was required by Descartes; it is factual evidence that counts, not self-evidence.

Moreover, prediction is to be based on exceptionless laws; causes, in other words, are to be understood as implying generalizations. Newton goes on to make this point when he states his second 'Rule.' It should be noted that it begins with the word 'therefore,' indicating that Newton takes it to be a straight-forward inference from the first. Here is the rule:

> Therefore to the same natural effects we must, as far as possible, assign the same causes.

Newton gives several examples. One is 'the descent of stones in Europe and in America.' The first rule tells us that the assigned cause must be 'true,' that is, true as a matter of fact, and a sufficient condition for the effect. The first rule tells us, in other words, that when we assign a cause we need inductive evidence that supports the claim that the cause is in fact sufficient for the effect. But Galileo did provide such evidence. The second rule tells one to assume that these effects have the same cause unless one has reason, *empirical reason*, to suppose something to the contrary. That is, one will take it to be *not possible* to assign the same cause to the same effect only when there is observational or inductive evidence establishing that it is not possible. The second rule thus instructs us to base our inferences to causes or, what is the same, to laws on the basis of empirical or inductive evidence. Thus, the second rule tells one to assign the same cause to all motions directed towards a massive central body: descending stones towards the earth, the moon towards the earth, the circumjovial planets towards Jupiter, and the solar planets, including the earth, towards the sun. The force of gravity that moves the stone towards the earth should be assumed to be the relevant cause in each case, as Newton argues in the Scholium to Proposition IV, Theorem IV of Book III of the *Principia*.

'Rule III' specifically enjoins us to generalize from observation and experiment: 'The qualities of bodies, which admit of neither intensification nor remission of degrees, and which are found to belong to all bodies within reach of our experiments, are to be esteemed the universal qualities of all bodies whatso-ever.' Newton indicates that he takes this rule to be based on the fact that it is such evidence alone that could justify our assertions of matters of fact: 'For since the qualities of bodies are only known to us by experiments, we are to hold for universal all such as universally agree with experiments. He gives this example: 'We no other way know the extension of bodies than by our senses, nor do these reach it in all bodies; but because we perceive extension in all that are sensible, therefore we ascribe it universally to all others also.'

Newton comments that '[w]e are certainly not to relinquish the evidence of experiments for the sake of dreams and vain fictions of our own devising; nor are we to recede from the analogy of Nature, which is wont to be simple, and always consonant with itself.' In this comment, Newton makes two points. The first concerns hypotheses: we are not to give up affirmations well supported by experiments simply because someone can dream up a hypothesis that would allow one to deduce the falsity of the conclusion given the observational evidence. This point Newton elaborates in his 'Rule IV,' which states '[i]n experimental philosophy we are to look upon propositions inferred by general induction from phenomena as accurately or very nearly true, notwithstanding any contrary hypotheses that may be imagined, till such time as other phenomena occur, or by which they may either be made more accurate or liable to exceptions.' It may well be that contrary hypotheses can be *imagined*. Since they are imaginable, they are clearly possible, not contrary to reason; and since they are not contrary to reason, it follows that the proposition that is accepted is not self-evident. A general proposition, such as Newton's axioms, may thus be accepted on the basis of inductive evidence even if it does not conform to the rationalist rule that requires axioms to be self-evident.

The second point that Newton makes in his comment is rather different: in it he states that we are not to 'recede from the analogy of Nature, which is wont to be simple, and always consonant with itself.' Nature being 'always consonant with itself' one presumes to mean that nature conforms to the rule *same cause / same effect*. The rule, then, is that one should infer in conformity with this principle, and not conclude that the same cause will sometimes bring about one effect and sometimes another. That is, one should not so conclude unless of course there is evidence for such a conclusion – and even then, one will so conclude only if one allows that when Y is sometimes followed by X and sometimes not, it is because there is another factor Z such that Y which is Z brings about X while Y which is not-Z fails to bring about X. For that is what is required by the 'same cause / same effect' rule (as Hume clearly saw when discussing this point – cf T I.III.xii).[69]

Newton's rules are certainly Baconian in spirit. Pemberton, in his *A View of*

Sir Isaac Newton's Philosophy (1728),[70] places both Newton's method and his science within a Baconian context. Pemberton refers to Bacon's 'admirable treatise, intitled NOVUM ORGANON SCIENTIARUM' and indicates that Bacon 'has there likewise described the true method, which ought to be followed' (*Newton's Philosophy*, 5). He refers to the *New Organon* (I.xix) for the distinction between a pseudo-science which makes 'a hasty transition from our first and slight observations on things to general axioms, and then to proceed upon those axioms, as certain and uncontestable principles, without farther examination,' on the one hand, and, on the other, a method 'which [Bacon] observes to be the only true one,' and which requires one 'to proceed cautiously, to advance step by step, reserving the most general principles for the last result of our inquiries' (*Newton's Philosophy*, 5). Pemberton indicates that 'by a just way of inquiry into nature' – that is, conforming to Bacon's method – 'we could not fail of arriving at discoveries very remote from our apprehensions' (12). He then goes on to remark: 'But what surprizing advancements in the knowledge of nature may be made by pursuing the true course in philosophical inquiries; when those searches are conducted by a genius equal to so divine a work, will be best understood by considering Sir Isaac Newton's discoveries' (13). Pemberton thus clearly connects Newton's method with that of Bacon. There is little doubt that this was also the view of Newton. For Pemberton, in his Preface, remarks that Newton 'also approved of the following treatise, a great part of which we read together' (a2).

But if Newton's rules are Baconian in spirit, and though they are clear statements that the method of science is *inductive*, they are none the less certainly not *clear* statements of the methods of *eliminative* induction. They are more directed at the attempt to use hypotheses of the Cartesian sort to criticize inductively justified conclusions than they are directed at locating the precise logic of the experimental situation. Their treatment of the crucial 'same cause / same effect' rule is indicative. This 'same cause / same effect' rule is, as Hume pointed out, crucial to the workings of the eliminative mechanisms in the logic of experiment if we are to conclude an affirmative after a series of negatives, to use Bacon's phrase. Yet Newton does not explicitly include it as one of his 'Rules,' referring to it as a sort of corollary to the more general principles of induction. This is no doubt correct: it is, indeed, something that we know by experience. As Hume put it, when commenting on this principle in his statement of the 'rules by which to judge of causes,' '[t]his principle we derive from experience, and is the source of most of our philosophical reasonings. For when by any clear experiment we have discovered the causes or effects of any phenomenon, we immediately extend our observation to every phenomenon of the same kind, without waiting for that constant repetition, from which the first idea of this relation is derived' (T 174). The point here, however, is not that the principle is correctly seen by Newton to be based upon experience, but rather that he does *not* see that it is central to the logic of experiment. In this respect, Newton's 'Rules of Reasoning in Philosophy,' as statements of the logic of experiment and, more generally, of inference in the natural sciences, fall short of the statement that we find in Hume. In particular, Newton's rules are less than explicit on the need for falsification and elimination, and on the need for the 'same cause / same effect' principle to justify the move from elimination to affirmation, from negatives to an affirmative. Hume was no doubt familiar with Newton's rules, if not from Newton himself, then perhaps from Pemberton, or Chambers's *Cyclopaedia*, or the books in Steuart's class library. But Hume was clear in his statement that what is essential is the method of elimination. For this he presumably needed a source other than Newton or Newton's popularizing followers.

It turns out that Bacon himself might have been such a source. Bacon's *Novum organum* and *De augmentis scientarum* were in Steuart's class library.[71] Moreover, Bacon's writings, the *New Organon* in particular, were translated into English and published in 1733, just prior to Hume's setting out to write the *Treatise*.[72] This was, in fact, the first translation into English of the *New Organon*, and it is quite possible that Hume had read it, or at least was aware of its publication. Still, maybe not: for, in his early memoranda,[73] he mentions no work later than 1731. But then, he mentions no work on methodology in those memoranda, and we do know that he read at least some works that dealt with this topic, since he recommended Malebranche's *Recherche* to the Chevalier Michael Ramsey to be read if the latter wanted to prepare himself to read the *Treatise*. As Hume put it, Ramsey should

read once over la Recherche de la Verité of Pere Malebranche, the Principles of Human Knowledge by Dr Berkeley, some of the more metaphysical Articles of Bailes Dictionary; such as those [on] Zeno and Spinoza. Des-Cartes Meditations would also be useful but don't know if you will find it easily among your Acquaintances. These books will make you easily comprehend the metaphysical Parts of my Reasoning and as to the rest, they have so little Dependence on all former systems of Philosophy, that your natural Good Sense will afford you Light enough to judge of their Force and Solidity.[74]

There is a striking omission in the case of Locke, but perhaps Hume felt that he did not need to recommend this author to a colleague who was reasonably well read in English philosophy and literature. In any case, Hume might well have read other methodological works that are, like the *Recherche*, not mentioned in the early memoranda. However, even if Hume did read the 1733 edition of the

New Organon, it remains true, as we have noted, that Bacon's own statement therein of the rules of eliminative induction, while accurate, is by no means as concise as Hume's. In fact, nowhere in the 1733 edition of Bacon do we find such a neat statement of the rules for the experimental method.

However, there is attached to this edition of Bacon's works an Appendix to the *New Organon* by the translator, Peter Shaw. Shaw, like Bacon himself, makes clear the eliminative nature of the method, the importance of falsification in moving one in the direction of causal knowledge. But again like Bacon himself, Shaw does not provide a concise statement of the rules in the way that Hume does.

Shaw is anxious, however, to connect Bacon's work with that of the Royal Society, and to illustrate its impact upon the practice of the new science. Shaw provides both a general reference and also more specific references to the works of various members of the Royal Society as providing other statements of the experimental method. If we turn to those works, however, they tend very often to be more programmatic than Bacon himself.

To mention one that Shaw does not note, there is Sprat's *History*, for example. This work, which was in Steuart's class library,[75] while lauding the role of Bacon and making clear the view of the members of the society that they were following the method of Bacon, nowhere states anything like the methods of eliminative induction.

Sprat is, in fact, critical of Bacon's actual practice. He tells us of Bacon that '[h]is rules were admirable: yet his *History* not so faithful, as might have been wish'd in many places, he seems rather to take all that comes, then to choose; and to heap, rather, then to register.'[76] This is not entirely fair: after all, we do not criticize the architect for not being a good bricklayer. In any case, when Sprat does tell us the nature of Bacon's method, it remains sufficiently vague that the whole point of eliminative induction, the element of falsification, is lost:

True Philosophy must first of all be begun on a scrupulous, and severe examination of particulars: from them, there may be some general Rules, with great caution drawn: But it must not rest there, nor is that the most difficult part of the course: It must advance those Principles, to the finding out of new effects, through al the varieties of Matter: and so both the courses must proceed orderly together; from experimenting to Demonstrating, and from demonstrating to Experimenting again.[77]

As for the specific references, these are to Boyle and Hooke in particular, as well as Locke and Newton.

In the case of Boyle, if we look at his programmatic announcements, we get the flavour of Baconianism as we do in Sprat, but we get no neat statement of rules. In fact, in the Appendix to Bacon's *New Organon*, Shaw states about Boyle that he 'has given us a particular Account of the Method he pursued, in his Philosophical Enquiries; which plainly appears to be formed upon the Model of the Lord Bacon; and is no other than a loose and imperfect kind of Induction ...' (567). This is just.

In the case of Hooke, the reference is to his *Posthumous Works*, and in particular to the essay on 'A General Scheme, or Idea of the Present State of Natural Philosophy, and How Its Defects May Be Remedied by a Methodical Proceeding in the Making Experiments and Collecting Observations.' Here is another statement of how to go about doing experiments. It includes long recommendations on the use of instruments, on the virtue of diligence, and so on. It is, in fact, a rather verbose statement. It does make clear that the gathering of data is for purposes of elimination. But, although Hooke does promise us a set of rules that will make the procedures of science akin to those of algebra, this portion of the work was left undone. He might have found it possible to give a concise statement of the eliminative methods in this part of the essay – though, given the prolix nature of what we do have, I would guess that that would be improbable. Hooke gives the usual complaint about past thinkers that 'we may find them even to wrest those few Experiments and Observations they had read, or collected, and to endeavour rather to adapt them to their Hypothesis, than to regulate their Thoughts by them, esteeming their own Understandings to be the Mine of all Science, and that by pertinacious ruminating, they could thence produce the true Image and Picture of the Universe' ('A General Scheme,' 4). In any case, what Hooke has us aiming at is the discovery of 'constant and ... necessary Effects, which are produc'd by the working Power ...' (45). To discover these he urges that 'the first way of discovering them, is by observing the Method or Progress of Nature in generating, increasing, weakening or destroying the same Property in divers States of the same Bodies' (48). Here we have a statement of the joint method of agreement and difference, though it is mixed up with the related method of concomitant variation. Hooke then goes on to state the method of agreement. This is followed by a rule designed to have us note conjunctive causes (49). Then, a little later, we get the method of difference (50). Some of the examples are nice, some messy, and throughout there is a good deal of wisdom from someone who has clearly mastered the practice of doing science. However, for our purposes it is equally clear that we are not going to find in Hooke a statement of the rules of eliminative induction that could be an immediate source of Hume's statement of those rules. Hooke is, in fact, less perspicuous on this point than is Bacon himself. We find no clear statement that the conclusions are affirmations of general regularities, nor any statement that the conclusions follow, given the eliminative data, only if we take for granted the principles of determinism and limited variety, principles which, as Hume makes clear in his statement of his fourth 'rule by which to judge of causes,' are themselves based on experience.

It would seem, then, that in the English tradition there is no statement of the rules of the experimental method as neat and as concise as Hume's prior to that of Hume himself.

But as we saw, Hume himself recommended to the Chevalier Michael Ramsey that, if the latter wished to prepare himself to read the *Treatise*, then certain books were important, and the first of these that Hume mentions is Malebranche's *Recherche*. In this work, a whole 'Book,' the sixth, is devoted to 'Method,' and 'Part Two' of this book opens with a chapter entitled 'Rules to be observed in the search after truth.' Perhaps, then, it is here that Hume found his 'rules by which to judge of causes.' In fact, as it turns out, Hume did not find them here. We have therefore to take seriously Hume's comment that the text was recommended with regard to the 'metaphysical Parts of my [Hume's] reasoning'[78] and not the methodological.[79]

In fact, Cartesian rules for method were as much 'in the air' as the Newtonian. Rules of this sort were provided by such well-known rationalists as Descartes, Arnauld, and Malebranche. Moreover, we know that Hume read Malebranche's *Recherche*, which, like several of Descartes's works, was in Steuart's class library.[80] In addition, the Port Royal Logic (*La Logique de Port-Royal*) of Arnauld and Nicole was a well-known work, one that was also likely to have been known to Hume. He therefore is unlikely to have missed the methodological norms proposed by Descartes and his followers. But, while it is certainly true that Hume was influenced by Cartesian and, in particular, Malebranchian occasionalism in formulating his doctrine of causation as regularity, it is also true that the rules proposed by these philosophers had little to do with the logic of truth that the new empirical science required. These philosophers all recognized that a logic of truth was indeed necessary, but the logic of truth that they proposed was a rationalist logic of truth, and made sense as a strategy only in the context of a rationalist metaphysics of essences.

It is in the context of his essentially Aristotelian metaphysics that we must situate Descartes's rule of method. The first and basic rule of this method states that one who wishes to inquire into truth ought 'never to accept anything as true if [he or she] did not have evident knowledge of its truth ...'[81] Arnauld and Malebranche followed Descartes in advancing rules of method of this sort. For us the point is that this is a rule for discovering truth by examining one's ideas; it is a rule for discovering the truth of things by turning to the ideas we have of those things, that is, the forms of things that are present innately in the mind. *It is not a rule for the empirical investigation of matter-of-fact regularities.* So this rule, though very much 'in the air,' would hardly be a source for Hume's 'rules by which to judge of causes and effects.'

We find in the Malebranche that Hume recommended to Chevalier Ramsey a restatement of this Cartesian method of clear and distinct ideas. 'The general rule concerning the subject of our studies,' he writes in the *Recherche*, is 'that we must reason only on the basis of clear ideas' (*Search after Truth*, 44). This he uses to attack Aristotle and the Aristotelians: Aristotle's 'followers have trouble understanding that these words [such as 'faculty'] signify nothing, and they are no more learned than before just because they are heard to say that fire dissolves metals because it has a dissolving faculty; and that man does not digest because he has a weak stomach or because his digestive faculty is not performing its functions well' (443). Malebranche uses this principle to establish that 'no body, large or small, has the power to move itself' (448). Indeed, 'not only are bodies incapable of being the true causes of whatever exists: the most noble minds are in a similar state of impotence' (449). That is, given this rule, the occasionalism follows: '*natural* causes are not causes; they are only *occasional* causes that act only through the force and efficacy of the will of God' (449). 'Only since Descartes' (444) have we been able to recognize how the metaphysics of the scholastics violated the basic rule of method, and only since him have we grasped the truth about the world: 'Descartes followed it [the rule] exactly enough in this system of the world' (488). The aim of reason is to discover the truth that consists in the relations of things or of ideas: 'we can generally say that in all questions we search only for the knowledge of some relations, be they relations between ideas or between things and their ideas' (489). We must form clear and distinct ideas of the subject matter with which we are concerned so that we will be able to compare them and 'thus to recognize their unknown relations' (489). However, we in fact find that what we obtain are not things clear and distinct enough to be self-evident, but merely *hypotheses*. Thus, Malebranche proposes 'an explanation of magnetism' (498ff) to illustrate his points. He begins with the principle that it is a 'law of nature that bodies move each other when they collide,' and therefore, he goes on, 'we should ... try to explain the motion of the magnet by means of some body that comes in contact with it' (499). Needless to say, this involves the introduction of purely hypothetical particles, unobserved entities, entities for which there is no observational warrant to justify claims to their existence. Indeed, Malebranche argues, 'To discover these tiny bodies, we need not open our eyes and approach this magnet; for the sense will impose upon reason, and we shall perhaps judge that nothing emanates from the magnet because we see nothing emanating from it' (499). Again, he explains the 'cause of the movement of our members' (502ff) in terms of, again hypothetical, animal spirits, small pores, and so on.

The point is twofold. *First*: these explanations are excluded by the fourth Newtonian rule against hypotheses, and, more generally, by Bacon's insistence – in which he was followed by Newton and Hume – that if that to which we appeal when we explain is merely hypothetical, then we have not yet succeeded in explaining: a hypothesis is acceptable as explanatory only after it has been confirmed and has therefore ceased to be a *mere* hypothesis.

But *second*, it is also clear that there is here nothing that comes close to being an *empirical* logic of truth, a set of 'rules by which to judge of causes' where *causes* are taken to be *matter-of-empirical-fact* regularities. So Hume could not have obtained *his* 'rules by which to judge of causes' from Malebranche. It really was the metaphysics that Hume was recommending to the Chevalier Ramsey, and not the rules of method that Malebranche defends.

In fact, Descartes is rather better on this point than is Malebranche. Malebranche nowhere seems to recognize that there might in fact be competing hypotheses about the relations among ordinary objects, that these objects may be too complicated for us to be able to establish by pure reason, a priori, which of these hypotheses is true, and that we must in such cases rely upon experiment to move us towards the truth by eliminating those hypotheses which are false. The order that he proposes is this: 'First I tried to discover in general the principles or first causes of everything that exists or can exist in the world' (*Discourse*, 143). For this purpose, Descartes considered only God, and from his being he was able to derive certain principles. Moreover, he 'derived these principles only from certain seeds of truth which are naturally in our souls.' Next, he went on to '[examine] the first and most ordinary effects deducible from these causes' (ibid.). From these basic metaphysical considerations, he was able – or so he believed – to derive explanations of the nature of the stars, minerals, water, fire, and so on. However,

when I sought to descend to more particular things, I encountered such a variety that I did not think the human mind could possibly distinguish the forms or species of bodies that are on the earth from an infinity of others that might be there if it had been God's will to put them there ... the power of nature is so ample and so vast, and these principles so simple and so general, that I notice hardly any particular effect of which I do not know at once that it can be deduced from the principles in many different ways; and my greatest difficulty is usually to discover in which of these ways it depends on them. I know no other means to discover this than by seeking further observations whose outcomes vary according to which of these ways provides the correct explanation.[82]

The metaphysical principles that depend upon God's being, and can (as Descartes thought) be derived therefrom, provide a framework, the framework of the mechanical philosophy; but the ordinary features of many things are too complex for us to be able to discern the logical steps that lead from God's being to the regularities describing these complex phenomena: there are, in fact, many hypotheses capable of conforming to the laws of nature insofar as I, a finite being, can grasp the logical structure of reality. We have, therefore, to resort to a second-best means, one that falls short of the certainty of the a priori method. This is the method of observation, and, Descartes makes clear, *its purpose is elimination*, elimination of the false from among the many hypotheses that seem to me to be, on the one hand, equally consistent with the fundamental laws, and, on the other hand, equally incapable of being deduced from those first principles.

The point is that Descartes, unlike his follower Malebranche, recognizes that there is a place in the methodology of science for the mechanisms of elimination based on the self-evident principles of the mechanical philosophy functioning as a set of principles of determinism and limited variety. The place is, to be sure, one that is second best. Descartes would prefer to make the patterns self-evident by deducing them from first truths about God. But he cannot do that, so he settles for second best, the use of observation to eliminate the false from among the range of possible alternatives admitted by the first principles.

However, even though Descartes recognizes the role of elimination, and, in fact, the role of the principles of determinism and limited variety, it is also true that he does not form these things into a set of 'rules by which to judge of causes.' Descartes is happy enough to give a set of rules for dealing with a priori truth. He does not give such a set of rules for his logic of *empirical truth*, the logic of elimination. So Hume, while he might have learned from Descartes the importance of elimination, did not learn from him the rules of eliminative induction.

There are other sources, however, from which Hume might have discovered his set of rules for the logic of truth. Bacon was, in fact, a widely read author,[83] not only in England but on the continent, and others could well have picked up the idea of a logic of experiment from reading his works, more easily perhaps than they could have thought up the idea of elimination as the key ingredient in the logic of experiment. Descartes might have been one of these. But there were others, too, who might have got the ideas from Bacon. There are, in particular, the methodological treatises published by the members of the Royal Academy in Paris. Both von Tschirnhaus and Mariotte published such works. In fact, Hume might have got the reference to von Tschirnhaus from Shaw's Appendix to Bacon's *New Organon*, since Shaw mentions him as one of the leaders in the attempts to formulate and put into practice the new methodology of empirical science.[84] But here again, while we do find statements indicating that their authors are aware of the importance of falsification and elimination, we do not find any statement of the rules as clear as that of Hume. We have no direct evidence that Hume ever read von Tschirnhaus's methodological essay, *Medicina mentis*, but it was among the books in Steuart's class library.[85] As for Mariotte's *Essai de logique*, while it too was in Steuart's library,[86] it is equally true that we have no direct evidence that Hume ever read his essay on methodology. None the less, it was Mariotte who discovered the blind spot in the eye,[87] and Hume makes reference to this fact of vision in his discussion of space in the *Treatise*;[88] so, as Hume was familiar with his work on the eye, it is entirely possible that Hume was also familiar with his methodological writings.

Von Tschirnhaus's essay *Medicina mentis* was first published in 1687, and a second edition was published in 1695.[89] In the second section of this essay, von Tschirnhaus inquires 'De quelle manière l'on peut avancer toujours sans commettre d'erreurs, en découvrant un nombre sans cesse croissant de vérités nouvelles' (*Médecin de l'esprit*, 91). The greater part of the discussion is taken up with reasoning in mathematics. Only then does von Tschirnhaus turn to matters of empirical fact. '[C]oncernant les êtres imaginables il faut apprendre cela par des expériences exécutées avec soin, qui devront d'ailleurs être organisées suivant des règles précises' (106). Here the crucial rule of inference is as follows: 'Il faut ... après cela, s'enquérir par des expériences de ce qui conformément à cette règle, est seul, et nécessairement, indispensable à la génération de telle ou telle chose: *est nécessairement indispensable à la génération d'un être tout ce en l'absence de quoi, toutes choses égales d'ailleurs, l'être à engendrer ne prend pas naissance*' (107). Here we have another statement of something that looks like a rule of eliminative induction. But it is certainly not as clearly stated as the method of elimination is stated in Bacon's own *New Organon*. We find no clear statement that the conclusions are affirmations of general regularities, nor any statement that the conclusions follow, given the eliminative data, only if we take for granted the principles of determinism and limited variety. Von Tschirnhaus certainly could have influenced Hume; but, equally, his essay seems not to be the sort of direct source for Hume's statement for which we have been looking.

The same is true of Edme Mariotte's *Essay de logique* (Paris 1678, with a second edition published as part of Mariotte's *Œuvres*, The Hague 1717).[90] Mariotte certainly had some interesting things to say concerning the methodology of science.[91] They are things that are part and parcel of the empiricist philosophy. In other words, he clearly recognizes that causal propositions are general statements of matter-or-fact regularity. Thus, in his Rule XXXVIII he states that

'[I]es propositions sensible universells, comme, *l'eau étient le feu, les hommes de l'Europe sont blancs*, dependent des particulieres, et ne sont connue vrayes que par elles, et sont fausses, lors qu'un particuliere est contraire' (*Essai*, 27). This states the regularity account of causation, and, moreover, recognizes that such propositions, being general, are capable of falsification. Such falsification is essential to the workings of the mechanisms of eliminative induction. As Bacon saw, these mechanisms move through negatives to affirmatives. The point to be noted is where this statement occurs in Mariotte's presentation. In fact, it occurs *after* Mariotte has enunciated eliminative rules. In Mariotte's presentation, the crucial 'Principes et propositions fondamentales, pour établir les sciences des choses naturelles' are Rules XI and XII. The former states: 'Si une chose estant posée il s'ensuit un effet, et ne l'estant point, l'effet ne se fait pas, toute autre chose estant posée: ou si en l'estant, l'effet cesse; et estant toute autre chose, l'effet ne cesse point: cette chose là est necessaire à cette effet, et en est cause' (19). Rule XII then states: 'Si deux choses estant posée il se fait un effet, et que l'une produise l'effet, et l'autre le reçoive; celle qui ne souffre point de changement, est celle qui produit l'effet' (19). Here again we get statements of eliminative principles that are close to the sort of thing that Hume said. But logically they are out of place. One can use these rules to infer causes only if causes are understood as regularities. Given this logic, the statement of the regularity position should occur *prior* to the eliminative rules. Since it does not, and, indeed, since Mariotte clearly fails to see any deep connection between the regularity view and the use of elimination, it follows that Mariotte's understanding of the logic of truth that Bacon developed is at best rudimentary, and perhaps confused, relative to the clarity of Hume's understanding.

Hume's statement of the 'rules by which to judge of causes and effects' is in fact far superior to that of Mariotte. The relevant points are the same as in the cases of Hooke and von Tschirnhaus. In the first place, Hume clearly locates the statement that the causal principles at which science aims are matter-of-fact regularities *prior* to the statement of the eliminative rules. And in the second place, Hume clearly indicates that the eliminative mechanisms will not lead to an affirmation except in the context of his rule four, the 'same cause / same effect' principle, that is the principles of determinism and of limited variety. Mariotte, like both Hooke and von Tschirnhaus, does not comprehend these points clearly.

Hume did not invent his 'rules by which to judge of causes and effects' out of nothing. They are, after all, the rules of Bacon's logic of induction. They were very much 'in the air'; various versions of them can be found in many places, as we have seen. Yet Hume's statement compares in its clarity only with that of

Bacon, and is far more concise than the latter. But further, Bacon had not seen the need for a systematic argument to establish the regularity account of causation. His methodological rules require that argument, if the principles that they establish, namely, regularities, are to be accepted as genuinely constituting science. Locke had started that argument, but it was Hume who first established, on the one hand, that the rationalist alternative that had come down from Aristotle through Aquinas to Descartes was either meaningless or contradictory, and, on the other hand, that the alternative, the regularity account, was the only reasonable account of causation. Furthermore, unlike Locke, Hume was able to place in the context of an empiricist metaphysics and epistemology the rules of the new logic of truth that was required by the new account of causation and of science. It is difficult not to conclude that Hume, as a young philosopher about to write the *Treatise*, himself thought through the method of the new science, and that his 'rules' were the product of this process, not merely lifted from the writings of some previous thinker. In other words, it is hard not to conclude that these rules clearly illustrate Hume's genius as a methodologist of empirical science.

Study Six
Causation and the Argument A Priori for the Existence of a Necessary Being

Demonstrative science was long thought possible in the case at least of the existence of God. Long after Locke had given up the notion of a demonstrative science for most, if not all, of the nature that we know in ordinary experience, including human nature, he continued to defend the notion that we can demonstrate this one existential fact, that God exists.[1] Samuel Clarke agreed with Locke on this issue, restating the argument in his Boyle Lectures.[2] But, of course, it was not just empiricists who continued to defend the argument. For, the very same argument was laid out by Locke's Aristotelian critic John Sergeant,[3] on the one hand, and by Locke's rationalist critic Leibniz, on the other.[4] Everyone seemed to be in agreement that, on this point at least, demonstration of matters of fact is possible. More strongly, the argument makes it reasonable to think that in order to understand any ordinary matter of fact, in the end one must turn to God as the necessary being who accounts for all contingent beings. God is inevitably part of the explanation scheme. It was Hume who challenged all this.[5] This challenge was, of course, part of his more complex challenge to the whole notion of causation as necessary connection.

The argument is not difficult to state. This, then, is where we shall begin. Moreover, as we shall see, it is, within its limits, a good argument, valid if not sound, though to put it that way fails to do justice to the fact that its defect is a matter of being part of an approach to explanation that is at its deepest and most crucial point simply mistaken. That was Hume's point: the whole idea upon which the argument was based was either meaningless or contradictory, and, in either case, not worthy of being reckoned part of the philosophy of any reasonable person. But the argument should not be dismissed. What is important is to see the assumptions concerning cause and explanation upon which it is based and which make it reasonable to conclude that a necessary being exists. What must be dismissed as irrational is not the argument itself, nor, in fact, its conclusion, but the concept of causation and of explanation upon which it is based.

Since the argument was a strong point of the traditional notion of a demonstrative science – Sergeant uses it as one example of demonstrative syllogistic – it is clear that much of the doctrine of traditional logic as capable of providing a demonstrative science is tied up with the notion of causation and of explanation which provides the argument with all its plausibility. Thus, if that framework of causation and explanation is undermined, as Hume did in fact undermine it, then what disappears with it is everything that made plausible the traditional notion of a demonstrative science and, at the same time, made plausible the traditional doctrines about the nature and power of logic or syllogistic.

In fact, if one pursues the traditional notion of explanation, then it is hard indeed to avoid the conclusion that a necessary being exists. God as an ultimate explanatory principle is an integral feature of the traditional metaphysics of explanation: the chain of explanation points to and ends with God. But once the metaphysics that supports the traditional patterns of explanation goes, then so does the the necessity of including God as an explanatory principle. It is not surprising, then, that with the emergence of the new science, God becomes an unecessary hypothesis. With Hume's critique of the traditional substance ontology, science is freed from the necessity of incorporating God into the world.

At the same time, of course, new concerns arise: just what should the new empiricist metaphysics that goes with the new science say about existence? Philosophers such as Hume struggled, as we shall see, with these issues.

I: The Argument A Priori for the Existence of God

The argument is a simple one: as Leibniz expresses it, 'since there are contingent beings, which can only have their final or sufficient reason in the necessary being, [therefore there exists] a being that has the reason of its existence in itself' ('Monadology,' § 45). Descartes expresses the same point succinctly when he states in the *Rules for the Direction of the Mind*[6] that 'from the fact that I exist I may conclude with certainty that God exists' (*Rules*, 1:45). Here is Locke's version:

man knows by an intuitive certainty, that bare nothing can no more produce any real being, than it can be equal to two right angles. If a man knows not that non-entity, or the absence of all being, cannot be equal to two right angles, it is impossible he should know any demonstration in Euclid. If therefore we know there is some real being, and that nonentity cannot produce any real being, it is an evident demonstration, that from eternity there has been something; since what was not from eternity had a beginning; and what had a beginning must be produced by something else.

Next, it is evident, that what had its being and beginning from another, must also have all that which is in, and belongs to its being, from another too. All the powers it has must be owing to, and received from, the same source. This eternal source then of all being must also be the source and original of all power; and so this eternal being must be also the most powerful. (*Essay*, IV.x.3–4)

Clarke puts it in more or less identical terms:

it is Absolutely and Undeniably certain, that *Something has Existed from all Eternity* ... For since Something Now Is; 'tis manifest that Something always Was: Otherwise the Things that Now Are, must have risen out of Nothing, absolutely and without Cause: Which is a plain Contradiction in Terms. For, to say a thing is produced, and yet that there is no Cause at all of that Production, is to say that Something is *Effected* when it is *Effected by Nothing*, that is, at the same time when it is *not Effected at all*. Whatever Exists, has a Cause of its Existence, either in the Necessity of its own Nature; and then it must have been of it self Eternal: Or in the Will of some other Being; and then That Other Being must, at least in the Order of Nature and Causality, have Existed before it. (*Discourse*, 1:9–10)

Sergeant includes the argument as an example of demonstrative syllogistic science at work in his *Method to Science*. Nothing occurs unless there is a cause: 'since Nothing can do Nothing, it follows that Nothing can be done, unless there be something that Does or Acts, that is unless there be an Efficient Cause.' (*Method*, 272). Now, 'Existence is no ways Intrinsical to any Created Ens; either Essentially, or as an Affection springing out of it's Essence' (304). Hence, 'All Created things have their Existence from something that is Extrinsical to them' (305). In other words, '[t]here must be some Uncreated Cause that gives Existence to all Created Entities' (306), and, moreover, '[t]his uncreated Cause of all Existence must be Self-Existent; that is, his Essence must be his Existence': 'For, were his Essence Indifferent to Existence or Existence Accidental to him and not Essential, he would need Another Cause to give him Existence, for the same reason Creatures do, and, so He would not be Uncreated' (306). Finally, we have Hume's statement of the argument:[7]

The argument, replied Demea, which I would insist on, is the common one. Whatever exists must have a cause or reason of its existence; I find only one argument employed to prove, that the material world is not the necessarily existent Being: and this argument is derived from the contingency both of the matter and the form of the world.

'Any particle of matter,' it is said [Hume here refers to Clarke – though the quote is not exact], 'may be conceived to be annihilated; and any form may be conceived to be altered. Such an annihilation or alteration, therefore, is not impossible.' (*Dialogues*, Part IX)

The argument, then, can be stated in apparently straightforward and simple terms. But if the argument can be stated simply, it must also be recognized that, when its structure is laid out, it is a structure that requires the support of the considerable metaphysical scaffolding that comes with it. In particular, what has to be recognized is that the traditional argument a priori is located within the context of the traditional scheme for explaining things.

Thus, Socrates is in prison. This event is followed by his not going to Thebes and by his drinking the poison. This sequence is in itself unstructured; so far as we can tell from our sense experience of these events, there is no connection. But, in fact, upon the traditional account, there is a connection. Or, at least, provided that we accept the principle that for every event there is a sufficient reason, then there is a connection. But this was accepted as a matter of course. As Leibniz puts it, in our thinking about the world, our reason is governed by the principle of contradiction on the one hand and, on the other hand, by the principle 'of sufficient reason, by virtue of which we consider that we can find no true or existent fact, no true assertion, without there being a sufficient reason why it is thus and not otherwise, although most of the time these reasons cannot be known to us' ('Monadology,' § 32). The explanation involves placing the event in question within the context of an underlying entity the connections to which provide a unifying structure. Again as Leibniz puts it, 'diversity must involve a multitude in the unity or in the simple' (§ 13). This unifying structure is provided by a unifying exercise of power by the underlying substance. This power is understood in Aristotelian terms. Leibniz puts the point this way: 'The action of the internal principle which brings about the change ... can be called appetition' (§ 15). Or, as we saw Clarke put it, whatever exists has a cause of its existence 'either in the Necessity of it own Nature ... Or in the Will of some other Being ...' (*Discourse*, I:9–10). The unifying exercise of power provides a necessary connection that links the events in the sequence. What is explained is the upshot of the exercise of power. This event that is explained consists of the substance both having a certain property and not having certain other properties: the sufficient reason explains 'why it [the thing] is thus and not otherwise,' as Leibniz put it. We explain why the substance is this way and is not a variety of other ways. This event that is explained is explained by relating it to a being, namely, the way the substance antecedently is, or to the way another substance is. Thus, the explanatory principle is this: for every event there is a cause with at least as much reality, that is, at least as much being, as the effect. No being can come from non-being.

Crucial to the argument is the notion that a being which is not something is thereby limited, and that limited being is contingent being; its existence is contingent upon something's non-being, perhaps, indeed, its own self not being in some way. If this non-being were in fact to be, then the being that is or exists would not be. Thus, for a limited being we can conceive of its being but also conceive of its not being. It is this that constitutes its contingency. Since the thing can be conceived as not being, it follows that it can be conceived apart from its existing, from the way it is. The entity and its existence are thus separable. Thus, as Sergeant puts it, 'the Notion or Nature of ens and of Existence in Creatures, (and consequently of Essence and Existence) are Distinct' (*Method*, 302). This distinction exists, Sergeant argues syllogistically, because

Every Notion of which Existent and not-existent may be truly predicated is Different from the Notion of Existent; But
The Notion of Ens (in its First and Proper Signification, taken for an Individual Substance) is a Notion of which Existent and not-existent may be predicated; therefore
The Notion of ens (thus understood) is different from the Notion of Existent; and, consequently, the Notions of essence and Existence are also distinct. (302)

It is the major which is crucial. With respect to this premise he further argues that '[t]he major is ... Evident; For, when we say [Petrus est] or [Peter is existent] were the notion of the Predicate [Existent] the same with [Peter] the Subject, the Proposition would be (in sense) formally Identical, and the same as 'tis to say, [what's Existent is Existent] Wherefore, when we say [Petrus non est] or Peter is non-Existent, Peter Signifying the same as Existent, it would be the same as if we said, what's Existent is not Existent, which is a contradiction' (303). And the argument goes on from there: 'Existence is no ways Intrinsical to any Created Ens; either Essentially, or as an Affection springing out of it's Essence' (304). From this it follows that '[a]ll Created things have their Existence from something that is Extrinsical to them' (305). Ergo, God. Limited being is dependent being.

The contrast to limited being is unlimited being. Since limited being is limited by non-existence, by its not being in certain ways, it follows that there is no non-being attached to an unlimited being. Unlimited being thus contains all being. Moreover, unlimited being has therefore to exist non-contingently: its being is not contingent upon the non-existence of some other being. Unlimited being, since its being is non-contingent, is necessary being. Since there is nothing which unlimited being is not, it is not possible for there to be a being which, were it to exist, would make it that the unlimited being did not exist. As Leibniz put it in the 'Monadology': 'that is why the ultimate reason of things must be in a necessary substance in which the diversity of changes is only eminent, as in its source' (§ 38).

Since this being contains all being within it and is the source of all diversity, there can be only one such being, only one unlimited and necessary being. Again as Leibniz put it, '[s]ince this substance is a sufficient reason for all this diversity, which is utterly interconnected, there is only one God, and this God is sufficient' ('Monadology,' § 39). To put it another way, the structure of an unlimited being cannot derive from another being, for that other being is something which it is not, contrary to the supposition that it is unlimited. The structure must therefore derive from itself: an unlimited being is self-existent. And, since unlimited beings are necessary beings, necessary beings are self-existent.

Further, if this being of the unlimited being does not derive from another being, neither does it derive from nothing. An unlimited being is therefore a being that just always is: an unlimited being is an eternal being.

The argument a priori for the existence of a necessary being is thus the argument that the existence of a contingent being implies necessarily the existence of a non-contingent or necessary being. But to recognize the premise, on the one hand, that there are contingent or limited beings, and to recognize the conclusion, on the other, that there is a necessary being, is not yet to understand just how the inference goes that is supposed to lead us from the one to the other.

So, just how does this inference go?

To begin to get a grasp of the structure of this inference, consider a series of contingent beings. Each member of the series is explained by reference to its predecessor. But the whole series itself is contingent: since there are beings which are not – that is why each member of the series is contingent – there are other series that could be but are not. So the whole series is contingent.

Since the whole series is an event, it requires explanation, and, in particular, an explanation by reference to something that is.

This could be an explanation in terms of another contingent being. But now we have another event that requires explanation.

We now introduce a crucial principle: *There can be no infinite regress of causes*. We must therefore arrive at an event which both explains the series of contingent events *and also* explains itself. But the being that explains its own being is a necessary, unlimited, eternal, self-existent being.

This necessary being is therefore a being which structures itself, and in structuring itself structures the whole: in causing itself, the necessary being causes all contingent being, both insofar as the latter is and insofar as it is not.

Leibniz puts it this way in the 'Monadology': 'since all this detail involves nothing but other prior or more detailed contingents, each of which needs a similar analysis in order to give its reason, we do not make progress in this way. It must be the case that the sufficient or ultimate reason is outside the sequence or series of this multiplicity of contingencies, however infinite it may be' (§ 37). Clarke puts the point this way: 'whatever Exists, must either have come into Being out of Nothing, absolutely without Cause; or it must have been produced by some External Cause; or it must be Self-Existent' (*Discourse*, 1:15). Nothing can come from nothing, as he argued, and there can be no infinite regress of external causes. So the ultimate cause must be something which is self-existent. Or, as we saw Sergeant put it, the 'uncreated Cause of all Existence must be Self-Existent; that is, his Essence must be his Existence,' for, 'were his Essence Indifferent to Existence or Existence Accidental to him and not Essential, he would need Another Cause to give him Existence, for the same reason Creatures do, and so He would not be Uncreated' (*Method*, 308).

Since power and essence coincide, it is clear that God as an unlimited being, as a being who contains all being, both its own being and all other being, within itself, has sufficient power to guarantee His own existence: nothing can prevent the necessary being from existing, which, of course, is precisely why it is a *necessary* being. Leibniz expresses the point this way:

Thus God alone (or the necessary being) has this privilege, that he must exist if he is possible. And since nothing can prevent the possibility of what is without limits, without negation, and consequently without contradiction, this by itself is sufficient for us to know the existence of God a priori. We have also proved this ... a posteriori since there are contingent beings, which can only have their final or sufficient reason in the necessary being, a being that has the reason of its existence in itself. ('Monadology,' § 45)

Crucial to the whole argument is the thesis that there can be no infinite regress of external causes. Clarke puts the point this way: 'An infinite Succession ... of merely Dependent Beings, without any Original; Independent Cause; is a Series of Beings, that has neither Necessity nor Cause, nor any Reason at all of its Existence, neither within it self nor from without ...' (*Discourse*, 1:13). But why is this so?

Why can there be no infinite regress? There is something in the whole scheme that is held by those who accept it to preclude such an infinite regress. What is there about the explanation scheme that precludes the possibility of an infinite regress of causes?

The crucial feature is the fact that the relevant notion of cause is that of a cause working for an end. We have seen clearly that the notion of cause that is used in this argument is the notion of a cause that *produces for an end*. It is this that establishes that there can be no infinite regress.

For each cause, the end is the form or structure of events that comes to be. The cause has as its aim or end that form or structure. All structure, in other words, is teleological structure. Now, the ultimate cause can bring about means that are intermediate to the ultimate end. The structure of the series is dependent upon the end determined by the ultimate cause. But a means without an end is contradictory. The very notion that the causal order is teleological implies that there can be no series the structure of which is not determined by an ultimate end. An infinite regress would be a series of means, each of which brings about its successor, but where there is no overall structure. There can therefore be no infinite regress: the notion of structure excludes it, that is, the notion of structure that is part of the explanation scheme in which the argument a priori is located. Upon this account of structure, of causal structure, the sequence of causes must terminate in an ultimate cause which determines the end.

A causal series that is ordered in this teleological way is what Aquinas[8] called a series that is *per se* ordered.[9]

In efficient causes it is impossible to proceed to infinity 'per se' – thus, there cannot be an infinite number of causes that are 'per se' required for a certain effect; for instance, that a stone be moved by a stick, the stick by the hand, and so on to infinity. But it is not impossible to proceed to infinity 'accidentally' as regards efficient causes; for instance, if all the causes thus infinitely multiplied should have the order of only one cause, their multiplication being accidental, as an artificer acts by means of many hammers accidentally, because one after the other may be broken. It is accidental, therefore, that one particular hammer acts after the action of another; and likewise it is accidental to this particular man as generator to be generated by another man; for he generates as a man, and not as the son of another man. For all men generating hold one grade in efficient causes – viz. the grade of a particular generator. Hence it is not impossible for a man to be generated by man to infinity; but such a thing would be impossible if the generation of this man depended upon this man, and on an elementary body, and on the sun, and so on to infinity. (*Summa*, FP, Q46, A2, RObj 7)

Given that the causal structure is, upon the explanation scheme, one that is ordered teleologically, it follows directly that there can be no infinite regress of causes. Thus, it follows that a series of contingent causes always implies a necessary, self-existent being as the ultimate cause: without such a being no structure would be possible. Aquinas sees this clearly: the existence of God is 'shown from the unity of the world':

For all things that exist are seen to be ordered to each other since some serve others. But things that are diverse do not harmonize in the same order, unless they are ordered thereto by one. For many are reduced into one order by one better than by many: because one is the 'per se' cause of one, and many are only the accidental cause of one, inasmuch as they are in some way one. Since therefore what is first is most perfect, and is so 'per se' and not accidentally, it must be that the first which reduces all into one order should be only one. And this one is God. (*Summa*, FP, Q11, A3)

It is equally clear in the statement of the 'fifth way' which is 'taken from the governance of the world.'

We see that things which lack intelligence, such as natural bodies, act for an end, and this is evident from their acting always, or nearly always, in the same way, so as to obtain the best result. Hence it is plain that not fortuitously, but designedly, do they achieve their end. Now whatever lacks intelligence cannot move towards an end, unless it be directed by some being endowed with knowledge and intelligence; as the arrow is shot to its mark by the archer. Therefore some intelligent being exists by whom all natural things are directed to their end; and this being we call God. (*Summa*, FP, Q2, A3)

In fact, Aquinas states the 'argument a priori' more or less as it was presented in the early modern period. Consider his first proof for the existence of God, the 'argument from motion.' This begins with the premise that there are things that are in motion: 'It is certain, and evident to our senses, that in the world some things are in motion.' Things that are in motion in this sense are contingent and limited. For, something can be in motion only if it is in potentiality to some state that it *is not*. The change itself occurs as a consequence of something that *is*, that is, something that is actual in some way or other.

Now whatever is in motion is put in motion by another, for nothing can be in motion except it is in potentiality to that towards which it is in motion; whereas a thing moves inasmuch as it is in act. For motion is nothing else than the reduction of something from potentiality to actuality. But nothing can be reduced from potentiality to actuality, except by something in a state of actuality. Thus that which is actually hot, as fire, makes wood, which is potentially hot, to be actually hot, and thereby moves and changes it. Now it is not possible that the same thing should be at once in actuality and potentiality in the same respect, but only in different respects. For what is actually hot cannot simultaneously be potentially hot; but it is simultaneously potentially cold. It is therefore impossible that in the same respect and in the same way a thing should be both mover and moved, i.e. that it should move itself. Therefore, whatever is in motion must be put in motion by another. If that by which it is put in motion be itself put in motion, then this also must needs be put in motion by another, and that by another again. But this cannot go on to infinity, because then there would be no first mover, and, consequently, no other mover; seeing that subsequent movers move only inasmuch as they are put in motion by the first mover; as the staff moves only because it is put in motion by the hand. Therefore it is necessary to arrive at a first mover, put in motion by no other; and this everyone understands to be God. (*Summa*, FP, Q2, A3)

This first cause is a being which is self-existent and in which essence and exis-tent are not distinct. It is, in other words, a non-contingent or necessary being.

[W]hatever a thing has besides its essence must be caused either by the constituent principles of that essence (like a property that necessarily accompanies the species – as the faculty of laughing is proper to a man – and is caused by the constituent principles of the species), or by some exterior agent – as heat is caused in water by fire. Therefore, if the existence of a thing differs from its essence, this existence must be caused either by some exterior agent or by its essential principles. Now it is impossible for a thing's existence to be caused by its essential constituent principles, for nothing can be the suffi-cient cause of its own existence, if its existence is caused. Therefore that thing, whose existence differs from its essence, must have its existence caused by another. But this cannot be true of God; because we call God the first efficient cause. Therefore it is impossible that in God His existence should differ from His essence. (*Summa*, FP, Q3, A4)

This necessary being, moreover, is perfect, lacking no perfections. In agreement with Leibniz, Aquinas argues that the necessary self-existent being contains within Himself (Herself, Itself) all being, its own being and all other being.

[S]ince matter as such is merely potential, the first material principle must be simply potential, and thus most imperfect. Now God is the first principle, not material, but in the order of efficient cause, which must be most perfect. For just as matter, as such, is merely potential, an agent, as such, is in the state of actuality. Hence, the first active prin-ciple must needs be most actual, and therefore most perfect; for a thing is perfect in pro-portion to its state of actuality, because we call that perfect which lacks nothing of the mode of its perfection. (*Summa*, FP, Q4, A1)

This indicates the last point that must be made. There is no infinite regress. The doctrine of explanation precludes it, the doctrine that the causal structure must in the end be teleological. But if the regress is precluded by the pattern of teleological causation, then we need to specify the end for which the ultimate cause strives. What is the end that God aims at in His (Her, Its) creative activity? What is the ultimate end, the ultimate form, to be achieved by the causal power of the unlimited necessary being? Leibniz makes clear what this ultimate end is: it is a striving for the best. As he puts it, for contingent truths, the 'principle is fitness [convenance] or the choice of the best' ('Monadology,' § 46). But the best consists in the maximization of being, 'perfection being nothing but the magnitude of positive reality considered as such, setting aside the limits or bounds in the things which have it' (§ 41). The world is the best of all possible worlds, the one with a maximum of being. As for God, 'here, where there are no limits, that is, in God, perfection is absolutely infinite' (§ 41).

Other philosophers make the same point in a variety of ways. Descartes disparages the idea that we can know God's purposes. These, he says, are of no use in physics.

When dealing with natural things we will ... never derive any explanations from the purposes which God or nature may have had in view when creating them and we shall entirely banish from our philosophy the search for final causes. For we should not be so arrogant as to suppose that we can share in God's plans. We should, instead, consider him as the efficient cause of all things; and starting from the divine attributes which by God's will we have some knowledge of, we shall see, with the aid of our God-given natural light, what conclusions should be drawn concerning those effects which are apparent to our senses.[10]

What we do know, however, is that God creates being, things that are. We know, further, that the laws God lays down for things are such that all possibilities are actual. Thus, discussing the laws of motion for material things, Descartes tells us that these 'cause matter to assume successively all the forms it is capable of assuming ...'[11] We know, therefore, that God as an all powerful being will bring into being as much being as is possible: that is what it means to say that all possibilities are actual. To be sure, these beings are all limited, and therefore contingent. Thus, for example, extended being is limited in its not being thinking being. Yet within these limits that render extended being contingent, all possibilities are actual. There is a necessary structure of essences that defines contingent or limited being; but within these limitations, God's end – in which, of course, He (She, It) achieves – is the maximization of being. As Guéroult has expressed it, Descartes 'rejects any penetration into God's counsels ...'; upon the Cartesian view, 'God cannot direct his will on anything other than being ...'[12] The metaphysics of Descartes is in important respects different from that of Leibniz, but in this respect they are agreed: that at which God aims in Her (His, Its) creative act is the maximization of being. It has been said that Descartes's world is a 'world without ends.'[13] We have seen in Study One that there is an important sense in which this is true: Descartes strips the world of bodies of the active powers that the Aristotelians had located there. At the same time, however, the point is misleading, as we also saw: activity retains an essential place in the Cartesian scheme, it is just that it is now all located in the Deity. Nor do we know much about the purposes for which God is acting: the ends of the Deity are largely mysterious. However, at the same time we do know *something* about those ends, namely, that God makes the most perfect world that She (He, It) can produce. '[R]emembering God's infinite power and goodness,' he tells us, 'we must not be afraid of overestimating the greatness, beauty, and perfection of His works ...'[14] But each way of being is a perfection, so that this implies that the world has as much being as is possible. This is achieved by making all possible being actual. In knowing this about the ends of the Deity, we know very little about His (Her, Its) teleology. Yet, and this is the important point, that is all that we do need for the argument a priori to proceed.

The ultimate end, then, that defines the structure of the *per se* ordered series of causes is that of maximizing being. Indeed, what else would an unlimited being strive to do in striving to achieve not only its own being but also all other being?

This, then, is the argument a priori for God, the argument that purports to demonstrate His (Her, Its) existence. It is this argument that Hume criticizes, a criticism which is part of his more general criticism of the traditional doctrine of causation as involving powers and necessary connections.

In this case, however, there are special complications. In particular, the argument purports to take *existence* to be a property of things, a property that some things have contingently and other things have necessarily, or, rather, one thing has necessarily. This requires some attention to be given to the three concepts of *existence*, *necessity*, and *contingency*. If, therefore, we want to see exactly how Hume criticized this aspect of the old view of science as *scientia*, as capable of demonstration, then we have to attend to Hume's views in particular on these topics, and the extent to which they pick up themes from the contemporaries whom he was criticizing.

Specifically, we have to look at Hume's views on the concept of *existence*. This is, however, one of the more obscure parts of Hume's philosophy. Cummins has done a valuable service simply by trying to unravel some of the puzzles; it is still more valuable for shedding as much light as it does on the issues.[15] There are, none the less, problems with the interpretation that he develops, and I would like to bring out some of these, and, moreover, develop them in such a way as to try to illuminate both the standard argument for the existence of God as a necessary being and Hume's critique of that argument.

I will address three questions.

a) What, more precisely, is the (abstract) idea of existence?
b) What is the notion of 'necessity' as Hume would use it in connection with the notion that something exists 'necessarily,' or, contrarily, 'contingently'?
c) Does it follow from the fact that whatever thing we conceive we conceive as existent that every thing necessarily exists?

The second question is, of course, relevant to how Hume criticizes those who hold that it is possible to demonstrate the existence of a deity. Indeed, it is precisely because Hume wishes to criticize those who hold that contingent being implies necessary being that we must put Hume's account of necessary being in the context of the position of those whom he is criticizing. There is no point in attributing to Hume a position that simply ignores what his purported opponents were saying; otherwise, one would be making of Hume someone who simply missed his mark. That is always a possibility, of course, but one should begin by assuming that it is most unlikely.

The third question is also crucial. As we shall see, it would seem that Hume's views on the concept of existence entail that every being exists necessarily. In the context of a distinction between contingent being and necessary being it would be difficult, to say the least, if it turned out that every being is a necessary being. So we must pay careful attention to this issue in our reading of Hume and of his critique of the notion of a necessary being and of the possibility of demonstrating the existence of such a being.

Moreover, in order to answer these questions, we will have to say something about Hume's doctrine of abstract ideas. But on all these issues, Hume is often not so much inadequate as incomplete; and I shall suggest that a number of points can be illuminated by looking at the historical context that Hume himself could take for granted. In any case, we will need a good deal of that context if we are to gain an adequate understanding of the traditional argument for God as a necessary being. The upshot will be not that Hume's views are unproblematic, but only that they are perhaps less problematic than Cummins makes out, and at any rate adequate to the task which he set himself of criticizing the argument that the existence of a God could be demonstrated a priori.

II: Abstract Ideas

In many respects Hume's doctrine of ideas is fairly traditional. We think in terms of ideas. In particular, our capacity to think generally, and to use general terms, is a matter of having abstract ideas. In this, Hume follows the views of his predecessors such as Locke, Descartes, and the Port Royal logicians. Hume also adopts the principle that what is conceivable is possible. Again, he follows the same tradition of Locke, Descartes and the Port Royal logicians. Where Hume differs from his predecessors is in his account of what precisely an abstract idea is.

For Hume's predecessors, one forms an abstract idea by separation. Thus, for Locke, forming ideas by abstraction consists in 'separating them from all other ideas that accompany them in their real existence' (*Essay*, II.xii.1). This sort of separation is the mechanism by which 'the mind makes the particular ideas, received from particular objects, to become general' (II.xi.9). The Port Royal logicians make the same point: abstraction occurs with respect to 'choses ... composées' when one 'les considérant par parties, et comme par les diverses faces qu'elles peuvent recevoir.' In abstraction what is thus known is considered separately from the whole; either one considers 'les parties séparément' in the case of 'parties intégrantes,' or one 'peut séparer les choses en divers modes,' or, finally, 'quand une même chose ayant divers attributs, on pense à l'un sans penser à l'autre' (*La Logique*, I.v).[16] The former occurs when, for example, we separate in thought the parts of the body. This is the simplest form of abstraction. Berkeley has no problems with this notion: 'I will not deny I can abstract, if that may properly be called *abstraction*, which extends only to the conceiving separately such objects, as it is possible may really exist or be actually perceived asunder' (*Principles*, sec. 5).[17] Neither does Hume have problems with this notion; it is the basis of his – and Berkeley's – view that ideas are either simple or complex and that some of the complex ideas are a result of our combining simple parts in ways that are not found in reality, as, for example, in our idea of the New Jerusalem (T I.I.i, 2–3).[18]

It is the other two sorts of abstraction that are more problematic. Consider the first of these two forms of abstraction. We begin with simple ideas of sensation, such as white (*Essay*, II.iii.1). One forms the abstract idea of white by separating the property white from the other qualities with which it is conjoined in the white particulars, such as snow or milk (II.xxi.73), that are given in sense experience. The resulting idea is distinct from other ideas, such as blue or heat (II.xii.1). The example given by Nicole and Arnauld is that of the geometers who explore the idea of three-dimensional space by considering first length alone, then area, and finally volumes or solids (which is the same, for these Cartesians as 'corps,' body) (*La Logique*, I.v). This is one form of separation. By this means one forms abstract ideas of species. But whatever is red is coloured. Here we have not only the abstract idea of a species but also an abstract idea of a genus. The method of forming the abstract ideas of genera is rather different. The mind, 'so to make other yet more general ideas, that may comprehend dif-

ferent sorts [species] ... leaves out those qualities that distinguish them ...'
(*Essay*, III.vi.32). That is, one forms the abstract idea of a genus from the ideas
of several qualities by separating the genus from the species. Thus, one forms
the idea of being coloured by separating the property of being coloured from
such properties as white, red, and so on, which are, of course, themselves
abstract ideas. One forms the abstract idea of an equilateral triangle by separating this property from the particular equilateral triangles.

Que si je passe plus avant, et que ne m'arrêtant plus à cette égalité de lignes, je considère seulement que c'est une figure terminée par trois lignes droites, je me formerai
une idée qui peut représenter toutes sortes des triangles. Si ensuite, ne m'arrêtant point
au nombre des lignes, je considère seulement que c'est une surface plate, bornée par
des lignes droites, l'idée que je me formerai pourra représenter toutes les figures rectilignes, et ainsi je puis monter de degré en degré jusqu'à l'extension. (*La Logique*, I.v)

Thus, one forms the abstract idea of extension, or of being extended, by separating the property of extension from the property of triangle or circle or some
other figure, that is, from these abstract ideas. These abstract general or universal ideas signify many particulars by signifying properties in those particulars.
But these properties do not exist independently of those particulars. These ideas
do not, therefore, represent or signify things that can exist as such, independently of particulars. 'general and universal ... belong not to the real existence
of things ... their general nature being nothing but the capacity they are put into
by the understanding, of signifying or representing many particulars' (*Essay*,
III.iii.11). Or again,

Quoique toutes les choses qui existent soient singulières, néanmoins, par le moyen des
abstractions que nous venons d'expliquer, nous ne laissons pas d'avoir tous plusieurs
sorts d'idées, dont les unes ne nous représentent qu'une seule chose ... et les autres en
peuvent également représenter plusieurs, comme lorsque quelqu'un conçoit un triangle
sans y considérer autre chose, sinon que c'est une figure à trois lignes et à trois angles;
l'idée qu'il en a formée peut lui servir à concevoir tous les autres triangles. (*La
Logique*, I:vi)

In general, one forms the abstract idea of a specific property by separating
that specific property from the concrete particulars presented to one; and one
forms generic abstract ideas by separating the generic property from the
specific.

This view is challenged by Hume, using, as we have noted in Study One, an
argument first deployed by Berkeley in his *Principles* (Introduction, secs 8–10).

The three propositions

a) (Necessarily) whatever exists is particular,
b) Whatever is possible in thought is possible in reality,
c) Abstract ideas are formed by separating specific and generic properties from existing particulars,[19]

are inconsistent. Accepting (a) and (b) requires one to reject (c). As Hume puts it, "'tis a principle generally receiv'd in philosophy, that every thing in nature is individual, and that 'tis utterly absurd to suppose a triangle really existent, which has no precise proportion of sides and angles. If this therefore be absurd in *fact and reality*, it must also be absurd *in idea*; since nothing of which we can form a clear and distinct idea is absurd and impossible' (T I.I.vii, 19–20). Berkeley and Hume conclude: so much the worse for the traditional doctrine of abstract ideas. Again as Hume puts it, '[n]ow as 'tis impossible to form an idea of an object, that is possest of quantity and quality, and yet is possest of no precise degree of either; it follows, that there is an equal impossibility of forming an idea, that is not limited or confin'd in both these particulars. Abstract ideas are therefore in themselves individual, however they may be general in their representation' (ibid., 20).

Berkeley leaves it at that. But this leaves him without any account of how words come to be general, since the tradition understood this in terms of words being associated with abstract ideas. Hume goes beyond Berkeley in attempting to provide a positive account of how words come to be general, that is, in effect, a radically revised account of abstract ideas, one that is not subject to the criticism that demolishes the traditional doctrine. In particular, Hume allows that we do in fact have ideas – clearly, *abstract* ideas – of such things as *existence*: 'There is no impression nor idea of any kind, of which we have any consciousness or memory, that is not conceiv'd as existent; and 'tis evident, that from this consciousness the most perfect idea and assurance of *being* is deriv'd' (T I.I.vi, 66). In holding this position Hume follows Locke, who also holds that existence is an idea that is 'suggested to the understanding, by every object without, and every idea within' (*Essay*, II.vii.7). Before Locke there was Descartes, who holds that '[e]xistence is contained in the idea or concept of every single thing, since we cannot conceive of any thing except as existing.'[20] But Hume, unlike Locke and the Cartesians, does not hold that we form this idea of existence by separation. Rather, he proposes a different thesis concerning the formation of abstract ideas, an alternative account that does not face the difficulties that he and Berkeley raised against the doctrine of abstraction by separation.

The problem to be solved is how an idea, that is, an image, that is particular becomes abstract, or, in Locke's terms, how it acquires 'the capacity ... of signifying or representing many particulars' (*Essay*, III.iii.11). It cannot signify them by their all being present to the mind when the particular idea that represents them all is present, since 'the capacity of the mind be not infinite' (T 18). Hume's solution turns on arguing that the ideas and impressions that an abstract idea signifies 'are not really and in fact present to the mind, but only in power' (T 20). This capacity is one that is acquired or learned through a process of association. Thus, Hume's positive account of abstract ideas depends upon his associationist psychology.[21] Just as our notion of causation is understood in terms of association based on the relation of contiguity, so abstract ideas are understood in terms of association based on the relation of resemblance.[22] Ideas (images) and impressions that resemble each other in a certain respect, say in being red or in being coloured or in being extended, will through the mechanisms of association become so associated in thought that any one of them can introduce another member of the set. At the same time, a word (that is, a sound or mark) comes to be associated with ideas and impressions insofar as they are associated with one another via some resemblance relation. Thus, 'red' comes to be associated with all members of the resemblance class of red impressions and ideas, 'coloured' comes to be associated with all members of the resemblance class of coloured impressions and ideas. And so on. When we encounter an impression or contemplate an idea (image) which is a member of the some resemblance class, that idea or impression introduces the word – the general term – that refers indifferently to each and every member of the resemblance class.

When we have a resemblance among several objects, that often occur to us, we apply the same name to all of them, whatever differences we may observe in the degrees of quantity and quality, and whatever other differences may appear among them. After we have acquired a custom of this kind, the hearing of that name revives the idea of one of these objects, and makes the imagination conceive it with all its particular circumstances and proportions. But as the same work is suppos'd to have been frequently applied to other individuals, that are different in many respects from that idea, which is immediately present to the mind; the word not being able to revive the idea of all these individuals, only touches the soul, if I may be allow'd so to speak, and revives that custom, which we have acquir'd by surveying them. They are not really and in fact present to the mind, but only in power; nor do we draw them all out distinctly in the imagination, but keep ourselves in a readiness to survey any of them, as we may be prompted by a present design or necessity. The word raises up an individual idea, along with a certain custom; and that custom produces any other individual one, for which we may have occasion. (T 20–1)

The mechanisms of association provide the explanation of how a word becomes general. The abstract idea just *is* the members of the resemblance class *qua* habitually associated with each other, or, more accurately, the associated resemblance class insofar as the general term has become associated with it. The parts of the idea, those impressions and ideas that fall under it, are the members of that class – though to be sure when the idea is before the mind, instancing a use of the general term, those parts are present not actually but *only potentially or dispositionally*, ready to be recovered through association as the occasion permits or requires.

When one judges that, say, whatever is extended is coloured, then one will have before the mind two abstract ideas, one the idea of being extended and the other the idea of being coloured, and, as in the tradition, these will be joined in making the affirmative universal judgment. Since these abstract ideas as such are habits or dispositions, they are of course as such not present to consciousness; rather, what is before the mind are particular ideas that are parts of the two abstract ideas, that is, particulars that have, through associations based on resemblance, come to be connected in thought with the use of general terms. Thus, it will be two particular ideas that are joined in the judgment, one a particular idea associated with the general term 'coloured' and the other a particular idea associated with the general term 'extension.' Or, even more likely, it will be *one* particular idea that is before the mind. For, a particular idea that is in the resemblance class of coloured objects will also be a particular that is in the resemblance class of extended objects. That means that both abstract ideas can be represented in consciousness by one and the same particular idea, with that idea associated with the two general terms. The judgment that whatever is extended is coloured will not consist of a compound idea, made up of two particular ideas, but will, rather, consist of a single idea without distinct parts, this single idea representing in consciousness the two abstract ideas of being coloured and being extended.

Hume explicitly recognizes this last point when he argues that the traditional division of mental acts into *conception*, *judgment*, and *reasoning* is mistaken. Conception is supposed to consist in the survey of one or more ideas, judgment in the joining or separating of two ideas, and reasoning in the joining or separating of ideas by means of intermediary ideas. Hume points out:

For *first*, 'tis far from being true, that in every judgment, which we form, we unite two different ideas; since in that proposition, *God is*, or indeed any other which regards existence, the idea of existence is no distinct idea, which we unite with that of the object, and which is capable of forming a compound idea by the union. *Secondly*, As we can thus form a proposition, which contains only one idea, so we may exert our reason without employing more than two ideas, and without having recourse to a third to serve as a medium betwixt them. We infer a cause immediately from its effect; and this inference is not only a true species of reasoning, but the strongest of all others, and more convincing than when we interpose another idea to connect the two extremes. (T I.III.vii, 96n1)

The point about inference and reasoning is, of course, but a restatement of the standard Lockean critique, examined in Study Two, of the Aristotelian idea that all scientific inference is syllogistic. But for present purposes what is important is Hume's claim that we can similarly have a proposition or judgment 'which contains only one idea.' The example that Hume uses is instructive, too. This example is *God is*. Here we have the idea of *God*; since God is a particular, this must be a particular idea. In the proposition *God is* we join this particular idea to the abstract idea of *existence*. That abstract idea will be represented in consciousness by a particular idea. This particular idea will not be another, distinct, particular idea that we attempt to join to the idea of *God*. It will, rather, be the same idea of *God*. That is, the judgment that *God exists* will appear in consciousness as a single rather than a compound idea.[23]

This departure from the traditional doctrine about the nature of judgment does not reflect a rejection of the traditional doctrine that judgment consists of joining and separating abstract ideas, but rather is a consequence of Hume's rethinking of the nature of abstract ideas, and his construal of them as associational habits represented indifferently in consciousness by any member of a resemblance class of ideas.

Hume's radical departure from the tradition does not consist in his rejection of the doctrine of abstract ideas but rather in the account that he gives of those ideas. This account is based on two points: one, the existence of different relations of resemblance, and two, the mechanisms of association.[24]

III: The (Abstract) Idea of Existence

We can now ask what, according to Hume, is our idea, that is, our abstract idea, of existence.

Hume tells us that 'there is no impression nor idea of any kind, of which we have any consciousness or memory, that is not conceiv'd as existent' (T I.II.iv, 66). As we have seen, Hume here follows Locke, who holds that *existence* is an *idea* that is 'suggested to the understanding, by every object without, and every idea within. When ideas are in our minds, we consider them as being actually there, as well as we consider things to be actually without us; which is, that they exist, or have existence' (*Essay*, II.vii.7), and Descartes, who holds that 'Existence is contained in the idea or concept of every single thing, since we cannot conceive of anything except as existing.'[25] This is the thesis which, as Cummins points out ('Hume,' 65), Hume uses as an unstated premise in his argument that there is no distinct impression of existence. Locke unfortunately adds nothing to this bald statement. Equally unfortunately, this leaves real problems. These are what Hume aims to address in his discussion of existence.

The doctrine of Locke and Descartes is that abstraction is a result of separation. It follows that the abstract idea of existence is obtained by separating existence from some existent thing. But Locke also holds that what is possible in thought is possible in reality. So it would seem to be possible to separate in reality a thing from its existence. It would seem possible to have in reality things that do not exist! Equally, if there is a separable idea of existence, then one will be able to confer existence on all sorts of things for which we have no reason to think that they exist, as Berkeley saw. In his defence of idealism, Berkeley points out that the defender of material objects may retreat from the claim that the external cause of our sensations is a material substance to the weaker claim that it is an 'unknown *Somewhat*' devoid of positive characteristics of the sort we know, something that is 'inert, thoughtless, indivisible, immovable, unextended, existing in no place.' Berkeley replies that if this be the concept of matter, then it is no different from the concept of nothing, and can have no more explanatory power than the latter (*Principles*, sec. 80). Berkeley then goes on to consider the possibility that his opponent will claim that the concept of matter that he advances is not that of nothing, for 'in the foresaid definition is included what doth sufficiently distinguish it from nothing – the positive abstract idea of *quiddity, entity*, or *existence*' (*Principles*, sec. 81). If existence were an entity that non-separately accompanied all other entities that one experienced but that could be separated in thought, then the world could be populated with all sorts of unacceptable entities.[26] Berkeley thus recognizes the dangers of having a separable idea of existence![27]

Hume makes the same sort of point. Judgments to the effect that a thing exists cannot, he argues, consist of joining two separable ideas, one of the thing and one of existence. He offers two reasons, one concerning the nature of our idea of existence, and the second concerning the absurd consequences of supposing that judgments of existence are of this sort. He concludes that a judgment of existence therefore consists not of joining the idea of a thing to another idea but simply of affirming the thing, construing affirming as a species of (cognitive) mental act (T, Appendix, 623).

Hume's first argument is that '[w]e have no abstract idea of existence, distinguishable and separable from the idea of particular objects. 'Tis impossible, therefore, that this idea of existence can be annex'd to the idea of any object, or from the difference betwixt a simple conception and belief' (ibid.). Hume, to provide himself with the premise he needs, is here relying upon the argument that Cummins is addressing – of which more in a moment. Note, however, that Hume is not here denying that we have an abstract idea of existence, only that we have an abstract idea of existence that is separable from the idea of particular objects or things. It is this, I shall be suggesting, that is the conclusion Hume wishes to draw from the argument that is Cummins's concern.

For our immediate purposes, however, it is Hume's second argument that is relevant; it is this that parallels Berkeley's argument against a separable idea of existence, by appeal to the absurd consequences that would result if one supposed otherwise. 'The mind has the command over all its ideas, and can separate, unite, mix, and vary them as it pleases; so that if belief consisted in a new idea, annex'd to the conception, it wou'd be in a man's power to believe what he pleased' (T, Appendix, 623–4). One could believe whatever absurdities one wanted, including the entities to which Berkeley objects.

Berkeley does allow that the traditional defenders of abstract ideas will claim, as does Locke, to have such an idea of existence. But that idea, like all other abstract ideas, is subject to Berkeley's critique. 'I own, indeed, that those who pretend to the faculty of framing abstract general ideas do talk as if they had such an idea, which is, say they, the most abstract and general notion of all: that is to me the most incomprehensible of all others' (*Principles*, sec. 81). *Existence* is just not the *sort* of entity that one could run across separately or even separably existing in experience: 'how ready soever I may be to acknowledge the scantiness of my comprehension, with regard to the endless variety of spirits and ideas that may possibly exist, yet for any one to pretend to a *notion* of Entity or Existence, *abstracted* from *spirit* and *idea*, from perceived and being perceived, is, I suspect, a downright repugnancy and trifling with words' (sec. 81).

However, where Berkeley rejects abstract ideas, Hume does not. To the contrary, he has his own doctrine of abstract ideas, and is prepared to allow that there is indeed an abstract idea of existence. But he also wishes to avoid the sort of problem to which Berkeley directs our attention; that is the thrust of the second of the two arguments we have just noted. So the question for Hume is this: what sort of abstraction is involved in forming the idea of existence? Locke does not tell us. Is it the unproblematic kind of abstraction in which an integral part is separated from a whole? Or is it the other sort of abstraction in which a mode or genus purports to be separated from a thing? The argument that Cummins examines in detail aims to show that it is not the former sort of abstraction: it is 'not deriv'd from a distinct impression, conjoin'd with every perception or object of our thought ...' (T I.II.vi, 66).[28] It is, rather, the other sort of abstraction, that obtains when we form the abstract idea of a species apart from particular things, or a genus apart from a species. Only, of course, for Hume this sort of abstraction is not a matter of separation but one of association based on resemblance. For, the species is *not* separable from the particular nor the genus from the species. What we have, therefore, is an 'abstract idea of existence,' but one which is *not* 'distinguishable and separable from the idea of particular objects' (T, Appendix, 623). Thus, 'Whatever we conceive, we conceive to be existent' (T 67).

However, this does not solve all the problems. For if, as Hume concludes, 'Whatever we conceive, we conceive to be existent,' then it would seem, as Cummins points out ('Hume,' 70), that we must take Hume to hold that everything necessarily exists, which hardly fits with the customary, and textually well-supported, view that for Hume no matter of fact or existence can be logically true. In order to deal with this issue, it is necessary to go on to see more precisely what *is* the abstract idea of existence.

Begin again with Locke. The latter contrasts ideas that come by one sense, such as our ideas of colours or of sounds (*Essay*, II.iii.1), with those that come by several senses, such as extension (II.v), with those that come by reflection, such as perception or volition (II.vi.2), with, finally, those that come by both sensation and reflection (II.vii.1). Among the last are the ideas of pleasure and pain (II.vii.2) but also duration and, of present interest, *existence* (II.vii.1, 7). The point to be noticed is that, in spite of the Kantian injunction that we have appropriated that existence is not a predicate, here it is in fact treated as such, parallel to such ordinary predicates as colours, shapes or forms of thought. This is, of course, part of a long tradition. The Port Royal Logic takes a *thing* (*chose*) to be that which 'l'on conçoit comme subsistant par soi-même'; it is what 'l'on appelle autrement substance.' The *manner of a thing* (*manière de chose*) 'étant conçu dans la chose, et comme ne pouvant subsister sans elle, la détermine à être d'une certaine façon.' This manner of being can be understood as either a mode, an attribute, or a quality (*La Logique*, I.ii). The Logic does not elaborate on the distinction between mode, attribute, and quality, but we find the relevant point in Descartes's *Principles of Philosophy*:[29]

By *mode* ... we understand exactly the same as what is elsewhere meant by an *attribute* or *quality*. But we employ the term *mode* when we are thinking of a substance as being affected or modified; when the modification enables the substance to be designated as a substance of such and such a kind, we use the term *quality*; and finally, when we are simply thinking in a more general way of what is in a substance, we use the term *attribute*. Hence we do not, strictly speaking, say that there are modes or qualities in God, but simply attributes, since in the case of God, any variation is unintelligible. (*Principles*, 1.1.56)

In created things, the attributes are, as it were, the modes that do not vary: 'in the case of created things, that which always remains unmodified – for example existence or duration in a thing which exists and endures – should be called not a quality but an attribute' (ibid.). The Cartesians thus include existence among the modes and attributes of things. As for Locke, for the Cartesians existence is a predicate, alongside extension, which is the principal attribute of body, alongside thought, which is the principal attribute of mind, alongside particular shapes and motions, which are modes of bodies, and alongside imagination and volition, which are modes of minds (1:1.53).

Descartes speaks of thought and extension as having different modes; these are the species under the genus. 'There are various modes of thought such as understanding, imagination, memory, volition, and so on; and there are various modes of extension, such as all shapes, the positions of parts and the motions of the parts' (1.1.65). Locke speaks in the same way. Thus, volition and judging are 'modes of thinking' (*Essay*, II.xix.1, 2); and among the 'simple modes of space' is 'each different distance' as well as 'figure' and the 'vast number of different figures' (II.xii.4, 5). To these Locke adds the 'modes of sounds' and the 'modes of colours,' the various sensory modalities that the Cartesians tended to ignore (II.xviii). The modes of extension or of thinking are the modes of those things or substances which are extended or thinking. More generally, a thing falling under the genus will be modified by one of the species of that genus.

The modes are, in the language of Aristotle, those entities that are 'present in' a substance (*Categories*, 1a20–5).[30] A thing or substance is said to *be* according to what is present in it. Thus, if a thing has red present in it as a modification, then that thing *is* red, or if it is modified by a circular shape, then it *is* circular. The modifications of a thing constitute its *being*, that is, *the ways in which it is*. But there are restrictions on the way in which a thing is; it cannot be just anything at all. Thus, if a thing is of a certain genus but not another it will have modifications of the one sort and not of the other. As Aristotle put it, 'being falls immediately into genera' (*Metaphysics*, 1004a4–5). Within the specific metaphysics of Aristotle, different genera may share qualities – both Peter Rabbit and Socrates can be modified by whiteness; Locke holds the same view, though the two philosophers give different accounts of the essences or forms of things: for Aristotle they have a real basis in substances, whereas for Locke they are, so far as humans are concerned, nominal. For Descartes, in contrast, things which are of different genera cannot have modifications falling under a different genus: an extended thing has the modifications of extension but not those of thought, that is, of thinking things. In any case, however, species and genera indicate *ways of being*: in Aristotle's words, '"white" indicates quality and nothing further, but species and genus determine the quality with reference to a substance: they signify substance qualitatively determined' (*Cat.*, 3b18–20).

Now, the difference between a thing that is something and a thing that is nothing is, of course, that the former exists while the latter does not. What is to be noted is that what it means to say that something is nothing is that it has no being. 'Nothingness possesses no attributes or qualities,' says Descartes (*Principles*, 1.1.11). For Locke, 'Non-entity' is 'the absence of all being' (*Essay*, IV.x.3). Following this tradition, Samuel Clarke tells us in his Boyle Lectures that 'Nothing is That which has No Properties or Modes whatsoever. That is to say, 'tis That of which nothing can truly be affirmed, and of which Every thing can truly be denied' (*Discourse*, 1:16).[31] *Existence is the presence of being*, and *non-existence is the absence of being.*[32] To say that *x* exists is to say that *x has being*, that is, that

there is a (specific) quality that *x* has

or, equivalently,

there is a (specific) quality *f* such that *x* is *f*.

We have already seen that this is one of the basic points of the substance ontology.

But of course, to say that *x is extended* is to say that

there is a (specific) quality *f* such that *f* is a mode of extension and *x* is *f*

and to say that *x is a thinking thing* is to say that

there is a (specific) quality *f* such that *f* is a mode of thought and *x* is *f*.

Extension, thought, and so on, are the genera into which being 'immediately divides.' To predicate these of things is to assert that these things *are* in certain ways, that is, that there are qualities that are certain sorts present in those things. Similarly, to predicate being of things is to assert that these things *are*, though not that they are in any certain way; it is to assert, that is, that there are specific qualities present in those things while not asserting precisely what that specific sort is which those qualities are. *Being* or *existence* is thus the genus that can be predicated of all things, and includes all less comprehensive genera within it.

Thus, for the tradition within which Hume on this point is located, being or existence is quite reasonably taken, *pace* Kant,[33] to be a predicate: it is a genus parallel to such genera as extension or thought differing only in being that genus that transcends all others;[34] it is that genus which comprehends all things and includes within itself all lesser genera.

All things of which we are aware are qualified or modified; they all resemble each other in this respect. This resemblance relation is the basis of the association that constitutes the idea of existence.

We can conclude that Hume's argument to the conclusion, that 'Whatever we conceive, we conceive to be existent,' is part of his more general argument concerning abstract ideas. It is in fact, simply Hume's views about the latter applied to the case of existence.

But what about *things*? What is the idea of a thing? Well, as we have seen, the Port Royal Logic takes a *thing* (chose) to be that which 'l'on conçoit comme subsistant par soi-même'; it is what 'l'on appelle autrement substance' (*La Logique*, I.ii). A *thing* is thus something that has being, that is, exists, but which also has the further property of being independent: a thing is an independent existent. Hume follows Arnauld and Nicole in this: as he puts it, 'the definition of a substance is *something which may exist by itself*,' that is, separately from all other things (T I.IV.v, 233). But Hume holds that what is different or distinct is distinguishable and that what is distinguishable is separable (T I.I.vii, 18). So entities that are distinct are separable and therefore may exist by themselves; they are therefore substances, according to this definition. Now, Hume has also systematically argued that perceptions or impressions are independent of each other. He draws the appropriate conclusion. 'since all our perceptions are different from each other, and from every thing else in the universe, they are also distinct and separable, and may be consider'd as separately existent, and may exist separately, and have no need of any thing else to support their existence. They are, therefore, substances, as far as this definition explains a substance' (T I.IV.v, 233). The entities that satisfy this definition are impressions and images (ideas). The impressions are all qualified entities; we have no impression that does not have determinate or specific characteristics. The entities that are to be reckoned substances are therefore all qualified and therefore existent. They resemble each other in this respect. But, in addition, these entities also resemble each other in being independent or separable. It is this relation of resemblance that is the basis of the association that constitutes the idea of a thing.

Here, then, we have two abstract ideas. But note that every idea or impression that is a member of the resemblance class for the abstract idea of a thing is also a member of the resemblance class for the abstract idea of existence. Everything falling under the former idea also falls under the latter. More strongly, it would seem that the two are inseparable: 'Whatever we conceive, we conceive to be existent.' It follows that *necessarily* every thing exists.[35] Conversely, the idea of a *non-existent thing* is self-contradictory. Hume makes this point when he says that 'no two ideas are in themselves contrary, except those of existence and non-existence, which are plainly resembling, as implying both of them an idea of the object; tho' the latter excludes the object from all times and places, in which it is supposed not to exist' (T I.I.v, 15). The judgment that an object exists or is existent consists in joining the idea of an object to the abstract idea of existence (this judgment may well, as we saw above, consist of a single particular idea). In contrast, the judgment that an object is non-existent consists in separating the idea of an object from the abstract idea of existence. But the idea of an object will be represented in consciousness by a particular idea, and there is no particular idea that is not of some specific kind or other; a particular idea is itself an object with certain modifications. However, an object modified is an object that exists, since, as we have argued, to exist just is to be modified. So in the judgment of non-existence one is attempting to separate from the idea of existence an idea that is necessarily tied to the idea of existence. But that, as we said, is to try the impossible; it is self-contradictory.

But to say this is hardly yet to understand what Hume has in mind; it does seem odd that *necessarily* every thing exists. We still require some elaboration: what is the relevant sense of *necessity*?

IV: Necessity

According to Hume, 'no matter of fact is capable of being demonstrated' (T III.I.i, 463). This seems to conflict with the claim that we are examining that 'whatever we conceive, we conceive to be existent.' But the conflict is not so clear once we begin to separate the relevant notions of 'necessarily.' Descartes agreed with Hume that whatever we conceive, we conceive to be existent: in that sense, *necessarily* every thing exists. 'Existence is contained in the idea or concept of every single thing, since we cannot conceive of anything except as existing.' But then Descartes hastens to add that '[p]ossible or contingent existence is contained in the concept of a limited thing, whereas necessary and perfect existence is contained in the concept of a supremely perfect being.'[36] Thus, while necessarily every thing exists, not every thing is a necessary existent.

Where Hume disagrees with Descartes is in holding that there is one thing the existence of which can be demonstrated. This is what he is insisting upon when he asserts that 'no matter of fact is capable of being demonstrated.' To insist upon this is possible, while agreeing with Descartes that necessarily every thing exists in the sense that whatever thing we conceive we conceive it as existent.

To get at this point, let us return to the philosophers with whom Hume was disagreeing when he held that no matter of fact could be demonstrated, that is, philosophers like Descartes, Locke, and Samuel Clarke, who held that it is possible to demonstrate the existence of at least one being. Hume's case would be a silly one if his own notion of necessary existence did not fit the notion with which these philosophers were operating when they held that they could demonstration a priori the existence of a necessary being.

Thus, Clarke argues that ordinary things are not self-caused, but causally dependent upon certain preceding beings. But these latter are also causally dependent, and not self-caused. Such a chain of beings causally dependent upon other beings does not contain within itself something that could account for its own existence. If there is nothing that could account for its own existence, then

it was originally equally possible, that from Eternity there should never have existed a Succession of changeable and dependent Beings. Which being supposed; then, What is it that has from Eternity determined such a Succession of Beings to exist, rather than that from Eternity there should never have existed any thing at all? Necessity it was not; because it was equally possible, in this Supposition, that they should not have existed at all. Chance, is nothing but a mere Word, without any signification. And Other Being 'tis supposed there was none, to determine the Existence of these. Their Existence therefore was determined by Nothing; neither by any Necessity in the nature of the Things themselves, because 'tis supposed that none of them are Self-existent; nor by any other Being, because no other is supposed to Exist. That is to say; Of two equally possible things, (viz. whether any thing or nothing should from Eternity have existed,) the one determined rather than the other, absolutely by Nothing: Which is an express Contradiction. (*Discourse*, 1:13–14)

From this Clarke concludes that 'consequently ... there must on the Contrary, of Necessity have existed from Eternity, some One Immutable and Independent Being' (1:14).

Locke uses the same argument. We know by intuitive certainty, he tells us, that 'bare nothing can no more produce any real being, than it can be equal to two right angles.' He draws Clarke's conclusion: 'If therefore we know there is some real being, and that non-entity cannot produce any real being, it is an evident demonstration, that from eternity there must have been something; Since what was not from eternity, had a beginning; and what had a beginning, must be produced by something else' (*Essay*, IV.x.3).

The argument that Locke and Clarke here offer is that we cannot suppose that the chain of dependent beings is caused by nothing. For, as nothing has no properties, and therefore, in particular, no causal powers, it is contradictory to suppose that the chain is caused by nothing. It must therefore be caused by something. But, as Hume points out, this is to misrepresent the situation. What we have to suppose is *not* that

> this thing is caused by nothing

but *rather* that

(@)   there is no cause for this thing,

and *this* is *not* self-contradictory: 'when we exclude all causes we really do exclude them, and neither suppose nothing nor the object itself to be the causes of existence; and consequently can draw no argument from the absurdity of these suppositions to prove the absurdity of that exclusion' (T I.I.iii, 81). So much the worse for the argument of Locke and Clarke.

None the less, the conclusion that Locke and Clarke draw from their argument is that a *necessary being* exists. The argument has been from the 'Impossibility of every Thing's being dependent' (Clarke, *Discourse*, 1:15) to the existence of an 'Independent Being' (1:14). This independent being *exists necessarily*: 'That unchangeable and independent Being, which has existed from Eternity, without any external Cause of its Existence; must be Self-Existent, that is, Necessarily-existing' (1:14).

> For whatever Exists, must either have come into Being out of Nothing, absolutely without Cause; or it must have been produced by some External Cause; or it must be Self-Existent. Now to arise out of Nothing, absolutely without any Cause; has already been shown to be a plain Contradiction. To have been produced by some External Cause, cannot possibly be true of every thing; but something must have existed Eternally and Independently; As has likewise been shown already. It remains therefore, that That Being which has existed Independently from Eternity, must of Necessity be Self-existent. (1:14)

The point to be emphasized is that the notions of 'necessity' and, correspondingly, of 'contingency' are part of the context of *causal discourse*. The 'Necessarily-existing' is the 'Self-Existent' in the sense that it itself is the '*cause of its existence*,' and, indeed, more strongly, since the self-existent is eternal, *the necessarily-existent has the sort of causal power that creates its own being and can prevent any other cause from destroying that being. That is, the necessarily-existent has a certain special sort of causal power; and this causal power is such that it is a contradiction to suppose that a thing has it and yet does not exist.* In parallel, the contingently existent has the sort of causal power that can neither create its own being nor prevent all other causes from destroying that being. The contingently existent also has a certain sort of causal power, namely a power such that there is no contradiction in supposing that it has this power and yet does not exist. Thus, *the notions of necessity and contingency refer to different kinds of cause and causal power and not to different kinds of existence; those notions refer to a relation between a thing and its existence; but that relation is not merely that of a thing modified to its modes but rather causal.* Hence, to conceive of some thing as necessarily existent is to conceive it as existent and to conceive that existence as created by a sort of causal power the presence of which is logically incompatible with the non-existence of the thing; while to conceive of some thing as contingently existent is to conceive it as existent and to conceive that existence as created by a sort of causal power the presence of which is logically compatible with the non-existence of the thing.[37]

What we should note is that we have *two* notions involved in the idea of *necessary existence. One* is the notion of causal power. The *second* is the notion of a logical contradiction. The necessarily existent is a thing such that (1) its *causal power* is of a certain special sort, to wit, the sort of power in which (2) having that power is *logically incompatible* with that thing's not existing, that is, the thing is conceived as having a causal power as such that one cannot conceive a thing having that power as not existing. As Clarke puts it,

[the necessity of God's existence] must antecedently force it self upon us, whether we will or no, even when we are endeavouring to suppose that no such Being Exists. For Example: When we are indeavouring to suppose, that there is no Being in the Universe that exists Necessarily; we always find in our Minds ... some Ideas, as of Infinity and Eternity; which to remove, that is, to suppose that there is no Being, no Substance in the Universe, to which these Attributes or Modes of Existence are necessarily inherent, is a Contradiction in the very Terms. (*Discourse*, 1:15).

The same two notions, that of causal power and that of logical incompatibility, are, it is clear, also involved in parallel fashion in the idea of *contingent existence.* It is thus misleading to say, as Cummins does, that 'contingent connections are explicated in terms of separability, which amounts to nothing more than the possibility of existing apart' ('Hume,' 72).

Hume deals with the Locke–Clarke argument in Part IX of the *Dialogues on Natural Religion.* It is presented by Demea: 'What was it ... which determined something to exist rather than nothing, and bestowed being on a particular possibility, exclusive of the rest? *External causes*, there are supposed to be none.

Chance is a word without a meaning. Was it *nothing*? But that can never produce any thing. We must, therefore, have recourse to a necessarily existent Being, who carries the REASON of his existence in himself; and who cannot be supposed not to exist without an express contradiction' (*Dialogues*, 189). Again, as in Clarke, the causal context of the discourse is clear. The necessity of a necessary existent consists in its containing within itself the reason for its being, that is, given the context, the cause for its being, and continuing to be; it is a matter of the causal relation between the thing and its being.

It is Cleanthes who replies to this argument. 'Nothing is demonstrable, unless the contrary implies a contradiction. Nothing, that is distinctly conceivable, implies a contradiction. Whatever we conceive as existent, we can also conceive as non-existent. There is no Being, therefore, whose non-existence implies a contradiction. Consequently there is no Being, whose existence is demonstrable' (*Dialogues*, 189). In this context, when Hume asserts that 'whatever we conceive as existent, we can also conceive as non-existent,' it is clear that he is asserting that we can conceive no thing with causal powers that are such that they are logically incompatible with the non-existence of the thing. That this is Hume's point is clear from what he next goes on to suggest, namely, that there doesn't seem to be any reason why the material universe can't have the causal properties that logically exclude its non-existence: 'We dare not affirm that we know all the qualities of matter; and for aught we can determine, it may contain some qualities, which, were they known, would make its non-existence appear as great a contradiction as that twice two is five' (190). It is the *qualities* of things that make for their necessary existence, if there is such, that is, the *causal qualities*. But, of course, as Hume has just said, there is nothing in the concept of any thing, that is, no causal quality, such that having that quality logically excludes non-existence; the more general case against any necessary existence holds for a material necessary existent.

We would be able to demonstrate a priori the existence of a thing if we conceived its having a causal quality that logically excludes its non-existence. 'But it is evident, that this can never happen, while our faculties remain the same as at present' (*Dialogues*, 189). The ideas that we have are derived from our impressions, and, Hume is asserting, we have no impressions, nor, therefore, any ideas, of things with causal qualities that logically exclude the non-existence of those things. In other words, Hume's argument here is of a piece with his general case concerning causal relations, that they never imply any necessity.

Hume's point about causation, his notorious point, perhaps, is that (@)

there is no cause for this thing

is not self-contradictory but is in fact *quite conceivable*. Which is to say that he argues that the causal maxim, that whatever has a beginning must have a cause, is neither intuitive nor demonstratively certain. He reasons as follows: 'As all distinct ideas are separable from each other, and as the ideas of cause and effect are evidently distinct, 'twill be easy for us to conceive any object to be non-existent this moment and existent the next without conjoining to it the distinct idea of a cause or productive principle ... The actual separation of these objects is so far possible, that it implies no contradiction or absurdity' (T I.III.iii, 78–9). In the perspicuous notation of the later logical atomists, Hume is claiming that

$$(\exists y)[By \ \& \ \sim(\exists x)Cxy]$$

is not a contradiction.[38] Here, *By* means '*y* has a beginning' and *Cxy* means '*x* causes *y*.' At this point in his argument, Hume does not have to tell us more about what '*x* causes *y*' means. (The passage quoted is from T I.III.iii, and he does not tell us what '*x* causes *y*' means until I.III.xiv) All that he needs to know about the meanings in order for the argument to go through is that the concept of *beginning* or *to be after not having been* does not imply the concept of *causing* – which, surely, it does not.[39] As Hume says, the concepts, or, in his terms, the ideas, are distinct, that is, logically distinct, and so separable.

V: Separating Events and Their Causes

The problem, of course, is that it is precisely on this point that Hume's opponents dispute his case. Leibniz, Clarke, and company all argue that the principle of sufficient reason, if it is not true by the law of contradiction alone, then at least is true, and a priori true at that – understanding this principle to mean that events are to be explained by unifying them into a causal structure, a set of causal relations, of the sort that implies that a necessary being exists.

Hume replies to this sort of claim by arguing that such a notion of a causal structure – this notion that is required if we are to be able to think of or even conceive of a necessary being – is beyond our ideas. He considers the point that the defender of the argument a priori holds that the starting place of the argument is the contingency of ordinary, material things. The starting point is the recognition that '[a]ny particle of matter may be conceived to be annihilated; and any form may be conceived to be altered. Such an annihilation or alteration, therefore, is not impossible.' Such a destruction is supposed not to be conceivable with regard to the deity, who is a necessary being. That, at least, is the argument. However, Hume, or, in the context, what is the same, Cleanthes, continues: 'it seems a great partiality not to perceive, that the same argument extends equally to the Deity, so far as we have any conception of him; and that the mind can at least imagine him to be non-existent, or his attributes to be altered. It must be some unknown, inconceivable qualities, which can make his non-existence appear impossible, or his attributes unalterable: and no reason can be assigned, why these qualities may not belong to matter. As they are altogether unknown and inconceivable, they can never be proved incompatible with it' (*Dialogues*, 190). We have no idea of a necessary connection and no idea of a unifying power that requires us to believe that it is impossible that a deity under the standard description not be in some way or other that it is standardly conceived to be. In fact, our idea of the powers that guarantee the existence of a deity is sufficiently obscure that there seems no reason why it might not be something that is had by matter – to be sure, not matter considered as mere potentiality for being as it was thought of by the scholastics, but matter as it is taken to be by the new science, something massy, hard, and apparently inert. As Hume puts it, 'why may not the material universe be the necessarily existent being, according to this pretended explication of necessity? We dare not affirm that we know all the qualities of matter; and for aught we can determine, it may contain some qualities, which, were they known, would make its non-existence appear as great a contradiction as that twice two is five' (190). The only argument used to prove otherwise is the argument from the contingency of matter. But at best this proves there is a non-contingent being; it does not prove that this being is somehow non-material. For, it could do the latter only if it were self-evident that matter cannot be the active cause of its own being. But as Locke argued, and Stillingfleet was unwilling to concede, this is certainly not self-evident.

Moreover, the defenders of the argument a priori themselves admit that the concept is, let us say, a difficult one. It is the concept of a being that is *limitless*. It has, in other words, within itself all being; there is no being which it does not have, no property which is to be denied of it.[40] This would seem to mean that extension, for example, a positive property, is to be predicated of the deity; after all, to deny it of Him (Her, It) would be to limit Him (Her, It). But this cannot be, for that would be to give to the deity the form of extension which is the, or an, essential attribute of body. So it would seem we are committed to the Spinozistic view that God is material. Descartes drew a distinction. It does not matter that I do not grasp the infinite, or that there are countless additional attributes of God which I cannot in any way grasp, and perhaps cannot even reach in my thought; for it is in the nature of the infinite not to be grasped by a finite being like myself. With regard to our thought of God, of the necessarily existent infinite being, '[i]t is,' he says, 'enough that I understand the infinite, and that I judge that all the attributes which I clearly perceive and know to imply some perfection – and perhaps countless others of which I am ignorant – are present in God either formally or eminently. This is enough to make the idea that I have of God the truest and most clear and distinct of all my ideas' (*Meditations*, in Cottingham, 2:31). The distinction is that between a property being present in a substance literally or formally, on the one hand, and not literally but eminently on the other. Leibniz makes the same distinction. In his 'Principles of Nature and Grace, Based on Reason,' he tells us that

[t]his simple primitive substance must eminently include the perfections contained in the derivative substances which are its effects. Thus it will have perfect power, knowledge, and will, that is, it will have omnipotence, omniscience, and supreme goodness. And since justice, taken very generally, is nothing other than goodness in conformity with wisdom, there must also be supreme justice in God. The reason that made things exist through him, makes them still depend on him while they exist and bring about their effects; and they continually receive from him that which causes them to have any perfection at all. But the imperfection that remains in them comes from the essential and original limitation of created things. ('Principles,' § 9)

Here, then, we have another form of predication. There is being, on the one hand, and eminent being on the other. But what exactly is the latter? Certainly, it enables us to say that the deity contains all properties without falling into the necessity of literally attributing all those properties to the deity. It enables us to say that the deity is both extended and spiritual but also that He (She, It) is after all not extended. It also enables us to say that the deity is both square and round, that is, contains both these properties eminently. That is important, of course, since they cannot both be contained formally in either the deity or anything else, at least they cannot with consistency both be said to be (formally) in any substance. All this is convenient – it enables us to hold consistently that there is an unlimited being for which there is no positive property that it lacks – but it hardly tells us what this new form of being is. In fact, one can only say that the notion is obscure. And to explain the nature of the unlimited being in terms of it is to say that the notion of an unlimited being is equally obscure.

In short, the idea of a power is sufficiently obscure that it is equally possible to suppose that the material universe has all the power it needs to be itself a necessary existent and to suppose that the deity as ordinarily conceived is not a necessary existent.

Conversely, again as Hume argues, the ordinary concept of cause does *not* support the claim that the world is structured by causal powers of the sort that imply the existence of a necessary being. Thus, for example, 'in tracing an eternal succession of objects, it seems absurd to enquire for a general cause or first author. How can any thing, that exists from eternity, have a cause, since that relation implies a priority in time, and a beginning of existence?' (*Dialogues*, 190).

More importantly, the concept of explanation that is tied up with our ordinary concept of cause, that concept for which we have an idea adequate enough to allow us to get on in the world, does not require us to make the sort of inferences that the defenders of the argument a priori claim that we are forced to make. Thus,

[i]n ... a chain ... or succession of objects, each part is caused by that which preceded it, and causes that which succeeds it. Where then is the difficulty? But the WHOLE, you say, wants a cause. I answer, that the uniting of these parts into a whole, like the uniting of several distinct countries into one kingdom, or several distinct members into one body, is performed merely by an arbitrary act of the mind, and has no influence on the nature of things. Did I shew you the particular causes of each individual in a collection of twenty particles of matter, I should think it very unreasonable, should you afterwards ask me, what was the cause of the whole twenty. This is sufficiently explained in explaining the cause of the parts. (*Dialogues*, 190)

The defenders of the argument a priori hold that we are forced to seek a cause for each part of a series of contingent causes, and also a cause for the series as a whole. In fact, if our concept of cause is as they claim, then such an inference is indeed forced upon us, as we have seen. However, if we are thinking about explaining things in terms of our ordinary concept of cause, then there is no basis to the claim that in order to explain an event we must go beyond assigning in a series a set of antecedent causes and also assign a cause to the whole series of causes. Hume thus drives a wedge between our ordinary concept of cause and the concept that is needed if the causal argument is to lead to the conclusion that a deity as self-existent and necessary does exist.

Philo (= Hume) has earlier explained both the nature of our ordinary causal reasonings and the gap between, on the one hand, the conception of causation that is therein involved, and, on the other hand, the very different concept, far distant from this one, that is needed if we are to apply it to the case of a deity. Here is the distance: 'You might cry out sceptic and railler, as much as you pleased: but having found, in so many other subjects much more familiar, the imperfections and even contradictions of human reason, I never should expect any success from its feeble conjectures, in a subject so sublime, and so remote from the sphere of our observation' (*Dialogues*, 149). Here is our ordinary and unproblematic concept of causal inference: 'When two species of objects have always been observed to be conjoined together, I can infer, by custom, the existence of one wherever I see the existence of the other; and this I call an argument from experience' (ibid.). Here is the contrast between how we ordinarily use the concept and how we are required to use the concept of cause by the defenders of the argument a priori: 'But how this argument can have place, where the objects, as in the present case, are single, individual, without parallel, or specific resemblance, may be difficult to explain' (ibid.).

It is, indeed, a point that he, Philo (= Hume) repeats. Our concepts of cause, insofar as they are useful to us, are derived from experience. So long as we stay with this concept of cause we have no problems. But as soon as we start to extend the concept to very different cases, our concept becomes more and more distant from the experience in which it is rooted. It becomes to that extent less and less something that permits us to infer causes with any confidence.

That all inferences, Cleanthes, concerning fact, are founded on experience; and that all experimental reasonings are founded on the supposition that similar causes prove similar effects, and similar effects similar causes; I shall not at present much dispute with you. But observe, I entreat you, with what extreme caution all just reasoners proceed in the transferring of experiments to similar cases. Unless the cases be exactly similar, they repose no perfect confidence in applying their past observation to any particular phenomenon. Every alteration of circumstances occasions a doubt concerning the event; and it requires new experiments to prove certainly, that the new circumstances are of no moment or importance. A change in bulk, situation, arrangement, age, disposition of the air, or surrounding bodies; any of these particulars may be attended with the most unexpected consequences: and unless the objects be quite familiar to us, it is the highest temerity to expect with assurance, after any of these changes, an event similar to that which before fell under our observation. The slow and deliberate steps of philosophers here, if any where, are distinguished from the precipitate march of the vulgar, who, hurried on by the smallest similitude, are incapable of all discernment or consideration. (*Dialogues*, 147)

Nor, clearly, are the vulgar the only ones who are guilty: so are the theologians and the other defenders of the argument a priori.

Given the task of the *Dialogues*, we do not find the careful argument of the *Treatise*. The *Dialogues* aim instead to ease us almost imperceptibly into the view of the *Treatise* without, however, stating the latter in a bold and dramatic way, one that would warn the theologian prior to his being caught that there is a trap waiting for him. But we do find a clear view in the *Dialogues*. Our ordinary concept of cause derives from experience. This concept of cause will not support the inferences that are needed for the argument a priori for the supposed demonstrative argument for the existence of a self-existent and necessary God.

As for the concept that is needed to support such inferences, this is obscure and distant from our ordinary concept, far too semantically imperfect to support the inferences that are supposed to hang upon it. The concept is either meaningful and does not support the inferences required or it is basically meaningless and does not support the inferences required. In either case, it does not support the inferences required. Those inferences are made only because the two concepts of cause are confused with each other. The argument a priori requires the principle of sufficient reason, where a sufficient reason is a special sort of cause. The argument is basically confused, resting on a meaningless concept.

Such is the gentle argument of the *Dialogues*. But it is the same argument as that of the *Treatise*. In the latter, however, it is boldly and directly stated that the concept of cause as active power is simply meaningless, or, at least, either meaningless or self-contradictory.

In the *Treatise*, we find the basic structure of the argument a priori once again restated.

Matter, say they, is in itself entirely unactive and deprived of any power by which it may produce, or continue, or communicate motion: but since these effects are evident to our senses, and since the power that produces them must be placed somewhere, it must lie in the Deity, or that Divine Being who contains in his nature all excellency and perfection. It is the Deity, therefore, who is the prime mover of the universe, and who not only first created matter, and gave it its original impulse, but likewise, by a continued exertion of omnipotence, supports its existence, and successively bestows on it all those motions, and configurations, and qualities, with which it is endowed. (T 159)

But it requires a concept of cause that in fact has no place in any coherent view of the universe, one not based on meaningless ideas. Our ideas insofar as they are meaningful are derived from our experience of things. All are agreed that we do not experience among the ordinary things of the world the exercise of power in effecting necessary connections. 'We have established it as a principle, that as all ideas are derived from impressions, or some precedent perceptions, it is impossible we can have any idea of power and efficacy, unless some instances can be produced, wherein this power is *perceived* to exert itself' (T, 160). Those who keep the necessary connections, the powers that are needed for the argument a priori, invent sources for these ideas other than experience. 'Now, as these instances can never be discovered in body, the Cartesians, proceeding upon their principle of innate ideas, have had recourse to a Supreme Spirit or Deity, whom they consider as the only active being in the universe, and as the immediate cause of every alteration in matter.' (ibid.). However, since Locke, it is agreed that the alternative sources for our ideas do not exist: all our ideas, to be meaningful, must derive from experience. In that case, the concept of a deity is not going to be much good for us. We will have no idea of an active power or necessary cause, nor, *a fortiori*, one that we can use to describe the power of the deity. Or if we do use such a concept to describe the deity, then we have to admit that it is a concept that is empirically meaningless, without sense.

[T]he principle of innate ideas being allowed to be false, it follows, that the supposition of a Deity can serve us in no stead, in accounting for that idea of agency, which we search for in vain in all the objects which are presented to our senses, or which we are internally conscious of in our own minds. For if every idea be derived from an impression, the idea of a Deity proceeds from the same origin; and if no impression, either of sensation or reflection, implies any force or efficacy, it is equally impossible to discover or even imagine any such active principle in the Deity. Since these philosophers, therefore, have concluded that matter cannot be endowed with any efficacious principle, because it is impossible to discover in it such a principle, the same course of reasoning should determine them to exclude it from the Supreme Being. Or, if they esteem that opinion absurd and impious, as it really is, I shall tell them how they may avoid it; and that is, by concluding from the very first, that they have no adequate idea of power or efficacy in any object; since neither in body nor spirit, neither in superior nor inferior natures are they able to discover one single instance of it. (T 160).

If our idea of power or cause as necessary connection is indeed obscure, then it will be able to do little philosophical work for one; it is a useless concept so far as proving anything about the world, and in particular it will do little to support the proof for the existence of a deity. 'If we have really no idea of a power or efficacy in any object, or of any real connexion betwixt causes and effects, it will be to little purpose to prove that an efficacy is necessary in all operations.' In those cases we simply do not understand what we are saying or talking about: our discourse, while not ungrammatical, in fact lacks content: it pretends to be meaningful but is instead about nothing. 'We do not understand our own meaning in talking so, but ignorantly confound ideas which are entirely distinct from each other' (T 168). Our idea of necessity arises not from any objective impression but from a subjective tie: 'after a frequent repetition, I find that upon the appearance of one of the objects the mind is determined by custom to consider its usual attendant, and to consider it in a stronger light upon account of its relation to the first object. It is this impression, then, or determination, which affords me the idea of necessity' (T 166). What we do is, as it were, to spread this subjective feeling upon the objective entities that are cause and effect, and suppose that there is an objective necessity and objective power. Our idea of power as an objective necessary connection is, in fact, a confused idea, one that arises from a fusion of the idea of cause as constant conjunction and the idea of cause as a determination of the mind to infer the cause from the effect. Ignorantly confounding ideas 'is the case when we transfer the determination of the thought to external objects, and suppose any real intelligible connexion betwixt them; that being a quality which can only belong to the mind that considers them' (T 168). Our whole discourse about objective powers and objective necessary connections is thus either meaningless, because the ideas derive from no objective impression, or contradictory, attributing what is subjective to what we acknowledge to be not subjective.

It follows, of course, as the *Dialogues* emphasize, that if we do use this obscure, essentially meaningless notion of cause, then it is so weak a reed to lean upon that we can guarantee neither that God Him- (Her-, It-)self is self-existent nor that the material universe itself is not a self-existent entity.

It is, then, Hume's general critique of causal discourse that shows the argument a priori for the existence of a necessary self-existent being to be fundamentally unsound, presupposing for its plausibility a concept of causation that is simply meaningless. On the ordinary, meaningful concept of causation, it is clear that the concept of *beginning* or *to be after not having been* does not imply the concept of *causing*; these two abstract ideas are separable. After all, the abstract idea of *beginning* does not presuppose the abstract idea of *being preceded by something else*. For things to *resemble in respect to having a beginning* is for things to resemble each other in a way that is different from things to *resemble in respect to being preceded by something else*. But, since the abstract idea of *beginning* does not presuppose the abstract idea of *being preceded by something else*, neither, therefore, does it presuppose the abstract idea of *cause*, since '*x* causes *y*' implies that '*x* precedes *y*.' So long as these abstract ideas are separable, there is no idea that an event requires of necessity for its beginning that it be produced by an active power. Nor, therefore, do we have any (coherent, meaningful) idea of causation that will support the notion that events can be explained only by fitting them into a causal structure that requires for its whole existence the existence of a self-existent, necessary being.

The notion of a demonstrative science depends upon some such principle of causation as appears in the argument a priori for the existence of God, an unlimited or necessary being. This causal principle, or, as Leibniz called it, the principle of sufficient reason, was taken by all, or almost all, philosophers prior to Hume as simply self-evident. But in accepting that, these philosophers were accepting more; they were, in fact, accepting a whole framework of explanation, a framework which took for granted that in the end all causes are active powers and all causation is teleological. When Hume challenges the principle of causation so understood, that is, the principle of sufficient reason, then he challenges this whole framework. And when that framework goes, so does the whole notion of demonstrative science. Locke had started the process of demolishing that framework, but it was Hume who completed the task, and put in its place a new standard of scientific explanation, a new framework for the causal understanding of the world, one in which all explanation is conjectural and in which there is no place for the notion that one can demonstrate matters of fact. With Hume science has replaced *scientia*.

But – there is usually a 'but' – Hume's argument rests upon his doctrine of abstract ideas, and there are real problems with that doctrine. Perhaps, then, it is Hume's doctrine of abstract ideas that is at fault here, rather than the concept of cause. In order to see whether this is so, it is important that we look at some of the problems with Hume's doctrine. It is to this task that we must now turn. Perhaps not surprisingly, it will turn out that whatever problems there are with Hume's account of abstract ideas, those problems do not affect the basic correctness of his account of causal inference nor his critique of the traditional notion of cause that alone provides the plausibility of the argument a priori for the existence of a necessary being, commonly known as God.

VI: Conceiving Things as Existent

'The idea of existence,' Hume tells us, 'is the very same with the idea of what we conceive to be existent. To reflect on any thing simply, and to reflect upon it as existent, are nothing different from each other. That idea, when conjoin'd with the idea of any object, makes no addition to it. Whatever we conceive, we conceive to be existent' (T I.II.vi, 66–7). If the argument that we have given so far is correct, then what this means is that whenever we conceive a thing we also conceive that there is a quality which is present in it.

It is evident that Hume takes this to be a logical truth. But this should not surprise the reader of the *Treatise*: he has already told us as much. Thus, earlier in the *Treatise* Hume has mentioned the notion that we sometimes have impressions which *apparently* have no precise or exact degree of quality or shape. But this is due not to the intrinsic nature but from our incapacity to view them clearly; to suppose otherwise is to fall into a contradiction: 'The confusions, in which impressions are sometimes involv'd, proceeds only from their faintness and unsteadiness, not from any capacity in the mind to receive any impression, which in its real existence has no particular degree nor proportion. That is a contradiction in terms; and even implies the flattest of all contradictions, *viz.* that 'tis possible for the same thing both to be and not to be' (T I.I.vii, 19).

Hume's 'to be' here, of course, ties in with the notions of existence and predication in the way that we have suggested: to assert that there is such an impression is to assert at once that it exists, that is, that there is a determinate mode present in it, and also that there is no determinate mode present in it. Which is, as Hume says, a contradiction.

But if it is indeed a logically necessary truth that '[w]hatever we conceive, we conceive to be existent,' then does it not follow, as Cummins suggests ('Hume,' 70), that existence is a necessary predicate of every thing? And how can that be reconciled with Hume's other claim that no matter of fact is necessary, that the existence of every thing is contingent?

The answers to these questions turn upon what one means by *existence* and by *necessity*. We have laid much of the groundwork; we can bring these points together by looking in a bit more detail at his argument concerning causation, quoted above. For these purposes, it is not the argument itself that concerns us (though it *is* a sound enough argument), but rather the nature of the ideas that argument asserts to be distinct and separable. These ideas have to be *abstract* ideas. Hume establishes the *logical possibility* of a thing that begins to exist but is not caused by *separating* the ideas of *beginning* and *caused*. These ideas are clearly abstract. Thus, to establish that a truth is not necessary it suffices to separate two abstract ideas. Conversely, to establish that a truth is necessary it suffices to show that two abstract ideas are inseparable.

Two abstract ideas will be separable just in case that the habit that associates particulars into the one idea is different from the habit that associates particulars into the other. Since these habits are based on relations of resemblance, the two ideas will be separable just in case that the relations of resemblance on which they are based are different, that is, *just in case that the modes or qualities in things that ground these resemblance relations are distinct*. Thus, the abstract ideas of red and white are separable because red and white as modes of things are distinct.

Let me say right now that there are real problems with Hume's notion of resemblance. Exact resemblance may not create too many difficulties, but inexact resemblance certainly does. The former is a two-term relation but the latter is polyadic. Hume notes that '*[b]lue* and *green* are different simple ideas, but are more resembling than *blue* and *scarlet*' (T, Appendix, 637). This shows that the relation to which he is here appealing is four-termed – x resembles y more than z resembles w – and, as Butchvarov has shown, such relations present insurmountable problems to the nominalist.[41]

Somewhat less problematic is the fact that resemblances themselves fall into classes. *Red* and *white* are both *colour*-resemblances, as opposed to *sound*-resemblances and *shape*-resemblances. The red and white resemblances resemble each other in being colour resemblances, and so on. Resemblances create abstract ideas; these ideas will be different, since the resemblances are distinct. But that distinctness among resemblances is compatible with those resemblances also resembling. These latter resemblances among resemblances will also be capable of generating abstract ideas. Specifically, the relation among resemblances of being colour-resemblances will generate the associational habit that constitutes the *generic* abstract idea of being coloured. This abstract idea will be different from both the abstract idea of red and the abstract idea of white, since it is based on a different resemblance relation. It is in fact based on the more general resemblance relation that consists, in effect, of the specific relations taken disjunctively:

x resembles y in being coloured

$= x$ resembles y in being white *or* x resembles y in being red *or*
...

The generic mode of resemblance is thus not separable from specific modes of resembling, nor, therefore, in Hume's (traditional) terminology, is it distinct. None the less, it is a more general relation of resemblance and, as a consequence, the individual ideas and impressions that are, via association, part of each of those specific abstract ideas will also be parts, via association, of the more generic abstract idea. The general term connected with the generic abstract idea will apply to more individuals than does the term connected with the specific abstract idea. None the less the generic idea will be inseparable from the specific ideas that fall under it, since, insofar as the generic mode of resemblance is not distinct from the specific modes, we will never be able to from the idea of an individual that falls under the generic idea but not under one of the specific ideas. Hence, the abstract idea of red is inseparable from the abstract idea of being coloured. But if the idea of F is inseparable from the idea of G, then F and G are necessarily connected: it is a necessary truth that all F are G. Hence it follows that, for example, the inference from being red to being coloured is necessary.

One can have other cases, of a more traditional sort. Thus, consider the abstract ideas of 'bachelor' and 'male.' The abstract idea of a bachelor is based on one resemblance relation, which itself is a simple conjunction of two resemblance relations, one of which is the relation that is the basis of the abstract idea of male. By virtue of the definition of the idea of bachelor, in other words, the relation that is the basis of the idea of bachelor is not distinct from the relation that is the basis of the idea of male. It follows that the abstract idea of bachelor is inseparable from the abstract idea of male, and therefore that the inference from being a bachelor to being a male is necessary.

This account of the necessity of the inference from being red to being coloured requires us to hold that we can have *abstract ideas* that are *different* but *inseparable*. Two abstract ideas will be *different* if the ways of resembling upon which they are based are different. The abstract ideas will be *separable* if the ways of resembling are distinct and therefore can form ideas (= images) of things that are the one but not the other. Two abstract ideas will be *inseparable* if the ways of resembling upon which they are based are inseparable, for example, by virtue of rules of definition or by one resemblance in question being a resemblance among resemblances. Inseparability is the criterion of *logical necessity*.

Here we are talking about ideas that are *different* yet *inseparable*. This would seem to conflict with the well-known Humean claim 'that all ideas, which are different, are separable' (T I.I.vii, 24). But this particular thesis applies to *things*, and abstract ideas are not things. The dictum applies to impressions and ideas in the sense of images, that is, particular ideas. But an abstract idea *is not a thing*; it is rather, a *habit* or *disposition*. It is thus compatible with Hume's dictum to hold that two abstract ideas can be different yet separable.

It is this sort of thing that is the basis for the Humean account of the distinction of reason. Briefly, where abstract ideas are different but inseparable, we have a distinction of reason. On this basis, Hume can hold that there is a distinction of reason between a thing and existence – these two ideas are different – while also holding that the idea of a thing and the idea of existence are inseparable.

Once again it is easy to fit this into the tradition in which Hume is situated. The distinction of reason (*distinctio rationis ratiocinatae*) is defined by Suarez as arising

from inadequate concepts of one and the same thing. Although the same object is apprehended in each concept, the whole reality contained in the object is not adequately represented, nor is its entire essence and objective notion exhausted, by either of them. This occurs frequently when we conceive an object in terms of its bearing on different things, or when we represent it in the way we conceive these different things. Hence such a distinction invariably has a foundation in fact, even though formally it will be said to spring from inadequate concepts of the same thing. (*Disputation* VII.19)[42]

Descartes describes the distinction of reason in much the same terms: the distinction of reason 'is a distinction between a substance and some attribute of that substance without which the substance is unintelligible; alternatively, it is a distinction between two such attributes of a single substance. Such a distinction is recognized by our inability to form a clear and distinct idea of the substance if we exclude from it the attribute in question, or, alternatively, by our inability to perceive clearly the idea of one of the two attributes if we separate it from the other' (*Principles*, 1.1.62). We have a distinction of reason with respect to a certain object, on the one hand, when the object is an inseparable unity of certain parts, and, on the other hand, when we can form two different ideas of the object based on those parts. Each of those ideas will be inadequate to the object in that neither represents the object in its total complexity, though each will be true 'as far as it goes.' The parts of the object are inseparable; they are therefore not real parts. None the less, there is in reality a foundation for this distinction. As Suarez puts it,

> things said to be thus distinct are real entities, or rather a single real entity conceived according to various aspects. The same is evident from the fact that reason does not produce the entities it thus distinguishes, but merely conceives things which are not distinct as though they were distinct. Hence it is not the objects distinguished but only the distinction itself that results from the reasoning.
>
> Nevertheless the mind does not err in thus distinguishing, because it does not assert that things conceived in this manner are distinct in fact, but simply and without composition – that is, without affirmation or denial – conceives them as distinct through precisive abstraction whereby it effects, as it were, this type of distinction. If later it predicates such reflection or composition of objects so conceived, it does not affirm simply, but only in a qualified manner, that is, according to the viewpoint of the mind, that they are distinct. (*Disputation* VII, 19–20)

The examples that Descartes gave of entities distinguishable only by reason are those between a thing and its attributes and between two attributes of a thing. Suarez gives different examples: 'Peter, man, animal, and other like predicates, as they really are in Peter, are not distinct in objective fact ... but only by the reasoned reason' (60).

It is this distinction of reason that Hume can make between the species and the genus. We have the idea of the genus applying to the species, but not wholly exhausting it. It is, therefore, an inadequate idea of the species. None the less, there is a foundation in reality for distinguishing the species from the genus. This foundation consists in the different resemblance relations which are the basis for the two abstract ideas. Of course, where Hume has different but inseparable abstract ideas, Descartes and Suarez have different and separable abstract ideas; they describe the distinction of reason in terms of 'separating attributes in thought.' But Suarez also holds that 'it is impossible for the same thing to be altogether disjoined and separated from itself in the real order; an open contradiction is involved, since no greater union can be thought of than total identity in nature. This is not merely union, but unity. As it is impossible for the indivisible to be divided, so it is impossible for what is wholly the same to be separated from itself' (*Disputation* VII, 41). And this, as Hume and Berkeley argue, can't be held jointly with the two further theses that these philosophers held, that what is possible in thought is possible in reality and that the abstract concepts are separable in thought. The result is the Humean notion that one can have two abstract ideas that are at once different but inseparable.

A similar account can be given for other truths traditionally reckoned to be necessary. Thus, for example, consider the truth that whatever is extended is coloured, the synthetic a priori truth that Hume, like Berkeley, uses to attack the Cartesian notion that ordinary objects are in reality substances whose essence is extension and which are wholly uncoloured. Hume argues that extension is, in Locke's terms, a 'modification of [a] simple idea.' It is 'made up only of that simple Idea of an Unite repeated'; 'simple modes' are '[r]epetitions of this kind joined together' (*Essay*, II.xiii.1). For Hume, '[t]he table before me is alone sufficient by its view to give me the idea of extension. This idea, then, is borrow'd from, and represents some impression, which this moment appears to the senses. But my senses convey to me only the impression of colour'd points, dispos'd in a certain manner' (T I.II.iii, 34). Thus, for Hume, the extended is constituted by coloured points, and whatever resembles anything in being extended must also resemble it in being coloured. It follows that being extended is not distinct from being coloured. On the other hand, the abstract ideas of being extended and being coloured are different ideas, since any extended thing is a multiplicity of coloured things; the idea of extension contains the idea of multiplicity that is absent from the idea of colour. Still, though the abstract ideas are different, they are not separable. It follows that it is a necessary truth that whatever is extended is coloured.

In this way Hume is able, at least if we don't probe his nominalism too deeply, to account for the necessity of what later have come to be called 'synthetic a priori' truths.

Among these truths is the proposition that whatever is a thing exists. The abstract idea of a thing is that of something qualified that is independent; the abstract idea of existence is that of something qualified. Clearly, the resemblances upon which the two abstract ideas are based are two. Yet, equally clearly, the two resemblances are inseparable. So the proposition that whatever is a thing exists is a necessary truth: 'Whatever we conceive, we conceive to be existent.'

But what about a particular thing, say Peter? Peter is a *whole*. This whole is a *unity.* Its parts are the various modifications that constitute it; Peter is *the individual that consists of the various qualities that modify him.* Suppose Peter is red and square. The abstract idea of red will apply to Peter, as will the abstract idea of square. Both these abstract ideas will contain the idea of Peter as parts; the associations based on the appropriate resemblance relations will ensure this. The abstract idea of red will thus be inseparable from the idea of Peter, though, of course, the two are different ideas. Similarly, the abstract idea of square will be different from the idea of Peter but inseparable from the latter. Neither of these will give an adequate idea of Peter, who is not just red but also square. There is, therefore, a distinction of reason between Peter and red and Peter and square. More generally, there will also be a distinction of reason between a thing like Peter and its existence. Hume's position here follows that of Suarez, who holds that 'existence and essence are not distinguished in the thing itself, even though the essence, conceived of abstractly and with precision, as it is in potency, be distinguished from actual existence ...' (*Disputation* XXXI.52).[43] Hume thus far follows the tradition.

Cummins suggests ('Hume,' 77) that for Hume there is no distinction of reason between a thing and its existence. He quotes Hume: 'Our foregoing reasoning concerning the *distinction* of ideas without any real *difference* will not here serve us in any stead. That kind of distinction is founded on the different resemblances, which the same simple idea may have to several different ideas. But no object can be presented resembling some object with respect to its existence, and different from others in the same particular; since every object, that is presented, must necessarily be existent' (T I.II.vi, 67). Cummins understands this to imply that there is no distinction of reason between a thing and existence, from which he concludes that for Hume there is no (abstract) idea of existence, or at least no idea of existence different from the idea of a thing. But the text is hardly conclusive; in fact it should be read in quite a different way. The question to be asked as we attempt to understand its import is this: what is the 'stead' that is supposed to be served by invoking the distinction of ideas (that is, reason)? The context makes clear: it will not serve in the 'stead' of 'opposing this,' to wit, the thesis that '[w]hatever we conceive, we conceive to be existent.' That is, it will not serve to establish that it is possible to conceive something that does not exist. What Hume is claiming is that we cannot appeal to the distinction of reason between a thing and existence to establish that we can form the idea of a thing that does not exist. That would require us to separate the idea of a thing from the idea of existence. But this we cannot do. Cummins suggests that this is because there is no distinction of reason between the idea of a thing and the idea of existence. But this is not what the passage says; it does not assert that there is no difference between the idea of a thing and the idea of existence but only that these ideas are inseparable. Normally where we draw a distinction of reason between the inseparable parts of an impression or idea – for example, between its colour and its shape (T I.I.vii, 25), say, the red colour and square shape present in the particular Peter – then we can form the idea of a thing which is red but not square, that is, which falls under the one abstract idea and not the other. One can do this because the two abstract ideas are separable. One cannot do this in the case of a thing and its existence, to form the idea of a thing that does not exist, because the abstract ideas, while different, are not separable.

However, if there is a distinction of reason between a thing and its existence, or between a thing and its modes, then it follows directly that a thing, Peter, is logically inseparable from his modes, and his existence. Necessarily Peter is red, and necessarily Peter exists. This is the problem that concerns Cummins, for it does seem to conflict with Hume's other doctrine that all matters of fact are contingent.

But it is essential to note that in Hume, as in his predecessors, the discussion of necessary versus contingent matters of fact takes place as we have seen in the context of *causal discourse*.[44] To say that a thing exists necessarily is to say that it has causal powers of a certain sort, while to say that it exists contingently is to say that it has causal powers of a different sort. Hume's argument is that no thing has causal powers of the former sort. *This argument stands.* Hence, the conclusion that Peter is necessarily red or necessarily exists is quite compatible with the claim that Hume wishes to defend against Descartes, Locke, and Clarke, that there are no necessary beings. One cannot appeal to Hume's claim that there are no necessary existents to argue that Hume cannot also say that a particular necessarily has the modes that it has, or necessarily exists.

Still, is there not something odd about asserting that Peter is necessarily red? Russell once argued that this was so, as part of a more lengthy argument against the idealists.[45] One can agree that his case against idealism is correct while not agreeing on this particular point. This argument concerned the claim of Bradley and company that a particular is a collection of qualities – that it is a whole of which its qualities are the parts. Thus, the particular Peter on this account is a whole consisting of the parts red and square. Various problems of individuation surround this account of particulars, and in part determine Russell's response. The main point, however, is that a particular is a whole of which its qualities are parts. This is the non-substantialist analysis of ordinary things that was first defended by Berkeley – inspired in part by Ramus's logic, as we argued in Study Three, above. However, if

Peter = {red, square}

where

(a)   { - - - , ...}

asserts that

(a′)   - - - is with ...,

then

>Peter is red

is the same as

>red ∈ Peter

>>= red ∈ {red, square}

>>= red is in the thing that is (=) {red, square}.

This is naturally taken to be an analytic truth or true by definition, and therefore necessary. But this is of little significance, about as significant ontologically as any proposition true by definition. What is important are the statements of the form (a) = (a′), the statements that represent the facts that certain qualities are together with each other, thereby forming the wholes that are the particulars. What is important is not that Peter is red but that red is with square to form Peter. If it is necessary that Peter is red, it is *not logically necessary* that red is with square. And, of course, it is not logically necessary that red be with square since the abstract idea of red is separable from the abstract idea of square.
   If it is not logically necessary that

>{red, square}

then in that sense if

>Peter = {red, square}

then neither is it logically necessary that

>Peter exists.

But have we not also said that a thing exists necessarily? Indeed we have, following Hume. However, in following Hume, the sense of 'exist' we have used when we assert that any thing is necessarily an existent thing is this: to say a thing exists is to say that there is quality present in it. Thus, what we are saying is that it is a necessary truth that

(E)   For any thing, there is a quality present in it.

This principle (E) is what has been called the principle of exemplification,[46] that there is no particular that is not qualified in some way or other. This cuts pretty deep, as deep as the anti-Platonist principle that there are no (simple) qualities that are not exemplified in at least one particular. Deep ontological truths of this sort are reasonably characterized as necessary. Having done so, however, we have to specify the sense of 'necessary.' Hume's account of abstract ideas provides us with such an account, and upon it (E) is indeed necessary. At the same time, it also implies, as we have just seen, that no particular, considered as a whole with its qualities as parts, necessarily exists, in the sense that it is not logically necessary that the qualities which, *qua* together, constitute a particular should exist together.

In short, when we say that it is necessary that Peter exist we are asserting about Peter an implication of (E), the principle of exemplification. When we say that it is not necessary that Peter exist we are asserting about Peter that there is nothing about the qualities that, *qua* together, constitute him that makes it logically necessary that they exist together. We have after all not landed ourselves, or Hume, in a contradiction.[47]

Conclusion

After making his illuminating remarks about Hume's argument concerning the inseparability of the idea of existence from the idea of a thing, Cummins ends by finding the doctrine still mysterious. What I have been arguing is that, if we locate Hume in the tradition of the substance philosophy, then we will be able to find the doctrine less mysterious than Cummins supposes. To say that, however, does not imply that it is free of all difficulties. In particular, I think that Hume's nominalism raises tremendous problems. On this view, all qualities of things are as particular as the things of which they are qualities. Generality is to be accounted for in terms of resemblance among these qualities – perfect particulars, as they have been called. As Russell has pointed out, this hardly solves the problem of universals, since resemblance itself must be taken to be a universal.[48] Moreover, there are real problems with respect to imperfect resemblance, as I have indicated. There are also problems with respect to resemblances among resemblances. Finally, we should note that Hume accepts the traditional doctrine that all relational statements have to be understood as, objectively, about non-relational facts.[49] This applies in particular to resemblance: this relation too, will have to be reduced to non-relational foundations. But, as this relation is a universal, the foundations into which it is analysed must be universals – contrary to the nominalist supposition that all qualities of things are as particular as the things of which they are qualities. In other words, while there is more sense in Hume than Cummins perhaps allows, there are also many problems that may in the end lead us to reject some of his basic ontological claims. Fortunately, none of these difficulties affects Hume's analysis of causation; but that is another story.[50]

Study Seven

Descartes's Defence of the Traditional Metaphysics[1]

Descartes opens his *Discourse on Method*[2] with a quotation from Montaigne's essays: 'Good sense is the best distributed thing in the world: for everyone thinks himself so well endowed with it that even those who are the hardest to please in everything else do not usually desire more of it than they possess' (*Discourse*, 1:111).[3] This reminds the reader that the edifice of knowledge, previously accepted as secure, had been challenged by Montaigne, and that the point of any new method should be to put an end to that challenge. As the discourse proceeds, Descartes clearly accepts the challenge to the senses that Montaigne raised in the 'Apology for Raimond Sebond.' He wants to discover an indubitable starting point for the edifice of knowledge. He agrees with Montaigne that the experiencing cannot justify the claim that the object of experience exists. The senses err, and we have dreams which we have subsequently come to discover to be false. Thus, no claim made by the senses that certain objects of sense experience exist should, upon the sense experience alone, be accepted as true. Moreover, people also make mistakes in reasoning. How, then, is even reason to be trusted? Descartes, however, goes on to propose that there is, for all that, a starting point for building the edifice of knowledge which escapes the critique of the senses. This is a truth that is beyond the critique of the senses because it makes a claim that has nothing to do with the objects of sense. Moreover, it is a truth that is so clear and distinct that it is indubitable. But that criterion of clarity and distinctness is obviously one which, if followed rigorously, will prevent one from falling into error. In this truth, then, one will find a guarantee against errors in reasoning. This truth that yields the criterion of truth and is an instance of that criterion is of course the well-known *cogito ergo sum*. From the fact that I think in any way – from the fact that I have experiences of sense objects, or from the fact that I reason – it clearly and indubitably follows that I exist. Whatever thinks must exist; it is clear that I think; therefore I exist.

Gassendi was later to raise the question as to the difference between *cogito ergo sum* and *ambulo ergo sum*: 'I think therefore I am' and 'I walk therefore I am.' Descartes's reply was that, as inferences, they were of a piece. The connection between walking and existing is as clear and distinct as the connection between thinking and existing. In both cases, the connection is an instance of the general rule that properties must be in substances. As Descartes puts it, 'aiming to destroy the arguments which led me to judge that the existence of material things should be doubted, you ask why, in that case, I walk on the earth, etc. This obviously begs the question. For you assume what had to be proved, namely that it is so certain that I walk on the earth that there can be no doubt of it' (*Objections*, 2:250). What is crucial is the premise, and this in fact makes for a vast difference between the two inferences: in the case of the *cogito*, the premise 'I think' is certain, whereas in the case of the *ambulo*, the premise 'I walk' is uncertain: its certainty has been cast into doubt by the critique of the senses. In the *Discourse* it is the certainty of this premise together with the certainty of the connection that is the starting point of the Cartesian reconstruction of the edifice of knowledge required by the challenge of Montaigne.

Descartes then introduces the proposition that 'if I exist then God exists.' This is another example of a self-evident or clear and distinct proposition. In the *Rules for the Direction of the Mind* Descartes had introduced it baldly as the proposition that 'I am, therefore God exists' (*Rules*, 1:45), the necessity and self-evidence of which is self-evident; but, by the time he came to write the *Discourse*, he clearly felt it to be important that some intermediate steps be inserted to make the self-evidence of this proposition more evident to his readers. In any case, from the knowledge that he now has that he himself exists, Descartes deduces the further knowledge that God exists. Then, from this further fact he can deduce the basic laws of the world. Geometry is a science that studies the connections among certain ideas, but there is nothing about these facts that guarantees that there are objects that fall under these ideas. It turns out, however, that the ideas of geometry are in this respect different from the idea of God. For, as we contemplate that idea, we see that it is such that it guarantees the existence of an object that falls under it. Moreover, we can see that this God is a good guy who would not deceive us, would not provide us with ideas that were, like those of geometry, clear and distinct and yet false. God is also all-powerful, and He is therefore a being who will guarantee that all our clear and distinct ideas, among them those of geometry, are true. We thus discover a genuine certainty rooted in the being of God for the basis laws that describe the real structure of the world, that is, the laws of geometry and also, upon Descartes's view, the laws of physics.

The strategy seems clear. The method of doubt clears the mind of all things that are less than certain – merely probable. The mind can then find ideas which have an internal structure – clarity and distinctness – which demands assent, an assent that is necessitated by that internal structure of the ideas and cannot be resisted. As long as the mind can focus on such an idea and such an idea alone, it will be guaranteed truth. We have before the mind two ideas, that of an *A* and that of a *B*. We infer the second from the first. In order to do this, we need to know two things. We need to know, first, the necessary connection between an *A* and a *B*. This connection guarantees that if something does indeed fall under the idea of *A* then there will necessarily be something falling under the idea of *B*. But we also have to know, second, that something does indeed fall under the idea of an *A*. We need to know not just that there truly is a necessary connection between the premise and the conclusion but also that the premise is true. In our terminology, we need to know not only that the argument is valid but that it is sound. As Descartes's reply to Gassendi makes clear, in the case of *cogito ergo sum*, we are certain that the premise is true in a way in which we are not certain that the premise of *ambulo ergo sum* is true.

This is a plausible strategy. But there are two problems with it. One of these Descartes did not see because he had every right to ignore it: it is not a fair criticism. The other criticism he did see and was later to take into account in the *Meditations*. The first of these criticisms is that the conclusion of the *cogito* attributes existence to something, and existence is not a predicate. This, I shall argue, is not a reasonable criticism. The second criticism is that the mere fact that we seem to have a criterion of truth in the clarity and distinctness of ideas is no guarantee that we do have a criterion.

Montaigne had already raised this issue in the 'Apology for Raimond Sebond.' Any attempt, he argued, to specify a criterion of knowledge, to specify some feature of our beliefs or perceptions that is a sure sign of truth, would have either to be circular or lead to an infinite and vicious regress. But this presupposes that one can, in fact, question the criterion, that one can raise a doubt about it and demand that it be proven. Descartes's original notion seems to have been that if the criterion is such that it cannot be doubted, then Montaigne's argument simply fails. In the *Discourse*, it seemed to Descartes that if he could find one truth the certainty of which derived from a property which made doubt impossible, then he would have replied adequately to Montaigne. This, of course, is not an adequate reply, even though the criterion that is proposed is not one that admitted any doubt. But to question that criterion is to try to doubt what cannot be doubted. It was Descartes's genius to see that it was, in fact, possible to raise a doubt about what cannot be doubted. More deeply, however, when he raised that doubt he was able to show that the seeds of scepticism were located at the very core of the traditional picture of the world. That ontology had such a structure that it called itself into question. The possibility of not being, and therefore of doubt, is at the very core of being. What had to be done to re-erect the traditional edifice was to develop a *metaphysical* argument that would show that being itself is such that it excludes the very possibility of its destroying itself.

It is clearly the aim of the first of Descartes's *Meditations* to take Montaigne's scepticism as deep as it could go. But for Montaigne the problem is primarily epistemological. What Descartes realizes is that the scepticism can be deepened, taken to a greater extreme than Montaigne had realized it could be taken, if we make the problem ontological, if we can locate a threat to the possibility of knowledge at the very core of ontology. But that would require not just an epistemological solution to the problem, but a metaphysical solution.

In the *Meditations*, Descartes deepens the scepticism of the *Discourse* by introducing the evil genius. The dream problem had been introduced in the *Discourse*, and this was part of the systematic questioning of all propositions that we affirm on the basis of sense experience. But the dream problem does not touch the necessary propositions of arithmetic and geometry; as Descartes points out in the First Meditation, these propositions retain their characteristic unchallengeability even in dreams. And, indeed, in the *Discourse* Descartes does not challenge them. But in the *Meditations* Descartes does challenge these propositions: he shows us how to doubt them. This is the hypothesis of a powerful being who sets out systematically to deceive us about the truth. Since this evil genius is powerful, it is within her power to make even self-evident propositions false. These propositions may well be such that it is impossible that we doubt them. None the less, from the hypothesis of the evil genius it follows that those propositions are false. So long as I have not established with certainty that the hypothesis of the evil genius is false, I must admit that it is possible that it is true, and therefore also possible that the propositions that I take to be self-evident are false. In this way, I can raise a doubt about those propositions that are such that it is impossible to doubt them.

In fact, the evil genius calls into question the *cogito* itself. Like the propositions of arithmetic and geometry, the *cogito* is clear and distinct and therefore self-evident. But all such self-evidence is challenged by the hypothesis of the evil genius. This includes the *cogito*. As Descartes puts it in the Third Meditation,

[b]ut what about when I was considering something very simple and straightforward in arithmetic or geometry, for example that two and three added together make five, and so on? Did I not see at least these things clearly enough to affirm their truth? Indeed, the only reason for my later judgement that they were open to doubt was that it

occurred to me that perhaps some God could have given me a nature such that I was deceived even in matters which seemed most evident. And whenever my preconceived belief in the supreme power of God comes to mind, I cannot but admit that it would be easy for him, if he so desired, to bring it about that I go wrong even in those matters which I think I see utterly clearly with my mind's eye. Yet when I turn to the things themselves which I think I perceive very clearly, I am so convinced by them that I spontaneously declare: let whoever can do so deceive me, he will never bring it about that I am nothing, so long as I continue to think I am something; or make it true at some future time that I have never existed, since it is now true that I exist; or bring it about that two and three added together are more or less than five, or anything of this kind in which I see a manifest contradiction. And since I have no cause to think that there is a deceiving God, and I do not yet even know for sure whether there is a God at all, any reason for doubt which depends simply on this supposition is a very slight and, so to speak, metaphysical one. (*Meditations*, 2:25)

Here the crucial sentence is this: 'I spontaneously declare: let whoever can do so deceive me, he will never bring it about that I am nothing, so long as I continue to think I am something; or make it true at some future time that I have never existed, since it is now true that I exist; or bring it about that two and three added together are more or less than five, or anything of this kind in which I see a manifest contradiction.' Descartes clearly includes the *cogito* alongside the propositions of arithmetic and geometry as those that are called into question by the hypothesis of the evil genius.

It immediately follows that the role of the *cogito* in the *Meditations* cannot possibly be the same as the role that it has in the *Discourse*. In the *Discourse* it has the role of a certain and certainly true premise for the start of a series of deductions. Because it is true and because all the deductions from it proceed on equally true premises, the *cogito* functions as a first truth and foundation upon which the edifice of knowledge is to be erected. But it cannot have that role in the *Meditations*, since it is now among the propositions that can be doubted. In the *Meditations* there must be a very different foundation for knowledge, a very different starting point for the rebuilding of the edifice of knowledge, the edifice that Montaigne had challenged and which, Descartes had discovered when he discovered the hypothesis of the evil genius, was an even more rotten structure than Montaigne had supposed. Many commentators take the role of the *cogito* in the *Meditations* to be identical to its role in the *Discourse*. Popkin is such a one.[4] Chevalier is another.[5] Kenny,[6] Williams,[7] Margaret Wilson,[8] and Guéroult[9] join the chorus.[10] Guéroult puts the interpretation this way: 'And so, according to the analytic order, one begins with the certain knowledge of my self, which, as the first truth for the subject (Cogito), is for me the first principle; in turn this first item of knowledge makes possible the knowledge of the existence of God, that is to say, the knowledge that the idea of perfection has an objective validity, which in turn makes possible, according to their respective limits, the knowledge of the objective validity of clear and distinct ideas, and then the knowledge of the objective validity of obscure and confused ideas.'[11]

Unfortunately for this common reading, Descartes clearly tells us in the Third Meditation what he does not allow in the *Discourse*, that the *cogito* can be doubted. The role of the *cogito* in the *Meditations* cannot, therefore, be that of providing a foundation for knowledge, a clear and certain truth from which we can proceed on to other truths. To be sure, the *cogito* in the *Meditations* is somehow a starting point; after all, it does appear in the Second Meditation. But it does not follow from the fact that it is a starting point that it is a first truth. It is this difference between the *Meditations* and the *Discourse* which we have to become clear upon.

But what, then, is the foundation of knowledge in the *Meditations*? Descartes tells us in the lines immediately following the passage just quoted. 'But in order to remove even this slight reason for doubt,' Descartes says 'as soon as the opportunity arises I must examine whether there is a God, and, if there is, whether he can be a deceiver. For if I do not know this, it seems that I can never be quite certain about anything else.' It is God's existence, not his own existence, which is Descartes's first truth in the *Meditations*. Descartes's task as he sees it is to provide the benefits of science, its fruits. These fruits depend, however, upon the trunk, which is physics, and the trunk in turn depends upon the roots. '[J]ust as it is not the roots or the trunk of a tree from which one gathers the fruit, but only the ends of the branches, so the principal benefit of philosophy depends on those parts of it which can only be learnt last of all' (*Principles of Philosophy*, 1:168). Descartes's task is to help produce the fruits by cultivating the roots of all science. The roots of that science are to be found in God: 'I have noticed certain laws which God has so established in nature, and of which he has implanted such notions in our minds, that after adequate reflection we cannot doubt that they are exactly observed in everything which exists or occurs in the world' (*Discourse*, 1:131). God is the root not only of the being of the world and its structure, but also of our capacity to know that being as so structured. Indeed, the very possibility of our knowing anything depends upon our knowing that our knowledge derives from the infinite being of God. '[I]f we did not know that everything real and true within us comes from a perfect and infinite being then, however clear and distinct our ideas were, we would have no reason to be sure that they had the perfection of being true' (1:129). The notion that there is a sense in which the existence of God is the central truth of the *Meditations* is not one that is surprising. But to say this is not yet to say too much: it is necessary to uncover the structure of the argument of the *Meditations* in order to see exactly what this means.

I: Cartesian Scepticism: The Ontological Roots of the Demonic Challenge

In the First Meditation, Descartes argues that our senses sometimes deceive us, and that we ought therefore to repose no trust in them. O.K. Bouwsma has argued in response[12] that if our senses deceive us they also provide the tools for correcting themselves. If I cannot distinguish a real flower from a paper flower on the basis of the colour and shape – that is, by sight – then I can tell the difference by means of another sense, touch: paper flowers and real flowers feel different. Descartes further argues that there is the possibility of an evil genius who so manipulates the world that everything of which I am aware by means of my senses is as a dream, false when considered as a true picture of the state of reality. Bouwsma further argues that such an evil genius is impossible. For the distinction between the real and the fake is something that is found *within* the world of experience: we characterize something as real – real flowers, say – only in contrast to something, also within experience, that we characterize as fake – paper flowers, for example. But if the difference between the real and the fake is found within experience, then it is not possible to characterize *all* of what we experience as fake. The hypothesis of the evil genius does not make any sense; it is meaningless given our use of the concepts of 'fake' and 'real.' Bouwsma in his own way characterizes the ordinary illusions of the world this way, contrasting them with the universal illusion supposedly raised by the demon: 'The dog over the water dropped his meaty bone for a picture in the water. Tom, however, dropped nothing at all.' The problem is words: 'The word "illusion" is a trap.' It seduces us into thinking that the evil genius could do what in fact cannot be done – cannot be done because the very idea, and so the problem, is unintelligible. '[The evil genius] was deceived in boasting that he could deceive, for his confidence in this is based upon an ignorance of the difference between our uses of the words, "heavens," "earth," "flowers," "milky," and "illusions" of these things, and his own uses of these words.'[13]

How, then, do we distinguish the fake from the real, the dreams and the fantasies from the solid and substantial? I look at the flower, and from what I see it seems to be that it is a flower. For to be a flower involves more than how it appears visually to me. Flowers have to feel a certain way to the touch. Some even have certain tastes. Smells are important, too. Certainly, flowers endure through time. So when, on the basis of what I see, I declare that the object is a flower, I am in effect making an inference from what is presented to me in my visual experience. The point about this particular thing, which appears visually to be a flower, is that when I touch it, I feel that it is paper. The inference from what appeared to me visually was mistaken. It was, after all, not a flower, but a fake, something made of paper but so made as to appear visually as a real flower appears. In the case of a second thing which also appears visually as a flower appears, it turns out that when I infer from the way it visually appears to how it would feel if I touched it, to what it will look like tomorrow, and to how it will smell when it sits in a vase in my living-room, all these inferences turn out to correct: it feels like a flower, it smells like a flower, it endures like a flower, or, rather, of course, it does not endure, in the way in which a real flower does not endure, where a fake paper flower does endure. I am deceived by the fake flower by virtue of making certain inferences that turn out to be mistaken. I am not deceived by the real flower, because the inferences that I make turn out to be correct. What is crucial are the inferences. A real flower is characterized by a certain pattern that holds among the aspects of the flower of which I am aware in various forms of sense experience. This structure that is objectively there in the case of real flowers I suppose to be there in the case of the paper flower on the basis, however, of only a restricted range of sensory experience; but in this I am mistaken. In the case of the real flower, there is a *coherence* among the various experiences which I have. This coherence *among* the experiences which I have reflects the objective structure of the world that I experience. But in the case of the fake flower, the coherence that I expect is not there. So I have a way in which we distinguish the real from the fake within experience: it is by means of a test of *coherence.*

The charge about our senses is that sometimes they deceive us. The conclusion is that we therefore cannot rely upon them as a source of knowledge. Bouwsma's reply is that we can, of course, discover that we are deceived in those cases where we are deceived. We do have a criterion for discovering such deception. This criterion is coherence. Therefore, the claim that they cannot provide us with knowledge is mistaken.

Interestingly enough, Descartes makes the same point. How do we distinguish dreams from realities? Coherence, he answers, in the Sixth Meditation:

Accordingly, I should not have any further fears about the falsity of what my senses tell me every day; on the contrary, the exaggerated doubts of the last few days should be dismissed as laughable. This applies especially to the principal reason for doubt, namely my inability to distinguish between being asleep and being awake. For I now notice that there is a vast difference between the two, in that dreams are never linked by memory with all the other actions of life as waking experiences are. If, while I am awake, anyone were suddenly to appear to me and then disappear immediately, as happens in sleep, so that I could not see where he had come from or where he had gone to, it would not be unreasonable for me to judge that he was a ghost, or a vision created in my brain, rather than a real man. But when I distinctly see where things come from and where and when they come to me, and when I can connect my perceptions of them with the whole of the rest of my life without a break, then I am quite certain that when I encounter these things I am not asleep but awake. And I ought not to have even the slightest doubt of their reality if, after calling upon all the senses as well as my memory and my intellect in order to check them, I receive no conflicting reports from any of these sources. (*Meditations*, 2:61)

Descartes accepts the same solution to the problem that our senses sometimes deceive us and to the dream problem as does Bouwsma. But Descartes clearly seems to think that there is a problem here where Bouwsma does not. The important difference is that Bouwsma argues that the demon hypothesis is meaningless, and therefore cannot challenge our ordinary practices in getting at the truth of the world. But equally clearly, Descartes disagrees: he would hardly have advanced the demon hypothesis if he thought it could be dismissed so easily. The question that Bouwsma never raises is why Descartes even thinks the hypothesis is meaningful. And this, of course, is the question that we must raise if we are to get a firm sense of what Descartes is about.

Now, the notion that coherence is the test of reality is one that many besides Bouwsma and Descartes have adopted. It is, for example, the same answer that Berkeley was later to give to the same problem. For Berkeley, all the things that are given to us in sensory experience are ideas in the mind. So, at least, he argues. But then we have to distinguish those experiences such as dreams which we declare to be not real and those such as experiences of cherry trees in the quadrangle of our college that we characterize as real. The things we experience are all part of the world of sensory experience, Berkeley declares. In that sense, both the experiences we have in dreams and the experiences we have when we look about the quad are equally existent as parts of the world given in experience. We distinguish the two parts of our experience on the basis of how they cohere with other parts of our experience, just as Descartes has distinguished dreams and reality. As Berkeley puts it in the *Principles*:

The ideas imprinted on the senses by the author of nature are called real things: and those excited in the imagination, being less regular, vivid, and constant, are more properly termed ideas, or images of things, which they copy and represent. But then our sensations, be they never so vivid and distinct, are nevertheless ideas, that is, they exist in the mind, or are perceived by it, as truly as the ideas of its own framing. The ideas of sense are allowed to have more reality in them, that is, to be more strong, orderly, and coherent than the creatures of the mind ...[14]

Earlier, Mersenne had made the same point.[15] We do, in fact, have the capacity to correct ourselves at points where the senses seem to deceive us.

Le sens commun est par dessus les sens exterieurs, auquel leurs operations abbouttissent commes les ligns de la circonference se terminent au centre afin qu'il juge de la difference sensible qui est entre la coleur, le son, l'odeur & les autres objects des sens externs; mais; l'entendement est par dessus les sens internes, & externes, c'est pourquoi il reçoit, & ramasse les operations des uns, & des autres, les unit en un point intelligible, & en juge en dernier ressort, de maniere qu'il reconoît, reprend, & corrige les fautes, & les abus qui pouroient estre arrivez par l'indisposition, ou l'incapacité des sens, comme nous voyons, lors qu'il conclud que le diametre du Soleil à plus d'un pied de Roy, encore qu'il ne paroisse pas si grand à l'oeil, & que la tour est quarrée, qui paroît ronde quand on la regarde de loin. [16]

On the basis of experience we can form 'maximes' which enable us to judge our experiences to be real or in need of correction. In fact, these maxims are often of sufficient certainty that we will use them as 'antecedens de nos demonstrations.'[17] Whatever the sceptic might in fact argue, experience does tell us that we can discover the truth and produce a reasonable science about the world that we experience.

There is, of course, a difference between the views of Berkeley and Mersenne. For Berkeley, the ideas of sense are given to us directly by God; there are no subsidiary secondary causes. For Mersenne, in contrast, there are such causes. It is just that we do not know them. With respect to these, Mersenne argues, the sceptic does have a point. He allows the sceptic the conclusion that 'nous ne sçauons rein de la substance, & du corps ... car nous ne pourrions rien appercevoir qui quelques accidens exterieur.'[18] But this does not really matter, argues Mersenne; we ought simply to rest content with a knowledge of the effects: 'c'est donc assez pour avoir la science de quelque chose, de sçavoir ses effets, ses operations, & son usage, par lesquels nous la distinguons de tout autre individu, ou d'avec les autres especes: nous ne voulons pas nous attribuer une science plus grande, ny plus particuliere que celle-là.'[19] As Mersenne says, this science, while limited, does suffice for our needs as a guide to action: 'ce peu de science suffit pour nous servir de guide en nos actions.'[20]

Descartes would no doubt agree. As we saw in Study One, he was perfectly willing to allow that there are areas of experience where the test of experience is the best that we can do. They are simply too complicated for the human intellect to grasp in detail. Thus, to cite the example which he gives in the *Discourse*, we need the test of experience to ferret out, as Harvey did, the fact that the blood is circulated by the heart. The patterns that we discover in experiments of the sort that Harvey carried out are known without a scientific knowledge of their causes. Yet, for all that, they are useful as we attempt to discover useful knowledge in medicine.

But if experimental tests of the sort that Harvey performed are to give us any degree of certainty about any hypothesis concerning the structure of experience, then, Descartes argues, we also need some framework in which to carry out our experiments and to limit the range of hypotheses that we have to test. This framework functions as a principle of determinism and of limited variety, telling us that there are patterns there to be discovered and telling us the general form which these patterns exhibit. If we had no such framework, then the hypothesis that survived our experimental tests could not be claimed to be true. Might not a hypothesis of some other form, for example, be true? Maybe we need an alchemist's hypothesis to explain the circulation of the blood. Certainly, Harvey's experiments had not eliminated such hypotheses. But Descartes's physical framework had. For Descartes, this framework is provided by the fundamental laws of physics, the laws that establish the mechanical nature of the material world. Thus, whatever pattern was going to hold, it would have to be one that conforms to the mechanical nature of the world. Since Harvey's hypothesis does, and the alchemist's does not, conform to the general pattern, and since Harvey's hypothesis has survived experimental tests that eliminated other alternatives such as Galen's, we can conclude, *given that we know that the framework is true*, that Harvey's hypothesis is true and the alchemist's is not. But these laws can be known to be *true* only if we have in fact uncovered their metaphysical roots which make them true. Only if we have uncovered these roots will we be able to ensure that the tree of knowledge bears *genuine* fruit that we can be sure will help us, as Mersenne puts it, as a guide to action.

It is not just a matter of not knowing the causes, as Mersenne suggests. It is, rather, that, however useful to action are the patterns that we do, in fact, in some sense know, these matters may none the less represent nothing but fakery. And if so, they are nothing to the mind that is engaged in the 'search after truth,' to take the title of Descartes's little dialogue on that subject.

We return, however, to Bouwsma's point: is the distinction between the fake and the real drawn *within* experience? If so, we do not have to worry, for coherence provides the criterion. Mersenne, however, clearly holds that the causes of things are outside the world of ordinary experience. This world of ordinary experience we can in fact cope with; the sceptic's arguments do not show that we cannot acquire enough such knowledge to get along quite well. What the sceptic's arguments establish, Mersenne is prepared to agree, is that we have no knowledge of the causes of what happens in the world of ordinary experience.

But this implies that the causes of what we experience are outside that world of experience. What we cannot know, Mersenne points out, are *substances*.

Bouwsma argues that the distinction between fake and real is one that is made *within* the world of ordinary experience. But the demon hypothesis makes sense only if there is a criterion by which the whole of the world of ordinary experience can be judged to be fake, only if there is a way in which the whole of the ordinary world, no matter what happens in it, can deceive me about reality. We do indeed need a contrast, as Bouwsma makes clear, between the fake and the real. But if the world of ordinary experience might as a whole turn out to be fake then we need a world *outside* that of ordinary experience to be the reality to which we appeal when we contrast it to the possibly fake world of ordinary experience. Bouwsma clearly assumes that there is no such world outside that ordinary experience to which the whole of ordinary experience can be compared. Bouwsma in fact recognizes this point, but rejects the position as unintelligible: the evil genius, if she succeeds in deceiving us, does so only because she 'has a sense denied to men.'[21] But if Bouwsma thinks that the existence of such a world is unintelligible to us, that is not the assumption of Mersenne. For Mersenne there are, in fact, entities outside the world of ordinary experience. These are the substances that are the causes of what happens in ordinary experience. And we may not have a 'sense denied to men,' but we have the capacity of reason to grasp a world beyond that revealed by the senses: if we do not have a special sense that gives us special sensible insight into a world beyond that revealed by our ordinary sensible insight, we do have rational insight.

Unlike Berkeley but like Mersenne, Descartes takes for granted that there are substances. This metaphysical framework he never questions. Berkeley, of course, had argued that the very notion of such secondary causes was metaphysical nonsense. That is why he can so readily agree with Bouwsma that the difference between the fake and the real, the dream and the waking, is a matter of coherence. The difference between Berkeley and Bouwsma is that Berkeley argued, where Bouwsma does not, that there is nothing outside ordinary experience in terms of which we can judge ordinary experience. But Descartes was of an earlier generation that could not conceive the possibility that the world of ordinary experience lacked metaphysical foundations in substances. He does not subject the notion of substance to the sort of critique which Berkeley and Hume were later to give it. He simply takes for granted that there are substances that are the causes of what we ordinarily experience, and – which is the crucial point – constitute a reality in terms of which ordinary experience can be judged to be real or fantasy.

What Descartes aims to establish is that he knows certain substances. In Bouwsma's terms, Descartes is prepared to argue that there is a world of substances outside the world of ordinary experience, and furthermore that we can know this world inhabited by the evil genius: the sense by which this world is known, or at least the rational intuition of substances by which this world is known, is after all *not* something, as Bouwsma put it, 'denied to men.' Descartes will prove that this is not denied to men. In this argument he takes for granted the basic intelligibility of the concepts of substance, essence, and so on.

Thus, in the *Discourse*, Descartes announces his conclusion that the soul is incorporeal this way: 'I realized that I was a substance whose whole essence or nature is simply to think and which does not ... depend on any material thing' (*Discourse*, 1:127). 'I realized that I was a substance ...': his claim is that he is a certain sort of metaphysical entity, a substance. Moreover, he also says of this substance that it has an 'essence or nature': not only is he a substance but he is a substance with a certain nature or essence, or, to use another term, form. Body, too, is a substance, and has an essence or form or nature. As he tells us in the *Principles,* we can consider thought and extension as 'constituting the natures of thinking substance and extended substance' (*Principles*, 1:215). Basically, he takes for granted the intelligibility of the traditional metaphysics of substance. That he never questions. What he questions is whether we can know the truth in a world of substances.

In order to see the nature of the problem that Descartes poses, it is necessary to recall certain features of the substance ontology.

Let us recall from Study One the structure of the substance ontology. Substances are individual entities. These entities have certain properties that are in them. Substances endure through time. The properties which are present in them may be different at different times. If there are such differences then the substance is said to change: change consists in a substance coming to have present in it properties that are different from those that were previously present in it. In addition, every substance has a nature or essence. This nature or essence determines which properties the substance has present in it. Thus, to use a non-Cartesian example, if the essence or nature of a substance contains the property of risibility, then if the property of hearing a joke is present in the substance then that property will be followed by the property of laughing. For, risibility is simply the property such that if a joke is heard then laughter follows:

x is risible if and only if, if a joke is heard then laughter follows.

In fact, upon the traditional view, risibility is an active power that *causes* the laughter to follow upon the hearing of the joke. If a joke is heard, then being risible *necessitates* that laughter follows. Risibility is a power that provides a necessary connection between the property of hearing a joke and the property of laughing. This is true in general: the nature or essence is a causal power that explains what happens in a substance by virtue of constituting necessary connections between the presence of the various properties that are in the substance over time. Thus, *the substance in virtue of its nature or essence determines the way in which the properties that are present in it are structured.* Or, to put it another way, *how the properties that a substance presents cohere is a function of the nature or essence of the substance.*

It is in this context that we have to place the notion of *truth.* Truth in its basic meaning for the substance ontology is not a matter of correspondence of thoughts with what is presented. Truth, rather, consists in the relation between the properties presented by a substance and its nature. Truth, in other words, is a *metaphysical* notion: truth is in the nature of things. Aristotle tells us that 'To say of what is that it is not, or of what is not that it is, is false, while to say of what is that it is, and of what is not that it is not, is true ...' (*Metaphysics,* 1011b 27–9).[22] But, as we have seen, Aquinas[23] tells us that

a house is said to be true that expresses the likeness of the form in the architect's mind; and words are said to be true so far as they are the signs of truth in the intellect. In the same way natural things are said to be true in so far as they express the likeness of the species that are in the divine mind. For a stone is called true, which possesses the nature proper to a stone, according to the preconception in the divine intellect. Thus, then, truth resides primarily in the intellect, and secondarily in things according as they are related to the intellect as their principle. (*Summa,* FP, Q16, A1)

Saying of what is that it is, is a rather complicated thing, in other words, in spite of its apparent simplicity. In the substance ontology, truth requires us to go to the heart of the ontology. According to Aquinas, the truth of a thing is a standard to which the thing bears a likeness. This standard is the form or species of the thing in the divine intellect. And this idea in the divine intellect is at the same time the principle of the thing, the nature or essence of the substance. The person who knows compares the object as presented to this standard. 'When ... [the mind] judges that a thing corresponds to the form which it apprehends about that thing, then first it knows and expresses truth' (FP, Q16, A2).

Within the substance ontology, the problem of truth consists of discovering whether the properties which a substance presents to one in ordinary experience in fact reflect the essence or form of that substance. The form defines what sort of substance the thing is. This form provides the standard by which we judge whether its appearances display its truth. If the appearances, the properties present in it, are those required by that essence, then it is presenting itself truly. But if those appearances, what we observe in ordinary experience, are other than those required by the essence or form, then it is not presenting itself truly. The essence or nature or form thus provides a standard *external* to the world of ordinary experience by which that world can be judged to be true or false, real or fake, real or merely a dream. Descartes in fact takes for granted the ontology of substances and the metaphysical account of truth that is part and parcel of that ontology. This framework provides him with the external standard that enables him to call into question the whole of ordinary experience. What vitiates Bouwsma's argument against Descartes is that he (Bouwsma) neglects precisely this ontological framework that Descartes takes for granted.

It is wrong to suggest that Descartes calls *everything* into question. Far from it. He takes for granted, for example, the intelligibility of ordinary discourse; he does not even raise the issue of whether ordinary discourse expresses concepts that make sense. Nor does he limit himself to ordinary discourse about ordinary things. It includes the metaphysical discourse of the substance ontology. Here, too, Descartes takes for granted that the concepts of substance, essence, or form, and so on, are intelligible. Berkeley was later, following Locke, to subject them to a devastating critique, challenging their very intelligibility. But Descartes does not raise that issue at all. His concern is not the intelligibility of the discourse of substances, but whether we can find out the truth about the substances which we are supposed to know.

In fact, Descartes not only accepts the intelligibility of the discourse of substances, he also takes for granted major parts of the epistemology that is part of the traditional substance metaphysics deriving from Aristotle.[24]

Consider once again Aquinas's account of truth. As we saw him say, 'when ... [the mind] judges that a thing corresponds to the form which it apprehends about that thing, then first it knows and expresses truth' (*Summa*, FP, Q16, A2). In fact, for Aquinas, the person who knows *abstracts* the essence or form – the species – from the thing judged about. This abstracted form is the standard of its truth, and judgment of truth consists in comparing the object to this standard. He follows that above statement with the remark that '[t]his it does [that is, the mind judges] by composing and dividing: for in every proposition it either applies to, or removes from the thing signified by the subject, some form signified by the predicate ...' (ibid.). When the mind is thus judging the thing, the form or nature of the thing known is *in the mind*. '[S]ince everything is true according as it has the form proper to its nature, the intellect, in so far as it is knowing, must be true, so far as it has the likeness of the thing known, this being its form, as knowing. For this reason truth is defined by the conformity of intellect and thing; and hence to know this conformity is to know truth' (ibid.). The thing that is known, or, at least, its essence or form, is thus literally in the mind of the knower. There is an identity of the knower and the known.

This identity of knower and known created by the presence of the form or essence of the known thing being present in the mind derives, of course, from Aristotle. According to Aristotle, 'it is *intuitive reason* that grasps first principles' (*Nicomachean Ethics*, 1141a7–8), or, equivalently, the indemonstrable knowledge which is the grasp of immediate premises is *rational intuition* (*Posterior Analytics*, 88b35–89a1). What inquiry aims at, then, is the rational intuition of the natures of things. But in what does such rational intuition consist? What is the ontology of the knowing situation? Already in Aristotle's time it was an old dictum to say that like knows like (*De anima*, 427a27), but in Aristotle this dictum takes on a fairly specific content. For Aristotle, the likeness amounts to an *identity*: in knowledge the mind is identical with its object (*De an.* 429a16, 429b20, 431a1). Since the object of knowledge is the form or nature, it follows that in knowledge the form that is in the thing known is also in the mind that knows it (431b30). But the form is not in the mind as a characteristic of it; for if it were, then the mind would *be* Socrates or Corsicus, that is, a man or a lump of earth, let us say, or in any case whatever substance it is that is known; but the mind is clearly not any of those. Thus, it is not the material object, only its form, which is in the mind. As the medievals were later to say, the form of the substance known is in that substance substantially but in the mind only intentionally. Thus, *rational intuition of a form consists in the form itself being literally in the mind.*

Descartes takes up this account of knowledge from Aristotle and Aquinas. He takes perception to fall under the category of thought; he makes this clear on the first page of the Third Meditation. Thoughts in general and perceptions in particular are modifications of the mind; they are acts of the mind. But thoughts and perceptions are also about, or as Brentano taught us to say, they 'intend,' certain objects. As Descartes puts it, thoughts and perceptions intend certain objects by virtue of *ideas* which inform them. An idea, then, is the form of an act of thought or perception; as Descartes puts it in his replies to the Third Set of Objections, 'by an idea I mean whatever is the form of a given perception' (*Objections*, 2:132). But an idea in this sense is not itself a content of thought that is apprehended by the mental act of which that idea is the form; we are *aware of* the idea but that idea is not the *object of perception*. Ideas, in other words, are not representative entities that intervene as it were between the act of the mind which is the perceiving and the object perceived. For, Descartes distinguishes two senses of the terms 'idea.' In one sense, it is a modification of the mind, and as such is something of which the mind is immediately aware. However, an act of thought or perception is also intrinsically connected to the object or body which it intends, by virtue of which connection the mind is aware of that object or body. We discussed this feature of Descartes's ontology of mind in Study One. This ontology of ideas is vital if we are to understand Descartes's argument in defence of the traditional substance metaphysics. Let us recall the details.

Descartes draws the relevant distinction between two senses of the term 'idea' in the 'Preface' to the *Meditations*: 'there is an ambiguity here in the word "idea." "Idea" can be taken materially, as an operation of the intellect, in which case it cannot be said to be more perfect than me. Alternatively, it can be taken objectively, as the thing represented by that operation' ... (*Meditations*, 2:7). This he elaborates in his reply to the Second Set of Objections:

> II. *Idea.* I understand this term to mean the form of any given thought, immediate perception of which makes me aware of the thought.
> III. *Objective reality of an idea.* By this I mean the being of the thing which is represented by an idea, in so far as this exists in the idea ... whatever we perceive as being in the objects of our ideas exists objectively in the ideas themselves. (*Objections*, 2:113)

Crucial here to the position of Descartes is the notion of 'objective being.' Descartes refers to actual being as 'formal' being or reality (*esse formale*): the formal being of a thing is constituted by its properties or modifications. This he distinguishes from the objective being of a thing. There is on the one hand the being of entities 'which possess what the philosophers call actual or formal reality' but also, on the other hand, 'in the case of ideas [there is] ... what they call objective reality'; the latter is 'the mode of being by which a thing exists objectively or representatively in the intellect by way of an idea' (*Meditations*, 2:28). This is 'the way in which its [the intellect's] objects are normally there' (*Objections*, 2:74). When a thing has objective being in the mind, that thing is not a mode or property of the mind, that is, it is not 'formally' in the mind; otherwise the mind would actually *be* the thing. None the less, though the thing that is objectively in the mind lacks, as such, formal reality, that thing 'imperfect though it may be, is certainly not nothing' (*Meditations*, 2:28) and is to be contrasted to things, or rather 'non-things' which are merely 'chimerical things which cannot exist' (2:30n). Thus, as Descartes put it to Caterus in the replies to the First Set of Objections, 'the idea of the sun is the sun itself existing in the intellect – not of course formally existing, as it does in the heavens, but objectively existing, i.e. in the way in which objects are normally in the intellect. Now this mode of being is of course much less perfect than that possessed by things which exist outside the intellect; but, as I did explain [in Meditation III], it is not therefore simply nothing' (*Objections*, 2:75). When Descartes distinguishes the thing – the sun, say – from the sun *qua* formally existing, the thing that is said to be in the mind, that is, the sun insofar as it does not formally exist, must be the essence of the thing. In other words, what is present to the mind when the thing is objectively in it is not this real object but its essence.[25] We therefore see that for Descartes, as for Aristotle and Aquinas, *the objective existence of things in the mind is constituted by the presence in the mind of the essence of the thing.*[26] But once we recognize this we also recognize that *the Cartesian account of perception is structurally that of Aristotle, in which the mind knows the object by virtue of having the form or essence of that object literally in it.*

To be sure, Descartes rejects the mechanisms by which Aristotle explained the presence of the form or essence in the mind. For Aristotle, there was the process by which the form was transported from the object known to the knowing mind. Descartes rejected this story, as is well known, to replace it by an atomistic account that fit with the new science.[27] This, though, is relatively trivial. Aristotelians were soon able to adapt Aristotle's account of cognition to an atomistic picture of reality.[28] But Aristotle also required the form to be abstracted from the objects given in sense. Descartes rejects this mechanism, too; that, after all, was the point of the wax example in the Second Meditation. This mechanism is replaced by the doctrine of innate ideas: all the forms or essences of things are native to the mind, and all we must do is retrieve them. But, for Aristotle, what is crucial to having a rational intuition of the thing known is for the form or essence of the thing to be present in the mind. *This central claim of the Aristotelian position is retained by Descartes. Descartes does not reject the central thesis of the Aristotelian account of knowledge in terms of rational intuition but only inessential claims about the mechanisms by which rational intuition is attained.*

What we have now to recall is that, within the metaphysics of substances, it is possible for a substance to have properties that are contrary to its form or essence. The classic example is that of heavy objects. Their nature or form requires them, according to Aristotle, to *be* at the centre of the universe, or at least to *be* as close thereto as they can get: it is in the nature of heavy objects to have this property. But, of course, many of them do not. If they do not, then they exist in a state that is contrary to their nature. If they do exist unnaturally, then their form or essence will bring it about that they move as directly as possible to be in such a way that they conform to their essence or form. This is free fall. But another thing or substance may cause them to move away from the centre of the earth, from their natural state of being. I may lift the stone up, and, so long as I hold on to it, it will remain away from its natural state of being. To be sure, it is striving to return to its natural state; that is the weight, the gravity, that I feel it exerting on my hand as I hold it above the surface of the earth. The stone will naturally strive to return to its natural state, and will in fact do so if I release it. But so long as I hold it up, its power is such that it cannot move itself as it is naturally inclined to do. I can do this because I can exercise a greater power, and restrain it from achieving its inclination to return to its natural state.

Substances, in short, can be in two different ways, either naturally or unnaturally. If they are naturally, then the properties that are present in them are those that truly express the nature or essence or form of the substance. If they are unnaturally, then the properties that are present in them are such that they do not truly express the nature or essence or form of the substance. Normally, a substance is in its natural state. But if there is a more powerful substance, then that substance can cause a less powerful substance to exist unnaturally.

Aquinas makes the distinction made previously by Aristotle between things which are natural and things which are unnatural or violent. '[A] thing is called natural because it is according to the inclination of nature,' he tells us, and 'we call that violent which is against the inclination of a thing' (*Summa*, FP, Q82, A1). Violent motion always involves a hindrance of a natural inclination to be in the natural way: 'that which has a form actually, is sometimes unable to act according to that form on account of some hindrance, as a light thing may be hindered from moving upwards' (FP, Q84, A3). Furthermore, violent motion needs an active cause to hinder the natural inclination: 'nothing violent can occur, except there be some active cause thereof. But a tendency to not-being is unnatural and violent to any creature, since all creatures naturally desire to be. Therefore no creature can tend to not-being, except through some active cause of corruption' (FP, Q104, A1, Obj 3). Now, sense organs are not deceived when it comes to their proper objects – or, rather, *not normally* deceived: they work satisfactorily unless something comes along to interfere with their normal operation. '[S]ense is not deceived in its proper object, as sight in regard to color; unless accidentally through some hindrance occurring to the sensile organ – for example, the taste of a fever-stricken person judges a sweet thing to be bitter, through his tongue being vitiated by ill humors' (FP, Q85, A6).

Aquinas makes this remark in the context of an argument why the intellect does not err. The conclusion to be drawn is that the intellect does not err unless accidentally through some hindrance. But the intellect knows things by virtue of those things being in the mind: 'intelligent beings are distinguished from non-intelligent beings in that the latter possess only their own form; whereas the intelligent being is naturally adapted to have also the form of some other thing; for the idea of the thing known is in the knower' (*Summa*, FP, Q14, A1). Thus, the intellect does not err with regard to the truth of things – unless accidentally through some hindrance. The question is: how do we know whether or not there is any such hindrance? The mind, after all, can be directly affected by God, as when he gives us dreams that foretell the truth. But in the same way the mind can be directly affected by demons:

[T]he outward cause of dreams is twofold, corporal and spiritual. It is corporal in so far as the sleeper's imagination is affected either by the surrounding air, or through an impression of a heavenly body, so that certain images appear to the sleeper, in keeping with the disposition of the heavenly bodies. The spiritual cause is sometimes referable to God, Who reveals certain things to men in their dreams by the ministry of the angels, according Num. 12:6, 'If there be among you a prophet of the Lord, I will appear to him in a vision, or I will speak to him in a dream.' Sometimes, however, it is due to the action of the demons that certain images appear to persons in their sleep, and by this means they, at times, reveal certain future things to those who have entered into an unlawful compact with them. (SS, Q95, A6)

No doubt such interventions in the natural workings of the world are in a way miraculous, where a miracle is understood as a systematic violation of everything that nature requires, or, rather, that is required by the totality of the natures of things. 'A miracle properly so called is when something is done outside the order of nature. But it is not enough for a miracle if something is done outside the order of any particular nature; for otherwise anyone would perform a miracle by throwing a stone upwards, as such a thing is outside the order of the stone's nature. So for a miracle is required that it be against the order of the whole created nature' (FP, Q110, A4). A systematic hindrance of the workings of the intellect could prevent it from grasping the truth of things. Such spiritual interference would violently prevent the intellect's natural working, and would prevent it from naturally achieving its natural cognitive end of the truth. 'God alone can do this [cause a miracle],' Aquinas concludes (ibid.). No doubt it would a very powerful spiritual being to systematically deceive us. But why could it not be a demon? So long as it was powerful enough it could cause our intellects to fall into error, to grasp forms that are not the truth of the world – as sense can sometimes fall into error even with respect to its proper objects and present to us images that incorrectly represent their causes.

Aquinas does not really see that there is a problem here. So far as he does, then he of course has the answer: God alone can do it and He won't. How do we establish it as true, however? This is the problem that Aquinas does not face. If there really are demons that could violently hinder the natural operations of the intellect itself, making it possible that the intellect systematically fails to grasp the truth of things, then how can we be sure of any proof that such demons do not exist because God alone can do such a thing and He won't? There is a challenge to knowledge claims that lies at the very heart of the substance ontology.

This challenge Aquinas quite fails to recognize. It is Descartes's genius to have recognized it and to have attempted to meet this challenge, this ontological challenge to any claim within the traditional substance ontology to know the truth of things.

Nor were other scholastics any more sensitive than Aquinas to the demonic possibilities lurking at the core of the Aristotelian metaphysics that they in effect took for granted. Thus, we might also consider Scotus.

'Truth,' according to Scotus,[29] 'implies a relation to an exemplar';[30] here he agrees with Aquinas before him. Of the exemplars, there are two: one created, one uncreated. The uncreated exemplar is the form in the mind of God, the eternal pattern for the thing or substance that He has created. The created exemplar is the form *in* the created substance. 'The created exemplar is the species of the universal created by the thing. The uncreated exemplar is the idea in the divine mind.'[31] The truth of the thing is in the first instance the form in the mind of God. It is, in the second instance, the created form in the substance. For, the form in the thing is the created image of the idea in God's mind. We can therefore come to know the truth of the thing by having an intuition of the form in the thing. '[T]he truths of things are known through their conformity to the intelligible species.'[32] This we obtain by abstracting that form from the thing as it is given to us in sense experience. Knowledge requires us to 'assume the existence of a species in the intellect';[33] this species is abstracted from the changing sense appearances which the thing presents to us: 'the object from which the exemplar is abstracted is itself mutable ...'[34] As for the sense object itself, it consists in a collection of separable properties that lacks within itself any intrinsic unity. '[T]he sensible thing outside causes a confused sense image, something with only an incidental unity in the faculty of imagination, which represents the thing according to its quantity, colour and other sensible accidents ... the sense image represents things only confusedly and according to the incidental unity ...'[35] When the form is grasped, it is separated in the mind from the sense appearances of the thing. '[B]y grasping just what things are of themselves, a person separates the essences from the many additional incidental features associated with them in the sense image.'[36] It is the form or essence in the thing that creates the unity out of these properties or accidents, and makes a substance out of them. 'It is only within the power of the few to attain the eternal reasons, because it is only the few that have an understanding of the essentials, whereas the many grasp things merely in incidental concepts ...'[37] The forms or natures of things produce regular patterns of behaviour. Scotus gives an example of how a fact that is known first by sense experience, as contingent and probable only, can come to be known scientifically, with demonstrative certainty.

[A]t times we experience [the truth] of a conclusion, such as: 'The moon is frequently eclipsed' ... Sometimes, beginning with a conclusion thus experienced, a person arrives at self-evident principles. In such a case, the conclusion which at first was known only by experience now is known by reason of such a principle with even greater certainty ... for it has been deduced from a self-evident principle. Thus for instance, it is a self-evident principle that when an opaque body is placed between a visible object and the source of light, the transmission of light to such an object is prevented. Now, if a person discovers by way of division that the earth is such an opaque body interposed between sun and moon, our conclusion will no longer be known merely by experience as was the case before we discovered the principle. It will be now known most certainly by a demonstration of the reasoned fact, for it is known through its cause.[38]

Again, there are facts that are never known through their causes. Scotus argues that when we know things only by experience, we have no notion of a necessary tie among events, a connection that is grounded in the natures of things. He points out that sometimes

we must be satisfied with a principle whose terms are known by experience to be frequently united, for example, that a certain species of herb is hot. Neither do we find any ... prior means of demonstrating just why this attribute belongs to this particular subject, but must content ourselves with this as a first principle known from experience. Now even though the uncertainty and fallibility in such a case may be removed by the proposition 'What occurs in most instances by means of a cause that is not free is the natural effect of such a cause,' still this is the very lowest degree of scientific knowledge – and perhaps we have here no knowledge of the actual union of the terms but only a knowledge of what is apt to be the case. For if an attribute is an absolute entity other than the subject, it could be separated from its subject without involving any contradiction. Hence, the person whose knowledge is based on experience would not know whether such a thing is actually so or not, but only that by its nature it is apt to be so.[39]

Nature, however, is uniform. Experience produces only probability, but we can raise this probability higher by invoking the self-evident principle that 'Quidquid evenit ut in pluribus ab aliqua causa non libera, est effectus naturalis illius causae ...':

(=)   Whatever occurs in a great many instances by a cause that is not free is the natural effect of that cause.[40]

This principle will make it certain that an object that for the most part produces certain effects is by its nature apt to do so.

In this sense, nature is uniform, according to Scotus. This is the meaning of the scholastic formula that 'Natura determinatur ad unum': 'The proposition that nature is determined towards a single end is to be understood not in the sense of being ordained to produce one and only one singular effect, but in the sense that it is ordained to a definite pattern of production – natural agents in this respect differing from voluntary agents which are not ordained to a definite pattern but rather determine themselves to one or the other of two opposite actions.'[41] Nature, or the nature of a thing, is the ultimate source of being, or the being of the thing, and the power that is the source of all its becomings, doings and beings. Hence the principle (=).

Given (=), we can infer that the effect which frequently follows a non-free cause is the natural effect of that cause. The regularities that we observe in the world we know by sense are thus rooted in the natures of things. Hence, even if a seeker after causes has observed only a restricted number of instances of a set of sensible appearances, he or she can still conclude that these have the same cause, and that this cause will produce the same effects always and everywhere (*semper et in omnibus*): 'As for what is known by experience, I have this to say. Even though a person does not experience every single individual, but only a great many, nor does he experience them at all times, but only frequently, still he knows infallibly that it is always this way and holds for all instances.'[42] We know this by virtue of (=), which is a self-evident proposition 'reposing in the soul.'[43] Observation of the world generates knowledge of this proposition – though, to be sure, 'the senses are not a cause but merely an occasion of the intellect's knowledge ...'[44]

This latter is an important point. Scotus considers the view that '[i]f an object is continually changing we can have no certitude about it ...'[45] Descartes was, as we well know, later to make much the same point, invoking the mutability of sensible appearances to challenge the notion that we could ever abstract the form or essence of a thing from its changeable appearances. This was the point of the wax example of the Second Meditation. Scotus argues, however, that the sense impressions which change are not the cause of our knowledge of the forms or essences of things. Rather, it is the form itself that is the cause of the knowledge. For, after all, it is in fact the form, and not the sensible appearances, that is the active power that accounts for and explains the effects of things. The point, of course, is that this form is immutable. '[I]t is not precisely this mutability in the object that causes the knowledge; it is the nature of this mutable object that does so, and this nature is immutable.'[46] The mutability of the sense object is therefore no objection to the claim that we cannot know necessary truths; to the contrary, we can indeed know such truths with 'infallible certainty,' Scotus argues.[47] The self-evident principles acquire their evidence through themselves, and do not depend upon the objects of sense. '[T]he intellect by its own power and in virtue of the terms will assent to [any such] proposition without the shadow of doubt. And it does not assent to this because it sees these terms verified in some things, as it does when it assents to the proposition "Socrates is white," because it saw the terms united in reality.'[48]

Once we know such truths in this way – as in (=) – then we can apply these to the events that are presented to us in sense experience and use the principles to strengthen our faith in what our senses tell us or even to correct what sense experience tells us. Thus, for example, *we correct* judgments of sight that imply that a building is as small as it appears to be, or that the stick in water that appears bent is really bent.[49] The self-evident principles upon which we rely in order to make these judgments hold for the intellect even when we dream and even for people who are mad. Scotus rejects the claim that 'as in the case of dreams or with madmen, the intellect could be so bound that it could not operate.' The problem in these cases is not that the intellect errs, but that 'it does not act.'[50]

It is clear, however, that given the further aspects of the Aristotelian framework with which Scotus is operating, this is not an adequate reply. This is what Descartes was to recognize when he raised the problem of the evil genius.

In order to justify our claims to knowledge, Scotus relies upon the principle (=). By virtue of it he can hold that 'it is recognized through experience and is certain that for the most part nature acts uniformly and in an orderly way.'[51] What is important, however, is that even causal facts that have demonstrative certainty are not invariable. Even though it is self-evident that

(=)   Whatever occurs in a great many instances by a cause that is not free is the natural effect of that cause

it is not only not self-evident that

As often as the cause is posited, its proper effect must also be posited

it is simply false. In this sense there is no absolute necessity to created or imperfect causes. 'To be able to produce an absolutely necessary effect does not pertain to the perfection of a secondary cause; in fact, there is no such thing (as a necessary effect) in a secondary cause ... for (the concept of) an absolutely necessary causation includes a contradiction.'[52] Natural laws, therefore, need not hold universally. Thus, even if there are regularities that hold 'for the most part' in the ordinary course of events, there are also *chance* causes. 'Natural causes, normally productive of certain effects, can sometimes have their normal operations prevented due to extrinsic interfering factors.'[53] It is true that when there is no interference, a natural cause will produce the effect at which it aims: 'Omnis causa sibi dimissa ... producit effectum cujus est per se' (Any natural cause, if left to itself, will produce its proper effect).[54] The production of the proper effect is an exercise of power, and the cause will exercise that power to its fullest so long as it is not prevented from doing so by some stronger power: 'Causa naturalis agit ad effectum suum secundum ultimum potentiae suae quando non est impedita' (A natural cause produces its effect to the utmost of its power so long as it is not prevented from doing so).[55] Thus, if we have

(=)   Whatever occurs in a great many instances by a cause that is not free is the natural effect of that cause

as a principle, then we also have

(==)   Every natural cause, unless prevented from doing so by some interfering factor, produces its effect necessarily.

That is why all observed effects attributable to secondary causes can be so attributed with only conditional necessity, not absolute necessity: a cause whose operation can be impeded, however unlikely such interference might be, is not a necessary cause.[56]

The point is that the principle (==) allows that there might well be chance causes that are operative, preventing a thing from producing the effect that it is naturally apt to produce. Once again we note the point that was examined in detail in Study One: *on the Aristotelian patterns, natural laws have exceptions.*

Once this is accepted, then we cannot count on our perceptual apparatus always to guarantee that its object exists. For, the natural cause of apprehension of things *could be* impeded in its normal operations, and therefore it could turn out that it produces within us a set of propositions which are false. Scotus proposes that this is not so.

In the first place, a necessary ontological proposition such as

(=)   Whatever occurs in a great many instances by a cause that is not free is the natural effect of that cause

is known infallibly. For, the knowledge is caused by the object known, and (=) applies to this cause too.

This proposition is known to the intellect even if the terms are derived from erring senses, because a cause that does not act freely cannot in most instances produce an effect that is the very opposite of what it is ordained by its form to produce ... if the effect occurs frequently it is not produced by chance and its cause therefore will be a natural cause if it is not a free agent. But this effect occurs through such a cause. Therefore, since it is not a chance cause, it is the natural cause of the effect that it produces.[57]

Since it is known infallibly through itself, it cannot be false. But this is merely circular: it cannot guarantee its own truth.

Secondly, Scotus argues that

the following truth reposes in the mind. 'A faculty does not err in regard to an object that is properly proportioned to it unless the said faculty is indisposed.' Now it is known to the intellect that the imaginative faculty is not indisposed during a waking state to such an extent that the sense image would represent itself as an object, for it is self-evident to the intellect that when it knows, it is awake, and that, consequently, the imagination is not bound [to rest in error] in a waking state as it is in sleep.[58]

But this again is circular. How do we know that the intellect is not indisposed with regard to the propositions before it? To be sure, they are self-evident. But, if it is indisposed, then it may be the case that even those propositions that are evident enough to move the intellect to assent may for all that turn out to be false. So long as natural laws can have exceptions, so long as faculties or powers can be indisposed and rendered improper in their actions by chance causes, then this rule will apply also to that power we call the intellect. So, although it is the proper action of the intellect to assent to truths, if it is, by chance causes, indisposed, then it will assent to a falsehood. One cannot appeal to a self-evident proposition to justify a claim that the intellect is not indisposed, since that proposition itself might be false and one to which assent is given only because the intellect is indisposed.

For Aristotle and the scholastics such as Scotus, there was a guarantee that one could acquire knowledge in the sense of incorrigible certainty. This guarantee was provided by the notion that there are certain 'natural tendencies' of the human mind towards the acquisition of knowledge. But for the defenders of the new science, whether empiricist or rationalist, the whole talk of powers is vacuous. In particular, therefore, talk of the power of abstraction is vacuous. But it is this 'naturalness' of this power of abstraction, the supposed fact that it is a feature of the human 'nature' or 'form' or 'essence,' that, in the end, is supposed to provide the guarantee for the whole of the Aristotelian system. 'By nature' all persons aim to know, the system claims. Coming to know is thus a *natural motion* which, like all natural motions, will be successful unless it is constrained. But such unnatural interventions, that is, being caused to abstract the wrong form, happen only incidentally. For what is natural is the norm; it is what happens *always or for the most part*. Hence, error, while it can occur, is unnatural, and the natural powers of the human mind guarantee that success, that is, knowledge of laws, is the usual outcome of induction. But what sort of guarantee is *this*? Even if we glomb onto, or have a rational intuition of, a nature, how do we know that we have abstracted the correct one? If our knowledge of the operation of natures is restricted to what they are 'apt' to produce, then our knowledge is not reliable until we have a criterion for discerning when they are acting reliably and when not. But if the basis for relying upon our knowledge of natures is the fact that natures for the most part are reliable in their operations, then we have no clear way to separate chance from natural implantations of propositions in the intellect.

Scotus cannot even argue that he knows by the nature of the intellect that the intellect by its nature is such that it *never* is indisposed. That would imply that there are some natural laws that are not subject to exceptions, that there are no powers that could interfere with, or impede these natural activities of the substance. Now, to be sure, there are some such laws that never have exceptions, for example, the laws that describe the motions of heavenly objects. But even these can be described universally as happening 'per causam naturalem ordinatam' – according to ordained natural causes, that is, predictably – only on the supposition 'quod causae naturales sibi dimittantur' – that is, only on the condition that it is left to itself – which is to say, only on the supposition 'quod per virtutem divinam non impediantur' – that is, the supposition that it is not interfered with by the divine omnipotent power.[59] God is not bound by the laws for the creatures He has brought into being, and may at any time suspend them. This would, of course, be a miracle. But, asks Descartes, might there not be a powerful being who could work the miracle of deceiving us even with respect to what seem like the most proper and certain of the objects of the intellect?

What has just been introduced, of course, is the Cartesian evil genius. The point is that Scotus, like Aristotle and Aquinas before him, has done nothing to ensure that such a demon does not exist. Scotus, like his predecessors, did not see that at the core of being there is a possibility lurking – the evil genius – that could call into question all possibility of gaining infallible knowledge of being. It was Descartes who was the first to see this problem. It lies at the heart of being within the Aristotelian ontology because, first, being is determined by the natures of things, and, second, there is no guarantee that the natures of things are always going to produce their proper effects. The problem is there within the Aristotelian ontology at the core of being because this ontology allows not only natural events but also violent or unnatural events – and provides neither a way of evidently discerning the difference nor, it seems, a way of guaranteeing that an unnatural course of events does not interfere with or impede the natural aim of the mind to know.

Descartes, of course, completely rejects the Aristotelian idea that the various objects presented in sense experience are separate substances. If they are separate substances, then some can exist naturally, some unnaturally. But for Descartes, ordinary things are aspects of a single substance, the one material substance that he calls body. For ordinary things, then, there is no distinction between existing naturally and existing unnaturally. There just is a single way of being. Observable things taken together constitute a single substance; they jointly form a single, unified whole.

On the substance ontology, if this single substance, this unified whole, is as it appears to be, then this whole truly expresses the essence or form or nature of body. It presents the truth of things: it is true, not false; it is real, not fake. This is so because it truly expresses the essence of body. The crucial point is that, while there is no unnatural/natural distinction to be found in the being of ordinary things, the material world as a whole as we perceive it could be wholly unnatural. This would be so if there were a powerful substance that could prevent the substance that presents to us the properties we perceive from presenting the properties that truly represent it. The unnatural/natural distinction presupposes that substances have power, and have it in different degrees, so that some substances can be more powerful than others and can prevent those others, if the more powerful is so inclined, from existing naturally or truly. And Descartes does, of course, allow that it is at least possible that there are some substances that are more powerful than others. It is, in fact, precisely that possibility that makes it possible to conceive the notion of an evil genius who sets out to mislead us systematically about the truth about the world that we know in ordinary experience.

The evil genius (if she exists) is a powerful substance. She is sufficiently powerful that she can bring it about that the substance that presents to me the properties that I observe in ordinary experience is such that those presented properties falsely express in a systematic way the way that substance is in its essence. What I am presented with in ordinary experience are properties that are structured as one would expect of a substance whose essence is extension. When I compare the idea of extension in my own mind with what is presented to me, I can recognize that the properties are presented as if they were the properties of a substance whose essence is given by that idea of extension. Yet perhaps this idea of extension that I have is not after all the essence of the substance that is presenting to me the properties I observe. Perhaps the substance that is presenting those properties to me is a substance with a different essence. Perhaps it is a thinking substance. Perhaps it is the evil genius herself, doing it directly. In that case, my ideas, which are the essences of things, present me with falsehood: they are not the essences of the things that are actually causing the ways in which things appear to me. '[T]he only reason for my later judgement that they were open to doubt was that it occurred to me that perhaps some God could have given me a nature such that I was deceived even in matters which seemed most evident. And whenever my preconceived belief in the supreme power of God comes to mind, I cannot but admit that it would be easy for him, if he so desired, to bring it about that I go wrong even in those matters which I think I see utterly clearly with my mind's eye' (*Meditations*, 2:25). In fact, maybe I am wrong about my own essence. The *cogito* gives me the essence of myself, but even that clear and distinct idea is also challenged by the evil genius. The evil genius calls into question the ideas in my mind, the essences of things. Are these essences the essences of the substances of whose properties I am aware in ordinary experience? But among the ideas in my mind is the idea of myself. Like all ideas, this has a certain objective reality. This objective reality consists in its being the essence of a thinking substance. Now, in ordinary experience I am presented with certain properties, the properties of thinking, doubting, and so on. These properties are presented as the properties of a substance whose essence is thinking. Yet the power of the evil genius is such that the substance that produces these properties may have some quite different essence. Maybe the properties that are presented to me in ordinary experience present me with falsehood. Maybe I am in reality something very different from a thinking thing. Maybe, in fact, I am the evil genius himself, deceiving myself about my own essence.

There is one very special case of unnatural being that we ought to consider: transubstantiation. Consider what Aquinas says about this.

During the Eucharist, when the priest raises the bread and wine and says the appropriate words, God miraculously changes the bread and wine into the body and blood of Christ. There is no visible change – the properties of the bread and wine remain unaltered. None the less, as Aquinas explains it, Christ 'is invisibly under the species of this sacrament, wherever this sacrament is performed' (*Summa*, TP, Q75, A1, RObj 2). What happens is that the substance changes in its essence: 'the substance of the bread is changed into the body of Christ ...' (TP, Q75, A3, RObj 2). 'The form ... is not changed into another form; but one form succeeds another in the subject; and therefore the first form remains only in the potentiality of matter' (TP, Q75, A3, RObj 2). The forms or essences of bread and wine are completely replaced by the forms of the body and blood of

Christ. To be sure, '[s]ome have held that the substance of the bread and wine remains in this sacrament after the consecration. But this opinion cannot stand' (TP, Q75, A2). Unnatural change can often be brought about by ordinary things. But the change that occurs during the Eucharist is a miracle; it cannot be performed by ordinary things. It is the doing of a superpowerful being. '[T]his change is not like natural changes, but is entirely supernatural, and effected by God's power alone' (TP, Q75, A4). God, being all powerful, and the source of all being, can effect the change where nothing else can.

[I]t is evident that every agent acts according as it is in act. But every created agent is limited in its act, as being of a determinate genus and species: and consequently the action of every created agent bears upon some determinate act. Now the determination of every thing in actual existence comes from its form. Consequently, no natural or created agent can act except by changing the form in something; and on this account every change made according to nature's laws is a formal change. But God is infinite act ...; hence His action extends to the whole nature of being. Therefore He can work not only formal conversion, so that diverse forms succeed each other in the same subject; but also the change of all being, so that, to wit, the whole substance of one thing be changed into the whole substance of another. And this is done by Divine power in this sacrament; for the whole substance of the bread is changed into the whole substance of Christ's body, and the whole substance of the wine into the whole substance of Christ's blood. Hence this is not a formal, but a substantial conversion; nor is it a kind of natural movement: but, with a name of its own, it can be called 'transubstantiation.' (TP, Q75, A4)

Now, it is evident to the senses that accidents of bread and wine remain. Moreoever, these accidents are perceived as having the structure of bread and wine. But 'the nature of the substance is known by its accidents' (*Summa*, TP, Q75, A3, RObj 2). Thus, when we perceive the accidents of bread and wine and perceive them as bread and wine, we have abstracted from these accidents the form or essence of bread and wine. By virtue of that form or essence being in the mind we have a rational intuition of bread and wine present before us. To be sure, bread and wine in their essential substance are not present. The accidents are, in fact, not in any substance at all; they no longer constitute the being of anything at all. For, if they were actually present in the substance that is there, then that substance would be, perhaps accidentally, white and doughy (in the case of the body of Christ) and red, liquid, and alcoholic (in the case of the blood of Christ), where, clearly, these things cannot be in these ways, even accidentally. The properties of bread and wine are present in a substance before the miraculous change; after the change they are present but not present *in* anything. This is truly unnatural. But God can effect this change, too.

[T]he accidents continue in this sacrament without a subject. This can be done by Divine power: for since an effect depends more upon the first cause than on the second, God Who is the first cause both of substance and accident, can by His unlimited power preserve an accident in existence when the substance is withdrawn whereby it was preserved in existence as by its proper cause, just as without natural causes He can produce other effects of natural causes, even as He formed a human body in the Virgin's womb, 'without the seed of man.' (TP, Q77, A1)

None the less, the accidents of bread and wine – these sensible appearances – are perceived as being in a substance; the forms of bread and wine are present in our intellect as we so perceive them. It is not by our senses or reason that we know that the substance of bread and wine have been replaced by the substance of the body and blood of Christ. It is by the suprarational faculty of faith. 'The presence of Christ's true body and blood in this sacrament cannot be detected by sense, nor understanding, but by faith alone, which rests upon Divine authority' (TP, Q75, A1).

If we did not have that faith it could be said that we were being deceived when we are in the presence of the consecrated 'bread' and 'wine.' We perceive the sensible appearances of bread and wine. Our reason, through the rational intuition of the being of these things, tells us that we are in the presence of bread and wine. But our reason here is wrong. In spite of our rational intuition of the essence of bread and wine, the essences of what is before us are very different essences. Our senses deceive us. Or, rather, a very powerful being has so changed things that neither our senses nor our reason can detect that what is given in ordinary experience is not the truth of the thing. The truth is very different. The truth is that what is before us is really, or essentially, the body and blood of Christ. We cannot by our natural faculties, including the faculty of reason, determine the truth of things in this case. That is what a very powerful being has been able to achieve.

Descartes's evil genius is a powerful being who is assumed to be capable of doing world-wide what God actually does locally in bread and wine when the ceremony of the Eucharist is performed. If God does it in the case of the bread and wine, why cannot He, or some other powerful being, do the same thing with everything? Why cannot there be a powerful being who will so alter the essences of things that sensible appearances of things and the rational intuitions I have of those things are all, in fact, false?

Aquinas never tries to answer this question. Descartes does. That he was able to formulate and try to answer this ontological question is the mark of his powerful metaphysical genius.

The hypothesis of the evil demon challenges the claim that I know the truth of anything, whether of material things or body or of myself. It shows that at the ontological core of the traditional substance metaphysics, there is the possibility that the truth of being can systematically elude us. The problem is not epistemological, but ontological. So long as there is the ontological possibility of there being a powerful substance that can make it such that all substances be in an unnatural state – so long as there is the ontological possibility of there being a powerful substance that can make all substances such that the properties present in them do not correspond to the essential nature of the substance – then the truth or reality of the things of which I am aware will be beyond me. The ontological possibility at the core of the traditional substance ontology must be removed if I am to be certain of the truth of anything, if I am to know that the ideas or essences of which I am certain really do present the truth of the world of ordinary experience. This ontological problem requires an ontological solution.

That ontological solution to the ontological problem of how we can know the truth of the world of ordinary experience is God.

II: The *Cogito* of the Meditations: The Demonic Challenge

In the *Meditations* the *cogito* is not the first truth that we know, as it is in the *Discourse*. It none the less plays an important role.

In the First Meditation Descartes attends to objects other than his own self. They are objects that are seen – horses, tables, chairs, his own body – and thought about – the structure of the perceived world, the structure constituted by the arithmetical and geometrical relations among things. The consciousness that is aware of these things is characterized in a certain way, namely, by the feature of attending to, or, more generally, thinking of these things. At the beginning of the Second Meditation, there is a dramatic shift of attention. The meditating consciousness now attends not to bodies or to body but to the attending itself: the consciousness as attending or thinking becomes the object of attention.

The attending to material objects, upon the account of ideas accepted or taken for granted by Descartes, is this. The attending consciousness is modified in a certain way. This modification is an idea. The formal reality of this idea is that of a property. This idea also has a certain objective reality. This objective reality is the object that is presented. In fact, the idea just is the essence of the object presented, that essence *qua* existing intentionally in the mind and not *qua* existing substantially formally as a substance.

Upon the shift of attention, consciousness is a consciousness of consciousness. This consciousness, therefore, has present in it an idea. This idea has a certain formal reality, namely that of a mode or property. It also has a certain objective reality. This objective reality is that of a substance. It has this objective reality by virtue of its being the essence of the presented substance. This essence is in fact that of a thinking thing. The meditating consciousness thinks of the thinking thing by virtue of the essence of that thinking thing characterizing the consciousness as an idea.

With this idea of a thinking thing before its mind, the meditating consciousness makes the inference 'I think, therefore I am.' There is the one idea, that of a thinking thing. And there is a second idea, that of an existing thing. These ideas are necessarily connected to each other, and it is this connection that the mind perceives when it makes the inference from the one to the other.

To see better what is involved in this inference, we have to spend a little while on the notion of what it is for a substance to exist or to be.

For this we have to go back to the roots of the substance ontology in Greek philosophy and the philosophy of Aristotle. Charles Kahn points out in an important essay[60] that, with regard to the notion of predication in Greek philosophy, 'for the philosophical usage of the verb, the most fundamental value of *einai* when used alone (without predicates) is not "to exist" but "to be so," "to be the case," or "to be true."'[61] He elsewhere elaborates this by distinguishing three aspects of 'is' or 'being' which are fused in the classical usage but which we are inclined to distinguish: '1) there must be a denotation, an existing subject which we are talking about or cognizing (we might compare Wittgenstein's *object*); 2) there must be a predication or saying-something about this subject (compare Wittgenstein's *sense*: a possible situation presented and affirmed); and 3) the state of affairs which we assert must itself be actual or 'existing,' if the cognition is true.'[62] This sense of the verb is described by Aristotle as the 'strictest' sense: 'being and non-being in the strictest sense are truth and falsity' (*Metaphysics*, 1051a35–6). Thus, a substance is made actual by virtue of having accidents or properties present in it; it *exists* or *has being* just to the extent that there are properties present in it; its *existence* or *being* is constituted by *what it is*, by *what is predicated of it*.

Aquinas takes up the substance ontology of Aristotle, and makes the same point as the philosopher. The individual, a substance having a certain form – for example, this man – 'is said to be a suppositum, because he underlies [*supponitur*] whatever belongs to man and receives its predication' (*Summa*, FP, Q2, A3). This subject is made actual by virtue of its having accidents predicated of it; for the substance to exist is for the substance to have being, that is, for it to have certain accidents predicated of it. '[A] subject is compared to its accidents as potentiality to actuality; for a subject is in some sense made actual by its accidents' (FP, Q3, A6). Furthermore, as Aquinas also indicates, *being* and *unity* coincide. The substance is an individual thing that is distinct from the accidents present in it and with respect to which it *is*: 'Although the universal and particular exist in every genus, nevertheless, in a certain special way, the individual belongs to the genus of substance. For substance is individualized by itself; whereas the accidents are individualized by the subject, which is the substance; since this particular whiteness is called "this," because it exists in this particular subject' (FP, Q29, A1). The unity of the individual derives from the substance. The collection of properties or accidents which characterize the individual are unified into the set of characteristics of a particular individual precisely because each of them inheres in the substance. The unity of any ordinary thing – a die, for example – derives from the substance in which all the properties of the thing inhere. The details of the structure or unity derive from the essence or form of the substance which is the cause of that unity.

Thus, as we have recognized already in Study One and Study Six, for a substance to exist or be is for it to be in some way or other. If *a* is a substance, then *a* exists provided that it is in some way or other. Let *F* be a property, some specific sort of accident. If it is true that

(*)   *a* is *F*

then *a* exists: it exists because it *is* in a certain way, namely, specifically it is *F*. So, to make the statement *tout court* that

       a exists

or

       a is

without committing oneself to what specifically is the property present in it, is to say that

(**)   there is a property *f* such that *a* is *f*

where *f* is, as we would say, functioning as a variable.

The contrast between (*) and (**) is this: (*) ascribes a *determinate* property to a; (**), in contrast, ascribes a *determinable* property to a.[63] The relation between the property that (*) ascribes to a and the property that (**) ascribes to a is the relation between

(@) x is red

and

(@@) x is coloured.

Let us represent 'red' by

($) R.

Then we know that

R is a colour

or, if we represent the second-order property of being a colour by

C

then we know that

(^) R is C.

Then 'x is coloured' is the predicate:

($$)   there is an f such that f is C and x is f.

The relation between the determinable (@@) and the determinate (@) is given by the logical relation between ($$) and ($). The relation between a property like

(+)   thinking specifically such and such

and

(++)   existing

is precisely this relation. If we take the truth (^) to be a truth stating a relation among universals, and therefore to be a truth which is in some sense necessary, then we should also take the corresponding truth

(^^)   thinking specifically such and such is a property (or mode)

as necessary. The necessity of (^) makes the relation between ($) and ($$) one of necessity; it is an entailment relation. Similarly, the necessity of (^^) makes the connection between (+) and (++) to be one of necessity, an entailment. Thus, within the substance ontology the inference from

> *a* is *F*

to

> *a* exists

is a necessary inference. Similarly, the inference from

> I think

to

> I exist

is a necessary inference.

Thus, *within the context of the substance ontology which Descartes accepts or takes for granted, the 'cogito' is a valid inference.*

Here it is not possible to say 'sound.' For, in the *Meditations* we cannot claim to know that the premise 'I think' is true. The evil genius has the power to make the substantial reality behind the appearances such that, although my thinking appears to be the expression of a substance whose essence is that of a thinking thing, it may be that that substance has a very different essence, a very different reality, a reality falsely presented by what appears. If, indeed, the substance behind the appearances is the 'I' of the *Meditations*, then that I is busy deceiving itself about its real being.

In fact, not only is it not possible to say that the inference is sound, it is also not possible, strictly speaking, to say that it is valid. For, even this necessary connection among my ideas, self-evident as it is in its clarity and distinctness, may well turn out to be false. The evil genius challenges even this connection among my ideas, that is, among the essential forms of things as these essential forms exist in the consciousness of the meditating self.

There have been many discussions of the Cartesian *cogito*, from Huet to Hintikka. Some of the discussion turns on whether we ought to construe the *cogito* as a syllogism or as something more akin to an enthymeme, with a suppressed premise.[64] The charge of begging the question in the *cogito* had been raised already by Gassendi. Descartes reported this objection in this way: 'The author of the Rejoinders [Gassendi] will have it that when I say *I think therefore I am*, I am presupposing the major premiss: *what thinks, is*, and have thus already embraced a prejudice.'[65] The point of the objection is that, as an inference, the *cogito* requires as a major premise 'whatever is a thinking thing is an existing thing.' But this general metaphysical premise, like the truths of geometry, is called into question by the demon hypothesis. Hence, Descartes cannot claim the *cogito* as a first truth. What Descartes objects to is the characterization of the dictum as a syllogism. As he puts the point in his replies to the Second Set of Objections,

> when we become aware that we are thinking things, this is a primary notion which is not derived by means of any syllogism. When someone says 'I am thinking, therefore I am, or I exist,' he does not deduce existence from thought by means of a syllogism, but recognizes it as something self-evident by a simple intuition of the mind. This is clear from the fact that if he were deducing it by means of a syllogism, he would have to have had previous knowledge of the major premiss 'Everything which thinks is, or exists'; yet in fact he learns it from experiencing in his own case that it is impossible that he should think without existing. It is in the nature of our mind to construct general propositions on the basis of our knowledge of particular ones. (*Objections*, 2:100)

Descartes's reply, note, does *not* say that the *cogito* escapes the demon. Instead, all that Descartes does is argue that the inference is not a syllogism. This latter point is, of course, quite reasonable. If there is a necessary connection between *thinking* and *existence* in the abstract, then there is also a necessary connection between them in the concrete.[66] In particular, then, the necessary connection between the fact represented by 'I think' and the fact represented by 'I exist' is discernible in those two facts, prior to our having to formulate the general premise that whatever thinks exists. Descartes elsewhere makes the same point about discerning necessary connections in particular facts: 'whatever we demonstrate concerning figures or numbers necessarily links up with that of which it is affirmed. This necessity applies not just to things which are perceivable by the senses but to others as well. If, for example, Socrates says that he doubts everything, it necessarily follows that he understands at least that he is doubting, and hence that he knows that something can be true or false, etc.; for there is a necessary connection between these facts and the nature of doubt' (*Rules*, 1:45).[67] It is the necessary connection between the forms of being that is the object of intuition in the *cogito*.[68] It *is* an inference, though not a syllogism.

That the *cogito* involves an inference has been denied. Thus, Hintikka has so argued in his important paper 'Descartes' *Cogito*: Inference or Performance?'[69] On his view, if the *cogito* were construed as an inference, then it would be trivial. For, Hintikka argues, if the premise is

(1)   *Fa*

then it would be trivial to add that

(2)   *a* exists

where he construes the latter as

(3) $(\exists x)(x = a)$.

The inference, to be sure, is valid. However, if *Fa* is to be meaningful, then *a* must be meaningful and, in particular, must denote an individual. But that is the same condition for (3) to be significant. So the inference from (1) to (3) is trivial, since the conclusion will already be contained in the premise.

To put it another way, the statement that 'I exist' – that is, assuming that I am (= I am identical with) *a*, '$(\exists x)(x = a)$' – is 'self-verifying'and its denial is 'existentially inconsistent.' This is not to say that such a sentence is literally inconsistent; to the contrary, not only is it consistent but its truth is contingent. Instead, '*p* is existentially inconsistent for the person referred to by "*a*" to utter if and only if the longer sentence

            p; and *a* exists

is inconsistent (in the ordinary sense of the word).'[70] Others, such as Ayer, [71] Mackie,[72] and Nahknikian[73] have made much the same point about the self-verifying nature of 'I exist.' Hintikka goes beyond this point, however. For, as he rightly indicates, what is crucial is the *performatory* nature of existential inconsistency and of self-verification. 'The inconsistency (absurdity) of an existentially inconsistent statement can in a sense be said to be of *performatory* (performative) character. It depends on an act or "performance," namely on a certain person's act of uttering a sentence (or otherwise making a statement); it does not depend solely on the means used for the purpose, that is, on the sentence which is being uttered. The sentence is perfectly correct as a sentence, but the attempt of a certain man to utter it assertively is curiously pointless.'[74]

Hintikka argues that what is important about the *cogito* is the fact that when I assert that

(4)   I think that I exist

what I am doing is expressing the thought that is the *performance* that makes

(5)   I exist

self-verifying; that is, verified by the speaker *or thinker* on this occasion of its use. Now, what is self-verifying in the performance is the conclusion (5). In that sense the certainty of the *cogito* can be found there, directly and solely in the conclusion. Yet I can *know* that that conclusion is true only if I am *conscious of* the facts that do actually verify it, conscious of the performance (4) that yields its self-verification. That is why we require not only 'I exist' but also 'I think.' I can assert (5) only if I am in fact conscious of thinking, when I am aware of my own awareness. But that consciousness of thinking which is a necessary condition for asserting (5) is a consciousness of the fact that (4) represents, the fact that makes it true. The 'I exist' of the *cogito* thus derives its peculiar certainty from the fact that it is both, as it were, self-verifying and also is verified by the fact of which the use of 'I think' expresses an awareness. We need, upon this account, not just the self-verifying (5) but also the

(6)   I think

that expresses the consciousness of consciousness that is the performance in which (5) verifies itself. We need, in other words, not just (5) but

(7)   I think, therefore I am.

As Hintikka puts it, 'The function of the word *cogito* in Descartes' dictum is to refer to the thought-act through which the existential self-verifiability of "I exist" manifests itself.'[75] The point is not that (7) is an *inference*, Hintikka is arguing, but that the awareness expressed by the 'premise' (6) or, at least, (4) constitutes the grounds for the conclusion ('therefore') in the sense that it is an awareness of the act or action that is the performance that verifies the 'conclusion.'[76]

There is an important point to what Hintikka says here. This is the point that in the *cogito*, I am aware of an awareness. A consciousness of something presupposes a consciousness of that consciousness. In the First Meditation, we are attending to bodies or body. This attention is a feature of consciousness: in other words, there is a consciousness of the consciousness of body. When the shift of attention occurs at the beginning of the Second Meditation, our attention is directed at the consciousness of body. That attention is a consciousness of the consciousness of body (or whatever). But this consciousness of the consciousness of body itself presupposes a consciousness. The consciousness of the consciousness of body is an immediate awareness of the consciousness of body. It is there when I attend to body, but it itself becomes an attending only when, in the shift of attention represented by the *cogito*, it itself becomes the object of an immediate awareness. In attending to the thinking that (6) represents, I become aware of the consciousness that always makes conciousnesses conscious. This is how Hintikka thinks that (5) is self-verifying, and how, therefore, the *cogito* acquires its peculiar certainty.

This point, that whenever I am conscious of something I am aware of that consciousness, is a very important point. But it does not have quite the significance that Hintikka attributes to it.

There are several things that are problematic about Hintikka's account of the *cogito*. In the first place, it proposes that the point of the *cogito* is to secure the certainty of a first truth. But, as we have seen, this is, in fact, not the point of the *cogito* in the *Meditations*, where, given the introduction of the demon, it could not so function.

This being said, it is also true that in terms simply of the phenomenology of uttering 'I think' Hintikka captures an important feature of how this phrase differs from, say, 'ambulo' ('I walk'). The *cogito* in part acquires its peculiar force that makes it attractive as a first principle precisely because it is, in the sense in which Hintikka makes clear, 'self-verifying' – a point that had also been made earlier by Ayer and Nahknikian. But to be important in this sense, and to be certain in this sense, does not suffice to create the ontological certainty that what I am conscious of in ordinary experience expresses the truth of things, the essential and substantial truth that lies behind the structure of properties that is represented to me in ordinary experience. Hintikka, like Ayer and Nahknikian before him, ignores this *ontological* point. In effect, they all take the Cartesian concerns to be purely epistemological, rather than grounded in ontology.

This points to the second problem of Hintikka's version of the *cogito*: it does not tie in with the substance ontology that Descartes clearly takes for granted. In particular, when Hintikka construes '*a* exists' as the formula (3)

$$(\exists x)(x = a)$$

he misses the point about the meaning of 'exists' in the substance ontology. This is the point that to say that a substance exists is to say that there are properties which it has. Hintikka's (3) does not bring this out at all.

In the third place, again with regard to the substance ontology that Descartes presupposes, what must be proven is that the 'I' that thinks is a substance, an entity that endures through time. In the First Meditation, the 'I' which is attending to body is an immediate awareness of the consciousness of body. In the Second Meditation, this immediate awareness itself becomes the object of an immediate awareness; that is, the 'I' of the First Meditation becomes the object of an immediate awareness. This latter is the 'I' of the Second Meditation. But there is a distinction between the two 'I's: even on Hintikka's account they are at least temporally distinct. But the two 'I's are two consciousnesses: they have different objects. Guéroult, in fact, denies that the two consciousnesses are distinct. They are, rather, one act spread slightly over time with different degrees of self-consciousness within it at different moments: '... the difference between the degrees of consciousness of a single thought do not make it different thoughts. Thus my belief, with or without the express consciousness that I believe, remains the same belief.'[77] The belief does indeed remain the same. But my consciousness of that belief is different from my consciousness of the consciousness. The two consciousnesses have different objects, and are therefore characterized by different ideas, ideas that differ in their form, in their objective reality. These two 'I's are therefore distinguishable.

But the *cogito*, if it is to do the job of a first truth within the substance ontology that Descartes accepts, must give us the truth about a substance: the 'I' about which it talks must be a substance. Hence, if the 'I' as a substance is to be discovered in the *cogito*, then the *cogito* must establish the substantial identity of these two 'I's. It fails to do this. It fails to do this on Hintikka's account, though it should do so if the *cogito* is to function, as Hintikka suggests that it should function, as a foundational first truth.

In fact, on Descartes's own account of the *cogito*, it fails to establish that there is a thinking substance. What I am according to the *cogito* is a thinking thing, but that thinking thing may in fact be entirely non-substantial: my reality may be very different. The thinking that is my consciousness may have no substantial reality, and the 'I' that is aware of an 'I' may have no substantial continuity. This is, in effect, the point that Huet made when he transformed the 'I think, therefore I am' into 'I may have thought, therefore perhaps I was':[78] the thinker that thinks the 'I am' exists *subsequently* to the thinker who thinks 'I think.' That is why we have after 'I think' an 'I was,' and why the consciousness that thinks this 'I was' can think of its basis only as an 'I thought.' Moreover, since there is a real distinction between the temporal parts of the self, those parts are separable, and the one could exist without the other. The existence of the second part therefore cannot guarantee the existence of the first part, even if the former thinks about the latter. The 'I think' cannot even be 'I thought'; it can only be 'I may have thought.' The thinker thinking that 'I was' can only know with probability, not necessity, that the thinker who previously thought 'I think' actually did exist. The inference must therefore be no stronger than 'I may have thought, therefore perhaps I was.' What this demolition of the *cogito* depends upon is the temporal separability of the events that constitute the inference within the mind of the meditator. The discovery of this point was hardly new with Huet, however. For, Descartes himself draws our attention to the crucial premise in the Third Meditation. He is considering the fact that so far as he can see, he is a substance. Now, a substance endures through time. But there is nothing in the concept which he has thus far acquired of himself – nothing in the essence and the powers which he has by virtue of that essence – which guarantees that he can conserve himself in existence. It might, in fact, be the case that the 'I' of one moment is different from the 'I' of the next moment. '[I]t does not follow from the fact that I existed a little while ago that I must exist now, unless there is some cause which as it were creates me afresh at this moment – that is, which preserves me. For it is quite clear to anyone who attentively considers the nature of time that the same power and action are needed to preserve anything at each individual moment of its duration as would be required to create that thing anew if it were not yet in existence' (*Meditations*, 2:33). Descartes is thus aware of the problem that Huet raises, that Guéroult denies, and that Hintikka does not notice. It shows once again that Descartes is well aware of the sceptical arguments of the sort that can be raised with regard to the *cogito*. It follows, as we have already concluded, that the *cogito* cannot provide a first truth on which to build the edifice of knowledge.

The *cogito* has a different role.

Here is how it is introduced in the *Meditations*.

I have convinced myself that there is absolutely nothing in the world, no sky, no earth, no minds, no bodies. Does it now follow that I too do not exist? No: if I convinced myself of something then I certainly existed. But there is a deceiver of supreme power and cunning who is deliberately and constantly deceiving me. In that case I too undoubtedly exist, if he is deceiving me; and let him deceive me as much as he can, he will never bring it about that I am nothing so long as I think that I am something. So after considering everything very thoroughly, I must finally conclude that this proposition, I am, I exist, is necessarily true whenever it is put forward by me or conceived in my mind. (*Meditations*, 2:16)

He has previously introduced the evil genius as a hypothesis, a supposition, and not as a proposition that is affirmed as true. 'I will suppose therefore that not God, who is supremely good and the source of truth, but rather some malicious demon of the utmost power and cunning has employed all his energies in order to deceive me' (2:15). He is now taking what had started as a hypothesis and actually asserting it: '*there is a deceiver ...,*' Descartes says. But asserting it in this way does not mean asserting it as true. Rather, it is merely being taken as the starting point in a series of inferences, that is, being taken *as if* it were true. And from this premise now being taken as if it were true, Descartes draws a conclusion. It goes like this.

> Given that there is a deceiver who is deceiving me, it follows that I am being deceived. But to be deceived is to affirm as true what I am thinking about when it is false. Hence, if I am being deceived, then I am thinking. Hence, (given that there is a deceiver,) it follows that I am thinking. But from this, that has now been established, that I am thinking, it follows that I exist.

Thus, *Descartes is concluding that he exists from the hypothesis that there is an evil genius who is deceiving him.* In other words, in the *Meditations, the 'cogito' takes its place as an inference from the hypothesis of the existence of the evil genius who is devoting her energies to deceiving one.*

Descartes's ultimate aim is to prove the immortality of the soul. What he goes on to do in the Second Meditation is to provide the materials for such a proof. But, as he tells us in the 'Synopsis,' the proof is not complete.

[S]ince some people may perhaps expect arguments for the immortality of the soul in this [the second] section, I think they should be warned here and now that I have tried not to put down anything which I could not precisely demonstrate. Hence the only order which I could follow was that normally employed by geometers, namely to set out all the premises on which a desired proposition depends, before drawing any conclusions about it. Now the first and most important prerequisite for knowledge of the immortality of the soul is for us to form a concept of the soul which is as clear as possible and is also quite distinct from every concept of body; and that is just what has been done in this section. A further requirement is that we should know that everything that we clearly and distinctly understand is true in a way which corresponds

exactly to our understanding of it; but it was not possible to prove this before the
Fourth Meditation. (*Meditations*, 2:9)

What he needs in order to demonstrate the immortality of the soul is a clear and
distinct conception of the soul as distinct from the body. This he proceeds to
give in the Second Meditation. Descartes has in his mind the idea of himself as
a thinking thing. This concept of himself is clear and distinct; he must therefore
affirm its existence. This is the conclusion of the *cogito*. But he has already, in
the First Meditation, pointed out that he can deny the existence of body. But if I
can deny the idea of X while affirming the idea of Y, then the two ideas are separable: that is the criterion of separability. It follows that there is nothing that
attaches to me of necessity other than my thinking. All other characteristics I
perceive clearly and distinctly as separable from the thinking that constitutes
my existence. It is thus this form and this form alone that constitutes my
essence: *sum res cogitans*.

At last I have discovered it – thought; this alone is inseparable from me. I am, I exist –
that is certain. But for how long? For as long as I am thinking. For it could be that
were I totally to cease from thinking, I should totally cease to exist. At present I am not
admitting anything except what is necessarily true. I am, then, in the strict sense only a
thing that thinks; that is, I am a mind, or intelligence, or intellect, or reason – words
whose meaning I have been ignorant of until now. But for all that I am a thing which is
real and which truly exists. But what kind of a thing? As I have just said – a thinking
thing. (2:18)

But this could not constitute a proof until it has been proven that what is clear
and distinct is true. For, the proof that the soul is distinct from the body and is
therefore metaphysically capable of surviving the body depends upon clearly
and distinctly perceiving that thoughts and all other characteristics are separable. But as Descartes tells us in the 'Synopsis,' he is unable to prove that what is
clear and distinct is true until after the Fourth Meditation. It therefore follows
that the aim of the *Meditations* of establishing the immortality of the soul cannot be accomplished until after the Fourth Meditation. This, of course, fits with
what we have been arguing, that the *cogito* does not constitute a first piece of
knowledge foundational for all other knowledge. Whatever Descartes is doing
at this point it is not what Euclid was doing, taking for granted certain self-
evident premises and drawing conclusions from those premises by means of
self-evident steps of inference. Nor, for the same reason, is he doing what the
author of the *Discourse* was doing.

  None the less, Descartes also tells us in the 'Synopsis' that he is following the pattern of argument of the geometers: 'the only order which I could follow was that normally employed by geometers, namely to set out all the premises on which a desired proposition depends, before drawing any conclusions about it.' Just what is this geometrical method if it is not the axiomatic method of the geometers?

III: The Method of Analysis: From the Meditator to the Deity by the Way of Ideas

In order to see what the Cartesian geometrical method is, it is necessary to recall that the Greeks proposed *two* methods for proceeding in geometry. One was the method of *synthesis* and the other the method of *analysis*.

Descartes, in his replies to the Second Set of Objections, points to the origins of these methods in the ancient world. 'It was synthesis alone that the ancient geometers usually employed in their writings. But in my view this was not because they were utterly ignorant of analysis, but because they had such a high regard for it that they kept it to themselves like a sacred mystery' (*Objections*, 2:111). The synthetic method is that which we find in Euclid's *Elements*, and it was indeed the favoured method of the ancients for the presentation of their results in geometry. But the method of analysis was well known. It received its clearest statement – at least, it is the clearest statement that has come down to us – in the 'Treasury of Analysis' of Pappus. Here is Pappus's statement concerning the nature of the analytic method: '*Analysis* ... takes that which is sought as if it were admitted and passes from it through its successive consequences to something which is admitted as the result of synthesis: for in analysis we assume that which is sought as if it were already done, and we inquire what it is from which this results, and again what is the antecedent cause of the latter, and so on, until by so retracing our steps we come upon something already known or belonging to the class of first principles ...'[79] In contrast, 'in *synthesis*, reversing the process, we take as already done that which was last arrived at in the analysis and, by arranging in their natural order as consequences what before were antecedents, and successively connecting them one with another, we arrive finally at the construction of what was sought ...' (Heath, *Greek Mathematics*, 400–1). Pappus distinguishes two kinds of analysis, theoretical and problematical. What he says about the former is also worth noting. 'In the *theoretical* kind [of analysis] we assume what is sought as if it were existent and true, after which we pass through its successive consequences, as if they too were true and established by virtue of our hypothesis, to something admitted: then (a), if that something admitted is true, that which is sought will also be true and the proof will correspond in the reverse order to the analysis, but (b), if we come upon something admittedly false, that which is sought will also be false.' (ibid.).

The method of analysis is likely clear to anyone who has taught logic. In logic we ask our students to construct formal proofs. Given certain premises, $P_1, P_2, ..., P_n$, students are required to construct a formal proof that some conclusion C follows validly from those premises. The problems are usually posed in such a way that the students know that such a proof does exist: their task is to solve the puzzle by discovering the proof that they know is there for them to discover. The formal proof consists in a series of steps that lead from the premises to C, each step justified by one of the rules of logical deduction that are used to define the set of formal proofs. We often give instructions to students that go something like this:

> There are two ways for searching out formal proofs. You can begin from the premises and work forward towards the conclusion. Or, you can begin from the conclusion and try to work backwards towards the premises.

The first way of trying to find a proof is the method of synthesis, the second way is the method of analysis. But if analysis is the method of discovery, it is not the method of presentation. The method of presentation is that of synthesis. In the method of analysis, one takes the conclusion C as given. It is accepted as a hypothetical starting point. Deductions are then made from it of exactly the same sort that are made from premises given as true. The aim is to find a series of steps that one will be able to reverse so as to yield a formal proof of the conclusion from the premises. If such a proof is found, if the method of analysis does produce such a proof, then the conclusion will have been proven true. If, as Pappus suggests, the deductions made from the conclusion towards the premises can be put in reverse order to give a synthetic proof of the conclusion from the premises, then the steps must all involve logical equivalences or identities, but in any case must all be strictly convertible.

Not every analytic procedure needs to involve strictly convertible steps, however. For Pappus allows *reductio ad absurdum* to be a species of analysis. In this case we proceed from the conclusion taken hypothetically as given and proceed to deduce things from it. If these turn out to be contradictory or otherwise false, then we have grounds for rejecting the conclusion as false.

The Greeks distinguished three sorts of conclusion in geometry. The first is a theorem, which is a theoretical principle. The second is a problem, which requires a construction. The third is a porism, which establishes an existence claim.[80] Problems in algebra are essentially porisms. Consider the problem that we would represent by asking for the solution to the two equations:

(a) $a + b \doteq 20$

(b) $a - b = 14$

What we are trying to prove when we are solving these equations is the theorem that states that

(c)   There are unique numbers x and y such that $x + y = 20$ and $x - y = 14$.

We prove this theorem in the following way.

> Assume that there are numbers that satisfy (c). Call these numbers a and b. Then (a) and (b) hold. It follows that $a = 20 - b$ and that $a = 14 + b$, and therefore further that $b + 14 = 20 - b$. From this we infer that $2b = 20 - 14 = 6$. Hence $b = 3$. Since from (b), $a = 14 + b$, we can conclude that $a = 17$. Since $17 + 3 = 20$ and $17 - 3 = 14$, the conclusion (c) follows.

We have assumed the conclusion as a hypothesis when we denote the numbers a and b and take them as satisfying certain equations. We then move from this conclusion, that is, potential conclusion now taken hypothetically as a premise, to infer that $a = 17$ and that $b = 3$. These inferences are by a series of substitutions of identicals. These inferences are convertible. So, from $a = 17$ and $b = 3$ as premises we can reverse the inferences and deduce (a) and (b) as the conclusion we want.

   We took our conclusion for granted and, by a series of inferences, moved from that to a set of propositions known to be true that are then used as premises from which to deduce and justify the conclusion that we previously simply took for granted. The first part of this process was a process of analysis. By this process we arrived at a set of premises from which we can deduce the conclusion. These premises are known truths. The inferences are now reversed, and we move from these premises to deduce the conclusion we want. This latter process is the synthesis. Nowadays, of course, in algebra we don't bother to do the synthesis after having completed the analysis. But we could.

   In the above algebraic example, the solutions to the equations we were looking at were discrete numbers. But we can also look at cases where the equations, as we say, define a continuous series of numbers. If such equations are considered geometrically, then what they define is not a set of discrete points but rather a curve or locus. The Greeks were interested in loci, but they could not adopt algebraic solutions of the sort that we use, since for the Greeks algebra deals with numbers and numbers are discrete units. There is an incommensurability, as it were, between geometry, which deals with continuous series, and algebra, which deals with discrete series. But the modern world did not share the sharp Greek distinction between the discrete and the continuous. We are therefore prepared to use algebraic means to deal with continuous curves or loci.

The discovery that algebra could be used to do geometry was, of course, made by Descartes.[81] The first presentation of this new analytic geometry was in the treatise on *Geometry* that was attached to the *Discourse on Method*.[82] The algebra which he develops is geometrically based; his primary terms are line segments, with different constructions representing different algebraic operations. We do not need to go into the details.[83] Suffice it to say that the method that Descartes uses is *analysis* in the sense of Pappus. Here is how he describes it:

if we wish to solve some problem, we should first of all consider it solved, and give names to all the lines – the unknown ones as well as the others – which seem necessary in order to construct it. Then, without considering any difference between the known and the unknown lines, we should go through the problem in the order which most naturally shows the mutual dependency between these lines, until we have found a means of expressing a single quantity in two ways. This will be called an equation, for the terms of one of the two ways [of expressing the quantity] are equal to those of the other. And we must find as many such equations as we assume there to be unknown lines.[84]

Descartes proposes that the method of analysis is the proper method for discovery, in contrast to the method of synthesis, which is the proper method for presenting material once it is known. It is therefore the method that he used in the *Meditations*. As he indicates, 'it is analysis which is the best and truest method of instruction, and it was this method alone which I employed in my Meditations. As for synthesis, which is undoubtedly what you are asking me to use here, it is a method which it may be very suitable to deploy in geometry as a follow-up to analysis, but it cannot so conveniently be applied to these metaphysical subjects' (*Objections*, 2:111). The analytic method is what Descartes used in the case of the *cogito*. A presentation according to the synthetic method would have involved a syllogism, in which 'I exist' is deduced from 'I think' by means of a major premise 'Whatever thinks, exists.' But that premise is not first in the order of discovery; to the contrary, the particular proposition 'I think, therefore I am' is the first in the order of discovery. The direct inference from 'I think' to 'I exist' by means of their necessary connection is the order of discovery, the order of the method of analysis. Descartes's denial that he is using the synthetic method of syllogistic is worth repeating: it makes clear that Descartes intends us to read the *cogito* as proceeding by the method of analysis.

[W]hen we become aware that we are thinking things, this is a primary notion which is not derived by means of any syllogism. When someone says 'I am thinking, therefore I am, or I exist,' he does not deduce existence from thought by means of a syllogism, but recognizes it as something self-evident by a simple intuition of the mind. This is clear from the fact that if he were deducing it by means of a syllogism, he would have to have had previous knowledge of the major premiss 'Everything which thinks is, or exists'; yet in fact he learns it from experiencing in his own case that it is impossible that he should think without existing. It is in the nature of our mind to construct general propositions on the basis of our knowledge of particular ones. (2:100)

But if it proceeds by the method of analysis, then there is no need to suppose that it is beginning with anything other than a hypothesis. That is, it is unnecessary to suppose that the 'I think' with which it begins is any more certain than any other hypothesis taken along the path of analysis towards a proposition whose truth may safely be taken as given – a truth whose certainty is beyond challenge by the demon.

This point may be applied more generally to the overall structure of the argument of the *Meditations*: the method of analysis begins by taking something as a hypothesis, treating it as given, and deducing other things from it, in the hope either of finding premises from which it might be deduced and thereby proved to be true, or finding known premises which contradict it, thereby proving it to be true. In fact, Descartes tells us clearly that this is precisely what he did in the *Meditations*:

I think I can easily get round this objection if I say that I have never denied that there are real accidents. It is true that in the Optics and the Meteorology I did not make use of such qualities in order to explain the matters which I was dealing with, but in the Meteorology ... I expressly said that I was not denying their existence. And in the Meditations, although I was supposing that I did not yet have any knowledge of them, I did not thereby suppose that none existed. The analytic style of writing that I adopted there allows us from time to time to make certain assumptions that have not yet been thoroughly examined; and this comes out in the First Meditation where I made many assumptions which I proceeded to refute in the subsequent Meditations. Further, it was certainly not my intention at that point to establish any definite results concerning the nature of accidents; I simply set down what appeared to be true of them on a preliminary survey. (*Objections*, 2:173)

And indeed, as we have seen, this is precisely how the *Meditations* proceeds: it begins with the hypothesis of the evil genius, and deduces other things from that hypothesis taken as given. In the first inference from this hypothesis Descartes, or, rather, the 'I' of the *Meditations*, deduces that he or she thinks, then further deduces that he or she exists, and then, still further, also deduces that his or her essence is that of a thinking thing. But, again given the nature of the analytic method, none of these is yet proven to be true; all that we know is that they follow from the hypothesis from which we started and which we are still taking hypothetically.

Guéroult has argued quite correctly that the method of the *Meditations* is the method of analysis. He has also argued quite correctly that this method involves the deductive method of classical geometry. As Guéroult puts it at one point, Descartes's 'demonstrations proceed everywhere from the spirit that moves Euclid ...; they can only be grasped by those who have a sense of mathematical demonstrations ... The *Six Meditations* are only the metaphysical answer to the fifteen books of Euclid's *Elements*.'[85] But this implies that the *Meditations* begin, like Euclid's text, from premises that are taken certainly to be true. To take the method of analysis to be deductive after the fashion of Euclid, to take Descartes to be attempting to provide *demonstrations* of the truth of certain propositions, requires Descartes to begin from truths which are known incorrigibly to be true. And so Guéroult has to read Descartes as starting from the *cogito* understood as providing a first truth.[86] Guéroult is right in his recognition that the steps taken by the meditating self are deductive, deductive in the sense of Euclid – but also in the sense of Pappus, which is, of course, the same sense as that of Euclid. The point is that where Euclid provides demonstrations that start from truths, Pappus allows one to begin from hypotheses. If, then, we take the method of analysis to be not only deductive – as Guéroult affirms – but also a method of discovery that proceeds from hypotheses rather than known truths – as Guéroult does not recognize – then we are not required to make more of the *cogito* in the *Meditations* than Descartes himself makes of it: Descartes allows that, given the demon hypothesis, the *cogito* must be taken like all other clear and distinct propositions, as a mere hypothesis, and, contrary to what Guéroult suggests, his method, taken from Pappus and not Euclid, allows him to do this.

Descartes carries on the process of analysis in the Third Meditation, the discussion here taking up where the Second Meditation ended.

The Second Meditation ended with a discussion of how we come to know the truth about the world as it appears to us. The wax example is used to establish, on the one hand, that we understand the wax in terms of our clear and distinct idea of extension, and, on the other hand, given the argument from sense variability that Descartes adapted from Montaigne and indirectly from Sextus, that this idea that describes the timeless structure of the world could not have been obtained by abstraction from the variable materials of sense experience. Descartes here locates the criterion that he employs to describe the truth of the world as we ordinarily experience it. It is the same criterion that he has employed to discover the truth about himself. In the case of body, the truth is given by the form or nature or essence that is in his mind, the clear and distinct idea of extension. In the case of himself, he has discovered that the truth about himself is given by his essence as that of thinking thing. The truth about things in both cases is given by clear and distinct ideas.

The Third Meditation begins by reminding us of this point, and then going on also to remind us that the hypothesis of the evil genius calls into question our clear and distinct ideas – including, as he here points out, the *cogito*, the clear and distinct idea that one has of oneself as a thinking thing. Although Descartes has discovered the criterion that he uses to discover the truth about the world of ordinary experience, the criterion used to discover the correct standard of coherence, he has not succeeded in showing that this criterion is reliable. To be sure, the doubt is slight. 'And since I have no cause to think that there is a deceiving God, and I do not yet even know for sure whether there is a God at all, any reason for doubt which depends simply on this supposition is a very slight and, so to speak, metaphysical one. But in order to remove even this slight reason for doubt, as soon as the opportunity arises I must examine whether there is a God, and, if there is, whether he can be a deceiver' (*Meditations*, 2:28). But, though the doubt is slight, in that he cannot conceive alternatives to what he is thus doubting, none the less it is a doubt that goes to the metaphysical core of reality. Given this metaphysically generated doubt, Descartes has yet to establish that the criterion of truth really does give the truth. He has not yet exorcised the demon.

The process of analysis that is taken up in the Third Meditation aims at exorcising this demon. What happens in the Third Meditation is that we have a movement towards God. Now, at this point the concept of God that he has in mind is one that is compatible with the demon hypothesis; indeed, the God that Descartes considers could be that demon. Thus, he speaks of 'the idea that gives me my understanding of a supreme God, eternal, infinite, immutable, omniscient, omnipotent and the creator of all things that exist apart from him ...' (*Meditations*, 2:28). In *this* idea there is nothing as yet that excludes God's being a deceiver: the idea as thus described has as yet no moral attributes, and this creator could thus also be a deceiver for all that *this* conception implies.

But one thing about *this* idea is clear and distinct: it is the idea of an infinite being. It therefore 'certainly has in it more objective reality than the ideas that represent finite substances' (*Meditations*, 2:28). This idea, like similar ideas of thinking and material substances, is an essence or form, the essence or form of the thing thought about *qua* literally in the mind of the knower, or, at least at this point, in the mind of the thinker. But among these ideas, some have within them more reality than others, and, specifically, the idea of God contains more reality within it than the idea of thinking or of material substance.

This introduces the notion of degrees of reality and, with that, degrees of perfection. The perfection of a thing consists, of course, in its existing in conformity to its form or essence: it is perfect, after its kind, just to the extent that it perfectly expresses in its being its essence or form. But some essences have within themselves a greater capacity for being. Such substances are therefore more perfect, and have more reality. In particular, of course, an infinite being, the first cause of all things, will have the highest degree of perfection and, consequently, of reality. As Aquinas puts it, 'the first active principle must needs be most actual, and therefore most perfect; for a thing is perfect in proportion to its state of actuality, because we call that perfect which lacks nothing of the mode of its perfection' (*Summa*, FP, Q4, A2). Note that the first principle is a first *active* principle. This is crucial, as we shall see below. Some who recognize the notion of degrees of reality and degrees of perfection fail to make the connection to activity.[87] But of that, more later. The present point is that, since the infinite God creates all things, the being of those things must be in Him: 'All created perfections are in God. Hence He is spoken of as universally perfect, because He lacks not ... any excellence which may be found in any genus' (FP, Q4, A2). Or, again, '[s]ince ... God is subsisting being itself, nothing of the perfection of being can be wanting to Him. Now all created perfections are included in the perfection of being; for things are perfect, precisely so far as they have being after some fashion. It follows therefore that the perfection of no one thing is wanting to God.' (ibid.). Of course, many of these ways of being will be present in the infinite being, God, eminently rather than formally: the infinite being will not be extended – extension is not in God formally; but if extended being does exist, then it is created by God and extension must therefore be in God, that is, in God eminently. Thus, degrees of reality correspond to the amount of being produced by different essences. The idea of an infinite being, an infinitely powerful being – this essence – , is an essence which has within it the capacity to produce all being, and therefore among all our ideas has the highest degree of reality.

Descartes now introduces a causal principle that is evident by the natural light of reason, a clear and distinct proposition with regard to relations among ideas.

[I]t is manifest by the natural light that there must be at least as much reality in the efficient and total cause as in the effect of that cause. For where, I ask, could the effect get its reality from, if not from the cause? And how could the cause give it to the effect unless it possessed it? It follows from this both that something cannot arise from nothing, and also that what is more perfect – that is, contains in itself more reality – cannot arise from what is less perfect. And this is transparently true not only in the case of effects which possess what the philosophers call actual or formal reality, but also in the case of ideas, where one is considering only what they call objective reality. (*Meditations*, 2:28)

With these Descartes is going to move from the idea of himself and from the idea of God to the existence of God.

But he has first to establish one more point with regard to the idea of God: *he must establish that we do in fact have this idea.* All that Descartes has done so far is *claim* that he has such an idea. It is possible that someone will claim that they have a different idea of God. But he must convince the meditator who is sharing the process with Descartes – the 'I' of the *Meditations* is supposed to be not only Descartes but the personally involved reader – Descartes must convince this other meditator that he or she has the same idea as he (Descartes) has. He is going to do this in a way that links the idea of God which he is now going to argue that we have with the causal principles that have just been introduced.

The relevant principles are, of course, necessary. Descartes distinguishes connections among ideas that are necessary and those that are contingent.

[T]he conjunction between ... simple things is either necessary or contingent. The conjunction is necessary when one of them is somehow implied (albeit confusedly) in the concept of the other so that we cannot conceive either of them distinctly if we judge them to be separate from each other. It is in this way that shape is conjoined with extension, motion with duration or time, etc., because we cannot conceive of a shape which is completely lacking in extension, or a motion wholly lacking in duration ... The union between such things, however, is contingent when the relation conjoining them is not an inseparable one. This is the case when we say that a body is animate, that a man is dressed, etc ... (*Rules*, 1:45)

It is the necessary connections that are traced out in any deduction, and, in particular, in the deductions that are part of the analytic process of the *Medita-*

tions. These necessary connections include the causal relations that Descartes has now introduced. In the *Rules* he tells us a number of important things about these connections that help us to understand what he is about in the *Meditations*.[88]

Specifically, in Rule VI, he asks us to note that 'everything, with regard to its possible usefulness to our project, may be termed either "absolute" or "relative" – our project being, not to inspect the isolated natures of things, but to compare them with each other so that some may be known on the basis of others' (*Rules*, 1:21). That thing is absolute which 'has within it the pure and simple nature in question; that is, whatever is viewed as being independent, a cause, simple, universal, single, equal, similar, straight, and other qualities of that sort. I call this the simplest and the easiest thing when we can make use of it in solving problems' (ibid.). In contrast, 'the "relative" ... is what shares the same nature, or at least something of the same nature, in virtue of which we can relate it to the absolute and deduce it from the absolute in a definite series of steps' (ibid.). The natures or ideas can be arranged in an order from the more complex to the simpler and finally to the absolute. These connections lead us from the least term to the absolute term by a series of steps. '[G]iven the last term we should be able to reach the one that is absolute in the highest degree, by passing through all the intermediate ones' (ibid.). Relative and absolute depend upon where you are in the series, and a term that is absolute with respect to a certain relative may itself be a relative with respect to something more absolute. What we have to discover is that which is 'absolute in the highest degree' (1:22). He illustrates what he has in mind:

For example, the universal is more absolute than the particular, in virtue of its having a simpler nature, but it can also be said to be more relative than the particular in that it depends upon particulars for its existence, etc. Again, certain things sometimes are really more absolute than others, yet not the most absolute of all. Thus a species is something absolute with respect to particulars, but with respect to the genus it is relative ... Furthermore, in order to make it clear that what we are contemplating here is the series of things to be discovered, and not the nature of each of them, we have deliberately listed "cause" and "equal" among the absolutes, although their nature really is relative. Philosophers, of course, recognize that cause and effect are correlatives; but in the present case, if we want to know what the effect is, we must know the cause first, and not vice versa. Again, equals are correlative with one another, but we can know what things are unequal only by comparison with equals, and not vice versa, etc. (ibid.)

On this account, as Descartes says, the species is relative to the genus, which is more absolute. Thus, doubting is relative to thinking, which is more absolute, and this in turn is relative to being, which is simpler and more absolute. This, of course, is the order of the *Meditations*. Again, the cause is more absolute than the effect; the cause and effect are related terms in a series, and the effect as the relative leads us towards the cause. Similarly, the equal is more absolute than the unequal; they, too, are related terms in a series, and the unequal as the relative leads us towards the equal. Further, in each case the absolute is simpler than that which is relative to it: the relative is derived from the absolute by adding a restriction that limits the absolute in some way. Thus, the genus is simpler than the species; the latter is derived from the former by adding the specific difference. The species thinking is relative to the genus being, and is a limitation of the genus to a restricted class. Similarly, the cause is absolute relative to the effect; the latter exists as a limitation of the former. Thus, I exist as a thinking thing; my cause will contain this reality within it, plus more. In this way, my being is a limitation of the being of my cause. And finally, the unequal is a limitation of the perfectly equal.

Now, in order to judge that something is *not red* it is necessary to have the concept of *red*, and, more generally, in order to judge that a concept is *not F* it is necessary to know what *F* is. And so, in order to judge that something is unequal it is necessary to know what it is to be equal. This is true with regard to each of the relative-absolute series that Descartes mentions. In order to judge of the limited that it *is* limited it is necessary to know the absolute of which it is the limitation. Thus, in order to know of an effect that it is an effect and therefore limited, it is necessary to know the cause of which it is the limitation. Similarly, in order to know of a species that it is a species, it is necessary to know the genus of which it is the limitation. In each case, the thing which is limited is limited because something is absent that is in the absolute. The genus, the determinable, contains all the species within it. The species is a limitation of the genus which excludes the other species. The effect is a limitation of the cause which excludes the other effects the cause is capable of producing. In thus excluding something, the limited is always the negative. If the limited is the absolute to which it is relative it is also the limitation and therefore the negation of that absolute.

This sort of reasoning was first developed by Plato in that most remarkable of philosophical documents, the dialogue called the *Phaedo*. Socrates argues that all that we experience by sense are imperfect equalities. Yet we judge all of them to be imperfect. We must therefore have the concept of perfect equality. But by hypothesis we do not have this concept from ordinary sense experience: all that we ever know by sense experience are imperfect equalities. The concept of perfect equality must therefore be a priori. Descartes uses the same form of argument to establish that we must have the idea of God as a perfect being.

When Margaret Wilson writes that 'Descartes claims to find in himself, among his other ideas, the idea of an infinitely perfect being,'[89] she misses the point completely. She gives the impression of Descartes rummaging around in his intellect and finding more or less by accident that he has this idea. Others may rummage around and find some other idea of God. But this is far from the correct story. Descartes provides an *argument* why he *must* have the idea of a perfect being, and similarly why anyone else who has followed him this far in the life of meditation must also have the same idea. Under these circumstances one simply cannot argue that finding the idea is accidental or that one has some other idea. One *may*, in some sense of "may," have some other idea of a supreme being – after all, Descartes himself has introduced the idea of an immensely powerful evil genius. But whatever other idea of a powerful being it is that one has, one *also* has, because one *must* have, the same idea of a supreme being as Descartes has.

We have an idea of God. This we already introduced in connection with the hypothesis of the evil genius. What we have now to note about this idea is that it is not merely a negative idea. It is not the idea of a being which requires the absence of something. To the contrary, as an infinite being, it contains all being within it; it is limited in no way.

Nor can it be said that this idea of God is perhaps materially false and so could have come from nothing, which is what I observed just a moment ago in the case of the ideas of heat and cold, and so on. On the contrary, it is utterly clear and distinct, and contains in itself more objective reality than any other idea; hence there is no idea which is in itself truer or less liable to be suspected of falsehood. This idea of a supremely perfect and infinite being is, I say, true in the highest degree; for although perhaps one may imagine that such a being does not exist, it cannot be supposed that the idea of such a being represents something unreal, as I said with regard to the idea of cold ... I understand the infinite, and that I judge that all the attributes which I clearly perceive and know to imply some perfection – and perhaps countless others of which I am ignorant – are present in God either formally or eminently. (*Meditations*, 2:31)

More importantly, from what I have discovered, it follows that we *must* have this idea of an infinite, and infinitely perfect being. I have discovered that, so far as I can tell, my essence is that of a thinking thing. I discovered this through an examination of other ideas. In particular, I distinguished myself from extended being: the two ideas, and therefore the two essences, so far as I can tell, are separable. But from this it follows that I am a limited being. But if I know that I am a limited being, then I must have the concept of a being without limitations, a being that contains within it all being. This idea of an infinite being, a being that includes all being within it, and is therefore supremely perfect, is an idea the presence of which is required by the idea of myself as a limited being.

But perhaps I am something greater than I myself understand, and all the perfections which I attribute to God are somehow in me potentially, though not yet emerging or actualized. For I am now experiencing a gradual increase in my knowledge, and I see nothing to prevent its increasing more and more to infinity. Further, I see no reason why I should not be able to use this increased knowledge to acquire all the other perfections of God. And finally, if the potentiality for these perfections is already within me, why should not this be enough to generate the idea of such perfections?

But all this is impossible. First, though it is true that there is a gradual increase in my knowledge, and that I have many potentialities which are not yet actual, this is all quite irrelevant to the idea of God, which contains absolutely nothing that is potential; indeed, this gradual increase in knowledge is itself the surest sign of imperfection. What is more, even if my knowledge always increases more and more, I recognize that it will never actually be infinite, since it will never reach the point where it is not capable of a further increase; God, on the other hand, I take to be actually infinite, so that nothing can be added to his perfection. And finally, I perceive that the objective being of an idea cannot be produced merely by potential being, which strictly speaking is nothing, but only by actual or formal being. (2:32)

This last statement about causation shows that the argument for the existence of the idea of an infinite being is part of the general causal argument for the existence of an infinite, and infinitely perfect, being. There are two arguments. The first proceeds from the fact that I have within me the idea of an infinite being. This idea is a form or essence. But it is also a limited form of the essence. It is a mode or property of my mind. So it *is* but *is in a limited way*. As a limited being, it presupposes an unlimited being which is its cause. This limited being could be caused to be in any way, and in particular could be caused to be as a mode or property of my mind, only by a cause that contains such an idea or essence within itself. But it is the essence of an infinite being. Nothing could cause such an essence to be in any way unless it contained that essence within itself. Therefore an infinite unlimited being exists as the cause of my having the idea of such a being within me.

Then, secondly, there is the argument from my own existence. I am a limited being. But any limited being presupposes unlimited being as its cause. Therefore, if I exist, God exists.

These arguments have as their conclusion the proposition that God exists. We have arrived at that conclusion by tracing out necessary connections as required by the method of analysis. Descartes makes a couple of points with regard to necessary connections that we should note. '[T]here are many instances of things which are necessarily conjoined, even though most people count them as contingent, failing to notice the relation between them: for example the proposition, "I am, therefore God exists," or "I understand, therefore I have a mind distinct from my body." Finally, we must note that very many necessary propositions, when converted, are contingent. Thus from the fact that I exist I may conclude with certainty that God exists, but from the fact that God exists I cannot legitimately assert that I too exist' (*Rules*, 1:45). In the Third Meditation Descartes takes up the proposition that 'I am, therefore God exists.' In the *Rules* it is taken for granted that the self-evidence of its necessity is clear. In the *Discourse* he had elaborated to a certain extent upon this argument. But he soon realized that more was needed. Shortly after the publication of the *Discourse* he wrote in a letter to Vatier (22 February 1638) that additional inferences were needed if he was successfully to explain and defend this crucial inference. 'I supposed that certain notions which habitual reflection had made familiar and evident to me would necessarily be so to everyone – for example that our ideas could not receive their form or their being except from some external objects ... [and] could not represent any reality or perfection not present in those objects' (3:86). Descartes is here referring in particular to the argument from the idea of an infinite being to the existence of such a being. Even the presentation of this argument in the *Meditations* was found difficult by some. Thus, the Thomist Caterus writes in the First Set of Objections that the idea of an infinite being, like the idea of any being, insofar as it has a certain objective reality, 'requires no cause; for objective reality is a pure label, not anything actual. A cause imparts some real and actual influence; but what does not actually exist cannot take on anything, and so does not receive or require any actual causal influence. Hence, though I have ideas, there is no cause for these ideas, let alone some cause which is greater than I am, or which is infinite' (*Objections*, 2:67). In fact, however, on the traditional account, the objective reality is not just something external to the mind but an actual feature of the mind. It *is* the essence *qua* in the mind, and the mind having just *this* characteristic and not some other must be accounted for; this, too, requires a causal explanation. This is the point that Descartes labours to make in his reply to Caterus:

Notice ... that he [Caterus] is referring to the thing itself as if it were located outside the intellect, and in this sense 'objective being in the intellect' is certainly an extraneous label; but I was speaking of the idea, which is never outside the intellect, and in this sense 'objective being' simply means being in the intellect in the way in which objects are normally there. For example, if anyone asks what happens to the sun through its being objectively in my intellect, the best answer is that nothing happens to it beyond the application of an extraneous label which does indeed 'determine an act of the intellect by means of an object.' But if the question is about what the idea of the sun is, and we answer that it is the thing which is thought of, in so far as it has objective being in the intellect, no one will take this to be the sun itself with this extraneous label applied to it. 'Objective being in the intellect' will not here mean 'the determination of an act of the intellect by means of an object,' but will signify the object's being in the intellect in the way in which its objects are normally there. By this I mean that the idea of the sun is the sun itself existing in the intellect – not of course formally existing, as it does in the heavens, but objectively existing, i.e. in the way in which objects normally are in the intellect. Now this mode of being is of course much less perfect than that possessed by things which exist outside the intellect; but, as I did explain, it is not therefore simply nothing. (2:74)

This response is entirely to the point. So not even the *Meditations* filled in all the details that were needed. None the less, the *Meditations* did add some intermediate steps that are in neither the *Rules* nor the *Discourse*, steps that are added in order to make the necessary connection, as it were, more evident. These intermediate steps all involve necessary connections: it is necessary connections among ideas that lead the meditator from his own existence to God. This is the movement of thought in the Third Meditation, where we find two arguments for the existence of God.

But there is one important difference between the *Meditations* and the earlier works. In the earlier works, these causal proofs for the existence of God were taken as indeed proving the existence of this infinite being. In the *Meditations*, however, they cannot be so construed. It should, in fact, be clear that these proofs do *not* establish the existence of God. For, we do not yet know that they are sound arguments. These proofs proceed from premises that are not yet known to be true. The one proof has as its premise the proposition that 'I have the idea of an infinite being within me,' and this in turn is inferred from the fact that I have previous established that I am a thinking, and therefore a limited, being. The other has as its premise the proposition that 'I, a finite being, exist.' These premises in turn are derived from previous steps in the process of analysis, following from the original premise, which is itself a mere supposition – the premise that the evil genius exists. This supposition calls into question the truth of all necessary propositions, however. The inferences, by the method of analysis, are these

> Suppose the demon exists as a deceiving thing, then I am deceived and therefore think, therefore I exist, therefore God exists.

The starting point is a hypothesis. Moreover, the principles that are involved in each case are necessary connections of various sorts – causal principles such as the principle that finite substances depend upon infinite substances, or principles of metaphysics such as the principle that accidents must be in substances. And these necessary truths, just like those of geometry and arithmetic, are called into question by the demon. These, too, are hypotheses. From hypotheses only hypotheses follow, and, moreover, that following itself, the apparent necessary connection, is itself only hypothetical. Analysis yields a proof only when it arrives at a premise that is known to be true. At that point, we either arrive at a truth that contradicts the starting hypothesis, in which case the hypothesis is rejected as false, or we arrive at a truth which will enable us to reverse the chain of inferences and prove the hypothesis true. But at this point we have no grounds for holding that we have arrived at a truth. What we need is a truth that can provide a starting point for inferences, one that establishes itself as true, and also establishes all the principles of inference upon which we have relied in constructing our chain of inferences according to the analytic method.

The set of inferences cannot simply be reversed, however, as Pappus suggests in his classic description of analysis. For, as Descartes pointed out in the *Rules*, as we have seen, certain relations carry a necessity in one direction and not in the other. An object x may be causally dependent as an effect upon another object y, while y is not causally dependent upon x. In that case we can deduce the existence of y from the existence of x, while we are unable to reverse the process and deduce the existence of x from the existence of y. And so, from the existence of a finite or limited being we can deduce the existence of an unlimited, infinite being. But from the existence of a unlimited being we cannot deduce the existence of a finite being. If I, as a finite being, exist, then God exists; in this way I am dependent for my existence upon the existence of God. But from the existence of God it does not follow that I exist. God is not dependent upon me for His existence. In this sense, God is independent of me.

Thus, arriving at a first truth will not lead directly to the establishment that I exist truly as I appear to exist, namely, as a thinking thing, a thing whose essence is to be thinking. But the latter is what Descartes wants to establish. Or, at least, it is a crucial part of what he wants to establish, namely, the immortality of my soul: since my essence is that of a thinking thing, I am distinct from my body and am therefore capable of surviving the disintegration of my body.

What Descartes does discover in the Third Meditation, however, is that the perfect being is no deceiver. This being is the infinitely powerful creator of all things. Because this being is infinitely powerful, it lacks nothing. Lacking nothing, it is unlimited; there is no defect or negativity about it. But it follows from this that God is no deceiver. As he puts it at the end of the Third Meditation, '[b]y "God" I mean the very being the idea of whom is within me, that is, the possessor of all the perfections which I cannot grasp, but can somehow reach in my thought, who is subject to no defects whatsoever. It is clear enough from this that he cannot be a deceiver, since it is manifest by the natural light that all fraud and deception depend on some defect' (*Meditations*, 2:35). Or, as he puts it again at the beginning of the Fourth Meditation, 'I recognize that it is impossible that God should ever deceive me. For in every case of trickery or deception some imperfection is to be found; and although the ability to deceive appears to be an indication of cleverness or power, the will to deceive is undoubtedly evidence of malice or weakness, and so cannot apply to God' (2:37). Thus, the idea of God implies that necessarily He is no deceiver. Thus, we can continue the inferences in our analysis, that if God exists, then he is no deceiver.

But from this it further follows that all my clear and distinct ideas are true. Since God is no deceiver, He could not create a spiritual substance that aims to know and, on the one hand, provide it with ideas the very form of which – their clarity and distinctness – commands assent while, on the other hand, making those ideas false.

Analysis has therefore led the meditator to the conclusion that *if* God exists, then all my clear and distinct ideas are true. It further follows, in particular, that mind is a thinking substance that is in fact separable from body. It is this that Descartes has set out to prove. He has thus deduced that the idea which he has of himself as a thinking substance is a true idea: the essence which this idea is does give the truth of the world as I am aware of it in ordinary experience. What I experience myself as is a thinking thing, and this which I experience is indeed a true expression of my inner essence. Descartes has thus located the truth of his being: he does truly exist as he appears to exist. But he has done this by deducing his existence as a thinking thing not from the existence of God but from the moral perfection of God, and the fact that the ideas that God gives me cannot, if they are clear and distinct and therefore command assent, be false.

Or rather, he has done so provided that Descartes can establish a first truth from which this might be inferred.

IV: God: The First Truth and the Foundation of All Knowledge

Given that God as an infinite and moral being exists, Descartes has solved his problems. He has established that his clear and distinct ideas must be true. This gives him both the separability of the soul from the body and also the criterion that he needs to establish as certain the laws of coherence that enable him to distinguish within experience between dreaming and waking experiences and between veridical and non-veridical perceptual experiences. These conclusions are drawn out in the Sixth Meditation.

Given the existence of a God that is infinite and moral, it follows that the demon can be exorcised. For, the powerful being who deceives is, by virtue of his being a deceiver, a being who has a defect. This being is therefore limited, and for that reason must have less reality than God. If there is, in fact, such an essence, God, being more real and therefore more powerful, can prevent it from existing, from coming into being and exercising its power to deceive us. In fact, God will have so to act. For, God is no deceiver: that means that He cannot create a world in which we are systematically misled as to the ontological truth of things. Since God will not allow us to be systematically misled by our clear and distinct ideas, He must prevent the existence of any demon who could systematically mislead us. The hypothesis with which we began is inconsistent with the existence of God. This is the conclusion to which the method of analysis has led us.

There is, of course, the fact of error. But, Descartes argues, this is compatible with the existence of a God who is infinite and not a deceiver. God has, in fact, given us the faculties that we need to discover the truth. These faculties are in themselves infallible, if we but use them correctly. Yet we certainly are not perfect beings, at least so far as we are cognizers: we do sometimes fall into error. This imperfection arises not from God but from another gift, another good thing, that He has given us, namely, free will. Our will as well as our ideas can determine us to judge. And sometimes it does so in a way that outstrips the evidence available. When that happens, it is possible that we fall into error. We can avoid error if we restrain the will, and allow our judgment to be determined solely by the clarity and distinctness of our ideas. Descartes deals with this problem in the Fourth Meditation. His solution is of a piece with the traditional solution to the more general problem of evil.

However, Descartes has exorcised the demon and established the true nature of the coherence of experience only if he has proven the existence of a God. One can infer the falsity of the demon hypothesis with which the analytic process began only if one knows it to be true that God exists. The proposition that our clear and distinct ideas are true is one at which we arrive as a step in the analytic process. When we reverse the analytic process, and begin from the final conclusion that God exists we prove that this proposition is true. This is what we need to justify the claim that the soul is separable from the body. But all this requires that it be true that God exists. And this has not yet been established: all that we have done is arrive at the proposition that God exists, *given* the hypothesis that the demon exists, the supposition from which the set of analytic inferences proceeds. What the inferences, so far considered, enable one to conclude is that there is an inconsistency among the steps in the chain of propositions in the set of inferences: the final step, that God exists, is inconsistent with the first step, that the deceiving demon exists. But we have not yet found a truth and have not determined which of these two propositions is true.

Frankfurt has argued that this suffices.[90] He reads Descartes as attempting to establish that so long as we accept the proposition that God exists, then reason will never lead us into an inconsistency or falsehood, and that this is all that Descartes either needs or wants. As Frankfurt reads Descartes, what is at issue in the Third Meditation is this: is it possible that our reason is in conflict with itself in the sense that it provides a distinct perception that its most distinct perceptions might be wrong? The meditator's task, upon this account, is to show that reason cannot consistently admit of such an idea. This will exclude the demon as inconsistent with the reason of clear and distinct ideas. This will not do, however. For, consistency is no guarantee of truth. Knowing that reason will never lead us into falsehood will not give us a truth from which to start in order to prove the separability of the soul from the body. Descartes, moreover, recognizes this. The *Meditations* opens with a clear statement of the aim both of the method of doubt and of the series of meditations as a whole: 'Some years ago I was struck by the large number of falsehoods that I had accepted as true in my childhood, and by the highly doubtful nature of the whole edifice that I had subsequently based on them. I realized that it was necessary, once in the course of my life, to demolish everything completely and start again right from the foundations if I wanted to establish anything at all in the sciences that was stable and likely to last' (*Meditations*, 2:12). 'Stable' and 'likely to last': these words indicate, surely, that Descartes is aiming at truth, at (re-)creating or rebuilding the edifice of *knowledge*, the structure of *known truths*. The truths he wants to build towards are those of the existence of God and of the immateriality of the soul. As he indicates in the 'Dedicatory Letter' to the philosophers of the Sorbonne, he proposes in the *Meditations* to *demonstrate* these propositions, to *prove* them:

I have always thought that two topics – namely God and the soul – are prime examples of subjects where demonstrative proofs ought to be given with the aid of philosophy rather than theology. For us who are believers, it is enough to accept on faith that the human soul does not die with the body, and that God exists; but in the case of unbelievers, it seems that there is no religion, and practically no moral virtue, that they can be persuaded to adopt until these two truths are proved to them by natural reason. (2:2)

In order to do this, Descartes needs in particular some truth on which to (re-)found the edifice of knowledge. For, of course, to establish any truth we need a truth as a starting point, and a reason that is merely consistent will not give us this: consistency does not yield demonstration. So Descartes both needs and wants truth, not just consistency.[91] Moreover, the issue is not just epistemological, as Frankfurt's reading suggests, but ontological. The problem is that not merely of keeping one's thought consistent but of eliminating the ontological possibility of systematic *illusion* and *deception*, that is, the systematic possibility of taking *falsehood* to be *truth*. The source of possible systematic error is not simply in reason understood as a process in the individual mind but is rather something that arises in the heart of the traditional substance ontology. The possibility of systematic error is not just a problem within us, but implies a rottenness at the very core of being. To exorcise this possibility requires an ontological solution, which in turn implies an error-free grasp of the truth of being. Mere consistency in reasoning will not yield that error-free grasp of the truth of being.

Now, the final step of our process of analysis has been the existence of God. But as a step in analysis His existence has not been established as true. If we are to get at the truth, we must assume that we have arrived at a proposition that is itself clearly true. If the process of analysis is to lead to truth we have to show that we have in fact arrived at a truth, and the process of analysis will not in fact show this.

However, as we have noted, Descartes makes it clear that God is his first truth. We should expect that, after the process of analysis ends in the proposition that God as an infinite and non-deceiving being exists, we will discover the truth of the proposition that God exists. And this we do: it is the point of the Fifth Meditation. The knowledge that God does indeed exist comes with the ontological argument as it is presented in this Meditation. The causal proofs do not establish this existence. The arguments of the Third Meditation depend upon causal principles for their validity, and these principles are called into question by the supposition that the demon exists. As Williams has put it, the relevant causal principles are 'scarcely luminous.'[92] By the end of this, the Third Meditation, the demon has not yet been exorcised. To be sure, it has been shown that we arrive at the concept of a God the existence of which is inconsistent with the existence of a deceiving demon. But we do not yet know which of the two propositions is true. So, if the process of coming to know is to be completed, we have to come upon knowledge that God exists, that the idea that we have of God is a true idea.

To establish the truth of this idea, we cannot rely upon any causal principles, at least not any ordinary causal principles that relate different sorts of things to each other. For, to repeat, these latter principles have been called into question by the demon that has not yet been exorcised. If we are to establish the truth of the idea of God, then we need a very different argument. This is provided by the Fifth Meditation: I refer, of course, to the so-called ontological argument for the existence of God.

Descartes has in fact made clear that this proof is the starting point of truth in the *Meditations*. If it was not clear to his first readers, he made it clear to others in his reply to the Second Set of Objections. Mersenne, the author of the Objections, has asked Descartes to put his argument in geometrical form, meaning in the form of a demonstration. Descartes explains that he has followed the order of demonstration in the *Meditations*. That is, he says, 'the reason for my dealing with the distinction between the mind and the body only at the end, in the Sixth Meditation, rather than in the Second' (*Objections*, 2:110). In other words, the *demonstration* of the mind-body distinction, while begun in the Second Meditation, is *completed* only in the sixth. The truth of the matter is not established until the end of the *Meditations*; in the Second Meditation, he is saying, the truth has not yet been demonstrated.

Descartes now proceeds, in his response to Mersenne, to distinguish analysis and synthesis: the geometrical method that he is following 'divides into two varieties: the first proceeds by analysis and the second by synthesis' (*Objections*, 2:110).

Analysis shows the true way by means of which the thing in question was discovered methodically and as it were a priori ...

Synthesis, by contrast, employs a directly opposite method where the search is, as it were, a posteriori (though the proof itself is often more a priori than it is in the analytic method). It demonstrates the conclusion clearly and employs a long series of definitions, postulates, axioms, theorems and problems. (2:111)

As for the method of the *Meditations*, 'it is analysis which is the best and truest method of instruction, and it was this method alone which I employed in my Meditations.' 'As for synthesis, which is undoubtedly what you are asking me to use here, it is a method which it may be very suitable to deploy in geometry as a follow-up to analysis, but it cannot so conveniently be applied to these metaphysical subjects' (ibid.). But, in order to satisfy the request of his interlocutor, Descartes proposes a brief demonstration of the same material using the synthetic method. He therefore provides, in synthetic form, 'Arguments proving the existence of God and the distinction between the soul and the body arranged in geometrical fashion' (2:113).

In the manner of Euclid, Descartes proceeds to give a series of definitions, postulates, and axioms. The crucial point is that 'PROPOSITION I' states that '[t]he existence of God can be known merely by considering his nature' (*Objections*, 2:117). This is the ontological argument. Thus, in the order of synthesis, the ontological argument is the first argument, not the causal arguments of the Third Meditation, which occur as Propositions II and III, the 'a posteriori' arguments from effect to cause (2:118). These are followed by the Corollary that 'God created the heavens and the earth and everything in them. Moreover he can bring about everything which we clearly perceive in a way exactly corresponding to our perception of it' (2:119). Finally, we are given Proposition IV: 'There is a real distinction between the mind and the body' (ibid).

This is the synthetic order of presentation, the order that reverses the analytic order. In the analytic order, we proceed from the proposition that mind and body are distinct to the truth of our clear and distinct ideas to the causal arguments for the existence of God to the ontological argument. The synthetic order reverses this. But why or when exactly does one reverse the analytic order? Precisely at that point where one has discovered a truth to anchor the deductive structure that one proposes to develop synthetically. The anchor at which Descartes arrives becomes the starting point of the synthetic process. This starting point for the system of knowledge, the truth upon which the other truths depend, is the existence of God, and this is established in the first place by the ontological argument.

Here, then, is the argument as it appears in the Fifth Meditation: 'it is quite evident that existence can no more be separated from the essence of God than the fact that its three angles equal two right angles can be separated from the essence of a triangle, or than the idea of a mountain can be separated from the idea of a valley. Hence it is just as much of a contradiction to think of God (that is, a supremely perfect being) lacking existence (that is, lacking a perfection), as it is to think of a mountain without a valley' (*Meditations*, 2:45). This is, in fact, a standard argument within the substance tradition. The idea is that a fully perfect being is a self-subsistent being.

We find the notion already in Plato,[93] together with the reliance upon hypotheses that we later find in Descartes. Plato complains that ordinary geometers are 'compelled to investigate from hypotheses, proceeding from these not to a first principle but to a conclusion' (*Republic*, 510b3). In contrast, the philosopher 'does not consider hypotheses as first principles, but as hypotheses in the true sense of stepping stones and starting points, in order to reach that which is beyond hypothesis, the first principle of all that exists' (511b3–5). The starting point of the inferences are hypotheses, but the truth of these hypotheses is discovered only when one reaches the first principle, which is no longer hypothetical, but evident truth, a truth which is the ground of all other truths. This ultimate principle is the Form of the Good, that Form which combines within itself the sum of all perfections. The existence of this Form is evident in itself but is also an existence that is not dependent upon any other thing: the Form of the Good is self-subsistent.

Aristotle, too, speaks of the existence of a perfect being[94] and, moreover, describes it as a perfect being. As such, its existence is necessary. But he does not take this existence to be self-evident. Rather, he uses a causal argument to justify the inference to this perfect and necessary being. His is, of course, a substance ontology. He takes for granted that ordinary things have the metaphysical status of substances. More strongly, he takes as his starting point, not mere hypotheses, as Plato requires, but rather what he takes to be *knowledge* of certain ordinary facts about these substances, that they exist, that they are limited, and so on. Then, from these *truths* which he *knows*, Aristotle uses a causal argument to arrive at a first cause, one which is perfect and lacks nothing. Aristotle took the point from Plato. In his early, and now lost, treatise 'On Philosophy,' Aristotle, is paraphrased by Alexander following Simplicius, as saying that there must be a perfect being, that is, one which lacks nothing.

In general, where there is a better there is also a best. Since, then, among existing things one is better than another, there is also something that is best, which will be the divine. Now that which changes is changed either by something else or by itself, and if by something else, either by something better or by something worse, and if by itself, either to something worse or through desire for something nobler. But the divine has nothing better than itself by which it will be changed (for that other thing would then have been more divine), nor is it right for the better to be affected by the worse; besides, if it were changed by something worse, it would have admitted something bad into itself – and nothing in it is bad. Nor yet does it change itself through desire for something nobler, since it lacks none of its own nobilities; nor yet does it change itself for the worse, since not even a man willingly makes himself worse, nor does it possess anything bad such as it would have acquired from a change to the worse. This proof too Aristotle took over from the second book of Plato's Republic.[95]

He later more explicitly connects up this notion of a first, and perfect, being with a self-moving mover. And the existence of the latter is defended by appeal to a causal inference that all motion presupposes an unmoved mover to keep the moved things moving.

[I]f there is something which is capable of moving things or acting on them, but is not actually doing so, there will not be movement; for that which has a capacity need not exercise it. Nothing, then, is gained even if we suppose eternal substances, as the believers in the Forms do, unless there is to be in them some principle which can cause movement; and even this is not enough, nor is another substance besides the Forms enough; for if it does not act, there will be no movement. Further, even if it acts, this will not be enough, if its substance is potentiality; for there will not be eternal movement; for that which is potentially may possibly not be. There must, then, be such a principle, whose very substance is actuality. Further, then, these substances must be without matter; for they must be eternal, at least if anything else is eternal. Therefore they must be actuality. (*Metaph.*, 1071b12–22)

Lacking nothing, it is pure actuality, without potentiality. But that which is something only potentially might not, and might never be in that way: potentiality implies the possibility of not being. Pure actuality therefore contains no possibility of not being. A being that is purely actual is a necessary being.

Aquinas elaborates upon Aristotle's claims. Aquinas, of course, provides, as he sees it, following a long tradition, a correction to Aristotle: where Aristotle's necessary being was needed to move the universe, for Aquinas the necessary being is not just a mover but also a creator.[96] There are other moves, too, that Aquinas makes. In particular, something that is important for our purposes, he argues that God, as the first mover of things, must be the most perfect. 'Now God is the first principle, not material, but in the order of efficient cause, which must be most perfect. For just as matter, as such, is merely potential, an agent, as such, is in the state of actuality. Hence, the first active principle must needs be most actual, and therefore most perfect; for a thing is perfect in proportion to its state of actuality, because we call that perfect which lacks nothing of the mode of its perfection' (*Summa*, FP, Q4, A2). Among the perfections is existence:

Existence is the most perfect of all things, for it is compared to all things as that by which they are made actual; for nothing has actuality except so far as it exists. Hence existence is that which actuates all things, even their forms. Therefore it is not compared to other things as the receiver is to the received; but rather as the received to the receiver. When therefore I speak of the existence of man, or horse, or anything else, existence is considered a formal principle, and as something received; and not as that which exists. (FP, Q4, A2, RObj 3)

In fact, God cannot be distinguished from His existence; He is inseparable from it, and could not exist otherwise than He does exist.

[W]hatever a thing has besides its essence must be caused either by the constituent principles of that essence (like a property that necessarily accompanies the species – as

the faculty of laughing is proper to a man – -and is caused by the constituent principles of the species), or by some exterior agent – as heat is caused in water by fire.
Therefore, if the existence of a thing differs from its essence, this existence must be
caused either by some exterior agent or by its essential principles. Now it is impossible
for a thing's existence to be caused by its essential constituent principles, for nothing
can be the sufficient cause of its own existence, if its existence is caused. Therefore
that thing, whose existence differs from its essence, must have its existence caused by
another. But this cannot be true of God; because we call God the first efficient cause.
Therefore it is impossible that in God His existence should differ from His essence.
(FP, Q3, A4)

Since God could not exist in any way other than He does exist, He is a being
that exists necessarily.

Now, for Aquinas the existence of this necessary being can be demonstrated
on the basis of causal principles. These are the well-known 'Five Ways.' These
are demonstrations *a posteriori*, from effects to cause. But God's existence cannot be demonstrated a priori, by insight into his essence. In this sense, the existence of God is not self-evident. Aquinas therefore rejects any appeal to the
ontological argument.

He first distinguishes two ways in which a proposition can be self-evident.

A thing can be self-evident in either of two ways: on the one hand, self-evident in itself,
though not to us; on the other, self-evident in itself, and to us. A proposition is self-
evident because the predicate is included in the essence of the subject, as 'Man is an
animal,' for animal is contained in the essence of man. If, therefore, the essence of the
predicate and subject be known to all, the proposition will be self-evident to all; as is
clear with regard to the first principles of demonstration, the terms of which are common things that no one is ignorant of, such as being and non-being, whole and part, and
such like. If, however, there are some to whom the essence of the predicate and subject
is unknown, the proposition will be self-evident in itself, but not to those who do not
know the meaning of the predicate and subject of the proposition. (FP, Q2, A1)

Aquinas now argues that 'this proposition, "God exists," of itself is self-evident,
for the predicate is the same as the subject, because God is His own existence as
will be hereafter shown (Q3, A4). Now because we do not know the essence of
God, the proposition is not self-evident to us; but needs to be demonstrated by
things that are more known to us, though less known in their nature – namely,
by effects' (ibid.). Thus, the proposition that 'God exists' is indeed in itself self-
evident. But in order to recognize that self-evidence it is necessary that we
grasp the essence of God. This, according to Aquinas, we cannot do. The reason

is, of course, that we are incarnate beings who can grasp a form, have it present in the intellect or mind, only if we have abstracted it from sense appearances. We are able to form a sort of definite description of God, as the cause of certain effects which we *do* know, namely, sensible things; but we are not able to grasp His essence. 'To know that God exists in a general and confused way is implanted in us by nature, inasmuch as God is man's beatitude. For man naturally desires happiness, and what is naturally desired by man must be naturally known to him. This, however, is not to know absolutely that God exists; just as to know that someone is approaching is not the same as to know that Peter is approaching, even though it is Peter who is approaching ...' (FP, Q2, A1, RObj 1). The point is, of course, that as incarnate beings we can grasp the forms or essences of things only if we abstract those forms or essences from sensible appearances. But the deity is not a sensible entity. We therefore cannot grasp the essence of God. If, *per impossibile*, we could grasp the essence of God, then the proposition that God exists would be self-evident to us, and the ontological argument would be successful. What Aquinas is arguing is that the ontological argument is valid and sound, but that we as finite and incarnate beings cannot know it to be sound, since we cannot know what we need to know in order to know that the premise is true: we do not grasp the essence of God. In contrast, of course, God grasps His own essence. It therefore follows that the ontological argument works for God. It is not, therefore, that the ontological argument is unsound or invalid; it is just that, given our limitations, we are never in a position to find the argument useful in proving to ourselves that the conclu- sion is true.

Descartes, like Aquinas, holds that the argument is both valid and sound.[97] But he holds in addition that we are, contrary to Aquinas, in a position to use the argument. For, contrary to Aquinas, Descartes holds that we grasp the essence of God. Descartes can do this precisely because he differs from Aquinas on the source of our ideas. Where Aquinas and Aristotle hold that we grasp essences by abstracting them from sensible appearances, Descartes argues that our ideas, our grasp of the essences of things, does not derive in any way from sensible appearances. That is the point of the wax example.

So Descartes follows Aquinas in accepting the validity and soundness of the ontological argument. But he differs from Aquinas in arguing that, when we discover the idea of God within us, what we achieve is a grasp of the essence of God.

Interestingly enough, Caterus, in the First Set of Objections, raises Aquinas's objection to the ontological argument.

[St Thomas] considers an objection to this put by Damascene: 'The knowledge of the existence of God is naturally implanted in all men; hence the existence of God is self-evident.' His reply is that the knowledge that God exists is naturally implanted in us only in a general sense, or 'in a confused manner,' as he puts it, that is, in so far as God is the ultimate felicity of man. But this, he says, is not straightforward knowledge of the existence of God, just as to know that someone is coming is not the same as to know Peter, even though it is Peter who is coming. He is in effect saying that God is known under some general conception, as an ultimate end or as the first and most perfect being, or even under the concept of that which includes all things in a confused and general manner; but he is not known in terms of the precise concept of his own proper essence, for in essence God is infinite and so unknown to us. (*Objections*, 2:70)

Descartes's response is that we of course do not fully grasp the essence of the deity, but that this form is none the less present in the mind and that we have sufficient insight into its structure so that we can recognize that to this essence belongs necessary existence.

[I]n the case of the thing itself which is infinite, although our understanding is positive, it is not adequate, that is to say, we do not have a complete grasp of everything in it that is capable of being understood. When we look at the sea, our vision does not encompass its entirety, nor do we measure out its enormous vastness; but we are still said to 'see' it. In fact if we look from a distance so that our vision almost covers the entire sea at one time, we see it only in a confused manner, just as we have a confused picture of a chiliagon when we take in all its sides at once. But if we fix our gaze on some part of the sea at close quarters, then our view can be clear and distinct, just as our picture of a chiliagon can be, if it is confined to one or two of the sides. In the same way, God cannot be taken in by the human mind, and I admit this, along with all theologians. Moreover, God cannot be distinctly known by those who look from a distance as it were, and try to make their minds encompass his entirety all at once. This is the sense in which St Thomas says, in the passage quoted, that the knowledge of God is within us 'in a somewhat confused manner.' But those who try to attend to God's individual perfections and try not so much to take hold of them as to surrender to them, using all the strength of their intellect to contemplate them, will certainly find that God provides much more ample and straightforward subject-matter for clear and distinct knowledge than does any created thing. (2:81)

This suffices to know that God exists from a knowledge of His essence: 'I asserted that existence belongs to the concept of a supremely perfect being ... and this point can be understood without adequate knowledge of God' (2:82).

In this exchange, Descartes surely has not read Aquinas aright: for the latter, we do not have *any* grasp of the essence of God, nor, therefore, any grasp sufficient to enable us, not even the learned among us, to infer God's existence simply from our grasp of His essence. We may leave that aside, however. What is important for our purposes is that it is clear that for *Descartes* we do have present in us the idea of God, that this idea is the essence of the object known, and that our grasp of this essence is sufficient to recognize that this essence *must* exist.

The crucial point is that in the idea of God we have discovered one idea which turns out to be true, and, in fact, turns out to guarantee its own truth. The idea is an essence; this essence guarantees its own existence; in fact, it exists in a way that truly expresses that essence; hence, the idea we have of God is a true idea.

Thus, Descartes, as a consequence of his anti-abstractionism, or, what amounts to the same, as a consequence of his innatism, can hold that we can in fact use the ontological argument to establish that this idea is true, that this idea or essence considered as objectively real does really exist. As Descartes puts it in the *Principles*, 'the idea of a supremely perfect being is not an idea which was invented by the mind, or which represents some chimera, but that it represents a true and immutable nature which cannot but exist, since necessary existence is contained within it' (*Principles*, 1:198). It is the ontological argument that provides Descartes with a true idea, a truth that can provide the starting point for the reconstruction of the edifice of knowledge. *The point of the ontological argument is that we have in the idea of God an idea that establishes its own truth.*

Before turning to a more detailed examination of Descartes's use of the argument, however, we must turn aside a standard criticism. This is the claim that existence is not a predicate. This is an argument that has been used since Kant to reject the ontological argument: it is invalid because it commits the fallacy of treating existence as a predicate. A cute version of this objection is this: Consider the *R*-being. This is that being that has present in it any property that is named by an adjective that begins in English with the letter *R*. Thus, the *R*-being is rusty, rotund, rambunctious, riotous, reposeful, recreational, rhomboid, rousing, and, of course, *real*. But the *R*-being is, clearly, not real: the fact that we cannot create a being out of a collection of properties simply by adding the property of being real to the collection shows that *real* is no ordinary predicate.

The problem with this objection is that it does not take account of how the property of being or existence is to be understood within the substance ontology that Descartes takes for granted. We have already seen that Descartes quite reasonably takes existence to be predicate when he introduces the *cogito*, and that Hintikka is wrong in his suggestion that we need to read the *cogito* in a way that does not imply that existence is a predicate. And in the very same way as it is legitimate in the substance ontology to treat the being of a mental substance as a predicate, so it is legitimate to treat the being of God as a predicate.

The next point is that, in considering the ontological argument, it is necessary to distinguish two senses of 'necessary.' One we have already encountered. This is the sense of 'necessary' in which two concepts are necessarily connected, inseparable from each other. The other, to which we must now turn, is the sense in which existence is necessary or contingent. Descartes introduces the distinction in the tenth of the propositions that he lists as 'Axioms' for his synthetic demonstration of the argument of the *Meditations*: 'Existence is contained in the idea or concept of every single thing, since we cannot conceive of anything except as existing. Possible or contingent existence is contained in the concept of a limited thing, whereas necessary and perfect existence is contained in the concept of a supremely perfect being' (*Objections*, 2:117). He makes the same point in his statement of the ontological argument in the *Principles*:

The mind next considers the various ideas which it has within itself, and finds that there is one idea – the idea of a supremely intelligent, supremely powerful and supremely perfect being – which stands out from all the others. And it readily judges from what it perceives in this idea, that God, who is the supremely perfect being, is, or exists. For although it has distinct ideas of many other things it does not observe anything in them to guarantee the existence of their object. In this one idea the mind recognizes existence – not merely the possible and contingent existence which belongs to the ideas of all the other things which it distinctly perceives, but utterly necessary and eternal existence. Now on the basis of its perception that, for example, it is necessarily contained in the idea of a triangle that its three angles should equal two right angles, the mind is quite convinced that a triangle does have three angles equalling two right angles. In the same way, simply on the basis of its perception that necessary and eternal existence is contained in the idea of a supremely perfect being, the mind must clearly conclude that the supreme being does exist. (*Principles*, 1:197)

Because the idea of a supreme being contains within itself the fact that it exists necessarily, it guarantees the existence of its object, that is, guarantees its own truth.

In this presentation, Descartes makes a further connection that is important. It is that the supreme being is 'supremely powerful.' In fact, as we saw in Study Six, the existence of the supreme being is guaranteed precisely because it is supremely powerful. He elaborates on this point in his replies to the First Set of Objections, in a passage that deserves to be quoted in full. It makes clear the connection between the notion of necessary existence and unlimited power, on the one hand, and the connection between the notion of contingent existence and limited power, on the other. At the same time, it also makes clear what we have already seen, the connection between possessing a power and possessing a perfection.

[I]f I were to think that the idea of a supremely perfect body contained existence, on the grounds that it is a greater perfection to exist both in reality and in the intellect than it is to exist in the intellect alone, I could not infer from this that the supremely perfect body exists, but only that it is capable of existing. For I can see quite well that this idea has been put together by my own intellect which has linked together all bodily perfections; and existence does not arise out of the other bodily perfections because it can equally well be affirmed or denied of them. Indeed, when I examine the idea of a body, I perceive that a body has no power to create itself or maintain itself in existence; and I rightly conclude that necessary existence – and it is only necessary existence that is at issue here – no more belongs to the nature of a body, however perfect, than it belongs to the nature of a mountain to be without a valley, or to the nature of a triangle to have angles whose sum is greater than two right angles. But instead of a body, let us now take a thing – whatever this thing turns out to be – which possesses all the perfections which can exist together. If we ask whether existence should be included among these perfections, we will admittedly be in some doubt at first. For our mind, which is finite, normally thinks of these perfections only separately, and hence may not immediately notice the necessity of their being joined together. Yet if we attentively examine whether existence belongs to a supremely powerful being, and what sort of existence it is, we shall be able to perceive clearly and distinctly the following facts. First, possible existence, at the very least, belongs to such a being, just as it belongs to all the other things of which we have a distinct idea, even to those which are put together through a fiction of the intellect. Next, when we attend to the immense power of this being, we shall be unable to think of its existence as possible without also recognizing that it can exist by its own power; and we shall infer from this that this being does really exist and has existed from eternity, since it is quite evident by the natural light that what can exist by its own power always exists. So we shall come to understand that necessary existence is contained in the idea of a supremely powerful being, not by any fiction of the intellect, but because it belongs to the true and immutable nature of such a being that it exists. And we shall also easily perceive that this supremely powerful being cannot but possess within it all the other perfections that are contained in the idea of God; and hence these perfections exist in God and are joined together not by any fiction of the intellect but by their very nature. (*Objections*, 2:84)

It is the supreme power of God that makes his existence necessary, something that could not be otherwise, and guarantees that the idea of God is true.
    Again and again Descartes ties together the notion of power and the notion of

existing. It enters the *Meditations* with the demon: 'I will suppose therefore that not God, who is supremely good and the source of truth, but rather some malicious demon of the utmost power and cunning has employed all his energies in order to deceive me' (*Meditations*, 2:15). The same connection is made in the Third Meditation: 'there is a deceiver of supreme power and cunning who is deliberately and constantly deceiving me' (2:15). In the Third Meditation Descartes refers to his, or the meditator's, 'belief in the supreme power of God' (2:25). Later he states that if a being 'derives its existence from itself ... then it is clear ... that it is itself God, since if it has the power of existing through its own might, then undoubtedly it also has the power of actually possessing all the perfections of which it has an idea – that is, all the perfections which I conceive to be in God' (2:34).

In the reply to the First Set of Objections, Descartes makes the point that 'I do readily admit that there can exist something which possesses such great and inexhaustible power that it never required the assistance of anything else in order to exist in the first place, and does not now require any assistance for its preservation, so that it is, in a sense, its own cause; and I understand God to be such a being' (*Objections*, 2:78). The notion of necessary existence thus has a causal component: a necessary being is the cause of its own existence, and is sufficiently powerful that there could be nothing that could prevent it from exercising its causal power. *This is the ontological argument for the existence of God.* The thrust of the argument is not merely logical; the necessity is not merely logical necessity. The necessity is, rather, causal: a necessary being is one that cannot be prevented from exercising its unlimited power to give itself being. This is why so much of the discussion of this argument since Kant has so dramatically missed the mark.

Within the essentially Aristotelian framework that Descartes takes for granted, then, the ontological argument can plausibly be taken to be a sound argument. Descartes, I think, makes this clear.

The idea that we have of anything is the essence of that thing as it is in the mind. On the view of thought that Descartes has taken over from Aristotle, the mind when it thinks is literally identical with the object of thought: the essence of the thing *is* (= is identical with) the form of thought in the mind. This point is often misunderstood. Thus, Guéroult claims that, in the causal proof for the existence of God, we 'establish that the idea we have of him faithfully reveals his essence. Without the proof by effects, we would have in ourselves only a purely representative idea, not an essence.'[98] This, however, is mistaken. The idea just *is* an essence. As such it constitutes the objective reality of an idea the formal reality of which is that of a property or modification of the mind. Our idea is not distinct from God's essence, re-presenting the latter in the mind.

Truth is not a relation between my idea on the one hand and the essence on the other, the former somehow picturing more or less adequately the latter. There is no question of picturing the essence, since the essence itself is the objective reality or form of the idea. The question of truth is not one of correspondence but rather one of whether the being, if any, of this essence expresses the essence truly. So long as we hold firmly to the fact that Descartes accepts the Aristotelian notion of rational intuition, that this consists of an essence being literally in the mind, then we will not be tempted to accept Guéroult's claim that a proof is required to establish that the idea is a faithful picture of the essence.

The idea of God, then, is the form or essence of God *qua* existing in the mind. Now, upon the substance ontology, this form or essence is a power that aims to bring that essence into existence, that aims to give to that essence the being appropriate to it. The more that the essence contains, the more reality it will have if it exists, and the more powerful it will be. Conversely, the more limited a being is, the less powerful it will be. Now, most forms or essences are limited. They therefore are limited in their power. A more powerful being could therefore prevent their existing truly, could prevent their having the being that is appropriate to them, or, indeed, could prevent them from being at all. That is why their existence is contingent. In contrast, a supremely powerful being must exist. A supremely powerful being will be such that no being could be more powerful, nor, therefore, could be such as to prevent it from existing truly according to its essence. The deceitful evil genius is powerful, by hypothesis, but by the same hypothesis is deceitful, and therefore has a defect, a limitation. Since it has a limitation, a more powerful being can be conceived, and can, if it exists, prevent the evil genius from existing. But nothing is more powerful than God – that is how He is conceived, a being without limitation, and a being therefore of infinite power. Such a being could not be caused to exist in another way by some other being: no being is more powerful than it. Its existence is therefore necessary.

The same point can be put in terms of perfections. A perfect being is a being that is not only perfect after its kind but a being which has *all* perfections: there is no property which it lacks. An essentially perfect being is one that contains within its essence the tendency to be all things. This tendency is an active power which will actualize itself unless prevented. It could be prevented from existing, that is, from being and therefore from being perfect, only by a more powerful being. But that would imply that the perfect being lacked some power which the other being possesses. But by hypothesis it is essentially perfect and there is therefore no power which it lacks. Lacking no power, there is no being more powerful. It is therefore not possible that it not exist, and not possible that it not exist truly, in the way appropriate to its essence. An essentially perfect being necessarily exists and necessarily exists perfectly: all perfections are present in it.

But to conceive a perfect being is for the essence of an essentially perfect being to be present in my mind. This means that when I conceive an essentially perfect being, I am conceiving a being that necessarily exists and, therefore, a being that does exist. When I conceive a perfect being I have in my mind an idea that guarantees its own truth.

The necessity of the existence of a unlimited being resides in the unlimited nature of the power of the essence of such a being: it is sufficiently powerful that it can, as it were, bring about and maintain its own being, sufficiently powerful that nothing could ever prevent the essence from existing truly. But that essence is an idea in my mind – the idea of God. This idea or essence is therefore the idea or essence of a being that necessarily exists, and necessarily exists truly. It is an idea that guarantees its own truth.

And in guaranteeing its own truth it guarantees, too, that all other clear and distinct ideas are true. This far the analytic process can be reversed. The first truth can establish this further truth, and also the truth that is the real goal of the argument of the *Meditations*, that the soul is distinct from the body.

In guaranteeing its own truth and the truth of the other essences which are in our mind, the idea of God removes the threat that there is something rotten at the core of being. It removes the ontological threat that alone creates the epistemological problem.

V: Critical Reflections

On the traditional account deriving from Plato and Aristotle, human reason is the capacity to grasp the reasons for things, that is, the forms or essences or natures that explain why things present to us the properties they do. Descartes raises the possibility that our faculty of reason is radically defective, radically incapable of grasping the reasons for things. This possibility lies not so much with reason itself as with the world that it is trying to grasp: perhaps the structure of the world is such that my reason cannot grasp it in a way that provides the truth about the world as I ordinarily experience it. The possibility that reason is radically defective lies not so much in reason itself, in the reasons of things that are in the mind, as in the reasons of things that are in the things themselves. The problem, apparently epistemological, is more deeply ontological, and the solution that Descartes offers is equally ontological: the God upon whom we depend is no deceiver and will allow no other being systematically to deceive us.

God is the foundation of all knowledge. This is, in fact, a claim that Descartes makes repeatedly. But the road to God is by way of the *cogito*. In the *Discourse* the *cogito* represents a first truth on the road to God, who then secures the basic principles of physics. But in the *Meditations* the demon, the evil genius, is introduced. The presence of this possibility calls into doubt the *cogito* that had hitherto in the *Discourse* been free of any doubt. This changes the role of the *cogito*, and the role of God.

The crucial text that makes this clear is in the Third Meditation:

[W]hen I turn to the things themselves which I think I perceive very clearly, I am so convinced by them that I spontaneously declare: let whoever can do so deceive me, he will never bring it about that I am nothing, so long as I continue to think I am something; or make it true at some future time that I have never existed, since it is now true that I exist; or bring it about that two and three added together are more or less than five, or anything of this kind in which I see a manifest contradiction. And since I have no cause to think that there is a deceiving God, and I do not yet even know for sure whether there is a God at all, any reason for doubt which depends simply on this supposition is a very slight and, so to speak, metaphysical one. (*Meditations*, 2:25)

This remark forces the reader to attribute a different status to the *cogito* in the *Meditations* than it had in the *Discourse*. It also forces one to rethink exactly what additional problem it was that came with the hypothesis of the demon: exactly why does the possibility of the evil genius render untenable the stance of the *Discourse* that the *cogito* constitutes a first known truth?

To be sure, not everyone takes at face value the passage that we have been discussing.

Thus, in his edition of the *Meditations*, F. Alquié comments that 'Ce texte semble mettre sur le même plan la certitude unique du *cogito* et des certitudes d'un autre type, telles les certitudes mathématiques. Rejette-t-il donc ce que nous a apporté la *Méditation seconde*, et nous laisse-t-il le choix entre un doubte universel, fondé sur l'hypothèse du Dieu trompeur, et l'acceptation de toutes les idées claires, sans privilège pour le *cogito*?'[99] Alquié gives us two possibilities, neither of which seems acceptable. There is, first, the possibility that all clear and distinct propositions, both mathematical and the *cogito*, are equal, and the latter has no privileged position. That seems unacceptable, given the argument of the Second Meditation, which does give some sort of privileged place to the *cogito*. This means that we seem to have to accept the second of Alquié's alternatives. This second is the possibility that there is a universal doubt from which there is no escape. Given Descartes's determination to escape scepticism, this, too, seems unacceptable. Alquié now refers us to Guéroult, *Descartes selon l'ordre des raisons*, chapter five. Now, Guéroult argues, as we have noted above, that Descartes begins his restoration of the structure of knowledge with the *cogito*, taking this to be the first of the truths that we know. The Third Meditation calls this reading into question. But, Guéroult goes on to argue, what Descartes is doubting in these remarks is not the *cogito* of the Second Meditation but rather the further proposition that is to be explored in the Third Meditation, that the *cogito* can successfully found the remainder of the sciences. As Alquié puts it, one must distinguish between the '*cogito actualisé*' and the '*cogito objectivé par rapport à moi, qui se situe dans l'ensemble des connaissances que non entendement a toujours réputés pour vrais ...*'[100] It is the latter, not the former, that is subjected to demonic doubt in the Third Meditation. In the order of reasons, we must go from the former to the latter, from the *cogito* simply present to mind but present as a known truth, to the *cogito* as a feature of the body of knowledge. The movement, found in the Third Meditation, will remove the doubt. The problem with this reading is that it still requires that the *cogito* have a privileged position as a first known truth, and it is precisely this that Descartes in fact denies: given the demon, it is in no stronger a position than any other clear and distinct proposition: whenever I think of the proposition, its clarity and distinctness make it impossible for me to doubt it, to do anything other than accept it; but none the less, the demon calls *every such* proposition into question: the *cogito* of the Second Meditation has no privileged position.

Nor is this the only passage where Descartes indicates that things that are clear and distinct in the Second Meditation are none the less not really proven until later in the *Meditations*. Thus, in the 'Preface' he considers the following objection to what he has written: 'From the fact that the human mind, when directed towards itself, does not perceive itself to be anything other than a thinking thing, it does not follow that its nature or essence consists only in its being a thinking thing, where the word "only" excludes everything else that could be said to belong to the nature of the soul.' He then answers this objection:

[In] that passage it was not my intention to make those exclusions in an order corresponding to the actual truth of the matter (which I was not dealing with at that stage) but merely in an order corresponding to my own perception. So the sense of the passage was that I was aware of nothing at all that I knew belonged to my essence, except that I was a thinking thing, or a thing possessing within itself the faculty of thinking. I shall, however, show below how it follows from the fact that I am aware of nothing else belonging to my essence, that nothing else does in fact belong to it. (*Meditations*, 2:7)

In the Second Meditation, he argues that thinking is his essence. For, since he can doubt that everything else exists but cannot doubt that he, as something thinking, exists, thinking alone is inseparable from himself. This is clear and distinct – as clear and distinct as the *cogito* itself. Indeed, it is precisely from the certainty of the *cogito*, the clarity and distinctness of this inference, that he is able to infer the inseparability of thought from himself. If the *cogito* were, in fact, a first truth, then Descartes would have to argue that directly after proving the *cogito* as a premise he thereupon succeeded in proving that his essence is thought. But what he says in the 'Preface' is that he has *not yet proven* that his essence is thought. To be sure, it *seems* to Descartes that thought alone is inseparable from his being. It seems so because it is clear and distinct. But it is not yet *proven* to be true. So mere clarity and distinctness are not yet established as the criteria of *truth*. Nor, therefore, could Descartes be taking the *cogito* to be true simply because it is clear and distinct.

This passage in the 'Preface,' in short, confirms what Descartes says in the Third Meditation, that the doubt raised by the demon infects even the *cogito*, making it impossible to take this principle as a first truth from which other truths can certainly be derived.

In fact, of course, there are other possibilities for reading the *Meditations*. One would be to ignore the passages in question, in particular that of the Third Meditation, and return to the reading which takes the *cogito* of the *Meditations* to be, like the *cogito* of the *Discourse*, a known and certain first truth. And, in fact, such a reading can obtain certain sorts of textual support. There are passages where Descartes seems to withdraw from doubting the *cogito*.

Thus, in the Fourth Meditation he tells us that 'since my decision to doubt everything, it is so far only myself and God whose existence I have been able to know with certainty' (*Meditations*, 2:39). This seems to say that at this point he has in fact *proved* both his own existence and that of God. It is necessary, however, to put it in context. Descartes is here worried about the fact that, because he errs at times, he is therefore imperfect; but could God not have created him perfect in this respect, that is, given him a reason that is error-free? Descartes suggests that one way to go about answering this question is by locating the part, that is, himself, in the context of a greater whole and, specifically, the universe as such. For, something that is itself imperfect may none the less through that imperfection contribute to a greater perfection in the whole. He then raises a possible objection to this line of reasoning. This objection turns on the structure of inferences in the series of meditations that he has been undertaking. For, in this line of reasoning the only entities that he has so far encountered are his own self and God, so that trying to place himself within the structure of the universe as a whole is simply not possible: he has as yet no idea about the other entities that would provide the context of the greater whole.

Here is the passage in more detail:

It also occurs to me that whenever we are inquiring whether the works of God are perfect, we ought to look at the whole universe, not just at one created thing on its own. For what would perhaps rightly appear very imperfect if it existed on its own is quite perfect when its function as a part of the universe is considered. It is true that, since my decision to doubt everything, it is so far only myself and God whose existence I have been able to know with certainty; but after considering the immense power of God, I cannot deny that many other things have been made by him, or at least could have been made, and hence that I may have a place in the universal scheme of things. (*Meditations*, 2:39)

What he is saying is that 'At this point in the discussion it would seem that at most I can talk about only two entities, and not any more.' This is compatible with his having doubted the *cogito*. More strongly, it is compatible with his *still* doubting the *cogito*. For, the process of removing the doubt may well require Descartes to look at himself in the context of a greater whole; and he must show how *this* is possible at this point – for it seems that it isn't, since he can at most talk about two entities.

Here is another relevant passage. It occurs in the *Conversation with Burman*.[101] Burman draws attention to the apparent circularity of Descartes's argument. 'It seems there is a circle. For in the Third Meditation the author uses axioms to prove the existence of God, even though he is not yet certain of not being deceived about these.' Descartes replies: 'He does use such axioms in the proof, but he knows that he is not deceived with regard to them, since he is actually paying attention to them. And for as long as he does pay attention to them, he is certain that he is not being deceived, and he is compelled to give his assent to them' (*Conversation*, 5–6). Of course, on the account that we have given above of the structure of the *Meditations*, there is no circularity, and we shall have more to say about this charge of circularity directly. But here the point concerns the argument for the existence of God in the Third Meditation. Descartes seems to be saying that there is no question about his taking clear and distinct axioms to be true. This seems to conflict with the reading we have given, based on the passage in the Third Meditation where Descartes apparently includes all clear and distinct propositions, including the *cogito*, as things that the demon hypothesis enables one to doubt: Descartes seems, in this remark to Burman, to suggest that he never, not even in the First Meditation, allowed the demon hypothesis to call into question those propositions which are clearly and distinctly conceived.

By way of comment, it is first necessary to note that a series of inferences

that proceeds on the basis of a hypothesis can be circular; it is not only demonstrative arguments that can be circular. The chain:

H, but A, and if H & A, then G; but why A? Well, if G then A

clearly includes a circularity. So the charge of circularity is possible even if the
chain proceeds from a hypothesis. What Descartes is saying is that the chain of
inferences proceeds on the basis of axioms that are clear and distinct. In laying
out this chain of inferences, he does not presuppose that the conclusion, that
God exists, is required for this use. This is quite correct: to use an axiom in the
chain of inferences treated as proceeding analytically, upon the basis of a
hypothesis, does not require Descartes to prove that God guarantees their truth.
All that he needs is that at each stage the inference is clear and distinct. This is
what he does. There therefore is, as Descartes maintains, no circularity. Moreover, to be clear and distinct is to be incapable of doubt: as long as we attend to
these propositions, they command our assent. And when they command our
assent, they render it impossible for us to conceive at that moment that we are
being deceived. Again, Descartes's remark about his procedure is quite to the
point. But note that this is compatible with the demon hypothesis providing us
with the means to doubt even those propositions where we cannot conceive that
we are deceived. Descartes's remark is not only to the point, but is also compatible with the reading that we have given: it is compatible with the claim that the
demon hypothesis enables us to doubt, albeit metaphysically, that which cannot
be doubted, that where we cannot conceive that we are being deceived.

Of course, in the end we cannot even conceive, at least not clearly and distinctly, the hypothesis that the evil genius exists. Thus, in the *Conversation with
Burman*, Burman cites the following passage from the Second Meditation,
emphasizing a qualification that Descartes makes concerning the hypothesis of
the evil genius: 'But what shall I now say that I am, when I am supposing that
there is some supremely powerful *and, if it is permissible to say so, malicious
deceiver*, who is deliberately trying to trick me in every way he can?' (*Meditations*, 2:18; italics added). Descartes comments that 'The restriction is added
here because the author is saying something contradictory in using the phrase
"supremely powerful and malicious," since supreme power cannot co-exist with
malice. This is why he says "if it is permissible to say so." '[102] Of course, Descartes at the point where this remark is made in his *Meditations* does not yet
know that the idea of a supremely powerful being who deceives is contradictory,
and therefore not clearly and distinctly conceivable. To be sure, he has formed
the hypothesis, and thus conceived it. But he has not conceived it clearly and distinctly. What the process of analysis will ultimately reveal, much later than this

stage of the investigation – in the Fifth Meditation – is that the hypothesis, is in fact, contradictory, and will therefore lead to the rejection of the hypothesis.

To return, however, to the point of these remarks. The suggestion we are looking at is that Descartes, in the conversation with Burman, asserts that his position, held in the *Meditations* from their beginning, is that clear and distinct propositions are not only indubitable but true. If this is, in fact, Descartes's position then there is no circle, no problem with the proof for God's eixstence in the Third Meditation, no problem with the *cogito* being dubitable. The argument of the *Meditations* will simply be that of the *Discourse*, and the interpretation of the *Meditations* that I have proposed is wrong. However, what we now see is that Descartes's remarks in the *Conversation with Burman* need not be read in this way, that there is a way in which they can be read which is consistent with the interpretation of the *Meditations* that I have been developing. Descartes's remarks in his *Conversation with Burman* thus do not establish that the reading that I have given is mistaken.

There are several other passages, besides the two cited, where Descartes seems to say that he never doubts either the *cogito* or, indeed, any other clear and distinct idea. Without attending to these passages one after another, let it simply be recorded that I think they, like the two passages just noted, can be reconciled with the claim that Descartes makes in the Third Meditation that the *cogito* can be doubted.

Those who wish to construe the *Meditations* along the lines of the *Discourse*, taking the *cogito*, and, perhaps, all other clear and distinct ideas also, to be a known and certain truth, will of course base their interpretation on passages such as those that I have just cited. They will then have to explain away the passage in the Third Meditation where Descartes clearly states, as we have seen, that he doubts the *cogito*. In contrast, if we take the latter passage seriously, as I have been suggesting that we must, then we have to, as it were, explain away those texts upon which the other interpretation relies, texts such as those just discussed. Either interpretation has certain texts that give it difficulty.

What recommends the interpretation that I have been developing is not simply the passage in the Third Meditation, however. There are four central points.

1) The introduction of the evil genius flows naturally from the substance ontology that Descartes takes for granted and is such that Descartes should, on that supposition, doubt the *cogito*.

2) The construal of the *cogito* as a hypothesis, and therefore something capable of doubt, rather than a known truth, is what is required by Descartes's claim that he is proceeding in terms of the method of analysis that he used to make his discoveries in geometry.

3) In the Replies to the Second Set of Objections, Descartes takes God's existence as established by the ontological argument to be the first truth, rather than God's existence as established by the argument that if any finite being exists then a necessary being exists as its cause, and, still more importantly, rather than his own existence as a thing separable from his body, that is, rather than the *cogito*.

4) The claim that the ontological argument is the first truth in the *Meditations*, fits, on the one hand, with the proposal that the *cogito* is a dubitable hypothesis rather than a first known truth in the *Meditations*, and, on the other hand, with the proposal that Descartes is using the method of analysis, which begins with hypotheses and moves towards the discovery of true premises.

These four points combine, I am arguing, to make acceptable the reading of the *Meditations* that I have outlined.

What is central to this reading is that it is God's existence and not the *cogito* which is the first and central truth in Descartes's philosophy. What this reading does, in effect, is take seriously the Cartesian notion that God is the foundation of all knowledge.

It has, however, been argued that, while Descartes *says* that God is the foundation of all knowledge, this is not in fact his true belief. This is the famous 'dissimulation thesis,' that Descartes really holds that the *cogito* is the foundation of knowledge and that when he proclaims to the contrary that the real foundation is the deity he is in fact dissimulating about his real beliefs.

This thesis has been defended by Hiram Caton, for one.[103] On his view, Descartes's system is developed in the *Discourse*, where he makes clear that after his discovery of his method he undertook physics for several years, only later returning to do metaphysics. Caton suggests that if Descartes was doing this then he clearly had no need to do any metaphysics in order to get on with the task of science. There are, therefore, two orders of development: 'According to one, mathematical foundations suffice for the sciences, and philosophy or metaphysics is superfluous; the other has it that God is the necessary foundation of the sciences, so that metaphysics must come between method and physics' ('On the Interpretation,' 233). Since Descartes could do physics without going on to God, it follows that the second is superfluous. Descartes's well-known contempt for scholasticism reinforces the same point. The presence of the second line of development is therefore misleading. Descartes is, in fact, dissimulating when he introduces this other pattern of development. Caton suggests that Descartes introduces it not because his science really needs it but rather in order to convince people that his image as a great metaphysician is justified. He quotes Descartes: 'But as I was honest enough not to wish to be taken for what I was not, I thought I had to try by every means to become worthy of the reputation that was given me' (126). and then comments that 'If his metaphysics is designed to save or enhance his reputation among the learned, it will necessarily address *their* opinions in such a way as to win their approval; in short, it will speak *ad hominem*' (235). This is why, according to Caton, Descartes introduces the 'language of the schools' when he introduces metaphysics in the *Discourse:* 'since I knew of some perfections that I did not possess, I was not the only being which existed (here, by your leave, I shall freely use some scholastic terminology), but there had of necessity to be some other, more perfect being on which I depended and from which I had acquired all that I possessed' (*Discourse*, 1:127). After all, Descartes was a strong opponent of scholasticism. Thus, he argued that '[n]or have I ever observed that any previously unknown truth has been discovered by means of the disputations practised in the schools. For so long as each side strives for victory, more effort is put into establishing plausibility than in weighing reasons for and against; and those who have long been good advocates do not necessarily go on to make better judges' (1:146). and also that '[i]n fact I expressly supposed that this matter lacked all those forms or qualities about which they dispute in the Schools, and in general that it had only those features the knowledge of which was so natural to our souls that we could not even pretend not to know them' (1:131). So, Caton proposes, we can safely infer that when Descartes introduces the scholastic metaphysics, he is introducing something he does not need for his real philosophy, but only to enhance his reputation: he is dissimulating.

We should, however, look carefully at these reasons. First, when Descartes objects to the disputations of scholasticism, he is not arguing against the substance metaphysics that had also been developed by the medievals. Second, when Descartes objects to the appeal to forms and qualities, he is objecting to medieval (and Aristotelian) physics, but from the fact that the physics is rejected it does not follow that Descartes also rejects the metaphysics of substance and essence. To reject one version of the metaphysics of substance and essence is not to reject all versions. Moreover, in the third place, there is no reason why the language of the schools should not be used in explaining the metaphysics: to reject the method of disputation is not to reject as useless all forms of discourse coming from the scholastics.

Finally, to be concerned about one's reputation does not imply that one must dissimulate. To the contrary, it suggests dissimulation in Descartes's case only if there are antecedent reasons for supposing that there need be no appeal to God or other forms of metaphysics in order to carry on the program in physics that Descartes was developing. But, in fact, already in the *Discourse* Descartes found it necessary to appeal to God in order to establish a basis for his physics.

Specifically, God must be introduced in order to secure the conservation of motion. Descartes claims self-evidence for the principles of physics. These principles include the axioms of Euclidean geometry. But these axioms do not give the full story about motion. In order to get laws of motion, Descartes has to add principles that cover motions, determining how they relate to each other. Among these principles is the law of inertia, for example. This principle is not found in Euclid. To introduce this principle, and other principles such as that of the conservation of motion, Descartes deduces them from the immutability of God. He indicates this in the *Discourse*: 'Further, I showed what the laws of nature were, and without basing my arguments on any principle other than the infinite perfections of God, I tried to demonstrate all those laws about which we could have any doubt, and to show that they are such that, even if God created many worlds, there could not be any in which they failed to be observed' (*Discourse*, 1:131). This deduction is explained more fully in the *Principles*: 'From God's immutability we can also know certain rules or laws of nature, which are the secondary and particular causes of the various motions we see in particular bodies. The first of these laws is that each thing, in so far as it is simple and undivided, always remains in the same state, as far as it can, and never changes except as a result of external causes' (*Principles*, 1:240). So, contrary to Caton, metaphysics must be introduced by Descartes if he is to establish the research program in physics that he worked upon for many years.

We may therefore safely conclude that Caton has failed to establish that Descartes has to be dissimulating when he introduces God and metaphysics into the *Discourse* and the *Meditations*.

Another defender of the dissimulation thesis is Louis Loeb.[104] Loeb suggests that the demon hypothesis is meant to establish that sense experience can err, and that the Second Meditation begins a process which shows that such error can be corrected by reason. But reason is epistemologically basic for Descartes, who, we are told, 'held that reason or the faculty of clear and distinct perception is epistemologically prior to sense perception, and indeed epistemologically basic.' Thus, for reason, '[t]here can be no question of this faculty's being susceptible to correction by some yet more basic cognitive faculty' ('Radical Dissimulation,' 262). There can thus be no role for the demon in the Third Meditation, as there is for her in the First Meditation. So Descartes's claim in the Third Meditation that the demon hypothesis undercuts both the *cogito* and all other clear and distinct ideas is simply an error, or, rather, a dissimulation with regard to Descartes's true view, which is, or has to be, the view that reason is epistemologically basic. Descartes's motive for such dissimulation, Loeb suggests, is his desire to introduce God into the picture in a way that will make his views attractive to the theologians at the Sorbonne or among the Jesuits. To be sure, God is not essential to Descartes's epistemology; but the dissimulation makes it seem that He is central. 'The effect of the misrepresentation is to suggest a specific, intimate connection between the existence of God and the possibility of human knowledge ...' (264). By thus creating the impression that God is central to any knowledge claims that we might make, Descartes can make his views appealing to the theologians, and gain for them a legitimacy that they otherwise would lack.

The strength of the interpretation that we have developed is that it takes seriously the challenge of the demon to the *cogito* – it allows Descartes to be here making a point that is perfectly reasonable – and that it takes seriously the Cartesian claim that God is indeed central to the possibility of humans knowing the truth of things. Loeb's reading requires us to attribute simple insincerity to Descartes, a deliberate attempt to mislead.

But Loeb arrives at his position by not examining in sufficient detail the reasons why the demon hypothesis succeeds in undermining knowledge. In order to see why the hypothesis succeeds, it is necessary to probe the ontological roots of the hypothesis, the ontological source of the doubts. Loeb does not examine these roots. He therefore does not recognize how it can be that the demon can challenge even clear and distinct ideas, those ideas that Loeb takes to be the foundational roots of the edifice of knowledge.

As a consequence, Loeb treats the problem facing Descartes as simply epistemological. But this will not do: the roots of the doubt *are* ontological. Failure to see the ontological roots of the doubt lead Loeb to searching for an epistemic rock bottom on which Descartes is supposed to found, or refound, the edifice of knowledge. Naturally enough, this epistemic rock bottom turns out to be clear and distinct ideas, of which the first is the *cogito*. However, since the problem is ontological, so is the solution: we solve the problem of knowledge not by finding some epistemic rock bottom, but by finding a point in reality that can guarantee the truth of things. This point in reality is, of course, as Descartes says, God.

Loeb would have been led into a deeper analysis, I suspect, had he not taken the notion of knowledge and of clear and distinct ideas more or less for granted. In fact, however, the notion of *reason* is not itself untainted. The reason of the empiricists is not the reason of the rationalists and the Aristotelians. It is important to recognize that the reason of Descartes is, like the reason of Aristotle, a grasp of the reasons of things, an intuition of their essences, and that this intuition is a matter of that essence or form being in the knowing mind. Once this is recognized, then we have to recognize the role of substances and essences in the philosophy of Descartes, and that truth is not merely 'correspondence' but has an ontological character. Loeb, like so many commentators on Descartes, fails to probe the nature of the reason that Descartes employs.

Another who develops the dissimulation thesis is Kenneth Dorter.[105] Dorter argues that there is a conflict between the science which Descartes really wanted to defend and the religion of the day. The science as materialistic threatens religion. In order to cover up the materialism of his science, Descartes introduced the various metaphysical elements that appear in the *Meditations*, including, of course, God. These elements are there in order to disguise the fact that the science Descartes aims to promote undermines the religion. But Descartes is clearly insincere in bringing in this material; he perhaps even meant for us to see through the disguise. But the insincerity is clear, since, on the one hand, the reasoning of the *Meditations* is patently circular, and, on the other hand, Descartes is well aware of the dangers of reasoning in a circle. In fact, in his dedicatory letter to the theologians of the Sorbonne, Descartes notes that '[i]t is of course quite true that we must believe in the existence of God because it is a doctrine of Holy Scripture, and conversely, that we must believe Holy Scripture because it comes from God; for since faith is the gift of God, he who gives us grace to believe other things can also give us grace to believe that he exists. But this argument cannot be put to unbelievers because they would judge it to be circular' (*Meditations*, 2:3). Since Descartes knows circular reasoning is bad, and since the *metaphysical* reasoning of the *Meditations* is circular, it follows that Descartes is insincere in introducing the latter, dissimulating his real, anti-religious views.

The problem of the Cartesian circle is well known. But is there a circle, as Dorter claims? We have, of course, offered an interpretation in which there is no circle. It will, however, pay to look at the issue in greater detail.

Descartes offers us a proof that our idea of God as all-powerful and no deceiver is true. If we can believe nothing for certain unless we have established that the idea of God is true, but must rely upon premises known to be true if we are to find the argument for God acceptable, then surely Descartes has reasoned in a circle and we have after all not yet solved the problem of the starting point of knowledge. This was the problem raised already by one of the earliest commentators on the *Meditations*, namely, as is well known, Arnauld, who put it this way in his (the fourth) Set of Objections to the *Meditations*: 'one further worry, namely how the author avoids reasoning in a circle when he says that we are sure that what we clearly and distinctly perceive is true only because God exists. But we can be sure that God exists only because we clearly and distinctly perceive this. Hence, before we can be sure that God exists, we ought to be able to be sure that whatever we perceive clearly and evidently is true' (*Objections*, 2:150). Descartes of course held that his argument was not guilty of circularity. He pointed to the distinction 'between what we in fact perceive clearly and what we remember having perceived clearly on a previous occasion' (2:171).

He then argued that 'we are sure that God exists because we attend to the arguments which prove this; but subsequently it is enough for us to remember that we perceived something clearly in order for us to be certain that it is true. This would not be sufficient if we did not know that God exists and is not a deceiver' (ibid.). Descartes reads Arnauld as claiming that the Cartesian argument justifies our faculties as reliable – including reason and memory – by relying on the existence of God, and then justifies the existence of God by relying upon our faculties. Descartes's response is that we do indeed justify our faculties as reliable by appeal to the existence of God. Then, having justified our faculties as reliable, we can quite reasonably rely upon them. In particular, we can rely upon them when we go through proofs in mathematics or metaphysics. But we do not rely in the same way upon our faculties to establish the existence of God. It is not true that in order to be sure that God exists we need to be sure that whatever we perceive clearly and distinctly is true. What is crucial here is that we have the idea of God present to us. We do not now *infer* that this idea is true. Rather, this idea – this essence – *in itself* guarantees its own existence, establishing itself as the creator, sustainer, and guarantor of all truth in the world. Having the idea, we have by and within that very fact the reason why the idea exists: it is not that clarity and distinctness guarantee the truth of the idea, but rather the fact that the idea is (= is identical with) the essence of a being that is essentially unlimited and therefore essentially unlimited in its power. Since clarity and distinctness do not guarantee that it is true that the entity presented by the idea exists, there is no circularity.

Another version of this criticism is sound against Guéroult's interpretation of the *Meditations*. Guéroult, too, like Arnauld, interprets the ontological argument as depending for its acceptability upon the prior acceptance of the premise that all clear and distinct ideas are true.[106] It presupposes, therefore, that this principle has been established. But this principle follows from the existence of God. God's existence must therefore have received a prior proof. This is the causal proof of the Third Meditation, according to Guéroult. In other words, the ontological argument presupposes the causal argument.[107] Guéroult can argue this because he has accepted the standard reading that Descartes uses the *cogito* in the *Meditations* as a first truth. On this reading it makes sense to hold that the *cogito* proves God's existence (as it did in the *Rules* and the *Discourse*), that this proves the truth of clear and distinct ideas, and that this in turn justifies the ontological argument. But, once one understands the analytic process as beginning from a hypothesis, and aiming at discovering a first truth by which to establish the other truths, it makes no sense to hold that the ontological presupposes the causal argument. Moreover, Guéroult has to take it for granted that the causal principle, and other mathematical and metaphysical principles, are true, secure beyond the doubt raised by the demon.[108] But dubitable they are: Descartes is clear that he holds that the demon hypothesis calls these into question, along with the reality of body. That is why Popkin can quite reasonably argue that, even if we grant the *cogito* as a first truth, Descartes can proceed no further. 'The discovery of one absolutely certain truth, the *cogito*, may overthrow the sceptical attitude that all is uncertain, but, at the same time one truth does not constitute a system of knowledge about reality.'[109] Further progress towards additional truths requires additional premises or principles of inference, and the method of doubt culminating in the demon hypothesis renders all such premises or principles dubitable. The demon calls the principles into question, and if the existence of God depends upon those principles, then that too is called into question. But if the existence of God is called into question, then one cannot appeal to that fact to exorcise the demon. Descartes has not escaped the circle. As Popkin puts it, Descartes ends up *sceptique malgré lui.*[110]

These criticisms do not apply to Descartes, however: he does not rely upon the principle that clear and distinct ideas are true in order to prove the existence God with the ontological argument. As we have seen, the Cartesian critique of the traditional metaphysics is not unreasonable. Nor is his solution to the problem any the less unreasonable: given the substance metaphysics that he takes for granted and the innatist, rather than abstractionist, account of knowledge within that framework, then a strong case for the Cartesian solution can be made. Descartes, in fact, makes it. In particular, the ontological argument for the existence of God as the source and guarantor of truth becomes very reasonable.

Critics often ignore this framework. Thus, for example, Kenny, in his discussion of Descartes, has provided an analysis of the ontological argument.[111] This analysis has as its starting point Meinong's notion that there are structures with various properties that are beyond being and non-being. Being itself is something that is, as it were, added to such a structure, and when it is, then that structure exists. Kenny suggests that these structures are akin to the objective realities which Descartes introduces. Furthermore, Kenny goes on[112] to introduce the Aristotelian point that, as Aquinas puts it, *omnia appetunt esse*: everything desires existence. '[E]verything naturally aspires to existence after its own manner ... [Moreover,] a natural desire cannot be in vain' (*Summa*, FP, Q75, A6). The second remark is crucial: unless something prevents a thing from actualizing itself, it will do so. Margaret Wilson wonders how something that is merely possible, such as God, could actualize itself, since that would require something non-real to bring into existence or cause something real, which seems to violate the Cartesian metaphysical principle that the cause contains within itself the reality of the effect. The merely possible cannot already contain the real and yet be merely possible. So, according to Wilson, Descartes falls into an inconsistency.[113] But this misses the point that within the Aristotelian framework the cause strictly speaking is always a power. The power contains the actuality within it potentially, and, as a power, actualizes that potentiality – that is, actualizes it unless prevented by some being which is more powerful. The Cartesian dictum about causal relations applies not to the relation between a power and its self-actualization but to the relation between an effect which is distinct from its cause, that is, where the power of one substance endows some other entity with existence. So, contrary to Wilson, Descartes falls into no inconsistency. As for Kenny's point with regard to this principle that everything desires existence, on his view we have to accept the ontological argument, that is, accept that being or existence will be among the properties predicated of the structure, provided that the desire for existence is antecedently included among the characteristics defining the structure. But it can actually cause existence to be predicated of the structure only if it, too, is among the properties predicated of the structure, in which case the structure already exists. Thus, Descartes's inclusion of it amounts simply to begging the question whether God exists: Descartes 'cannot prove God's existence from God's essence without begging the question.'[114]

There is some point to the analogy between Meinong's structures that are beyond being and the objective realities of Descartes. There are, however, three points that Kenny misses, and if these are added, then the conclusion that Descartes begs the question vanishes. First, the objective realities that Descartes introduces are indeed structures, but, unlike Meinong's structures, the Cartesian structures are also *powers*, they are active entities that bring about their own being. Desiring existence is not a further characteristic that is, as it were, added to the characteristics that define the essence; rather, in true Aristotelian fashion, the essence simply *is* a power, it simply *is* a form striving for being. Secondly, for Descartes, existence is not something that is a further property-like entity that is added to a structure. Rather, a structure or essence acquires being when it acquires a property: to have being or existence just *is* to have a certain specific property, and to have a specific property is what it is to be or exist. This is how the concept of being or existence is understood within the substance ontology. Third, the structure for Descartes is not, as it is for Meinong, something to which the mind is externally related. Rather, for Descartes the form or essence or structure is both external *and also* literally in the mind: there is an identity between the two.

This latter point is missed by many besides Kenny. Thus, E.J. Ashworth tells us that 'Descartes adopted a representative theory of sense perception by which ideas are a necessary intermediary between the mind and what is external to it, and he extended this theory to apply not only to sensory objects but to all external reality, including god and the eternal truths.'[115] According to E.B. Allaire, for Descartes '(a) There are two kinds of entities, *ideas* and *things*. (b) Things are *represented* to us by ideas *presented* to us.' Or, as Allaire rephrases his point, 'we are acquainted *directly* with ideas and *indirectly* (by means of ideas) with things'[116] On this view, the problem is to know that some of our ideas actually represent. As stated, the problem is insoluble, as Allaire suggests.[117] Any attempt to escape the veil of ideas by appeal to the clarity and distinctness of some of those ideas will inevitably be circular, as both Allaire and Ashworth make clear.[118] But Cartesian ideas do not constitute a veil: rather, when we have an idea we are in direct contact with the external object, or, at least, with its form or essence. For Descartes, then, *to have an idea is* already *to be outside the circle of ideas*. This issue is not how to get outside a veil of ideas; *the problem is, to the contrary, how to establish that the essences that are both present to us and also external to us constitute the* ontological truth *about the world external to us.*

What secures the latter, of course, is the discovery that one of these essences guarantees its own truth. That is why we need not only the identity of knower and known to solve Descartes's ontological problem of knowledge of truth but also the other two points, that essences simply are powers, simply are strivings for being, and that being or existence is not a property of things alongside other properties but simply is the having of some property or other.

Once these three points are made, we have added what is needed in order to make the ontological argument into something that is neither circular and nor question begging. But then, neither is the result Meinong. It is, quite differently, Aristotle. Kenny is sensitive to the Aristotelian tradition in which Descartes locates himself, but in the end he misreads Descartes because he (Kenny) fails to locate Descartes securely enough in that tradition.

Williams has argued that the project of using the method of doubt as a starting point in the quest for knowledge presupposes that knowledge is *possible*.[119] Indeed, it presupposes that one *knows* that such knowledge is possible. Otherwise the project would have as its goal something that is impossible, and is known to be impossible, and it is unreasonable to strive to attain a goal that one knows cannot be achieved. The real issue about the Cartesian metaphysics is not whether the ontological argument is valid or whether existence is a predicate: these things are so, given the substance metaphysics that Descartes, following Aquinas, takes for granted. The problem lies rather with precisely the latter, that which Descartes simply takes for granted, the framework itself: does it really make sense? Is knowledge of the sort to which it aspires really possible? This is the point that Kenny raises when he points out that Descartes does *not* doubt everything, contrary to what it seems he purports to do. As Kenny puts it, 'Descartes does not doubt that he knows the meaning of the words he uses to construct and resolve the doubts of the *Meditations*.'[120] In a way this is unfair. It was the traditional framework that Montaigne challenged, the framework that had come down to the modern world from Aristotle and through such medieval philosophers as Aquinas. So it was this framework that Descartes set out to defend. The question of the criterion was raised within this context: *given* the doctrine of substance and the account of knowledge which was part of that ontology, *what then* is the criterion of knowledge? It is this question that Descartes set out to answer, and to insist that he should have stepped farther back and called into question that framework that Montaigne accepted is hardly fair. To this extent, Kenny's comment is not just. None the less, it is also true that one can raise the question whether the very terms in which the problem is stated are intelligible. Is the framework really meaningful? Does the substance ontology make any real sense? Descartes does not undertake to answer this question. To this extent, Kenny's comment is entirely appropriate. It was up to later philosophers, however, to take up these issues. And, of course, take them up they did.

One of these was Simon Foucher, who offered some telling criticisms of the Cartesian philosophy in general and Malebranche's version in particular.[121] This discussion was itself caught up in the controversy between Arnauld and Malebranche as to the nature of ideas, that is, of Cartesian ideas.

Arnauld more or less followed Descartes in taking for granted the Aristotelian account of ideas as the forms or essences of things in the mind. Like Descartes, Arnauld also makes the object perceived to be itself in the mind that knows it, there in the mind only objectively: '*the perception* of a square indicates more directly my soul as perceiving a square and *the idea* of a square indicates more directly the square insofar as it is *objectively* in my mind.'[122] He also affirms the Aristotelian principle of the identity of the knower and the known: 'the *idea* of an object and the perception of an object [are] the same thing.'[123] He continues: 'an object is present to our mind when our mind perceives and knows it'; when I so conceive of a thing that 'thing is *objectively* in my mind ...'[124]

But there is for Arnauld, as for Aquinas and Descartes, another sort of existence for forms or essences besides the being they have formally and the being they have in the mind. They are also, in fact, the divine exemplars, the divine ideas in conformity with which God creates his creatures.

With reference to perception, however, what is crucial is the identity of the idea in the mind with the object known, or, more accurately, with the form of the object known. This requires one and the same entity, the form or essence, to be in two ways, once formally as a substance and once objectively as an idea. In the latter way of being, however, it is a form or characteristic of a mode of the mind. In respect of this form – which is shared by the substance, which is not mental, and the mode, which is mental – these two entities are *the same*: in respect of this form the two entities *exactly resemble* each other. Thus, the mind that knows and the thing known stand in a relation of exact resemblance. In fact, it is by virtue of that exact resemblance that the mind knows the object.

It was this point at which Foucher directed much of his criticism of the Cartesian position.[125] He argued at some length that the mind cannot resemble what is known. Foucher notes that Malebranche assumes from the beginning that mind and body are *toto caelo* distinct, that the one is spiritual and the other material and that they share no properties.[126] Foucher then argues that this supposition is incompatible with the supposition that ideas represent things without us.[127] He argues in particular that, if the external substance is literally in the mind, that is, if our ideas are for this reason *like* the things that are external to us, then that substance ought to have within us the same effects that it has without us.

[I]f our ideas represent these things to us, it is necessary that they cause the same effect in us that these things would if they were present, by causing us to know what they are in themselves, and not the ways of being that these same things would excite in us if they act through our senses. For as these ways of being are not *like* these things, following the acknowledgement of these philosophers, these ideas would not represent these things, but represent only their effects. So to represent these things as they are in themselves it is necessary that our ideas dispose us exactly as though the things were now in us and were *immediately* present to us. For this it is necessary that our ideas cause an effect in our soul that is at least *like* the effect the things would cause if they were really there, which these ideas could not do unless they were *like* the things. Otherwise, far from causing in us ways of being such as we would have if those things were immediately present to us, it is obvious that the ways of being the ideas would cause in us would be entirely different. From which it follows that they would not represent those things.[128]

The issue is ontological. How can one and the same entity be both a substantial form, on the one hand, and a characteristic of a property which is a property or mode of a substance? How can one and the same entity be, on the one hand, that of which properties are predicated, and also a property that is predicated of a substance? If it is the same entity, then it should have the same effects wherever it is. But it does not: it has one sort of effect when it exists without the mind, a different sort of effect when it exists as an idea. Huet was later to make much the same point. 'The *Form*, or *Image*, which comes from a Tree, is that a Tree?

And if it be not a Tree, can it resemble a Tree? ... the *Species*, or *Image*, of this Tree is different from the Tree in many Things. The Tree is visible, without Motion, and solid; its *Species*, or *Image*, is not visible, has no Consistence, and is very moveable, very thin, and very fluid.'[129] In effect, the view requires one and the same entity to occupy two very different ontological categories, in a way that makes it impossible to think of the entity in question as literally one and the same. As Berkeley was later to put the same point, 'Only an idea can be like an idea.' 'But say you, though the ideas themselves do not exist without the mind, yet there may be things like them whereof they are copies or resemblances, which things exist without the mind, in an unthinking substance. I answer, an idea can be like nothing but an idea; a colour or figure can be like nothing but another colour or figure. If we look but ever so little into our thoughts, we shall find it impossible for us to conceive a likeness except only between our ideas' (*Principles*, sec. 8). This is the point raised by Foucher and Huet, and it cuts very deeply. In fact, it points to an ontological incoherence at the heart of the traditional account of knowledge. That theory required one and the same form or essence to be in the mind of the knower as it was in the thing known. This was how the theory accounted for the mind's being able to get outside itself and into contact with the thing known: in knowing, the thing known was in the mind, and, since this same object was the one outside the mind, it turns out the knowing mind is outside itself without being outside itself. This view can be defended, however, only by making one and the same entity occupy two very different ontological categories, that is, in effect, only by making the ontology basically incoherent. So much the worse, then, for the traditional account of knowledge.

This criticism is telling against both Descartes and Arnauld – as well as against the whole earlier tradition coming via Suarez and Aquinas from Aristotle. Malebranche was, in fact, sensitive to it. He argued in detail that the ideas that we have of things could not be in the mind.[130]

Like the tradition of which he was a part, he held that the perception of ordinary things involved an intuition of the form or essence of the thing. For Arnauld and Descartes, this form or essence had a threefold being: in the thing as the substantial form, in the mind as the act by which that thing is known, and in the mind of God as the divine exemplar or pattern in terms of which He has created all things. On this account, to know the thing is at once to have the thing literally in one's mind and also to grasp the divine exemplar, the form that God intended the thing to have. Malebranche denied, however, that the form or essence could literally be in one's mind: what are in one's mind are modes or properties, and a form or essence is just that and not a property. There are two different ontological categories, and one and the same entity cannot be in both.

Malebranche refers to any 'sensations, imaginings, pure intellections, or simply conceptions, as well as its passions and natural inclinations,' as a 'thought, mode of thinking, or modification of the soul'; and the mind is aware of all these entities simply by virtue of their being in the mind, that is, by virtue of 'the inner sensation that it [the mind] has of itself [and therefore of its properties].'[131] Now, the human mind, 'being neither material nor extended,' is a 'simple indivisible substance without composition or parts.'[132] While we do perceive material objects, none the less 'we do not perceive objects external to us by themselves.' To the contrary, since we do sometimes perceive objects that do not exist, and none the less have something before the mind when those perceivings occur, it follows that there must be something more than and other than the physical object before the mind. This something is an idea of that object: 'for the mind to perceive an object, it is absolutely necessary for the idea of that object to be actually present to it ...'[133] Malebranche considers various possibilities to account for the presence of these ideas to the soul: they might be species transmitted, as Aristotle held, from the object to the perceiver; they might be produced by the soul itself; God might produce a new idea each time we perceive; or the ideas might be already in the soul. These he systematically rejects. In particular, he argues that neither creatures – either material objects or ourselves – nor God could or would create the infinity of ideas that would be needed if ideas were in the mind. All modifications of the mind are particular, where the concept of triangle or of extension are general, and include an infinite number of particulars under them.

[S]ince there is an infinite number of different figures, the mind must have an infinity of infinite numbers of ideas just to know the figures.
   ... it is clear the idea, or immediate object of our mind, when we think about limitless space, or a circle in general, or indeterminate being, is nothing created. For no created reality can be either infinite or even general, as is what we perceive in these cases.[134]

In particular, the idea of extension is not in the mind: 'The soul does not contain intelligible extension as one of the modes because this extension is not perceived as a mode of the soul's being, but simply as a being. This extension is conceived by itself and without thinking of anything else; but modes cannot be conceived without perceiving the subject or being of which they are modes.'[135] These ideas are the reasons for things; it is through them that we understand things. But this reason cannot be in the mind of the individual: 'It must be concluded ... that the reason consulted by all minds is an immutable and necessary Reason.'[136]

Malebranche thus concludes that the ideas which we know when we know things cannot be in the mind. Where Descartes and Arnauld ascribed a threefold sort of existence to ideas – in the thing known, in the mind of the knower, and in the mind of God – Malebranche has now removed one of those forms of existence.

Malebranche does not deny that in perceiving or knowing a thing we are aware of ideas. To the contrary, he continues to hold that position. Only, these ideas are not in our own minds. They are rather the ideas in the mind of God, the divine exemplars of things.

[I]n short, the mind clearly sees the infinite in this Sovereign Reason, although he does not comprehend it. In a word, the Reason man consults must be infinite because it cannot be exhausted, and because it always has an answer for whatever is asked of it.

But if it is true that the Reason in which all men participate is universal, that it is infinite, that it is necessary and immutable, then it is certainly not different from God's own reason, for only the infinite and universal being contains in itself an infinite and universal reason.[137]

This is the famous Malebranchian thesis that *we see all things in God*: 'the mind can see God's works in Him, provided that God will to reveal to it what in Him represents them.'[138] These ideas in God's mind that He uses to represent things to us, Malebranche holds, as it were 'touch' the mind but are not in it. '[T]hrough His presence God is in close union with our minds, such that He might be said to be the place of minds as space is, in a sense, the place of bodies.'[139]

The problem, of course, is that this new relation between a mind and what is known has no place within the traditional substance ontology. Even Malebranche explains it merely by means of a metaphor. In the substance ontology there are substances and there are properties (characteristics, modes). The latter are in the former. This is the only structural relation that one has in the ontology. With the introduction of the new relation of 'touching,' Malebranche has, in effect, argued that there are deep problems in the traditional framework that require radical changes if they are to be solved.

Descartes took for granted the traditional ontology of knowing, that is, epistemology, that was part of the substance tradition deriving from Aristotle. Thinkers such as Foucher and Huet criticized this traditional ontology of knowing as ontologically incoherent – 'only an idea can be [ontologically] like an idea,' they argued. Malebranche accepted this point, and thereby accepted that the traditional ontology was radically defective. His own remedy – that we perceive by means of ideas in God's mind that touch us when we see and know various things – leaves much to be desired: it requires a new structural relation that has no neat place within the traditional ontology.

Malebranche made another, equally crucial, breach in the traditional metaphysics.

On the traditional ontology, ordinary things are substances. These substances have forms or essences, and these forms or essences are active powers. Through their activities they establish or, better, constitute the necessary connections that hold among the perceivable events that we see. The rationalists criticized these doctrines. In particular, they held that among many of the qualities of which we are aware in ordinary experience there are no necessary connections. Thus, for example, besides the clear and distinct ideas that are present in our minds, as Descartes and Arnauld held, or present *to* our minds, as Malebranche held, there are also various *sentiments* that are *not* clear and distinct. Because they are not clear and distinct, reason has no insight into how they are connected to other things. In fact, precisely because the deity does not mislead, the fact that we are not aware of such connections means that there are no connections there but unperceived: God is no deceiver.

Here is the way Descartes puts it:

there is in me a passive faculty of sensory perception, that is, a faculty for receiving and recognizing the ideas of sensible objects; but I could not make use of it unless there was also an active faculty, either in me or in something else, which produced or brought about these ideas. But this faculty cannot be in me, since clearly it presupposes no intellectual act on my part, and the ideas in question are produced without my cooperation and often even against my will. So the only alternative is that it is in another substance distinct from me – a substance which contains either formally or eminently all the reality which exists objectively in the ideas produced by this faculty (as I have just noted). This substance is either a body, that is, a corporeal nature, in which case it will contain formally and in fact everything which is to be found objectively or representatively in the ideas; or else it is God, or some creature more noble than a body, in which case it will contain eminently whatever is to be found in the ideas. (*Meditations*, 54)

Since God is no deceiver, these ideas must come from an extended or corporeal substance.

I do not see how God could be understood to be anything but a deceiver if the ideas were transmitted from a source other than corporeal things. It follows that corporeal things exist. They may not all exist in a way that exactly corresponds with my sensory grasp of them, for in many cases the grasp of the senses is very obscure and confused.

But at least they possess all the properties which I clearly and distinctly understand, that is, all those which, viewed in general terms, are comprised within the subject-matter of pure mathematics. (ibid.)

As for the other properties, those which I do not see clearly and distinctly, while I cannot *know* that these are connected with material substance, I none the less can have a *firm belief*. In fact, such a belief is *natural* in the sense of inescapable. I naturally believe that I have a body, that there are other bodies around me, that these other bodies cause certain sensations in me, and in particular cause sensations of pain and pleasure.

My sole concern here is with what God has bestowed on me as a combination of mind and body. My nature, then, in this limited sense, does indeed teach me to avoid what induces a feeling of pain and to seek out what induces feelings of pleasure, and so on. But it does not appear to teach us to draw any conclusions from these sensory perceptions about things located outside us without waiting until the intellect has examined the matter. For knowledge of the truth about such things seems to belong to the mind alone, not to the combination of mind and body. (2:56)

But I am finite and am often forced to make inferences without having the time to check things meticulously by reason to find out the truth insofar as I can discover it. Error derives from my will and finite nature rather than from God – as before! 'It is quite clear from all this that, notwithstanding the immense goodness of God, the nature of man as a combination of mind and body is such that it is bound to mislead him from time to time' (2:61).

In this proof of the reality of the causes of such sentiments as those of pleasure and pain, of the colours, sounds, and tastes that I experience, two things have to be said. *First*, we do not *know* that these causes exist. It is a matter of belief. To be sure, it is a natural and inescapable belief; but it is a belief, none the less. This derives from the fact there is no perceivable connection between these sensations and the properties such as shape which bodies have by virtue of the essence of corporeal substance. Nor *could* such a connection exist. Since there is no perceivable connection, and since God is no deceiver, it follows that there *is* no connection.

Second, since there is no necessary connection, it follows that whatever connection is established is only a regularity, not genuinely causal. This regularity is maintained by God, because He is no deceiver. It is God's activity that is crucial, not His intellect. He simply acts to produce an effect on my soul when the body is in a certain state, or produces an effect on my body when my soul is in a certain state. Bodily states are the *occasion* for God's creating a certain state in my soul, and states of my soul are the *occasion* for God's creating a certain state in my body. *There are regularities among various qualities of things which are presented as separable and with no necessary connections among them; but there are no causal connections among events with these qualities because causal connections presuppose necessary connections. None the less, the regularities are causal in the sense of involving activity, not the activity of finite substances, but the activity of God.*

Malebranche appeals to the same sort of quality in things: the heat of the fire, the whiteness of the snow, the brilliant light of the sun.[140] 'It is easy,' he says, 'to understand that pleasure and pain, heat, and even colours are not modes of bodies, that sense qualities in general are not contained in the idea we have of matter ...'[141] These qualities are in the mind, not body, and have no apparent necessary connection with the bodies that are their apparent cause. Since there is no apparent necessary connection, there is no necessary connection, nor, therefore, any necessity in there being a cause which produces them. Their existence therefore in no way proves that there is a corporeal cause for them. However, 'it is at least possible that there are external bodies. We have nothing that proves to us there are not any, and on the contrary we have a strong inclination to believe there are bodies.'[142] This natural inclination comes from God, and we should, therefore, accept it – but that acceptance is by way of faith in God as a good being and not by way of evidence. 'God speaks to the mind and constrains its belief in only two ways: through evidence and through faith. I agree that faith obliges us to believe that there are bodies; but as for evidence, it seems to me that it is incomplete and that we are invincibly led to believe there is something other than God and our own mind.'[143] Again, the point is that there are regularities here but no causal = necessary connections.

Foucher objects that his opponents have proved either too little or too much. We have presented to our consciousness both extension and sentiments such as colour. We are naturally inclined to believe that both are characteristics of external objects. We either follow our natural inclination or not. If we do, then we must believe that colours are in corporeal objects. In this alternative, the Cartesians have proven more than they wanted to. But if we accept the Cartesian reasoning, then we must reject the natural inclination to believe that colours are characteristics of bodies. However, in that case we have no grounds to accept any argument based on our natural inclinations to believe. In particular, then, we have no grounds to believe that our consciousness of body as extended is true. In this alternative, the Cartesians have proved less than they wanted to. '[I]f it is possible that our ideas, which have nothing in them like what is in external objects, as he maintains, represent them to us nevertheless, why is it that the ways of being we receive through the senses do not represent to us the objects that produce them, even though these ways of being are not like those objects?'[144]

The point is that there are not two ways of being conscious of things, one way in terms of ideas such as those of extension and a second way in terms of sentiments such as those we have of colours, and so on. We cannot claim two ways of having ideas, as both Descartes and Malebranche try to have it – the former with modifications of consciousness some of which are essences of other things, some of which are not; and the latter with some ideas that merely touch our minds, some of which are literally in it.

There is another point that is equally important, however. It is that Descartes allows that there are systematic gaps in the rational structure of the world, gaps that are filled not with necessary connections but with the activities of the deity. Malebranche accepts this gappiness in the rational structure of the world. More importantly, he extends it further. *For Malebranche there is no causal activity in the world save that of God.* In fact, it will be recalled from Study One, it is not unreasonable to read Descartes as also holding this opinion: the only force or activity at work in the world is that of God. But Descartes is not as explicit on this point as Malebranche. We see here that there is some point to the dissimulation thesis: Descartes was always fairly careful never to express too clearly those positions where he was disagreeing widely with the scholastic Aristotelianism of the Jesuits.

With regard, in particular, to the soul, we have seen that Descartes allowed that there is no causal power in it that could preserve it in existence from moment to moment. The temporal parts are distinguishable and therefore separable. These parts being separable, there are no discernible necessary connections between them. Since God is no deceiver, where there are no discernible necessary connections, there are no necessary connections. We are therefore preserved in being not by our own powers but only by the power of God. Malebranche rightly concludes that we therefore have no clear and distinct idea of the soul as a continuing being. '[W]e have no *clear idea* of our soul, but only *consciousness* or inner sensation of it, and ... thus we know it much less perfectly than we do extension.'[145] In fact, Malebranche cannot conceive any power in any finite thing. 'There are many reasons preventing me from attributing to *secondary* or *natural* causes a force, a power, an efficacy to produce anything. But the principal one is that this opinion does not even seem conceivable to me ... whatever effort of mind I make, I can find force, efficacy, or power only in the will of the infinitely perfect Being.'[146] In particular, corporeal objects, like minds, have no power to move either themselves or others.

When I see one ball strike another, my eyes tell me, or seem to tell me, that the one is truly the cause of the motion it impresses on the other, for the true cause that moves bodies does not appear to my eyes. But when I consult my reason I clearly see that since bodies cannot move themselves, and since their motor force is but the will of God that conserves them successively in different places, they cannot communicate a power they do not have and could not communicate even if it were in their possession. For the mind will never conceive that one body, a purely passive substance, can in any way whatsoever transmit to another body the power transporting it.

... since I know that all the changes that occur in bodies have no other principle than the different communications of motion that take place in both visible and invisible bodies, I see that it is God who does everything, since it is His will that causes, and His wisdom that regulates, all these communications.[147]

The only active power is God. All *apparent* causes are mere occasions for God's activity in bringing about the cause consequent upon the effect. The lack of connection that Descartes attributes to the mind-body complex is extended by Malebranche to all complexes. There is a similar lack of connection among other entities that are apparently causally related. Not only are the mind-body connections mere regularities that provide occasions for God's activity, so are mind-mind relations and body-body relations. *All causal powers are withdrawn from objects which we know and replaced by the causal activity of the deity.*

The traditional doctrine of substance had the forms or essences doing three things. First, they were ideas in God's mind, the divine exemplars of things, the models that constitute the structures of things created. But these forms are, second, also the forms *in* things, the active powers by which things caused themselves, in natural motion, to be. Finally, third, forms or essences in the mind were the acts by which things are known. We have seen that Malebranche rejects the latter role for forms or essences. We now see that he also rejects the second role. 'Some philosophers prefer to imagine a *nature* and certain *faculties* as the cause of the effects we call natural, than to render to God all the honor that is due his power ...'[148] Malebranche, in contrast, prefers a 'God who knows all and who does all.'[149] Thus, in contrast to the threefold role of forms or essences in the substance tradition, there is but a single role for ideas in Malebranche's philosophy. Their single role is that of ideas as the divine exemplars. Through them God, on the one hand, touches our souls to make visible to us corporeal things external to our minds, and, on the other hand, creates the collections of properties that are structured in conformity to these ideas.

The traditional doctrine of substance had ordinary things, or, in the Cartesian scheme, corporeal substance, consisting of properties inhering in a substance the form or essence of which determined those properties to be in it. That determination was not merely logical but, more importantly, causal: the form or essence was an *active power*. Descartes removed that power from the essences of corporeal things: God alone had the power to move things. The form or essence remained, however, and so the traditional doctrine of predication remained: properties are predicated of a substance that has the relevant form or essence. But, as Malebranche came to see, this makes the essence or form *in* the substance redundant: if God, as it were, does all the causal work, then all that *He* needs is the divine exemplar. So the form or essence *in* things disappears.

What does this leave us with? As for knowledge, we have on the one hand the external thing and on the other hand the knowing mind. That by which the thing is known is not in the mind. Neither is it in the thing. It therefore hovers as an object that is situated *between* the mind and the thing known. Arnauld retains a direct knowledge of the thing known, at least in the way that Aristotle did: the form or essence of the thing known is literally in the mind of the knower. But this creates an ontological problem, as Foucher argued. In order to solve this problem, Malebranche moved the idea by which things are known out of the mind. For other, equally sound, metaphysical reasons he eliminated the form or essence as something *in* the thing known. The idea, as form or essence continues to make the object known, as previously, and continues to provide the structure of the thing known, as previously, but it does so only as a divine exemplar and therefore as something that lies, as we now see, *between* the knower and the known. In short, *Malebranche solves the ontological problem of knowing that confronts the older view of Aristotle, Descartes, and Arnauld, but only at the cost of a version of representationalism.* No doubt this position was not exactly where he wanted to end up: he repeatedly insists that we *do* perceive bodies.[150] But just as Descartes arguably ends up a sceptic *malgré lui*, so Malebranche ends up a representationalist *malgré lui*.[151] As Locke was to put it, '[a]ccording to his hypothesis of seeing all things in God, how can he know that there is any such real being in the world as the sun? Did he ever see the sun? No; but on occasion of the presence of the sun to his eyes, he has seen the idea of the sun in God, which God has exhibited to him; but the sun, because it cannot be united to his soul, he cannot see. How then does he know that there is a sun which he never saw?'[152]

We can also ask what is left of corporeal things, once the essence or form is removed? It is clear: *a corporeal thing is a congeries of properties.* For the tradition deriving from Aristotle, things were also congeries of properties, but they were also more: they had within them a form or essence, an active power, that accounted for the structure exhibited by this congeries. But for Malebranche, with the disappearance of all active powers save that of God, it follows that a thing is *nothing more than* a congeries of properties, where the structure is now accounted for by means of God's activities rather than by an internal form or essence.

If this sounds like Berkeley, then it should.[153]

Berkeley, sometimes for reasons that reflect Malebranche's, also rejects material or corporeal substance. As for Malebranche, ordinary things are congeries of properties, except that for Berkeley these properties are sensible characteristics rather than the properties of extension that they are for Malebranche. This reflects the move towards empiricism, the appeal to a principle of acquaintance (PA) that we find in Berkeley.[154] At the same time, however, Berkeley notoriously maintained that any adequate ontology needed not only the passive entities – sense impressions or ideas – but also active entities. He argued that any account of the order or structure that we observe among the sense impressions of which we are aware requires the introduction of an active entity. For the 'external' world, that is, the most coherent structures among the ideas of which we are aware, we need to introduce a deity to account for the perceived order. So Berkeley retains the Malebranchian deity as the active force in the world. He also retains finite minds as centres of activity.

Descartes and Malebranche, in fact, challenge this latter feature also. They both argue that the mind is a spiritual substance which is, when compared to extended or corporeal substance, simple. As Malebranche puts it, '[b]eing neither material nor extended, the mind of man is undoubtedly a simple, indivisible substance without composition of parts ...'[155] It is, however, as we have seen a substance that lacks any active power; it is unable to sustain its own existence over time; and its essence is, at least for Malebranche, not presented to us. So it, too, is a substance for which the traditional substantial form has disappeared. Like Malebranche's material substances, so also his mental substances: both are merely congeries of properties, with the structure of these into a self accounted for not by any activity intrinsic to the self but by the activity of God.

In these ways the philosophical positions developed by Descartes and Malebranche point towards the empiricist critique of rationalism. None the less, there are two important differences.

The first difference, clearly enough, is the appeal to PA. Consider Locke. He considers the possibility of knowledge of causes of the sort desired by the rationalists. For such knowledge to be possible in the case of such qualities as sweet, red, and so on, there must be necessary connections between these qualities, necessary connections that are established by the forms or essences or natures of the substances that cause these qualities to appear. These necessary connections must be both ontological, in the entities themselves, and epistemological, giving us, when in the mind, scientific knowledge of those entities. But, Locke argues, we grasp no such connections:

'tis evident that the bulk, figure, and motion of several bodies about us, produce in us several sensations, as of colours, sounds, tastes, smells, pleasure and pain, etc. These mechanical affections of bodies, having no affinity at all with those ideas, they produce in us, (there being no conceivable connexion between any impulse of any sort of body, and any perception of a colour, or smell, which we find in our minds) we can have no distinct knowledge of such operations beyond our experience; and can reason no otherwise about them, than as effects produced by the appointment of an infinitely wise agent, which perfectly surpasses our comprehensions ...[156]

Locke's appeal is to the empiricist's principle of acquaintance (PA).[157] Descartes and Malebranche also note that such connections are not presented to us. They also argue that such connections do not exist. That is why they have to offer a special theological case for the existence external to the mind of material causes for these ideas. But Descartes and Malebranche justify the *apparent lack of connection* as a *real lack of connection* by appeal to a non-deceiving God who would not in such a case mislead us by placing a real connection where – so far as we can, from our faculties, tell – there is none. That is how they achieve knowledge that the soul is indeed separable from the body. Locke does not make such an appeal. For him the detour by way of God is not required. To the contrary, if things are presented as *apparently unconnected* then he concludes directly by appeal to PA that they *really are* unconnected.

The second difference between the empiricist critique of the substance metaphysics and that of Descartes and Malebranche is that Descartes and Malebranche simply take for granted the intelligibility of the traditional discourse. They take for granted that the notion of *substance* is intelligible. It may be that the traditional metaphysics coming down from Aristotle by way of Aquinas has got some details wrong. It may be, as Malebranche held, that the traditional doctrine and even Descartes's version of it were terribly wrong about forms, essences, and ideas, about our perception of things, and about the ontological structure of things. But all this disagreement is still 'within the ring': it all takes the notion of *substance* to be intelligible. As well, these philosophers, whatever their disagreements among themselves, all take the notion of *necessary connection* as intelligible. And, finally, they all take the notion of *(unanalysable) causal activity* or *(simple) causal power* for granted as intelligible notions. It is precisely this assumption that the empiricists were to challenge.

This challenge is worked out most completely in Hume,[158] who systematically applies the PA to the statement and resolution of philosophical problems. PA, as he expresses it, states that all our ideas derive from impressions, that is, impressions of sense – including inner experience. In particular, he insists that PA be applied to the supposed idea of an objective necessary connection based on the causal efficacies or powers that are supposed, by philosophers, rationalists, and Aristotelians alike, to move things.

Ideas always represent their objects or impressions; and *vice versa*, there are some objects necessary to give rise to every idea. If we pretend, therefore, to have any just idea of this efficacy, we must produce some instance wherein the efficacy is plainly discoverable to the mind, and its operations obvious to our consciousness or sensation. By the refusal of this, we acknowledge that the idea is impossible and imaginary; since the principle of innate ideas, which alone can save us from this dilemma, has been already refuted, and is now almost universally rejected in the learned world. (T 157–8)

Although there has been much talk about such powers, it in fact turns out that in the end everyone agrees, from Aristotle to Descartes, that these powers and the necessary connections that they are supposed to produce are not given to us in sensible experience. 'The small success which has been met with in all the attempts to fix this power, has at last obliged philosophers to conclude that the ultimate force and efficacy of nature is perfectly unknown to us, and that it is in vain we search for it in all the known qualities of matter' (T 159). Hume specifically mentions the Cartesians in this context, drawing attention in a footnote to the philosophy of Malebranche. These philosophers are led to divest matter of all causal power or efficacy. They place all such power in the deity.

Matter, say they, is in itself entirely unactive and deprived of any power by which it may produce, or continue, or communicate motion: But since these effects are evident to our senses, and since the power that produces them must be plac'd somewhere, it must lie in the Deity, or that Divine Being who contains in his nature all excellency and perfection. It is the Deity, therefore, who is the prime mover of the universe, and who not only first created matter, and gave it its original impulse, but likewise, by a continued exertion of omnipotence, supports its existence, and successively bestows on it all those motions, and configurations, and qualities, with which it is endow'd. (T 159)

However, an appeal to PA soon shows that this solution to the problems of unanalysable powers and objective necessary connections will not do. '[I]f every idea be deriv'd from an impression, the idea of a deity proceeds from the same origin; and if no impression, either of sensation or reflection, implies any force or efficacy, 'tis equally impossible to discover or even imagine any such active principle in the deity. Since these philosophers, therefore, have concluded that matter cannot be endow'd with any efficacious principle, because 'tis impossible to discover in it such a principle, the same course of reasoning should determine them to exclude it from the supreme being' (T 160). Hume

concludes that discourse about causal powers is in fact *meaningless*: 'Thus, upon the whole, we may infer, that when we talk of any being, whether of a superior or inferior nature, as endow'd with a power or force, proportioned to any effect; when we speak of a necessary connexion betwixt objects, and suppose that this connexion depends upon an efficacy or energy, with which any of these objects are endowed; in all these expressions, *so apply'd*, we have really no distinct meaning, and make use only of common words, without any clear and determinate ideas' (T 162). PA excludes such language as ontologically meaningless. *There are no objective necessities. There are no unanalysable powers.*

With this critique, causal powers were eliminated. So, therefore, was the ontological argument for the existence of an infinitely perfect being. For that argument depends upon there being an intimate ontological connection between the notion of perfection and the notion of power. And once that argument goes, so does the Cartesian solution to the epistemological problem at the core of being as it was posed by the threat of the evil genius. The core of being, however, was only the core that was defined by the framework, established by the substance ontology. Once that framework goes, there is no epistemological threat at the core of being – in the relevant sense, there is no such being and therefore no such core. Nor is there any need to overcome the threat by reference to some transcendent force or power. With the critique of the framework, the Cartesian solution to the epistemological problem can no longer be maintained; but there is no longer any need for it, since the problem it proposed to solve has also disappeared.

Under a variety of ontological pressures, the traditional doctrine of causal powers and objective necessities was eventually subjected to a searching critique by the empiricists and by Hume in particular. It did not survive the critique. But this is the framework that Descartes took for granted. In the end, once his sceptical methods were turned on this framework, it did not survive. But when it went, so did the problem of knowledge to which it gave rise. The sceptical crisis created by Montaigne and deepened in the First Meditation presupposed the framework of the substance ontology for its very statement. Descartes attempted, with not unreasonable success, to solve that problem, that is, solve it within that framework. The solution could not withstand the empiricist critique. But that did not leave an epistemological question unanswered. To the contrary, the empiricist critique that eliminated the solution also eliminated the problem.

If Descartes had in fact pushed his sceptical arguments farther than he did in the First Meditation, pushed them to the point to which Hume took them, then he would have seen that he did not need to resort to God in order to justify coherence as the criterion for distinguishing within ordinary experience those things which are fake and those which are original, those things which are appearance and those which are real. Bouwsma, in effect, more or less saw this point: if we restrict ourselves to the world of ordinary experience, the world delineated by PA, then there can be no problem of an evil genius of the sort that Descartes attempted to pose. What Bouwsma did not see was that the framework of the substance ontology had to be subjected to a sceptical critique before the demon could finally be exorcised without appeal to a God or anything else beyond the world of ordinary experience. Bouwsma's case, in fact, presupposes the prior argument of Hume that the framework of the substance ontology was simply meaningless. What Bouwsma failed to show, and what Hume was able to show in detail, is how PA invalidates the framework that makes equally possible the threat of the evil genius, on the one hand, and the appeal to God to eliminate that threat, on the other. *Sic transit gloria Dei.*

Virgil could praise the philosopher-poet:

Felix qui potuit rerum cognoscere causas.

'Fortunate is he who can discern the causes of things.' Dryden translated these lines from the Second *Georgics* this way, making clear that the reference was to Lucretius:

Happy the Man, who, studying Nature's Laws,
Thro' known Effects can trace the secret Cause.
His mind possessing, in a quiet state,
Fearless of Fortune, and resign'd to Fate.[159]

A more recent translation puts it much the same way:

Blessèd is he whose mind had the power to probe
The causes of things and trample underfoot
All terrors and inexorable fate
And the clamour of devouring Acheron ...[160]

This has been preceded by an invocation of the poet's Muse, and a prayer that she will grant the poet's desire to know the causes of things. In Dryden's translation:

Wou'd you your Poet's first Petition hear,
Give me the Ways of wandring Stars to know:

The Depths of Heav'n above, and Earth below.
Teach me the various Labours of the Moon,
And whence proceed th' Eclipses of the Sun.
Why flowing Tides prevail upon the Main,
And in what dark Recess they sink again.
What shakes the solid Earth, what Cause delays,
The Summer Nights, and shortens Winter Days.[161]

In the modern version:

Teach me to know the paths of the stars in heaven,
The eclipses of the sun and the moon's travails,
The causes of earthquakes, what it is that forces
Deep seas to swell and burst their barriers
And then sink back again, why winter suns
Hasten so fast to plunge themselves in the ocean
Or what it is that slows the lingering nights.[162]

There is here a curiosity about the order of things, about why things go on as they do, about what the causes are of all the things that happen in the world. For Aquinas, however, this sort of curiosity was simply a sin.

To be sure, it was not a sin as evil as heresy. If, perchance, your curiosity took you that far into unbelief, then all was lost. You deserved to be turned over to the secular arm, and dispatched without the loss of blood. For Aquinas, heresy is a sin that is so awful that the heretic deserves death. As he puts it, heresy is a 'sin, whereby they [the heretics] deserve not only to be separated from the Church by excommunication, but also to be severed from the world by death. For it is a much graver matter to corrupt the faith which quickens the soul, than to forge money, which supports temporal life. Wherefore if forgers of money and other evil-doers are forthwith condemned to death by the secular authority, much more reason is there for heretics, as soon as they are convicted of heresy, to be not only excommunicated but even put to death' (Summa, SS, Q11, A3).

If heresy could be the sinful product of curiosity pursued for its own sake, that motive for studiousness, though not so heinous a sin as heresy, is a sin none the less. To be sure, 'we must judge differently of the knowledge itself of truth, and of the desire and study in the pursuit of the knowledge of truth. For the knowledge of truth, strictly speaking, is good, but it may be evil accidentally, by reason of some result, either because one takes pride in knowing the truth, according to 1 Cor. 8:1, "Knowledge puffeth up," or because one uses the knowledge of truth in order to sin' (*Summa*, SS, Q167, A1). But, especially, a simple desire to know the way the world is, that is, the world as we know it in sense experience, is a sin 'when a man desires to know the truth about creatures, without referring his knowledge to its due end, namely, the knowledge of God. Hence Augustine says (De Vera Relig. 29) that "in studying creatures, we must not be moved by empty and perishable curiosity; but we should ever mount towards immortal and abiding things"' (SS, Q167, A2). Here Aquinas quotes Augustine that 'concupiscence of the eyes makes men curious' (ibid.), and adds that 'since concupiscence of the eyes is a sin, even as concupiscence of the flesh and pride of life ... it seems that the vice of curiosity is about the knowledge of sensible things' (ibid.).

For Aquinas, one could study nature, but only if one had one's eye firmly on the fact that such a study should not be for its own sake but only because the facts so studied lead one inevitably to God: the only morally acceptable cognitive end is knowledge of God, and any attempt to study nature for its own sake misses this end and is therefore sinful.

But with the emergence of empiricism and the new science, *Sic transit gloria Dei*: with the demise of the substance metaphysics so came the demise of God as central to the cognitive enterprise: there was no God to be in this way central. Since discourse about such entities was essentially empty, such entities could not be the aim of any cognitive interest: there was nothing there in which to take a cognitive interest. Nor, therefore, could it any longer be a sin that one took a cognitive interest in natural causes with no further cognitive end in mind. One could explore the structure of nature for its own sake, without having constantly to look beyond this order to some being outside the order but somehow responsible for it and somehow there to ensure that we would be successful in our search.

We can see the cognitive end of trying simply to understand nature becoming as acceptable in the seventeenth century as it was to Virgil. In Milton's *Paradise Lost*, Adam asks Raphael about creation, asks if he may not explore it in order to understand it. To be sure, Adam is guarded: he indicates that he knows that what he will find will testify to the glory of God, make Him more worthy of adoration.

> If unforbid thou mayst unfold
> What wee, not to explore the secrets ask
> Of his Eternal Empire, but the more
> To magnify his works, the more we know.[163]

Independently of Adam's careful qualifications, Raphael allows the legitimacy of the desire to know – but not to know too much: curiosity is to be limited to the realm of ordinary experience, things that we can know by means of our senses.

> To ask or search I blame thee not, for Heav'n
> Is as the Book of God before thee set,
> Wherein to read his wond'rous Works, and learn
> His Seasons, Hours, or Days, or Months, or Years;
> This to attain, whether Heav'n move or Earth,
> Imports not, if thou reck'n right; the rest
> From Man or Angel the great Architect
> Did wisely to conceal, and not divulge
> His secrets to be scan'd by them who ought
> Rather admire.[164]

Like hunger, curiosity is a natural appetite which must be satisfied – though within limits – 'within bounds,' as Milton puts it.

> This also thy request with caution askt
> Obtain: though to recount Almighty works
> What words or tongue of Seraph can suffice,
> Or heart of man suffice to comprehend?
> Yet what thou canst attain, which best may serve
> To glorify the Maker, and infer
> Thee also happier, shall not be withheld
> Thy hearing; such Commission from above
> I have receiv'd, to answer thy desire
> Of knowledge within bounds; beyond abstain
> To ask, nor let thine own inventions hope
> Things not reveal'd, which th' invisible King,
> Only Omniscient, hath supprest in Night,
> To none communicable in Earth or Heaven:
> Anough is left besides to search and know.
> But Knowledge is as food, and needs no less
> Her Temperance over Appetite, to know
> In measure what the mind may well contain,
> Oppresses else with Surfeit, and soon turns
> Wisdom to Folly, and Nourishment to Wind.[165]

As for God Himself, He is

Unspeakable, who sit'st above these Heavens
To us invisible or dimly seen
In these thy lowest works, yet these declare
Thy goodness beyond thought, and Power Divine;[166]

In contrast to Aquinas, Milton is arguing that curiosity about matters of fact is permissible as an end. In pursuing it for its own sake we will, in fact, discover things that will reveal to an even greater extent the glory of God; but we seek the knowledge for its own sake, not simply because we aim to know God. In fact, unlike Aquinas, Milton does not allow that we can, in any intelligible sense, set out to comprehend God. We may seek knowledge only 'within bounds,' and God is beyond those bounds.

Aquinas had no shame in claiming to know a good many things about the Deity and His purposes. To be sure, God created things in order that there be being: 'In the very fact of any creature possessing being, it represents the Divine being and Its goodness. And, therefore, that God created all things, that they might have being, does not exclude that He created them for His own goodness' (*Summa*, FP, Q65, A2, RObj 1). But we know much more than this. With regard to the details of creation, we are told that

if we wish to assign an end to any whole, and to the parts of that whole, we shall find, first, that each and every part exists for the sake of its proper act, as the eye for the act of seeing; secondly, that less honorable parts exist for the more honorable, as the senses for the intellect, the lungs for the heart; and, thirdly, that all parts are for the perfection of the whole, as the matter for the form, since the parts are, as it were, the matter of the whole. Furthermore, the whole man is on account of an extrinsic end, that end being the fruition of God. So, therefore, in the parts of the universe also every creature exists for its own proper act and perfection, and the less noble for the nobler, as those creatures that are less noble than man exist for the sake of man, whilst each and every creature exists for the perfection of the entire universe. Furthermore, the entire universe, with all its parts, is ordained towards God as its end, inasmuch as it imitates, as it were, and shows forth the Divine goodness, to the glory of God. Reasonable creatures, however, have in some special and higher manner God as their end, since they can attain to Him by their own operations, by knowing and loving Him. Thus it is plain that the Divine goodness is the end of all corporeal things. (FP, Q65, A2)

Here we may contrast Descartes, who explicitly denies that we can have any insight into the purposes of the Deity: whatever His ends may in detail be, they remain a mystery to us. All that we know is that He is a creator: He brings things into being. He puts the point briefly: 'It is not the final but the efficient causes of created things that we must inquire into' (*Principles*, 1:202); and then at greater length:

When dealing with natural things we will, then, never derive any explanations from the purposes which God or nature may have had in view when creating them and we shall entirely banish from our philosophy the search for final causes. For we should not be so arrogant as to suppose that we can share in God's plans. We should, instead, consider him as the efficient cause of all things; and starting from the divine attributes which by God's will we have some knowledge of, we shall see, with the aid of our God-given natural light, what conclusions should be drawn concerning those effects which are apparent to our senses. (ibid.)

As Guéroult has expressed it, Descartes 'rejects any penetration into God's counsels ...'; upon his view, 'God cannot direct his will on anything other than being ...'[167] Descartes is keeping our ability to know the ways of God 'within bounds.'

With Hume the bounds upon what we might claim to know and where we may reasonably seek to satisfy our curiosity about the ways of the world are even more severely limited. Hume allows that we are moved by curiosity. He argues that even if we

suppose this curiosity and ambition should not transport me into speculations without the sphere of common life, it would necessarily happen that from my very weakness I must be led into such enquiries. It is certain that superstition is much more bold in its systems and hypotheses than philosophy; and while the latter contents itself with assigning new causes and principles to the phenomena which appear in the visible world, the former opens a world of its own, and presents us with scenes, and beings, and objects, which are altogether new. Since, therefore, it is almost impossible for the mind of man to rest, like those of beasts, in that narrow circle of objects, which are the subject of daily conversation and action, we ought only to deliberate concerning the choice of our guide, and ought to prefer that which is safest and most agreeable. And in this respect I make bold to recommend philosophy, and shall not scruple to give it the preference to superstition of every kind or denomination. (T 271)

Why recommend philosophy? Well, here we have the possibility of a method that will yield solutions to problems. As for speculations that go beyond the realm of philosophy, beyond the task of assigning new causes to the phenomena which appear in the visible world, while they are tempting, they are also useless and, because empirically meaningless, incapable of any reasonable resolution.

'While a warm imagination is allowed to enter into philosophy, and hypotheses embraced merely for being specious and agreeable, we can never have any steady principles, nor any sentiments, which will suit with common practice and experience. But were these hypotheses once removed, we might hope to establish a system or set of opinions, which if not true (for that, perhaps, is too much to be hoped for), might at least be satisfactory to the human mind, and might stand the test of the most critical examination' (T 272). Knowledge is now kept quite 'within bounds': those bounds are the visible world, the world of ordinary sense experience. Milton, like Descartes, had a world in which God was an essential ingredient. But the world in which they lived had a God that was in most respects much more limited, much more shadowy, than the deity in the world of Aquinas. A necessary being, to be sure, but for all that a being who is mysterious and inscrutable, beyond human capacities to comprehend.

As for the motive for the search after causes in the visible world, in the world of ordinary experience – that motive is simple curiosity. But there are other sentiments, too, that move one to search after matter-of-fact truth: there is the desire to help humankind come to know the foundations of knowledge; there is the love of the fame that might be the result of success in such researches. But all are natural, none sinful.

At the time, therefore, that I am tired with amusement and company, and have indulged a reverie in my chamber, or in a solitary walk by a river side, I feel my mind all collected within itself, and am naturally inclined to carry my view into all those subjects, about which I have met with so many disputes in the course of my reading and conversation. I cannot forbear having a curiosity to be acquainted with the principles of moral good and evil, the nature and foundation of government, and the cause of those several passions and inclinations which actuate and govern me. I am uneasy to think I approve of one object, and disapprove of another; call one thing beautiful, and another deformed; decide concerning truth and falsehood, reason and folly, without knowing upon what principles I proceed. I am concerned for the condition of the learned world, which lies under such a deplorable ignorance in all these particulars. I feel an ambition to arise in me of contributing to the instruction of mankind, and of acquiring a name by my inventions and discoveries. These sentiments spring up naturally in my present disposition; and should I endeavour to banish them, by attaching myself to any other business or diversion, I feel I should be a loser in point of pleasure; and this is the origin of my philosophy. (T 273)

Sic transit gloria Dei: with the empiricist critique of the substance ontology, God disappears from the ontological structure of the universe. With the disappearance of God goes the disappearance of the attitude that a curiosity about the ordinary world of sense experience is somehow sinful and unworthy of the proper human intellect. To the contrary, that world now becomes the only object of study. Discourse about God is not meaningless because God has become inscrutable; rather, such discourse is cognitively meaningless and God has therefore become unthinkable. Investigations into God will not 'stand the test of the most critical examination.' Curiosity is therefore properly to be restricted to the visible world, where we might not know for sure which causal principles are true but where we can at least expect that they 'might' stand that test. Nor is there any reason to condemn or denounce such curiosity on the grounds that there is a realm or world that is superior and therefore the more proper object of our cognitive interest.

With the empiricist critique of the substance metaphysics and of God, the mind has been freed to search after the truth about causes in the world of ordinary sense experience. The method is to hand, the method of empirical experiment, and we have been made free to use it to discover the causes of things.

Of course, with the demise of the substance ontology, there disappeared, too, the old notion of cause as involving objective necessary connections established through the exercise of simple unanalysable powers. The old science of *scientia* disappeared. But the disappearance of the old science could not automatically lead to a new science. For that to happen, for there to come to be a new science that limited itself to the world of ordinary experience, a new notion of cause, a new notion of explanation, a new notion of human reason had to be developed. Causation could no longer be the obtaining of objective necessary connections. Explanation could no longer be by appeal to forms or natures or essences. Reason could no longer be the grasp of such forms or natures or essences. It was the empiricists who developed and defended this new notion of cause, this new notion of explanation, and this new notion of human reason.

But they did not develop it *ex nihilo*; nor were the arguments they used to defend it without historical precedent. In fact, the history of the substance philosophy in the early modern period is a story in which its defenders, such as Descartes and Malebranche, gradually removed the necessary connections and active powers that are its central doctrine more and more away from any explanatory role in the occurrences of ordinary events and processes. This, that, and the other thing all possessed such powers, and we were supposed everywhere to recognize their exercise. But ordinary things gradually lost these powers, until, in the end, only God was able to effect causal changes. Where Aristotle and Aquinas saw the exercise of powers in things, Descartes and Malebranche saw things that were mere *occasions* for the exercise of the divine power. All we are ever aware of with regard to ordinary processes are patterns, matter-of-fact regularities. The real causal power lies forever beyond us, or at least almost beyond us. As our relationship to causal powers became more and more attenuated, it became more and more reasonable to think of regularities as what for science, in its ordinary practice, were the objects of cognitive interest. To be sure, ideally we could aspire to know objective necessary connections, and a concern with regularities was still a concern for the second best. But the second best became more and more the best that we ever could achieve. And when the second best turns out upon close examination to be the very best that we can achieve, then, relative at least to what we, as humans, *can* achieve, that really is *the best*.

This is the logic of the situation that we find in Locke. From there it is but a short step to the final critique of rationalism in Hume, the final demise of objective necessary connections. Hume is able to see clearly that not only is the grasp of regularity the best that we *can* do but that *there is nothing better that we might do*. So long as we allow, as Locke does, that we have the idea of, even if very few instances of, an objective necessary connection, then we have to allow that there is always the possibility that there is something to causation beyond regularity. But Hume argued that all such discourse is without cognitive content. Thus, *there is, AND CAN BE, nothing to causation beyond regularity*. Taking causation to be regularity is no longer a mere second best, something we can always, however vaguely, hope to surpass. Objectively, regularity is all there is *or could be* to causation. For Hume, when we discover that causation is regularity we have reached not just the limits of knowing, but the limits of being. We have not reached the limits of our knowledge beyond which stretches a realm of being that we cannot know. Rather, we have reached the limits of our knowledge because we have reached the limits of being. With regard to causation, there is nothing in being beyond regularity that we might aspire to know.

Causation thus comes to be redefined as matter-of-fact regularity. Explanation is similarly redefined as subsumption under such regularities. And reason in its turn is redefined as the coming-to-know such regularities. Reason is now understood not as the capacity to glomb onto or grasp forms or natures or essences but rather as the capacity to come to know matter-of-fact generalities. We exercise our reason when we conform to those patterns of inductive research that best serve the cognitive interest we have in trying to understand the world of ordinary experience, which is to say, our cognitive interest in coming to know the regularities that describe the world that we know through this experience. These norms of inductive research are Hume's 'rules by which to judge of causes and effects,' that is, the rules of eliminative induction.

Reason is now fallible. Reason is now restricted to the world of sense experience. But, at least, it is a reason that has a real object, not the illusions of the

Aristotelians and the rationalists. Above all, it is a reason that, as Hume said, allows us to 'hope to establish a system or set of opinions, which if not true (for that, perhaps, is too much to be hoped for), might at least be satisfactory to the human mind, and might stand the test of the most critical examination.' That is more than the Aristotelians and the rationalists ever attained.

Notes

Introduction

1 Cf P. Frank, *Philosophy of Science*, 1–20, 361–3.
2 I.B. Cohen, *The Birth of a New Physics*, 25.
3 Ibid., 117ff.
4 T. Kuhn, *The Copernican Revolution*, 83.
5 Ibid., 96.
6 S. Shapin, *The Scientific Revolution* (Chicago 1996).
7 F.C. Beiser, *The Sovereignty of Reason: The Defense of Rationality in the Early English Enlightenment*, x.
8 One should also mention Pierre Gassendi, whose importance has recently been demonstrated by T. Lennon in his *The Battle of the Gods and Giants: The Legacies of Descartes and Gassendi, 1655–1715*.

Study One

1 A. Arnauld, *The Art of Thinking*, trans. J. Dickoff and P. James, 167; unless otherwise noted references to the Port Royal Logic will be to this translation. See Francis Bacon, *New Organon*, ed. with an Introduction by F.H. Anderson, I.xv, xvi, lxiii.
2 Arnauld, *Art of Thinking*, 15, 235ff; Descartes, *Rules for the Direction of the Mind*, in *Descartes: Philosophical Essays*, trans. L.J. Lafleur, 150–1. (Also in *The Philosophical works of Descartes*, trans. Cottingham, et al., 1:11ff; hereafter cited as Cottingham, with volume and page number).
3 Bacon, *New Organon*, 12.
4 S. Gaukroger, *Cartesian Logic*.
5 L. Jardine, *Francis Bacon: Discovery and the Art of Discourse*.
6 Cf R. Brown, 'History versus Hacking on Probability,' argues (656) against the case

I once made for the emergence of modern science corresponding to the success of the empiricists in implementing their program against that of Aristotle. I made this case (in outline) in a 'Critical Notice' of I. Hacking *The Emergence of Probability*. History shows, Brown claims, a continuity where I claim a radical break exists. We shall discuss Brown's views briefly, in sections IV:3 and IV:4 below. Suffice it to say for now that Brown can think he sustains his case only because he fails to understand the difference between the sources of error and fallibility in Aristotle and in empiricism and the new science, with this failure deriving from his apparent incapacity to understand the central differences between these two philosophies. Superficial similarities – in terminology, for example – are paraded to support the claim for continuity, while the deeper differences are simply ignored. I suppose that the moral is that before one ventures to discuss philosophy one should attempt to learn some. By the way, as the 'Critical Notice' argues, I do not agree completely with Hacking, but none the less I would argue that Brown's criticisms of Hacking fail for the same reasons that his criticisms of my own case fail.

7 William Harvey, *The Circulation of the Blood and Other Writings*, trans K. Franklin, 176.

8 J. Weinberg, 'Induction.'

9 Descartes, *Discourse on Method, Optics, Geometry, and Meteorology*, trans. P.J. Olscamp, 41; Cottingham, I:136.

10 Descartes, *Discourse*, trans. Olscamp, 50; Cottingham, 1:142.

11 Bacon, *New Organon*, 37.

12 Unless otherwise noted, references are to the *New Organon*, by Book and Aphorism.

13 Ibid., 19.

14 The relevance of a clear grasp of the cognitive goals of science for understanding the nature of scientific explanation has been emphasized in F. Wilson, *Explanation, Causation and Deduction*, following G. Bergmann, *Philosophy of Science*, ch. 2. Though this relevance was clear to such early thinkers as Bacon and Descartes, more recent discussions, such as those of C.G. Hempel and P. Oppenheim, 'Studies in the Logic of Explanation,' have tended to ignore it. As a consequence, there have been a lot of unjustified attacks on the model of explanation given by Bacon and Descartes, that is, the so-called 'covering law model,' according to which explanation of particular facts consists in subsuming them under matter-of-fact regularities. *Explanation, Causation and Deduction* defends the traditional view of Bacon, Descartes, Hume, and Mill against many recent critics.

15 Bacon, *New Organon*, 35.

16 Ibid., 19.

17 Cf J.L. Mackie, 'Causes and Conditions.'

18 Cf Bergmann, *Philosophy of Science*, ch. 2. See also Wilson, *Explanation, Causation and Deduction* and *Laws and Other Worlds*.

19 If *whenever 'A' then 'B'* (e.g., whenever there is a fire then oxygen is present) then *B* is a *necessary (causal) condition* for *A*; and if *whenever 'C' then 'D'* (e.g., whenever sugar is in water then it dissolves), then *C* is a *sufficient (causal) condition* for *D*.

20 Cf Bergmann, *Philosophy of Science*, ch. 2.

21 Ibid.

22 Cf ibid., and Wilson, *Explanation, Causation and Deduction*.

23 Since there are factual limitations on measurement, it is *un*reasonable to aim to overcome these.

24 There may be other variables that do not affect the variables in which one is interested. For example, in computing the positions and velocities of the planets, one can ignore the colours of those objects.

25 Descartes, *Principles of Philosophy*, trans. V.R. Miller and R.P. Miller (Bk III, ¶ 47), 108 (Cottingham, 1:258). The curly brackets are used by Miller and Miller to indicate words that are in the French translation but not in the Latin, and for which they (Miller and Miller) believe that there is independent evidence that they reflect Descartes's own view.

26 Or rather, what we know since Einstein to be a good approximation to such knowledge.

27 Mackie, 'Causes and Conditions.'

28 Bergmann, *Philosophy of Science*, ch. 2.

29 See Wilson, *Explanation, Causation and Deduction*, for a detailed defence of this point, and discussion of many philosophers who fail to recognize this point.

30 Bacon, *New Organon*, 20.

31 Ibid.

32 Ibid.

33 D. Hume, *Treatise concerning Human Nature*, ed. L.A. Selby-Bigge, Book I, Part III, section xv. References are by book, part, and section, and/or by page, following a 'T.'

34 Cf F. Wilson, *Empiricism and Darwin's Science*, 29ff.

35 Descartes, *Discourse*, trans. Olscamp, 34; Cottingham, 1:131.

36 Ibid., 37; Cottingham, 1:134.

37 Ibid., 37–8; Cottingham, 1:134.

38 Ibid., 52; Cottingham, 1:144.

39 Bacon, *New Organon*, 21.

40 Descartes, *Discourse*, trans. Olscamp, 33; Cottingham, 1:131.

41 Ibid., 45; Cottingham, 1:139.

42 Ibid., 112–13; Cottingham, 1:172.

43 Descartes, *Discourse*, trans. Olscamp, 114ff.

44 Bacon, *New Organon*, 21.

45 To say this is not to say that one cannot accept as laws certain regularities that may

correctly be characterized as 'gappy'; see Wilson, *Explanation, Causation and Deduction*.

46 Bacon, *New Organon*, 19.

47 Ibid., 20.

48 Ibid., 19.

49 *Rules*, trans. Lafleur, 163; Cottingham, 1:20–1.

50 Ibid., 184; Cottingham, 1:37.

51 Arnauld, *Art of Thinking*, 24, 167–8.

52 Ibid., 167.

53 Ibid., 319.

54 Ibid.

55 Descartes, *Discourse*, trans. Olscamp, 38; Cottingham, 1:134.

56 For this tradition, see P. Reif, 'The Textbook Tradition in Natural Philosophy, 1600–1650' also L. Thorndyke, 'The Cursus Philosophicus before Descartes.'

57 G. Vlastos, 'Reasons and Causes in the *Phaedo*.'

58 R.G. Turnbull, 'Aristotle's Debt to the "Natural Philosophy" of the *Phaedo*.'

59 In this essay we shall be concerned exclusively with changes that occur *in* a substance, rather than with what Aristotle called 'generation' and 'corruption,' that is, changes in which a substance itself comes or ceases to be. The latter is not a central issue for the concerns that motivated the critics of the Aristotelian position; basically, if sense cannot be made of ordinary change, there is little possibility that sense can be made of substantial change. In fact, the rationalists in the early modern period – Descartes, Spinoza, and Leibniz – all rejected the notion that there could be change in the sense of generation and corruption.

60 Unless otherwise noted, all quotations from Aristotle are from *The Basic Works of Aristotle*, ed. R. McKeon.

61 The place of this Aristotelian teleology in late scholastic thought is discussed in Dennis Des Chene, *Physiologia: Natural Philosophy in Late Aristotelian and Cartesian Thought*, 20ff.

62 I have benefited from reading Richard Sorabji, *Necessity, Cause, and Blame*.

63 Cf A. Gotthelf, 'Aristotle's Conception of Final Causality.'

64 Aristotle, *The Generation of Animals*, in *The Complete Works of Aristotle*, ed. J. Barnes.

65 I shall, therefore, on the whole use 'form' and 'nature' interchangeably. For our purposes, though of course not for others, there is not much need to distinguish these.

66 A. Arnauld, *On True and False Ideas*, trans. E.J. Kremer, 46.

67 I discuss other aspects of the traditional doctrine of substance in Study Six, below.

68 See W.D. Ross, *Aristotle*, 171.

69 Ibid.

70 Ibid., 69.

71 Sorabji, *Necessity*, 171; cf Gotthelf, 'Aristotle's Conception of Final Causality,' 232, 234.

72 Cf Wilson, *Laws and Other Worlds*.

73 Sorabji, *Necessity*, 170.

74 The textbook tradition held that '*science* consists in certain, universal, and unchanging knowledge achieved through causal demonstration' (Reif, 'Textbook Tradition,' 21).

75 Again, for our purposes, the distinction between 'form' and 'essence' is not important.

76 See Russell's 'Lectures on the Philosophy of Logical Atomism,' in his *Logic and Knowledge*, ed. R.C. Marsh, 231.

77 Cf J. Hintikka, 'Necessity, Universality, and Time in Aristotle,' 111.

78 Ibid., 117–18.

79 Hume, *Treatise of Human Nature*, Bk I, Part III. Cf Wilson, 'Hume's Defence of Causal Inference,' and 'Hume's Defence of Science'; also *Laws and Other Worlds*.

80 Cf L.A. Kosman, 'Understanding, Explanation, and Insight in Aristotle's *Posterior Analytics*': 'If all Ks being L *could* explain this K being L, then surely, on the same grounds, if it is puzzling that this K is L, the fact that *all* Ks are L would be equally puzzling ...' (375). This states the Aristotelian argument against explanation in terms of the covering law model succinctly but correctly. It is, however, an inconclusive argument: for detail on this point, see Wilson, *Laws and Other Worlds*.

81 The textbook tradition insisted that the aim in natural philosophy is 'to attain *essential* knowledge of natural bodies and their properties' (Reif, 'Textbook Tradition,' 21).

82 Contrary to. M. Hocutt, 'Aristotle's Four Becauses.' Compare: 'The primary object of scientific understanding is a phenomenon which, as Aristotle claims, is necessary, for explanation is from the necessary and concerns that which could not be other than it is ...' (Kosman, 'Understanding,' 377).

83 Cf B. Brody, 'Toward an Aristotelian Theory of Scientific Explanation'; and Sorabji, *Necessity*, ch. 3.

84 'Prima facie ... understanding the why of something is not understanding that thing, but some other thing, namely its cause, that which is responsible for it being the case.

'Any account of what leads Aristotle to identify understanding something and knowing its causes must begin with the defeat of that prima facie expectation. For it must understand "cause" to refer *not* to something other than the entity in question, but to the entity itself under that description which reveals certain of its *kath auto* predicates ... The *why* in terms of which scientific understanding is defined is simply *the nature of the phenomenon in question* ... The asking a why question is thus an attempt to understand more fully the *nature* of the phenomenon being explained' (Kosman, 'Understanding,' 376).

85 The textbook tradition often ended up trying 'to resolve their difficulties by resorting to such pseudo-explanations as "occult" qualities or simply "nature"' (Reif, 'Textbook Tradition,' 21).

86 A. Arnauld, *On True and False Ideas*, trans. Kremer, 4.

87 D. Hume, *Enquiries concerning Human Understanding and concerning the Principles of Morals*, 25, 163.

88 This is why it is quite wrong to say, as James H. Lesher, 'The Meaning of *NOUS* in the Posterior Analytics,' does, that 'If ... we mean by "intuition" a faculty which acquires knowledge about the world in an *a priori* manner, then it will be inappropriate to think of the Aristotelian *nous* as intuition' (64). To be sure, the knowledge, according to Aristotle, is not innate, but acquired only after a certain amount of sense experience from which it is abstracted; but the *product* of that process of abstraction *is*, contrary to Lesher, a rational intuition of a necessary structure that is not given in sense as such.

89 '[B]y "possible" [Aristotle] understood more than merely "conceivable." He seems to have thought of "possibilities" as something not unlike "natural tendencies"' (Hintikka, 'Necessity,' 113).

90 There is a qualification that must be made here; see note 95, below.

91 Cf Lesher: 'This schema for the expansion of scientific knowledge, producing scientific syllogisms by interpolating the middle or causal factor until we have reached premises which no longer admit of further "packing," is well attested in Aristotle's writings and it brings out a feature of syllogistic reasoning which is sometimes neglected: that we employ and construct syllogisms on the way to first principles, as well as from them' ('Meaning of *NOUS*,' 56–7).

92 Cf Kosman, 'Understanding,' 378.

93 L. Bourgey, 'Observation and Experiment in Analogical Explanation,' 180.

94 These points are extensively discussed in Wilson, *Explanation, Causation and Deduction, Laws and Other Worlds*, and *Empiricism and Darwin's Science*.

95 See note 90, above.

96 For a discussion of the Aristotelian notion of 'Nature' in late scholastic thought, see Des Chene, *Physiologia*, ch. 7.

97 Contrary to what Sorabji, *Necessity*, 49, holds.

98 Aristotle 'seems to have thought of "possibilities" as something not unlike "natural tendencies." In some cases at least, they become actualised unless there is something to prevent them from doing so' (Hintikka, 'Necessity,' 113).

99 'On this [Aristotle's] view, a possibility is never realised only if there is *always* something which prevents it from being realised. But saying that something is always prevented from happening seems to be the same as to say that it is *impossible* for it to happen. Hence, on Aristotle's view every genuine possibility is bound to be actualised sooner or later' (Hintikka, 'Necessity,' 113).

100 Cf Wilson, 'Explanation in Aristotle, Newton and Toulmin.'

101 Ross, *Aristotle*, 77, suggests that 'There will exceptions to rules, but these exceptions will be according to rule.' But this is not just. To the contrary, exceptions are *not* subject to higher order generalizations. Ross's suggestion too easily assimilates Aristotle to modern science. Cf Sorabji, *Necessity*, 62.

102 See Wheelwright, *The Presocratics*, 54. For discussion, see G. Vlastos, 'Equality and Justice in Early Greek Cosmology,' and Charles Kahn, 'Anaximander's Fragment: The Universe Governed by Law.'

103 Cf F. Wilson, 'Mill's Proof That Happiness Is the Criterion of Morality,' and 'Hume's Cognitive Stoicism,' and, in greater detail, in *Hume's Defence of Causal Inference*, ch. 2.

104 'We said that to look for a causal explanation is to look for a description appropriate to a subject such that the subject under that description is per se the predicate in question. Such a description Aristotle terms ... the middle; it is a description which links subject and predicate both conceptually, as providing the intelligible source of explanation, and ontologically, as the real ground of the subject exhibiting the predicate' (Kosman, 'Understanding,' 379).

105 Cf Wilson, *Explanation, Causation and Deduction*, and *Laws and Other Worlds*.

106 Cf F. Wilson, 'The Lockean Revolution in the Theory of Science.'

107 Cf Lesher: 'experience provides us with principles which we then endeavor to structure within syllogistic form ...' ('Meaning of *NOUS*,' 58).

108 Kosman, 'Understanding,' 383. See also Lesher: '*nous* is not restricted to the grasp of first principles but exhibited whenever from a series of observations of particular cases we grasp the universal principle at work in each case ...' ('Meaning of *NOUS*,' 52).

109 Ross, *Aristotle*, 44.

110 Sorabji, *Necessity*, 200.

111 Cf Raymond, 'La Théorie de l'induction – Duns Scot précurseur de Bacon.'

112 Unless otherwise noted, references to Scotus are to the *Opus oxoniense*, I, dII, q4, as translated in Duns Scotus, *Philosophical Writings*, ed. and trans. Alan B. Wolter, 96–132. These references will be given as 'Wolter.' Scotus's various works are discussed in this volume, at xviiff. For an important discussion of Scotus's views on knowledge, see Peter C. Vier, *Evidence and Its Function according to John Duns Scotus*. Vier cites a number of Scotus's works beyond the passages in Wolter; references to these sources will be given as 'Vier.' References from Vier to the *Opus oxoniense* that are not in Wolter are referred to as *Oxon*. Other references taken from Vier are to *Quaestiones subtilissime in Metaphysicam Aristotelis*, referred to as *Metaphy*., and to *Quaestiones quodlibetales*, referred to as *Quodl*.

113 Duns Scotus, *Opus*, Wolter, 100.

114 Ibid., 128, 131.

115 Ibid., 100.

116 Ibid., 118.

117 Ibid., 128, 131.

118 'Est intellectus hujus propositionis: natura determinatur ad unum, non quidem ad unum producibile, unum inquam numero sive singulare; sed determinatur ad unum determinatum modum producendi, quia non est ibi principium indeterminatum respectu oppositorum sicut est voluntas' (Duns Scotus, *Quodl.* Q2, n10; quoted in Vier, *Evidence*, 142).

119 Duns Scotus, *Opus*, Wolter, 109.

120 Ibid., 109.

121 Ibid., 109.

122 Ibid., 108.

123 '[V]idet et certus est naturam ut in pluribus uniformiter agere et ordinate' (Duns Scotus, *Metaphy.* I, q4, n19; quoted in Vier, *Evidence*, 143).

124 Duns Scotus, *Opus*, Wolter, 110.

125 'Posse habere causatum simpliciter occessarium non est perfectionis in causa secunda; immo et hoc nulli causae secundae convenir ... Simpliciter enim necessario causate includit contradictionem et ideo hoc nulli causae secundae convenit' (Duns Scotus, *Oxon.* II, d1, q3, n12; quoted in Vier, *Evidence*, 146).

126 Duns Scotus, *Metaphy.* VI, q2, n7; quoted in Vier, *Evidence*, 144.

127 'Causa naturalis, licet ex se terminetur ad effectum, potest tamen impediri ... si ab extrinseco ponatur impedimentum' (Duns Scotus, *Metaphy.* IX, q14, n1; quoted in Vier, *Evidence*, 144).

128 Duns Scotus, *Metaphy.* VI, q2, n7; quoted in Vier, *Evidence*, 145.

129 Duns Scotus, *Oxon.* I, d3, q2, f27r, b; quoted in Vier, *Evidence*, 145.

130 'Licet aliqua connexio (effectus ad causam) sit secundum quid necessaria, nulla tamen simpliciter est necessaria, quia quaelibet dependet a prima quae contingenter causat ... similitur communitur secundae causae impedibiles, et causa impedibilis quantumcumque non impediatur non est necessaria' (Duns Scotus, *Metaphy.* V, q3, n5; quoted in Vier, *Evidence*, 144).

131 Duns Scotus, *Opus*, Wolter, 110–11.

132 Vier, *Evidence*, 139–40.

133 Peter Dear, *Discipline and Experience*, 11.

134 Ibid., 15.

135 Ibid., 19.

136 Ibid., 18.

137 F. Aguilonius, *Opticorum libri sex*, 215; quoted in Dear, *Discipline and Experience*, 19.

138 Duns Scotus, *Opus*, Wolter, 108.

139 Ibid., 116.

140 Ibid., 109–10.

141 Bacon, *New Organon*, 19.

142 P. Coffey, *The Science of Logic*, 2 vols, 2:33.

143 Charles B. Schmitt, 'Experience and Experiment: A Comparison of Zabarella's View with Galileo's in *De motu*.' 100–1.

144 Ibid., 127.

145 J.H. Randall, Jr, 'The Development of Scientific Method in the School of Padua.'

146 Bacon, *New Organon*, 37.

147 Ibid., I.xix.

148 Ibid., 20.

149 *Meditations*, in Descartes, *Philosophical Essays*, trans. Lafleur, 87–91; Cottingham, 2:20ff.

150 Descartes, *Rules for the Direction of the Mind*, trans. Lafleur, Rule II, 150; Cottingham, 1:11.

151 See S. Gaukroger, in his 'Introductory Essay' to his translation of Arnauld's *On True and False Ideas*, 4–7.

152 Thomas Aquinas, *Summa theologica*, trans. by the Fathers of the Dominican Province. References are to the First Part (FP) or First Part of the Second Part (FS) or Second Part of the Second Part (SS), by Question (Q), Article (A), Objection (Obj) by number, and, where appropriate, Reply to Objection (RObj) by number.

153 Cf Stillman Drake, *Galileo: Pioneer Scientist*.

154 Other theories of projectile motion deriving from the ancient world were equally metaphysical, or, to use a phrase of Michael Wolff, 'speculative.' This includes the influential theory of impetus deriving from Philoponus, as well as other ancient theories of motion such as Aristotle's. As Wolff emphasizes, none of them were empirical. There is, in fact, a discontinuity between the metaphysical tradition and the new science. Wolff puts the point this way: 'it is now rather difficult to regard the theory [of impetus] as a connecting link between Aristotelian physics and classical mechanics. For the words "connecting link" imply more than the mere temporal order of these theories. They are intended to point to the continuous evolution of one theory into another. But, if we keep in mind the non-empirical character of impetus theory, it is difficult actually to recognise such continuity.' See Michael Wolff, 'Philoponous and the Rise of Preclassical Dynamics,' 85–6.

155 We thus see that J.C. Pitt, 'Galileo: Causation and the Use of Geometry,' 187, is just wrong to suggest that Galileo was 'refining, for scientific purposes, the Aristotelian modes of causal analysis.' Galileo was in fact eliminating the whole idea of Aristotelian causation in favour of a very different concept of science.

156 See also Stillman Drake's 'Impetus Theory Reappraised.'

157 The symmetry thesis has been challenged by critics of the covering law model of explanation, but has been defended in detail in Wilson, *Explanation, Causation and Deduction*.

158 Unless otherwise noted, references are to Isaac Newton, *Mathematical Principles of Natural Philosophy and His System of the World*, trans. Andrew Motte, 1729, revised by Florian Cajori.

159 I shall ignore the necessary qualification that must be made since Einstein.

160 As in biological phenomena, for example; see Wilson, *Empiricism and Darwin's Science*.

161 Cf J. Herival, 'Newton's Achievement in Dynamics'; I.B. Cohen, 'Newton's Second Law and the Concept of Force in the *Principia*'; and R.S. Westfall, *Force in Newton's Physics*.

162 Newton undoubtedly understands this law in its explicit formulation in terms of momentary, though not instantaneous, forces – specifically, forces of impact that last for only a very short time. That, at least, is the force of the examples that he gives to illustrate this law (for which, see below). On the other hand, when it comes to application, he does apply it to the case of continuously acting forces, that is, forces whose action is instantaneous. Newton's implicit understanding of the second law as he formulates it is, therefore, that which the tradition has taken up, that is, our formulation (@). For details on this, see Cohen, 'Newton's Second Law.' In effect, what Newton does is generalize from the case of impact to the case of continuously acting forces; assuming as part of this generalization that the latter are limits of the former. See below.

163 Descartes, *Principles of Philosophy*, trans. V.R. Miller and R.P. Miller.

164 Cf W. Dray, *Laws and Explanation in History*.

165 Cf the discussion of 'narrative explanations' in T.A. Goudge, *The Ascent of Life*, and in Wilson, *Empiricism and Darwin's Science*.

166 Cf Wilson, *Laws and Other Worlds*.

167 Cohen, 'Newton's Second Law,' 152–3.

168 T. Kuhn, *The Structure of Scientific Revolutions*, 2nd ed.

169 Cf Wilson, *Laws and Other Worlds*, 70 ff, and *Empiricism and Darwin's Science*, sec. 2.3. The point is developed in detail in the final section of Wilson, *Hume's Defence of Causal Inference*.

170 Cf Wilson, *Empiricism and Darwin's Science*, 53f, 290ff.

171 Bacon, *New Organon*, I.civ.

172 These sorts of inference are explored in detail in Wilson, *Empiricism and Darwin's Science*.

173 Arnauld, *Art of Thinking*, 320.

174 Ibid., 321.

175 Ibid., 322.

176 Descartes, *Discourse*, trans. Olscamp, 16; Cottingham, 1:120.

177 Bacon, *New Organon*, I.civ.

178 This point is emphasized in the useful discussion of Bacon's views in Urbach, *Francis Bacon's Philosophy of Science: An Account and a Reappraisal*, 38ff.

179 See ibid., 45. H.G. van Leeuwen, *The Problem of Certainty in English Thought from Chillingworth to Locke*, while often illuminating, wrongly assimilates Bacon's pursuit of certainty to that of Descartes.

180 I. Newton, *The Correspondence of Isaac Newton*, ed. H.W. Turnbull, vol. 1, Newton's letter of 10 June 1672, quoting from the English paraphrase.

181 Bacon, *New Organon*, I.lxiii.

182 J. Rohault, *Rohault's System of Natural Philosophy*, trans. J. Clarke, with notes by Samuel Clarke, in 2 vols (London 1723).

183 London 1716.

184 London 1730.

185 Descartes, *Principles of Philosophy*, trans. Miller and Miller (Bk III, ¶ 56), 112; Cottingham, 1:259.

186 Ibid. (Bk II, ¶ 11), 44; Cottingham, 1:227.

187 Ibid. (Bk II, ¶22), 49; Cottingham, 1:232.

188 Ibid. (Bk III, ¶ 57), 112; Cottingham, 1:246. (The passage is not, however, translated in Cottingham.)

189 London 1670.

190 Des Chene, *Physiologia*, 311f.

191 I have warned some years ago against a similar misreading of Newton. See Wilson, 'Explanation in Aristotle, Newton and Toulmin.'

192 Descartes, *Principles*, trans. Miller and Miller (Bk II, ¶ 36), 58; Cottingham, 1:240.

193 Ibid.

194 This implies that Descartes was committed to an occasionalist account of the causation we observe in material things. I shall have more to say about Descartes's occasionalism in Study Seven, below. For a somewhat different reading of Descartes on this point, see Daniel Garber, *Descartes' Metaphysical Physics*, 299ff.

195 Descartes, *Principles*, trans. Miller and Miller (Bk II, ¶ 37), 59; Cottingham, 1:240–1.

196 For discussion, see Garber, *Descartes' Metaphysical Physics*, chs 7 and 8.

197 G.W.F. Leibniz, *Philosophical Papers and Letters*, 1:484–5.

198 P. Costabel, 'Newton's and Leibniz's Dynamics,' 121.

199 Gary Hatfield, 'Force (God) in Descartes' Physics,' 134–5. One should note, however, that the suggestion that minds, that is, immaterial *finite* substances, have Aristotelian powers is mistaken. See Study Seven, below.

200 M. Guéroult, 'The Metaphysics and Physics of Force in Descartes,' 198.

201 Descartes, *Principles*, trans. Miller and Miller (Bk II, ¶ 40), 61; Cottingham, 1:242.

202 Descartes, *Principles*, trans. Miller and Miller, 58n31. Alan Gabbey makes the same point; see his 'Force and Inertia in Seventeenth-Century Dynamics,' 25.

203 As Alan Gabbey has shown, Descartes was more or less halfway to the notion of 'reaction' that appears in Newton's third law. See his 'Force and Inertia in Seventeenth-Century Dynamics,' 28.

204 Guéroult, 'The Metaphysics and Physics of Force in Descartes,' 197.
205 Gabbey, 'Force and Inertia in the Seventeenth Century: Descartes and Newton,' 236.
206 Ibid.
207 Ibid., 327–8.
208 Leibniz, *Specimen dynamicum*, in *Philosophical Papers and Letters*, 2:714.
209 Ibid., 712.
210 See note 185, above.
211 Garber, *Descartes' Metaphysical Physics*, 297.
212 Ibid., 298.
213 Hume, T 167.
214 T. Hobbes, *Elements of Philosophy, the First Section, Concerning Body*, 121–2.
    Hobbes's views on causation and the correct method to discover causes are discussed further in Study Four, below.
215 Ibid., 128.
216 Ibid.
217 Hume, T 171.
218 Ibid., 311. For greater detail on Hume on dispositions, see F. Wilson, *Hume's Defence of Causal Inference*, ch. 1.
219 Hume, T 312.
220 Hobbes, *Elements of Philosophy, the First Section, Concerning Body*, 212.
221 Gabbey, 'Force and Inertia in Seventeenth-Century Dynamics,' 7.
222 Des Chene, *Physiologia*, ch. 10.
223 We shall deal at greater length with these issues in Studies Six and Seven, below.
224 Hume, T 159.
225 Ibid., 159–60.
226 Cf J. Yolton, *Perceptual Acquaintance from Descartes to Reid*; T. Lennon, 'The Inherence Pattern and Descartes's Ideas'; and Monte Cook, 'Descartes' Alleged Representationalism.'
227 See Arnauld, *On True and False Ideas*. For discussion of Arnauld, see S. Nadler, *Arnauld and the Cartesian Philosophy of Ideas*.
228 Descartes, *Meditations*, in *Philosophical Essays*, trans. Lafleur, 91; Cottingham, 2:26.
229 Descartes, *Objections and Replies*, in *The Philosophical Works of Descartes*, trans. J. Cottingham, Robert Stoothoff and Dugald Murdoch, 2:132.
230 Cf Arnauld, *On True and False Ideas*: 'When I attack *representative beings* as superfluous, I am referring to those which are assumed to be really distinct from ideas taken in the sense of perceptions. I am careful not to attack every kind of *representative* being or modality, since I hold that it is clear to whoever reflects on what takes place in his own mind, that all our perceptions are modalities which are essentially *representative*' (20).

231 Ibid., 25.
232 Ibid., 19.
233 Ibid.
234 *Meditations*, in *Descartes: Philosophical Essays*, trans. Lafleur, 68; Cottingham, 2:7.
235 Arnauld, *On True and False Ideas*, 26.
236 Descartes, *Objections and Replies*, Cottingham, 2:113.
237 Cf T.J. Cronin, *Objective Being in Descartes and Suarez*; Calvin Normore, 'Meaning and Objective Being: Descartes and His Sources.'
238 '[E]ssentiam creaturae, seu creaturam de se, et priusquam a Deo fiat, nullam habere in se verum esse reale, et in hoc sensu, praeciso esse existentiae, essentiam nonesse rem aliquam, sed omnio esse nihil ...' F. Suarez, *Disputationes metaphysicae, disputatio* 31, *sectio* 2, paragraph 1. Future references will be by *disputatio, sectio*, and paragraph.
239 '[R]ecipiendo veram entitatem a sua causa' (31.2.2).
240 'In hoc enim distinguuntur essentiae creaturarum a rebus fictitiis et impossibilibus ut chymera, et hoc sensu dicuntur creaturae habere reales essentias, etiamsi non existant' (31.2.2); see also 31.2.10.
241 '[E]sse, quod appellant essentiae ante effectionem, seu creationem divinam, solum est *esse potentiale objectivum*' (31.2.2).
242 'Ut autem vera esset scientia qua Deus ab aeterno cognovit, hominem esse animal rationale, quia illud esse non significat actuale esse et reale, sed solam connexionam intrinsicam talium extremorum; haec autem connexio non fundatur in actuali esse, sed in potentiali' (31.2.8).
243 '[H]oc esse cognitum non esse in illis aliquod esse reale intrinsecum ipsis' (31.2.1).
244 'Ab aeterno non fuisse veritatem in illis propositionibus, nisi quatenus erant objective in mente divina, quia subjective seu realitur non erant in se ...' (31.2.8).
245 'Essentia creaturarum tantum habent vel esse in causa, vel objective in intellectu' (31.2.11).
246 'Non [sunt] in se, sed objective tantum in intellectu' (31.2.10).
247 Cf Descartes, *Objections and Replies*, Cottingham, 2:293–4.
248 See Meditation III, in *Descartes: Philosophical Essays*, trans. Lafleur, 97ff; Cottingham, 2:29.
249 Descartes, *Objections and Replies*, Cottingham, 2:74.
250 Descartes, *Meditations*, trans. Lafleur, 98; Cottingham, 2:28.
251 Descartes, *Meditations*, trans. Lafleur, 100; Cottingham, 2:30.
252 Descartes, *Objections and Replies*, Cottingham, 2:75.
253 Arnauld, *On True and False Ideas*, trans. Kremer, 20.
254 Whether this constitutes 'direct realism,' as Nadler, *Arnauld and the Cartesian Philosophy of Ideas*, suggests is a further issue, probably not worth pursuing in depth.

255 See Gaukroger, 'Introductor Essay' to his translation of A. Arnauld, *On True and False Ideas*.

256 Cf John Sergeant, *Solid Philosophy Asserted against the Fancies of the Ideists* (London 1696). Cf Wilson, 'The Lockean Revolution in the Theory of Science.'

257 For the distinction between 'high' and 'low' sciences, see Hacking, *The Emergence of Probability*, 35.

258 Descartes, Meditation III, in *Philosophical Essays*, trans. Lafleur, 100; Cottingham, 2:30.

259 Arnauld, *On True and False Ideas*, trans. Kremer, 21.

260 Ibid., 180.

261 Ibid., 175.

262 See note 257, above.

263 R. Brown, 'History versus Hacking on Probability,' 669.

264 See notes 27 and 28, above.

265 Hacking, *Emergence of Probability*, 35.

266 Brown, 'History versus Hacking,' 668.

267 Ibid., 668.

268 Ibid., 669.

269 J. Locke, *Essay concerning Human Understanding*, Introduction, sec. 2. References are given by Book, chapter, and paragraph, or Introduction and section.

270 Sextus Empiricus, *Outlines of Pyrrhonism*, trans. R.G. Bury, Bk II, ch. xv, 204.

271 Ibid., II.xiv, 195ff.

272 Cf Ibid., I.xv, 175; II.xvi, 205.

273 Thomas Aquinas, *Summa theologica*, see above, note 152.

274 Duns Scotus, *Opus*, Wolter, 118.

275 Duns Scotus, *Metaphy.* VI, q2, n7; quoted in Vier, *Evidence*, 144.

276 In my 'Critical Notice' of Hacking's *Emergence of Probability*, I argue that Hacking, for all the brilliance of his insights, also does not do full justice to this point that the crucial break comes with the elimination of real essences and objective necessary connections.

277 For detail on Foucher, see the important book by R.A. Watson, *The Downfall of Cartesianism*.

278 R. Rapin, *Reflexions sur la philosophie ancienne et moderne* (Paris 1676).

279 Huet was a friend of both Foucher and the anti-Cartesian Jesuit Rapin, as well as of Leibniz; both Foucher and Huet were friends of Leibniz.

280 See R. Popkin, 'The High Road to Pyrrhonism.'

281 Paris 1694.

282 Amsterdam 1723.

283 Cf Wilson, 'Lockean Revolution in the Theory of Science.'

284 The quotations in French are from A. Arnauld, *La Logique de Port-Royal*, ed. Ch.

Jourdain. References are by part and chapter. The translation of Dickoff and James is not helpful on the areas with which we are now concerned; their translation obscures the presence throughout the discussion of abstraction of the whole notion of *separability*.

285 G. Berkeley, *Principles of Human Knowledge*. References are by numbered paragraph.

286 Arnauld, *Art of Thinking*, trans. Dickoff and James, 164.

287 J. Weinberg has drawn our attention to this logical argument against abstraction; see his essay on 'The Nominalism of Berkeley and Hume.'

288 Huet, *Traité philosophique de la foiblesse de l'esprit humain*, 197.

289 On associationism in general, see F. Wilson, *Psychological Analysis and the Philosophy of John Stuart Mill*.

290 Hume's associationist account of abstract ideas is discussed in detail in F. Wilson, 'Abstract Ideas and Other Rules of Language in Hume.'

291 London 1749, 1:291.

292 Ibid., 1:329.

293 I have discussed these in detail elsewhere; see F. Wilson, 'Abstract Ideas and Other Rules of Language in Hume,' and also F. Wilson, 'Hume and Derrida on Language and Meaning.'

294 D. Livingston, *Hume's Philosophy of Common Life*.

295 Hume, *Letters*, ed. J.Y.T. Grieg, 1:201.

296 P. Jones, *Hume's Sentiments*, 146.

297 Ibid., ch. 4.

298 Ibid., 146.

299 Ibid., 144.

300 A. Baier, *A Progress of the Passions*, 100.

301 This long argument has been examined in detail in F. Wilson, 'Hume's Defence of Causal Inference.'

302 See Baier, *A Progress*, 15–27.

303 These remarks by Hume occur in the 'Conclusion' to Book I of the *Treatise*. The structure of this passage has been examined in detail in F. Wilson, 'Is Hume a Sceptic with regard to Reason?' and *Hume's Defence of Causal Inference*, ch. 3.

304 Baier, *A Progress*, 200.

305 On reason as a virtue, see also ;rdal, 'Some Implications of the Virtue of Reasonableness in Hume's *Treatise*'; and F. Wilson, 'Hume's Defence of Science.'

Study Two

1 J. Maritain, *Formal Logic*, 222ff.

2 P. Coffey, *The Science of Logic*, vol. 1, ch. 8.

3 See L. Wittgenstein, 'On Logic, and How Not to Do It' (review of Coffey, *The Science of Logic*).

4 Cf I. Copi, *Introduction to Logic*, 7th ed., chs. 5, 6.

5 Cf Copi, *Introduction to Logic*, sec. 5.5.

6 P.F. Strawson, *Introduction to Logical Theory*, ch. 6, secs I, II.

7 Cf Geach, 'Distribution: A Last Word?' in *Logic Matters*, 62–4.

8 Cf Coffey, *Science of Logic*, 1:253–4.

9 G. Boole, *Laws of Thought*.

10 See J.W. Miller, *The Structure of Aristotelian Logic*, § 91.

11 See O. Bird, *Syllogistic and Its Extensions*, § 19.

12 As we shall see in the next section, this customary presentation needs to be slightly modified.

13 Cf J.N. Keynes, *Studies and Exercises in Formal Logic*, 2nd ed., 180. For a more recent statement, see S. Barker, *The Elements of Logic*, 69.

14 P. Geach, *Reference and Generality*.

15 S. Barker, *Elements of Logic*.

16 B. Katz and A. Martinich, 'The Distribution of Terms.'

17 The terminology 'AA$_1$' is from Keynes, *Studies and Exercises in Formal Logic*, 127.

18 See Peter of Spain, *Summulae logicales*, ed. I.M. Bochenski, sec. 6.11. Compare A.N. Prior, *Formal Logic*, 2nd ed., 110.

19 W. Kneale and M. Kneale, *The Development of Logic*, 272–3.

20 This idea is worked out in detail, but from a slightly different perspective, in W.T. Parry, 'Quantification of the Predicate and Many-sorted Logic,' sec. 2. The point is also made briefly in W.V.O. Quine, Review of Geach, *Reference and Generality*. Quine attributes it to Peirce, *Collected Papers*, vol. 2, art. 458.

21 Geach, 'Distribution: A Last Word?' 62–4.

22 See Miller, *Structure of Aristotelian Logic*, ch. 5.

23 Geach, 'Distribution,' 64.

24 Cf Bird, *Syllogistic and Its Extensions*, 52ff.

25 Miller, *Structure of Aristotelian Logic*, §§ 61–7.

26 Bird, *Syllogistic and Its Extensions*, 62.

27 Henry Aldrich, *Artis logicae compendius* (Oxford 1691). See also Henry Aldrich, *Artis logica rudimenta*, ed. with notes by H.L. Mansel, 4th ed. (London 1862).

28 Robert Sanderson, *Logicae artis compendium*.

29 Franco Burgersdijck, *Monitio logica, or an Abstract and translation of Burgersdiscius His Logick*, by a Gentleman (London 1697).

30 J. Maritain, *An Introduction to Philosophy*, trans. E.J. Watkin.

31 Aristotle, *Metaphysics*, Bk IX, 1051a35–6. References to Aristotle are to *The Complete Works of Aristotle*, ed. J. Barnes.

32 Thomas Aquinas, *Summa theologica*, trans. by the Fathers of the Dominican Prov-

ince. References are to the First Part (FP) or First Part of the Second Part (FS) or Second Part of the Second Part (SS), by Question (Q), Article (an A), Objection (Obj) by number, and, where appropriate, Reply to Objection (RObj) by number.

33 J. Maritain, *A Preface to Metaphysics*, 67.

34 Aristotle, *De interpretatione*, trans. J.L. Ackrill.

35 Aristotle's 'indefinite' nouns were characterized in Latin by Boethius as 'infinite' nouns. See Sir William Hamilton, *Lectures on Metaphysics and Logic*, 2:178.

36 Deborah Black, 'Aristotle's *"Peri hermeneias"* in Medieval Latin and Arabic Philosophy: Logic and the Linguistic Arts,' 41ff.

37 Ibid., 43. The reference is to Martin of Dacia, *Martinus de Dacia Quaestiones super librum Periermenius*, q22, 260.19–26.

38 Simon of Faversham, *Quaestiones super libro Perihermenias*, q4, 152.27–153.2; quoted in Black, 'Aristotle's "Peri hermeneias,"' 43.

39 As the Kneales put it, 'the treatment of the *Topics* is in some ways more open than that of the *Analytics*' (Kneale and Kneale, *The Development of Logic*, 37).

40 A similar caution about attempts to read ancient logic simply as modern formal logic, independent of any logic of truth, can be found in Charles Kahn, 'Stoic Logic and Stoic LOGOS,' 158–9.

41 I.M. Bochenski, *Ancient Formal Logic*.

42 The five Stoic 'indemonstrable' forms are these:
   1) If p, then q; but p; so q
   2) If p, then q; but not q; so not p
   3) Not both p and q; but p; so not q
   4) p or q; but p; so not q
   5) p or q; but not q; so p
   See Diogenes Laertius, *Lives of the Eminent Philosophers*, 7:80–1.

43 W. Hay, 'Stoic Use of Logic,' 148.

44 Sextus Empiricus, *Against the Logicians*, 297 (II, 113).

45 Cf J. Rist, 'Zeno and the Origins of Stoic Logic,' in J. Braunschweig, ed., *Les Stoiciens et leur logique*.

46 Diogenes Laertius, *Lives of the Eminent Philosophers*, 7:73.

47 Cicero, *De fato*, vi.

48 Sextus Empiricus, *Against the Logicians*, 463 (II, 430).

49 I. Mueller ('Stoic and Peripatetic Logic') is thus wrong when he suggests that, while for the Aristotelians logic was an *organon*, a logic of truth, for the Stoics in contrast logic was simply a logic of consistency (see 184–5). To the contrary, the Stoic logic was, like the logic of Aristotle, intended to be a logic of demonstration. Moreover, precisely because the indemonstrable inference forms were intended to be forms of demonstrations with the major premise a necessary truth, it follows, as Hay has stated ('Stoic Use of Logic,' 148), that the conditionals that form the premises of

hypothetical syllogisms have to be general truths. Mueller is therefore incorrect in his rejection (185) of Hay's position. M. Frede ('Stoic vs. Aristotelian Syllogisms') agrees with Hay against Mueller that the Stoics located their indemonstrable forms within a metaphysical or ontological context that is very different from that of the Aristotelians; see 7ff. So does J.S. Kieffer; see his Introduction to Galen, *Institutio logica*, 8.

50 On Chrysippus in particular, see Josiah B. Gould, *The Philosophy of Chrysippus.*

51 Cf A.A. Long, *Hellenistic Philosophy*, 125.

52 Sextus Empiricus, *Against the Logicians*, 383 (II, 275–6).

53 Quoted in Gould, *Philosophy of Chrysippus*, 163.

54 Gould, *Philosophy of Chrysippus*, 93.

55 Cicero, *De fato*, xiv.

56 Cf S. Sambursky, *The Physics of the Stoics*, ch. 3; Gould, *Philosophy of Chrysippus*, ch. 5.4.

57 As Gould, *Philosophy of Chrysippus*, puts it, the wise man 'is the one of all men who has made the truest generalizations about things which are advantageous and injurious ...' (173).

58 See A.A. Long, 'Language and Thought in Stoicism,' 95; also J. Rist, 'Zeno and the Origins of Stoic Logic,' 391–3.

59 Kieffer, Introduction to Galen, *Institutio logica*, 9.

60 For a discussion locating Stoic logic in the broader context of its metaphysics/ epistemology, see Kahn, 'Stoic Logic and Stoic LOGOS.'

61 Galen, *Institutio logica*, ch. 4.

62 Cicero, *De fato*, vi.

63 Ibid.

64 Aulus Gellius, *Noctes Atticae*, xvi.8.

65 Galen, *Institutio logica*, XIV.7–8, 47.

66 C.I. Lewis and C.H. Langford, *Symbolic Logic*, 2nd ed., ch. 6, sec. 2.

67 Cicero, *De fato*, viii, 16–17.

68 Benson Mates, *Stoic Logic*, 60–1.

69 Émile Bréhier, *Chrysippe*, 64.

70 Quoted in Gould, *Philosophy of Chrysippus*, 67.

71 Cicero, *De divinatione*, I.xvii.

72 Ibid., I.xiii.

73 Ibid., I.xiv.

74 Quoted in Gould, *Philosophy of Chrysippus*, 76–7.

75 Cicero, *De divinatione*, I.lvi.

76 Sambursky, *The Physics of the Stoics*, 65ff.

77 C. Normore, 'Medieval Connectives, Hellenistic Connections: The Strange Case of Propositional Logic.'

78 Aristotle, *De interpretatione*, trans. J.L. Ackrill, 18a18–26.

79 Cf K. Dürr, *The Propositional Logic of Boethius*, 11.
80 Cf Normore, 'Medieval Connectives, Hellenistic Connections.'
81 William of Ockham, *Summa logicae*, II.32. Cited and translated in Normore, 'Medieval Connectives, Hellenistic Connections,' 34.
82 Cited and translated in Normore, 'Medieval Connectives, Hellenistic Connections,' 34.
83 Ibid., 36–8.
84 Peter Abelard, *Sic et non*, a critical edition by B. Rover and R. McKeon.
85 Cf Jean Jolivet, *Abélard*, 73ff.
86 He taught at Padua from 1306 to 1316. For a discussion of the range of his interests in the context of the curriculum at the University of Padua, see Nancy Siraisi, *Arts and Sciences at Padua.*
87 Cf Nancy Siraisi, *Medieval and Early Renaissance Medicine*, 81.
88 Ibid., 81.
89 Ibid., 75–6.
90 Ibid.
91 Ibid., 80.
92 For an illuminating discussion of the method of the Talmudists, see Louis Jacobs, *Studies in Talmudic Logic and Methodology*, ch. 7.
93 Aristotle, *Topica*, 100a30.
94 John of Salisbury, *The Metalogicon*, trans. D.D. McGarry, Bk II, chs 9, 10.
95 Boethius, *De topicis differentiis*, trans. E. Stump; and *In Ciceronis topica*, trans. E. Stump. The former includes an important essay by Stump on Boethius's treatment of the topics.
96 On the difference between the way hypothetical syllogisms and categorical syllogisms are treated, see the essay by E. Stump in her edition of Boethius, *De topicis differentiis*. On hypotheticals, see 186ff; on categoricals, see 182ff.
97 See C.J. Martin, 'William's Machine.'
98 See O. Bird, 'The Tradition of the Logical Topics: Aristotle to Ockham'; O. Bird, 'The Formalizing of the Topics in Mediaeval Logic'; E. Stump, 'Topics: Their Development and Absorption into Consequences.' See also E. Stump, *Dialectic and Its Place in the Development of Medieval Logic.*
99 See E. Stump, 'Logic in the Early Twelfth Century.'
100 Gould, *Philosophy of Chrysippus*, 67–8.
101 Bernard of Clairvaux, *Epistolae*, in *Patrologia Latina*, vol. 182:355; quoted in Anders Piltz, *The World of Medieval Learning*, trans. D. Jones, 83.
102 Bernard of Clairvaux, *Epistolae*, 355; quoted in Anders Piltz, *The World of Medieval Learning*, 83.
103 Mordechai Breuer, 'Pilpul,' *Encyclopaedia Judaica.*
104 Quoted in Frederic Thieberger, *The Great Rabbi Loew of Prague*, 21.
105 Ibid.

106 E. Gilson, *The Mystical Theology of Saint Bernard*, trans. A.H.C. Downes, 64.

107 Descartes, *Rules for the Direction of the Mind*, in *The Philosophical Works of Descartes*, trans. Cottingham, et al., 1:11; hereafter cited as Cottingham with volume and page number.

108 Ibid., 1:12.

109 J. Locke, *Essay concerning Human Understanding* IV.xvii.6.

110 For such a suggestion, see L. Rose, *Aristotle's Syllogistic*, App. VII.

111 Amy Mullin has made this suggestion, as part of a general case that Descartes fails to live up to the critical standards that he proposes to defend, see her 'Masks and Accusations.'

112 Sem Dresden, *Humanism in the Renaissance*, trans. M. King, 76.

113 Ibid., 226.

114 Locke, *Essay* III.ix.10.

115 Descartes, *Rules*, Cottingham, 1:13.

116 See H. Baron, *The Crisis of the Early Italian Renaissance*. For another view, see J.E. Seigel, '"Civic Humanism" or Ciceronian Rhetoric?'

117 For European universities in general, see Sven Stelling-Michaud, *Les Universités Européennes du XIVᵉ au XVIIIᵉ siècle*. For English universities, see L. Stone, *The University in Society*.

118 Cicero, *De oratore*, I.lvii, 241.

119 Ibid., I.lvii, 241–2.

120 Cicero, *Lucullus*, II.iii, 7–8; quoted in L. Jardine, 'Lorenzo Valla and the Intellectual Origins of Humanist Dialectic,' 150.

121 Cf J. Glucker, '*Probabile, Veri Simile*, and Related Terms,' in J.G.F. Powell, ed., *Cicero the Philosopher*, 115–43.

122 Cicero, *De fato*, ii, 3.

123 Quoted in Jardine, 'Lorenzo Valla,' 151n46.

124 Cicero, *De finibus bonorum et malorum*, trans. H. Rackham, IV.iv, 8–10.

125 Cicero, *Topica*, II, 6.

126 Cf Jardine, 'Lorenzo Valla,' 154ff.

127 Cf Jill Kraye, 'The Philosophy of the Italian Renaissance,' in G.H.R. Parkinson, ed., *The Renaissance and Seventeenth-Century Rationalism*, vol. 4 of the *Routledge History of Philosophy*; for Valla, see p. 43–4.

128 Jardine, 'Lorenzo Valla,' 146.

129 Cf Jardine, 'Humanistic Logic.'

130 For example, A.J. Ayer, *The Problem of Knowledge*, 28–34.

131 This part of this study elaborates certain points of my earlier 'The Lockean Revolution in the Theory of Science.'

132 Antoine Arnauld and Pierre Nicole, *La Logique de Port-Royal*, with Introduction and Notes by Charles Jourdain. Translations are my own.

133 Jean-Claude Pariente, *L'Analyse du langage à Port-Royal*, ch. 9.

134 Jill Buroker, 'Judgment and Predication in the *Port-Royal Logic*.'

135 'Il est certain que nous ne saurions exprimer une proposition aux autres que nous ne servions de deux idées: l'une pour le sujet et l'autre pour l'attribut, et d'un autre mot qui marque l'union que notre esprit y conçoit.

  'Cette union ne peut mieux s'exprimer que par les paroles mêmes dont on se sert pour affirmer, en disant qu'une chose est une autre chose.

  'Et de là il est clair que la nature de l'affirmation est d'unir et d'identifier, pour le dire ainsi, le sujet avec l'attribut, puisque c'est ce qui est signifié par le mot *est*' (Arnauld, *La Logique*, 152).

136 '[I]l est impossible qu'une chose soit jointe et unie à une autre, que cette autre ne soit jointe aussi à la première, et qu'il s'ensuit fort bien que si A est joint à B, B aussi est joint à A, il est clair qu'il est impossible que deux choses soient conçues comme identifiées, qui est la plus parfaite de toutes les unions, que cette union ne soit réciproque, c'est-à-dire, que Leon ne puisse faire une affirmation mutuelle des deux termes unis en le manière qu'ils sont unis ...' (ibid., 154).

137 'J'appelle *compréhension* de l'idée, les attributs qu'elle enferme en soi, et qu'on ne peut lui ôter sans la détuire, comme la compréhension de l'idée du triangle enferme extension, figure, trois lignes, trois angles, et l'égalité de ces trois angles à deux droits, etc.' (ibid., 45).

138 For the best systematic discussion of the connections between the concepts of species/genus, determinate/determinable, and mode/attribute, see A.N. Prior, 'Determinables, Determinates, and Determinants.' These problems are discussed in a way that is remarkably unsophisticated by J. Woods and R. Thomason in an interchange in *Noûs*: see J. Woods, 'On Species and Determinates,' and R. Thomason, 'Species, Determinates and Natural Kinds.' Woods comes closer to discussing the relevant issues, but he uses a totally unexplicated notion of 'entailment' – it simply won't do to take this as an unanalysed primitive relation as Woods does; and, moreover, he takes the notion of adding a specific difference to be mere conjunction and a genus to be a mere disjunction of species. Thomason has even more severe problems, however. He takes the problem to be one of explicating the species/genus relationship. He therefore offers an axiomatics, with model-theoretic analyses of the logical structure. But he totally ignores the real problem deriving from the tradition, namely, that the connection between a species and its genus is taken to be *necessary*. Woods, though he takes 'entailment' to be an unanalysable notion, at least recognizes that this is the important philosophical issue. Whatever the technical problems that confront Woods's treatment, he at least recognizes what is of central philosophical importance. Thomason doesn't.

139 '[U]ne idée est toujours affirmée selon sa compréhension; parce qu'en lui ôtant quelqu'un de ses attributs essentiels, on la détruit et on l'anéantit entièrement, et ce n'est plus la même idée; et, par conséquent, quand elle est affirmée, elle l'est toujours selon tout ce qu'elle comprend en soi. Ainsi, quand je dis *qu'un rectangle est*

un parallélogramme, j'affirme du rectangle tout ce qui est compris dans l'idée du parallélogramme; car, s'il y avait quelque partie de cette idée qui ne convînt pas au rectangle, il s'ensuivrait que l'idée entière ne lui conviendrait pas, mais seulement une partie de cette idée: et ainsi le mot de parallélogramme, qui signifie l'idée totale, devrait être nié et non affirmé du rectangle. On verra que c'est le principe de tous les arguments affirmatifs' (Arnauld, *La Logique*, 153).

140 Pariente, *L'Analyse du langage à Port-Royal*, 270n.

141 George Boole, *Laws of Thought*.

142 John Sergeant, *Method to Science* (London 1696).

143 'Lors donc que la seule considération de ces deux idées ne suffit pas pour faire juger si l'on doit affirmer ou nier l'une ou l'autre, il a besoin de recourir à une troisième idée ... et cette troisième idée s'appelle *moyen*' (Arnauld, *La Logique* 160).

144 'Il faut donc que ce terme moyen soit comparé, tant avec le sujet ou le petit terme, qu'avec l'attribut ou le grand terme ...' (ibid., 160).

145 'Une vrai démonstration demande deux choses: l'une, que dans la matière il n'y ait rien que ce certain et indubitable; l'autre, qu'il n'y avait rien de vicieux dans la forme d'argumenter ...' (ibid., 296).

146 Sergeant, *Method to Science*, Preface, unpaginated, d4.

147 Not surprisingly, Sergeant notes that 'God himself has expressed his own Supreme Essence by this Identical Proposition, *Ego Sum qui Sum* ...' (*Method*, 145).

148 Descartes, *Rules*, Cottingham, vol. 1.

149 Descartes, *Meditations*, in *Descartes: Philosophical Essays*, trans. Lafleur, 91; Cottingham, 2:25–6.

150 Descartes, *Objections and Replies*, Cottingham, 2:132.

151 Cf Arnauld, *On True and False Ideas*: 'When I attack *representative beings* as superfluous, I am referring to those which are assumed to be really distinct from ideas taken in the sense of perceptions. I am careful not to attack every kind of *representative* being or modality, since I hold that it is clear to whoever reflects on what takes place in his own mind, that all our perceptions are modalities which are essentially *representative*' (20).

152 Descartes, *Meditations*, trans. Lafleur, 68; Cottingham, 2:7.

153 Descartes, *Objections and Replies*, Cottingham, 2:113.

154 See Descartes, Meditation III, trans. Lafleur, 97ff; Cottingham, 2:29.

155 Descartes, *Replies to the First Set of Objections*, Cottingham, 2:74.

156 Descartes, *Meditations*, trans. Lafleur, 98; Cottingham, 2:30.

157 Ibid., trans. Lafleur, 100; Cottingham, 2:31.

158 Descartes, *Objections and Replies*, Cottingham, 2:75.

159 Il est 'nécessaire que la matière soit trouvée pour la disposer ...' (Arnauld, *La Logique*, 210).

160 'J'appelle *étendue* de l'idée les sujets à qui cette idée convient; ... comme l'idée du triangle en générale s'étend à toutes les diverses espèces de triangle' (ibid., 45).

161 '[L]'affirmation mettant l'idée dans le sujet, c'est proprement le sujet qui détermine l'extension de l'attribut dans la proposition affirmative, et l'identité qu'elle marque regarde l'attribut comme resserré dans une étendue égale à celle du sujet, et non pas dans toute sa généralité, s'il en a une plus grande que le sujet; car il est vrai que les lions sont tous animaux, c'est-à-dire que chacun des lions renferme l'idée d'animal; mais il n'est pas vrai qu'ils soient tous les animaux' (ibid., 153).

162 Michael Tooley, 'The Nature of Laws.'

163 C.D. Broad, *Examination of McTaggart's Philosophy*, 1:51–3.

164 G. Bergmann, 'On Non-Perceptual Intuition.' See also H. Hochberg, 'Natural Necessity and Laws of Nature'; and 'Possibilities and Essences in Wittgenstein's *Tractatus*.'

165 F. Wilson, *Laws and Other Worlds*, ch. 2, sec. III.

166 J.R. Brown, *The Laboratory of the Mind*, 82ff.

167 As John Earman has put it, 'What remains to be worked out [on the view that causal relations are relations among universals] is the formal semantics of the entailment relation [that holds between the statement about universals and the matter-of-fact regularity]; whether this can be done consistently ... remains to be seen' ('Laws of Nature: The Empiricist Challenge,' 221n21). Earman puts the matter rather too cautiously: there is no reason to suppose that it can be done, no reason to think that one can find grounds to make plausible the claim that a second-order atomic statement about universals, '$R(F, G)$,' should entail a first order generalization, '$(x)(Fx \supset Gx)$.'

168 John Stuart Mill, *System of Logic*, 8th ed. Bk II, ch II, sec. 2.

169 D.M. Armstrong, *What Is a Law of Nature?*

170 See F. Wilson, *Explanation, Causation and Deduction*, sec. 3.6, for an extended discussion of another, rather different, defence of the idea of a primitive nomological connective.

171 D. Hume, *Enquiry concerning Human Understanding and concerning the Principles of Morals*, 25, 163.

172 For a discussion of these inferences in the context of empiricism, see F. Wilson, *Hume's Defence of Causal Inference*, ch. 1.

173 Those criticized by Wilson in his *Explanation, Causation and Deduction*, sec. 3.6, were equally unforthcoming with regard to our knowledge of the nomological connective which, in spite of that lack of knowledge, they none the less insist must exist. *Credo ut intelligam.*

174 Such as in A. Legrand, *Institutio philosophiae* (London 1672), 9.

175 Sergeant, *Method to Science*, 248ff. Cf T. White, *An Exclusion of Sceptics from All Title to Dispute: Being an Answer to the Vanity of Dogmatizing* (London 1665), 5, 8. (*The Vanity of Dogmatizing* is by J. Glanvill.)

176 London 1698.

177 Locke, *Essay*, I.i.4.

178 We are indebted to John Yolton for publishing these; see his 'Locke's Unpublished Marginal Replies to John Sergeant.'

179 Cf F. Wilson, 'Acquaintance, Ontology and Knowledge. 'Also F. Wilson, 'Bradley's Conception of Ideality: Comments on Ferreira's Defence.'

180 Cf Wilson, 'Acquaintance, Ontology and Knowledge.' See also F. Wilson, 'Was Hume a Sceptic with regard to the Senses?'; and 'Was Hume a Subjectivist?'

181 Cf F. Wilson, 'Weinberg's Refutation of Nominalism'; and Review of M. Mandelbaum, *The Anatomy of Historical Knowledge*. Also F. Wilson, 'Bradley's Contribution to Empiricism.'

182 J. Locke, *Some Thoughts concerning Education*, 199f.

183 Ibid., 200.

184 Cf F. Wilson, 'Logical Necessity in Carnap's Later Philosophy,' sec. 3.2 and also ch. 4.

Study Three

1 Aristotle, *The Complete Works*, ed. J. Barnes.

2 Thomas Aquinas, *Summa theologica*, trans. by the Fathers of the Dominican Province. References are to the First Part (FP) or First Part of the Second Part (FS), by Question (Q), Article (A), Objection (Obj) by number, and, where appropriate, Reply to Objection (RObj) by number.

3 Franco Burgersdijck, *Monitio logica, or an Abstract and Translation of Burgersdiscius His Logick*, by a Gentleman (London 1697). For an account of Burgersdijck's career, see Paul Dibon, *La Philosophie Néerlandaise au siècle d'or*, 1:90ff. Dibon discusses the teaching of logic at Leyden, 51ff, and deals with Burgersdijck's *Logic* in detail, 100ff.

4 J. Locke, *Some Thoughts concerning Education*, 199f.

5 It was used as a textbook at a number of dissenting academies; see H. McLachlan, *English Education under the Test Acts*, 301.

6 I discuss this in greater detail in F. Wilson, 'The History of Relations from Burgersdijck to Bradley.'

7 Locke, *Essay concerning Human Understanding*. References are to book, chapter and numbered paragraph.

8 For a clear exposition of the history of this doctrine of relations from the Presocratics through to Peirce, see J. Weinberg, *Abstraction, Relation and Induction*.

9 G. Berkeley, *The Principles of Human Knowledge*. References are by numbered paragraph.

10 Kenneth Winkler, *Berkeley: An Interpretation*, 1.

11 Ibid., 231.

12 London 1686.

13 Boyle does not exclude the use of final causes in science. Indeed, in his *Disquisition about the Final Causes of Natural Things* (1688), he allows that the best explanation for certain phenomena is to be found in terms of the final causes for which the Creator intended them. But the teleology here is external, rather than internal, by reference to metaphysical natures. On Boyle's view, the hypothesis of God is a hypothesis about the causes of natural phenomena, and God is a scientific entity alongside atoms and other minute parts and mechanisms to which our experiments and observations enable us to make inferences.

14 E. Halley, 'A Discourse concerning Gravity,' 5.

15 Berkeley, *Philosophical Commentaries*. References are to numbered entry.

16 I have explored some other aspects of this account of Berkeley on the structure of the world in F. Wilson, 'On the Hausmans' "New Approach." '

17 Rudolphus Agricola, *De inventione dialectica libri tres*.

18 In vol. 6 of Robert Sanderson's *Works*.

19 References will be to Pierre de la Ramée, *Dialectique* (1555), edition critique avec introduction, notes et commentaires de Michel Dassonville.

20 Walter J. Ong, *Ramus and Talon Inventory*.

21 Ibid., 179ff.

22 Ibid., 182.

23 Ibid., 184.

24 McLachlan, *English Education under the Test Acts*, 300.

25 See Constantia Maxwell, *A History of Trinity College Dublin, 1591–1892*, 149.

26 See J.W. Stubbs, *The History of the University of Dublin*, 146.

27 Professor Steven Daniel has informed me of this fact.

28 Descartes, *Rules for the Direction of the Mind*, in *The Philosophical Works of Descartes*, trans. J. Cottingham, et al., vol. 1.

29 A. Arnauld, *The Art of Thinking*, trans. J. Dickoff and P. James; *La Logique de Port-Royal* (Paris 1854).

30 'Il est nécessaire que la matière soit trouvée pour la disposer ...' (*La Logique*, 210).

31 W.J. Ong, *Ramus, Method, and the Decay of Dialogue*.

32 Ong refers to J. Peghaire, *Intellectus et ratio selon S. Thomas D'Aquin*.

Study Four

1 A.E. Taylor, *Thomas Hobbes*.

2 F. Copleston, *History of Modern Philosophy*, vol. 5 [*Modern Philosophy: The British Philosophers*], Pt 1 ['Hobbes to Paley'].

3 J.W.N. Watkins, *Hobbes's System of Ideas*.

4 Thomas Hobbes, *Elements of Philosophy, the First Section, Concerning Body.*
Unless otherwise noted, page references are to this work.

5 Cf J. Weinberg, 'Induction.'

6 Cf G.H. von Wright, *The Logical Problem of Induction,* 2nd ed.

7 *Descartes, The Philosophical Writings of Descartes,* trans. John Cottingham, Robert
Stoothoff, and Dugald Murdoch, 2:125.

8 Ibid.

9 For details, see Study One, above.

10 See Descartes, Meditation III, in *Descartes: Philosophical Essays,* trans. L.J.
Lafleur, 97ff; Cottingham, 2:29ff.

11 Descartes, *Objections and Replies,* Cottingham, 2:74.

12 Descartes, *Meditations,* trans. Lafleur, 98; Cottingham, 2:30.

13 Ibid., 100; Cottingham, 2:30.

14 Descartes, *Objections and Replies,* Cottingham, 2:75.

15 For an extended discussion of Antisthenes, and the evidence concerning his views,
see C.M. Gillespie, 'The Logic of Antisthenes.' Gillespie carefully distinguishes the
views of Antisthenes from the relativism of Protagoras. Gillespie also draws an interesting parallel between the views of Antisthenes and those of Hobbes.

16 All quotations from Aristotle are from *The Basic Works of Aristotle,* ed. R. McKeon.

17 See, for example, *Sophist* 251A–B and 252 C where views are stated without attribution but which seem to be the same as those attributed by Aristotle to Antisthenes.

18 See *Theaetetus* 202 Aff.

19 See J.S. Mill, *System of Logic,* Bk I, ch. II, sec. 5.

20 I have argued in Study One that the scheme as here adumbrated is clearly to be found
in Aristotle, and that it was the notion of science to which the rationalism of Descartes and Arnauld was intended to be a response.

It is clear that what Hobbes was taught when he was a student at Oxford was a
thoroughgoing diet of Aristotle. He rejected all this teaching, and later scoffed at the
university's 'Aristotelity' (*Leviathan,* 670) tricked out with the jargon of 'vain philosophy' (674). He argued, not without merit, that the 'natural philosophy [of the
Greeks] was rather a dream than science' (668), and dismissed the 'error of *separated essences*' (675), which he said was 'built on the vain philosophy of Aristotle'
(674), as the source of 'absurdities' (675) in natural philosophy, religion, and politics. In this condemnation of the Aristotelity of the universities he was joined by
other critics such as Clarendon and Bishop Butler (see Strickland Gibson, 'The University of Oxford,' 270).

Bachelors, as Hobbes was when he was a student, were expected to study, as part
of their course of instruction, natural philosophy. For the latter they were supposed to
read Aristotle's *Physics,* the *De caelo et mundi,* the *De meteoris,* and the *De partibus
naturalibus* or the *De anima* (see Mark H. Curtis, *Oxford and Cambridge in Transi-*

tion: 1558–1642, 91). The tutors would presumably add supplementary texts, as they did at Cambridge (*ibid.*, 111); in natural philosophy, these might include Toletus's *Commentaria ... in octo libros Aristotelis* and Giacomo Zabarella's *De rebus naturalibus libri XXX* (*ibid.*). Other textbooks were commonly available. For this tradition, see Reif, 'The Textbook Tradition in Natural Philosophy, 1600–1650,' and also L. Thorndyke, 'The Cursus Philosophicus before Descartes.' See also William T. Costello, *The Scholastic Curriculum at Early Seventeenth-Century Cambridge.* There were, of course, many variations in this tradition, but there was an essential Aristotelian core that the textbooks represented not inaccurately. The textbook tradition held, with Aristotle, that '*science* consists in certain, universal, and unchanging knowledge achieved through causal demonstration' (Reif, 'Textbook Tradition,' 21); and further, like Aristotle, the textbook tradition often ended up trying 'to resolve their difficulties by resorting to such pseudo-explanations as "occult" qualities or simply "nature"' (ibid.). It is these aspects of the Aristotelian tradition to which Hobbes, like Descartes, Arnauld, Locke, and the rest of the early modern tradition, objected. And it is, therefore, these aspects that I am here trying to describe briefly. Again, for greater detail, see Study One.

21 D. Hume, *Enquiries concerning Human Understanding and concerning the Principles of Morals*, 25, 163.

22 D. Hume, *Treatise of Human Nature.* All further references to the *Treatise* in this study are given in parentheses after a 'T.'

23 J.A. Robinson, 'Hume's Two Definitions of "Cause"' (cited in text as 'Hume's Two Definitions'; Thomas J. Richards, 'Hume's Two Definitions of "Cause"'; and J.A. Robinson, 'Hume's Two Definitions of "Cause" Reconsidered' (cited in text as 'Reconsidered'). The mentioned question has been raised by Robinson, 'Hume's Two Definitions,' 123, and 'Reconsidered,' 162.

24 A.C. Ewing, *The Fundamental Questions of Philosophy*, ch. 8, 'Cause.'

25 R. Fogelin, *Hume's Scepticism in the Treatise of Human Nature*, esp. ch. 4.

26 T. Penelhum, *Hume*, 54.

27 J.A. Robinson, 'Hume's Two Definitions,' 138–9.

28 Ibid.

29 See Norman Kemp Smith, *The Philosophy of David Hume*, 91–2.

30 As Hume puts it in the *Enquiries*, 'It is only when two *species* of objects are found to be constantly conjoined, that we can infer the one from the other ...' (*Enquiries concerning the Human Understanding and concerning the Principles of Morals*, 148; his italics). He makes the same point in the *Treatise* (T 87). Thus, D. Lewis is simply wrong when he suggests (see his 'Causation') that law-deduction is not essential to Hume's account of causation; and, for that matter, to Hume's account of the *justified* assertion of counterfactuals.

31 See Alan Hausman, 'Hume's Theory of Relations.'

32 J. Locke, *An Essay concerning Human Understanding*, References are to the sections of the Introduction or to book, chapter, and section.

33 N. Malebranche, *De la recherche de la vérité*, ed. G. Rodin-Lewis.

34 Cf F. Wilson, 'Acquaintance, Ontology and Knowledge.'

35 A relation considered philosophically is what is objective, *in* the things related; it is, Hume says, those properties objectively in things on the basis of which we 'think proper to compare' two ideas (T 13). When a relation is considered naturally, the things related are *associated* in the mind; the association moves the mind naturally from one relatum to the other. As Hume puts it, a relation taken naturally is that 'by which two ideas are connected together in the imagination, and one naturally introduces the other' (T 13).

36 See, for example, R.M. Chisholm, 'Law Statements and Counterfactual Inference.'

37 For a more adequate placing of Hume's ethical views see Páll Árdal, *Passion and Value in Hume's Treatise*, esp. ch. 9.

38 For example, *Treatise*, III.III.i, 582.

39 B. Blanshard, 'The Case for Determinism,' 26ff.

40 S.T. Coleridge, *Biographia Literaria*, 1:73, 86. For a discussion of Coleridge, see F. Wilson, 'The Ultimate Unifying Principle of Coleridge's Metaphysics of Relations and of our Knowledge of Them.'

41 For greater detail, see Wilson, *Hume's Defence of Causal Inference*, chs. 1 and 3.

42 H.A. Prichard, *Knowledge and Perception*, 184, suggests that Hume thus reduces all causal inferences to irrationality.

43 For a discussion of Hume's positive case against necessary connections, see F. Wilson, 'Acquaintance, Ontology, and Knowledge'; also Hausman, 'Hume's Theory of Relations.'

44 Hume, *Enquiries*, First Enquiry (XII.II), 160. See J. Lenz, 'Hume's Defense of Causal Inference'; also A.J. Ayer, *The Problem of Knowledge*, 75.

45 That is, without gathering more evidence, or enlarging the sample.

46 This is a bit too simple. Contrary generalities can be confirmed if we pick our samples correctly; clearly, we are presupposing a principle of total evidence. But the point remains, since even the total evidence is but a sample relative to total population – the population of the universe.

47 Compare the discussion of subjective and objective justification in G.E. Moore, *Ethics*, 118–21, where it arises in the context of a discussion of utilitarianism.

48 See T. Beauchamp and T. Mappes, 'Is Hume Really a Sceptic about Induction?'

49 Cf T. Beauchamp and A. Rosenberg, *Hume and the Problem of Causation*.

50 For discussion of the transition from *scientia* to empirical science, see F. Wilson, 'The Lockean Revolution in the Theory of Science'; and also 'Critical Notice' of I. Hacking, *The Emergence of Probability*.

51 Cf F. Wilson, *Hume's Defence of Causal Inference*.

52 For an extended discussion and defence of the Humean position, see F. Wilson, *Laws and Other Worlds.*

Study Five

1 D. Hume, *Treatise of Human Nature.* References are by book, part, and section, and/or by page, following a 'T.' For discussion of aspects of Hume's defence of causal inference as based on his rules by which to judge of causes and effects, see F. Wilson, *Hume's Defence of Causal Inference,* and also 'Hume's Defence of Science.'
2 Thomas Aquinas, *Summa theologica,* trans. by the Fathers of the Dominican Province. References are to the First Part (FP) or First Part of the Second Part (FS), by Question (Q), Article (A), Objection (Obj) by number, and, where appropriate, Reply to Objection (RObj) by number.
3 For a discussion of these, see G.H. von Wright, *The Logical Problem of Induction,* 2nd ed.
4 Cf B. Brody, 'Toward an Aristotelian Theory of Scientific Explanation.'
5 Cf Stillman Drake, *Galileo: Pioneer Scientist.*
6 Galileo Galilei, *Two New Sciences,* trans. with a new Introduction and Notes by Stillman Drake.
7 See also Stillman Drake's 'Impetus Theory Reappraised.'
8 See also R.H. Naylor, 'Galileo: The Search for the Parabolic Trajectory,' and J. MacLachlan, 'Galileo's Experiments with Pendulums: Real and Imaginary.'
9 R.H. Popkin, *The History of Scepticism from Erasmus to Spinoza.*
10 Descartes, *Meditations,* in *Descartes: Philosophical Essays,* trans. L. Lafleur, 91; also in *The Philosophical Works of Descartes,* trans. Cottingham, et al., 2:26; hereafter cited as Cottingham, with volume and page number.
11 Descartes, *Objections and Replies,* Cottingham., 2:132.
12 *Meditations,* trans. Lafleur, 68; Cottingham, 2:7.
13 See *Meditations,* trans. Lafleur, 97ff; Cottingham, 2:29.
14 Descartes, *Objections and Replies,* Cottingham, 2:74.
15 Ralph Cudworth, *True Intellectual System of the Universe* (London 1678); references are to the second edition (London 1743).
16 Joseph Glanvill, *The Vanity of Dogmatizing* (London 1661).
17 Joseph Glanvill, *Scepsis scientifica,* 2 vols. (London 1665).
18 Cf R.H. Popkin, 'Joseph Glanvill: A Precursor of Hume.'
19 N. Malebranche, *De la recherche de la vérité,* trans. T.M. Lennon and P.J. Olscamp as *The Search after Truth,* 449.
20 T. Sprat, *History of the Royal Society,* ed. J.I. Cope and H.W. Jones.
21 See B. Wood, 'Methodology and Apologetics: Thomas Sprat's *History of the Royal*

Society'; and M. Fisch and H.W. Jones, 'Bacon's Influence on Sprat's *History of the Royal Society.*'

22 For a discussion of scholastic Aristotelian physics as taught at Cambridge in the seventeenth century, see W.T. Costello, *The Scholastic Curriculum at Early Seventeenth-Century Cambridge*, 83–102. Costello gives an outline of the Aristotelian doctrines of Keckermann.

23 B. Keckermann, *Systema physicum*, 15; quoted in Costello, ibid., 86.

24 Aristotle, *Physics* (in R. McKeon, ed., *The Basic Works of Aristotle*), 192b14–16.

25 Ibid., *Physics*, 192b22–3.

26 W.D. Ross, *Aristotle*, 68.

27 Aristotle, *Physics*, 193b7–9.

28 Cf Aristotle, *De caelo*, 291b13; *Generation of Animals*, 741b1 (in *The Complete Works of Aristotle*, ed. J. Barnes).

29 Robert Boyle, *A Free Enquiry into the Vulgarly Receiv'd Notion of Nature* (London 1686.).

30 Boyle does not exclude the use of final causes in science. Indeed, in his *Disquisition about the Final Causes of Natural Things* (London 1688), he allows that the best explanation for certain phenomena is to be found in terms of the final causes for which the Creator intended them. But the teleology here is external, rather than internal, by reference to metaphysical natures. On Boyle's view, the hypothesis of God is a hypothesis about the causes of natural phenomena, and God is a scientific entity alongside atoms and other minute parts and mechanisms to which our experiments and observations enable us to make inferences.

31 John Locke, *An Essay concerning Human Understanding*. References are to book, chapter, and section.

32 John Stuart Mill, *System of Logic*, 8th ed., 137.

33 Cf Walter J. Ong, *Ramus, Method and the Decay of Dialogue*.

34 Franco Burgersdijck, *Monitio logica, or an Abstract and Translation of Burgersdiscius His Logick*, by a Gentleman (London 1697).

35 Cf W.S. Howell, *Logic and Rhetoric in England, 1560–1700*; also his *Eighteenth Century British Logic and Rhetoric*.

36 R. Sanderson, *Logicae artis compendium* (Oxford 1615; 2nd ed., Oxford 1618); and *Editio nova emendata* (Oxford 1841).

37 Sanderson *Logicae*, 1618 ed., 226; 1841 ed., 177.

38 John Wallis, *Institutio logicae* (Oxford 1687). References are to the edition in the *Opuscula quaedam miscellanea* appended to Wallis's *Operiaem mathematicorum*, vol. 3 (Oxford 1699). The pagination of *Opuscula* recommences at page 1, following 708 pages numbering the works that preceded. The *Institutio logica* begins at page 81 of this new series.

39 See Sergeant *Method to Science* (London 1696) and his *Solid Philosophy Asserted*

against the Fancies of the Ideists (London 1698). See also T. White, *An Exclusion of Sceptics from all Title to Dispute: Being an Answer to the Vanity of Dogmatizing* (London 1665). The latter is an Aristotelian answer to Glanvill's *Vanity of Dogmatizing.*

40 Isaac Watts, *Logick*, 2nd ed. (London 1726).

41 For an interesting discussion of Bacon as a methodologist, see J. von Leibig, 'Lord Bacon as a Natural Philosopher,' and his 'Induction and Deduction'; and also the anonymous reply, 'Lord Bacon an Imposter?' The latter elicited a response by Leibig, 'Was Lord Bacon an Imposter?' See also Otto Sonntag, 'Leibig on Francis Bacon and the Utility of Science.'

42 Julius Weinberg has argued convincingly that the eliminative methods were discovered by Bacon; see his essay on 'Induction.'

43 See Francis Bacon, *The Philosophical Works*, in 3 vols., trans. with notes by Peter Shaw (London 1733). The *New Organon* is in vol. 3. References to this work are to book and aphorism.

44 Cf J. Charles Robertson, 'A Bacon-Facing Generation: Scottish Philosophy in the Early Nineteenth Century.'

45 Dugald Stewart, *Philosophy of the Active and Moral Powers of Man*, vol. 2, in vol 7 of his *The Collected Works of Dugald Stewart*, ed. Sir W. Hamilton, 18ff.

46 *Encyclopaedia Britannica*, 5th ed, supp. vol. 2 (Edinburgh Constable 1817).

47 'Logic,' *Encyclopaedia Britannica*, 1st ed. (Edinburgh 1771), vol. 2.

48 W. Duncan, *Elements of Logic*. References are to the Scolar Press reprint, 1970.

49 London 1748.

50 'Logic,' *Encyclopaedia Britannica*, 3rd ed. (Edinburgh Constable 1797), vol. 10.

51 Tatham is famous (as these things go) for a two-hour sermon that he preached in 1802, about a dispute concerning the First Epistle of St John (verse 7), which concluded by leaving the topic to the learned bench of bishops 'who have little to do and do not always do that little.'

52 Edward Tatham, *The Chart and Scale of Truth* (Oxford 1793).

53 Thomas Belsham, *Elements of the Philosophy of the Mind, and of Moral Philosophy. To Which Is Prefixed a Compendium of Logic* (London 1801).

54 James Mill, Review of Thomas Belsham, *Elements of the Philosophy of the Mind, and of Moral Philosophy. To Which Is Prefixed a Compendium of Logic* (1802). Belsham was a unitarian minister, of considerable importance in the organizational development of the denomination. He also taught at the dissenting academies at Daventry and Hackney. (For his role in these academies, see H. McLachlan, *English Education under the Test Acts*, 156, 251.) The psychology he was defending was the associationism of Hartley and Priestley, which for all intents and purposes had become the basis of unitarian theology in Britain at the end of the eighteenth century. Both the logic and the psychology had been presented as lectures in the unitarian

academy where Belsham taught. Mill is critical of Belsham's psychology. At this stage, Mill was still very much the student of Thomas Reid and Dugald Stewart, and had not yet become persuaded of the truth of associationism.

55 J.W.F. Herschell, *Preliminary Discourse on the Study of Natural Philosophy*, 151ff.

56 A. Arnauld and P. Nicole, *La Logique de Port-Royal* (Paris 1854); *The Art of Thinking*, trans. J. Dickoff and P. James.

57 Francis Hutcheson, *Logicae compendium*. This was first published at Glasgow in 1756, as a posthumous edition of the text used in his logic course for many years.

58 Ephraim Chambers, *Cyclopaedia* (London 1728; 2nd ed. 1738). Further editions, 1739, 1741, 1746.

59 Michael Barfoot, 'Hume and the Culture of Science in the Early Eighteenth Century.'

60 Ibid., 158.

61 Ibid., 157.

62 Ibid., 158.

63 London 1722. See Barfoot, ibid., 157, 170n49.

64 London 1720. See Barfoot, ibid., 157, 170n49.

65 London 1719. See Barfoot, ibid., 157, 170n49.

66 Second edition, London 1726. See Barfoot, ibid., 157, 170n49.

67 For some aspects of the early history of Newtonianism in Scotland, see C.M. Eagles, 'David Gregory and Newtonian Science.'

68 Isaac Newton, *Mathematical Principles of Natural Philosophy and His System of the World*, trans. Andrew Motte, 1729, revised by Florian Cajori, 398–400.

69 'The vulgar, who take things according to their first appearance, attribute the uncertainty of events to such an uncertainty in the causes, as makes them often fail of their usual influence, though they meet with no obstacle nor impediment in their operation. But philosophers observing that almost in every part of nature there is contained a vast variety of springs and principles, which are hid, by reason of their minuteness or remoteness, find that it is at least possible the contrariety of events may not proceed from any contingency in the cause, but from the secret operation of contrary causes. This possibility is converted into certainty by further observation, when they remark, that upon an exact scrutiny, a contrariety of effects always betrays a contrariety of causes, and proceeds from their mutual hindrance and opposition. A peasant can give no better reason for the stopping of any clock or watch than to say, that commonly it does not go right: but an artizan easily perceives that the same force in the spring or pendulum has always the same influence on the wheels; but fails of its usual effect, perhaps by reason of a grain of dust, which puts a stop to the whole movement. From the observation of several parallel instances, philosophers form a maxim, that the connexion betwixt all causes and effects is equally necessary, and

that its seeming uncertainty in some instances proceeds from the secret opposition of contrary causes' (T 132).

70 Henry Pemberton, *A View of Sir Isaac Newton's Philosophy*, London 1728; repr. with an Introduction by I.B. Cohen (New York 1972).

71 Barfoot, 'Hume and the Culture of Science,' 159n22.

72 Bacon, *Philosophical Works*, see note 43. There had previously been published, in English, *The Novum Organum ... Epitomiz'd* (London 1676), but this epitome completely misses most of what is important and novel in Bacon, that is, the logic of experiment.

73 See E.C. Mossner, 'Hume's Early Memoranda, 1729–40: The Complete Text.'

74 The whole letter is reprinted in R.H. Popkin 'So, Hume Did Read Berkeley.'

75 Barfoot, 'Hume and the Culture of Science,' 157.

76 T. Sprat, *History of the Royal Society*, 36.

77 Ibid., 31.

78 See the passage cited by note 74.

79 For a discussion of the Hume–Malebranche connection, see Charles McCracken, *Malebranche and British Philosophy*.

80 Barfoot, 'Hume and the Culture of Science,' 158.

81 Descartes, *Discourse on Method*, in Cottingham, 1:120.

82 Ibid., 1:143.

83 See Macvey Napier, 'Remarks, Illustrative of the Scope and Influence of the Philosophical Writings of Lord Bacon,' for a thorough discussion of the impact Bacon had on subsequent thought, especially on the continent. Napier's 'Remarks' are in reply to an article in the *Quarterly Review* (1817), discussing Dugald Stewart's 'Dissertation First: [prefixed to the Supplemental Volumes of the Encyclopaedia Britannica] Exhibiting a General View of the Progress of Metaphysical, Ethical, and Political Philosophy, since the Revival of Letters in Europe,' in supp. vol. 1, to the 5th ed. of the *Encyclopaedia*. The *Quarterly* reviewer suggests that Stewart overestimated the significance of Bacon. See also the review of Stewart's 'Dissertation' in the *Edinburgh Review* (1816). The article in the *Quarterly* was by William Rowe Lyall; see Hill Shine and Helen Chadwick Shine, *The Quarterly Review under Gifford*, 55. The article in the *Edinburgh* was by James Mackintosh; see Walter Houghton, *The Wellesley Index to Victorian Periodicals*, 1:455.

84 See Bacon, *Philosophical Works*, Shaw, 'Appendix,' 562f.

85 Barfoot, 'Hume and the Culture of Science,' 158.

86 Ibid.

87 See John Milton Hirschfield, *The Académie Royale des Sciences 1666–1683*, ch. 8.

88 Hume, T I.II.i, 27. This connection was drawn to my attention by David Fate Norton.

89 For details of von Tschirnhaus's career and of the history of the text, see the 'Intro-

duction' by J.-P. Wurtz to his translation: E.W. von Tschirnhaus, *Médecin de l'esprit, ou préceptes généraux de l'art de découvrir* (Paris Editions Ophyrs 1980).
90 Edme Mariotte, *Essai de logique.*
91 Cf B. Rochot, 'Roberval, Mariotte et la logique'; Pierre Brunet, 'La Méthodologie de Mariotte'; and Ernest Coumet, 'Sur L'*Essai de Logique* de Mariotte: L'Établissement des sciences.'

Study Six

1 John Locke, *An Essay concerning Human Understanding.* References are to book, chapter, and paragraph.
2 Samuel Clarke, *A Discourse concerning the Being and Attributes of God*, 4th ed. (London 1716).
3 John Sergeant, *Method to Science* (London 1696).
4 G.W.F. Leibniz, 'Monadology,' in *G.W. Leibniz: Philosophical Essays*, trans. R. Ariew and D. Garber. Unless otherwise noted, all Leibniz references are to the translations of this volume. Concerning the demonstrability of God according to Leibniz, see also his *Theodicy.*
5 For a good discussion of Hume and of the context in which Hume was arguing, see M.A. Stewart, 'Hume and the "Metaphysical Argument *A Priori*."'
6 In R. Descartes, *The Philosophical Works of Descartes*, 3 vols, trans. J. Cottingham et al. The *Rules* is in vol. 1, and references will be given in parentheses as '*Rules*' followed by volume and page number. Other works in vol. 1 are the *Principles* and the *Discourse on Method*; the former will be referred to by volume, part, and section, while the latter will be referred to by volume and page number.
    The *Meditations* and *Objections and Replies to the Meditations* occur in vol. 2; these will be referred to in parentheses by title followed by volume and page number.
7 D. Hume, *Dialogues on Natural Religion*, ed. N. Kemp Smith, 2nd ed.
8 Thomas Aquinas, *Summa theologica*, trans. by the Fathers of the Dominican Province. References are to the First Part (FP) or First Part of the Second Part (FS) or Second Part of the Second Part (SS), by Question (Q), Article (A), Objection (Obj) by number, and, where appropriate, Reply to Objection (RObj) by number.
9 Cf G. Patterson Brown, 'Infinite Causal Regression.'
10 Descartes, *Principles of Philosophy*, Cottingham, 1:202.
11 R. Descartes, *Principles of Philosophy*, trans. V.R. Miller and R.P. Miller, 108; Cottingham, 1:258.
12 M. Guéroult, *Descartes' Philosophy Interpreted according to the Order of Reasons*, 2:237.
13 Dennis Des Chene, *Physiologia: Natural Philosophy in Late Aristotelian and Cartesian Thought*, ch. 10.

14 Descartes, *Principles*, trans. Miller and Miller, 85; Cottingham, 1:248.

15 P. Cummins, 'Hume on the Idea of Existence.'

16 A. Arnauld, *La Logique de Port-Royal*, ed. Ch. Jourdain (Paris Hachette 1854). References are by part and chapter.

17 G. Berkeley, *Principles of Human Knowledge*. References are by paragraph.

18 D. Hume, *A Treatise of Human Nature*. References are to book, part, section, and/or page, following a 'T.'

19 J. Weinberg has drawn our attention to this logical argument against abstraction; see his *Abstraction, Relation and Induction*.

20 Descartes, *Objections and Replies*, Cottingham, 2:117.

21 On associationism in general, see F. Wilson, *Psychological Analysis and the Philosophy of John Stuart Mill*.

22 Hume's associationist account of abstract ideas is discussed in detail in F. Wilson, 'Association, Ideas and Images in Hume.'

23 Note, by the way, that in the passage we have just quoted Hume does *not* deny that we have an idea of *existence*, that is, an *abstract* idea of *existence*. All that he is saying is that, in the judgment that *God is*, this abstract idea is not represented in consciousness by a particular idea that is *distinct* from the idea of *God*. Pears simply misunderstands this passage; see D. Pears, *Hume's System*, 34, 57.

24 A fuller account would have to refer to the role of language; see Wilson, 'Association, Ideas and Images in Hume.' For the radical nature of Hume's break with the tradition, see F. Wilson, Critical Review of Jones, *Hume's Sentiments*, and 'Hume and Derrida on Language and Meaning.'

25 Descartes, *Objections and Replies*, Cottingham, 2:117.

26 As Cummins ('Hume,' 75f) points out, the position of Dennis Bradford ('Hume on Existence') that there are two sorts of entities, those that exist and those that don't, simply makes no sense in the Humean context.

27 Cummins perhaps does not see the force of this; see 'Hume,' 67f.

28 The reconstruction (Cummins, 'Hume,') that culminates on 71 I find entirely persuasive.

29 In Descartes, *Philosophical Works*, trans. Cottingham, vol. 1.

30 Quotations are taken from *The Basic Works of Aristotle*, ed. R. McKeon.

31 S. Clarke, *A Discourse concerning the Being and Attributes of God, the Obligations of Natural Religion, and the Truth and Certainty of the Christian Religion*, 6th ed. (London 1725).

32 It is customary to make a sharp distinction for Aristotle between the 'is' of existence and the 'is' of predication. Clearly, I disagree: the 'is' of existence is to be understood in terms of the 'is' of predication. I see no good argument of a philosophical sort to back up the claim for two senses of 'is.' The justification is often made in terms of the 'absolute' and 'predicative' constructions in the philosopher's Greek.

Kahn has pointed out the difficulty of making a firm syntactical distinction between the two constructions, and has argued further that the 'absolute' construction ought not to be construed as making 'existential' claims as opposed to 'predicative' ones. Kahn's case is convincing. See C. Kahn, 'The Greek Verb "To Be" and the Concept of Being.'

33 And also David Pears; see his *Hume's System*, 34. For I. Kant, see his *Critique of Pure Reason*, trans. N.K. Smith, 550–7 ('The Impossibility of an Ontological Proof of the Existence of God').

34 And therefore, it *is* different from other predicates; none the less, it is still a predicate. Pears, *Hume's System*, 56, misses this point.

35 This necessary connection supports the subjunctive conditional to the effect that to conceive of something is to conceive of that thing as it would be if it were to exist. David Pears, *Hume's System*, makes this conditional central to his account; he does not notice that the subjunctive conditional is parasitic upon the general truth or necessary connection. For a discussion of the way in which subjunctive conditionals receive support from laws, see F. Wilson, *Laws and Other Worlds*. This discussion is explicitly Humean. One often has the impression, as one does in Pears, that reference to subjunctive conditionals is introduced without any serious attempt to make a connection with a Humean account of subjunctive conditionals. For a glance at the latter, see F. Wilson, *Hume's Defence of Causal Inference*, ch. I.

36 Descartes, *Objections and Replies*, Cottingham, 2:117.

37 I will argue in Study Seven that Descartes's account of the demonstration of a necessarily existent being is of a piece with that of Locke and Clarke.

38 Cf J. Weinberg, *Ockham, Descartes, Hume*, 143.

39 As Weinberg, ibid., ch. 10, has pointed out against the obscure criticisms of Miss Anscombe ('Hume Reconsidered,' 188) and Anthony Kenny (*The Five Ways*, 67).

40 That is, no positive properties. Privations are, of course, to be denied of it.

41 P. Butchvarov, *Resemblance and Identity*, 115ff.

42 F. Suarez, *On the Various Kinds of Distinctions*, trans. from *Disputationes metaphysicae, Disputatio VII*, by C. Vollert.

43 F. Suarez, *On the Essence of Finite Being as Such, On the Existence of That Essence and Their Distinction*, trans. from *Disputationes metaphysicae, Disputatio XXXI* by N.J. Wells.

44 Cummins fails to note this; cf his remarks in n27.

45 Cf B. Russell, 'The Philosophy of Logical Atomism,' in his *Logic and Knowledge*, ed R.C. Marsh, 204.

46 Cf G. Bergmann, 'Stenius on the *Tractatus*.'

47 This blurring together the necessity of the principle of exemplification with the necessity of other truths occurs elsewhere in the history of philosophy. See F. Wilson, 'The World and Reality in the *Tractatus*.'

48 B. Russell, 'On the Relations of Universals and Particulars,' 111–12.

49 Cf F. Wilson, 'Weinberg's Refutation of Nominalism.'

50 Causation is the one relation that *does* fit the Aristotelian account of relations that Hume adopts!

Study Seven

1 I have benefited from comments by Tom Lennon and R. Imlay. It was a paper of the latter that first stimulated my thinking concerning the ontological argument for God's existence, which, as I now see it, is crucial for correctly understanding the *Meditations*; see R. Imlay, 'Descartes' Ontological Argument: A Causal Argument,' and also his 'Descartes' Ontological Argument.'

2 In R. Descartes, *The Philosophical Works of Descartes*, 3 vols, trans. J. Cottingham et al. The *Discourse* is in vol. 1, and references will be given in parentheses by *Discourse* followed by the volume and page number. Other works in vol. 1 are the *Principles of Philosophy* and the *Rules for the Direction of the Mind*; these two works will be referred to in similar fashion.

The *Meditations* and *Objections and Replies to the Meditations* occur in vol. 2; these, too, will be referred to in parentheses by title followed by volume and page number.

Many of Descartes's letters have been translated in vol. 3, trans. J. Cottingham, R. Stoothoff, D. Murdoch, and A. Kenny; these will also be referred to in parentheses by volume and page number.

3 The quotation is from Montaigne, 'Of Presumption,' *Complete Essays*, Book II, ch. xvii. Cf E. Gilson, *René Descartes, Discours de la méthode: Texte et commentaire*, 81–4. The Cartesians, of course, hated Montaigne; see the discussion of Montaigne's style in N. Malebranche, *The Search after Truth*, trans. T.M. Lennon and P.J. Olscamp, 184ff.

4 R. Popkin states that 'The *cogito* functions not, as some of the critics claimed, as the conclusion of a syllogism ... but as the conclusion of doubt. Just by pushing scepticism to its limit, one is confronted with a truth that one cannot doubt in any conceivable manner. The process of doubting compels one to recognize the awareness of oneself, compels one to see that one is doubting or thinking, and that one is here, in existence.' The method of doubt is 'the cause rather than the occasion of the acquisition of knowledge.' This knowledge is constituted by the 'discovery of one absolutely certain truth' and results in the 'overthrow [of] the sceptical attitude that all is uncertain ...' (*History of Scepticism from Erasmus to Descartes*, 187).

5 J. Chevalier, *Descartes*, 218.

6 A. Kenny states that 'The *cogito* ... established as certain his [Descartes's] own existence' (*Descartes: A Study of His Philosophy*, 63; cf 40).

7 B. Williams states that, at the beginning of the Second Meditation, Descartes 'makes the reflection which brings the Doubt for the first time to a halt, and which sets him off in the opposite direction, on the path of positive knowledge' (*Descartes: The Project of Pure Enquiry*, 72); this reflection is the *cogito*, which makes it 'certain that he [Descartes] exists' (ibid., 102).

8 Margaret Wilson, *Descartes*, 61.

9 M. Guéroult, *Descartes' Philosophy Interpreted according to the Order of Reasons*, 1:27.

10 Here are some more examples. L.J. Beck writes that 'The force of the hypothesis of the Malignant Spirit breaks on the rock of the *Cogito*. The recognition of one truth as indubitably true, and self-evidently so, gives a rational conviction which is sufficient to destroy the hypothesis once and for all. The *Cogito* destroys the very basis of the postulate of an all-powerful deceiving being' (*The Metaphysics of Descartes*, 143).

   M. Versfeld states that 'His [the evil genius's] essence was to possess full powers of deception. Without that he is nothing. The evil genius, then, disappears with the affirmation of the *Cogito*' (*An Essay on the Metaphysics of Descartes*, 49).

   S. Tweyman places the *cogito* as the first truth, and 'once Descartes showed that the *Cogito* leads to the conclusion that veracious God exists he believed himself to have established the reliability of reason' ('The Reliability of Reason,' 136).

11 Guéroult, *Descartes' Philosophy*, 1:9–10.

12 O. K. Bouwsma, 'Descartes' Evil Genius.'

13 Ibid., 151.

14 G. Berkeley, *Principles of Human Knowledge*, sec. xxxiii.

15 Marin Mersenne, *La Verite des sciences contre les sceptiques ou Pyrrhoniens*. On Mersenne's 'mitigaged scepticism,' see R. Popkin, *History of Scepticism from Erasmus to Descartes*, ch. 7; and, from a somewhat different perspective, Dear, 'Marin Mersenne and the Probabilistic Roots of "Mitigated Scepticism."'

16 Mersenne, *La Verite des sciences*, 191–2.

17 Ibid., 193.

18 Ibid., 9.

19 Ibid., 14–15.

20 Ibid., 14.

21 Bouwsma, 'Descartes' Evil Genius,' 151.

22 Aristotle, *The Complete Works of Aristotle*, ed. J. Barnes.

23 Thomas Aquinas, *Summa theologica*, trans. by the Fathers of the Dominican Province. References are to the First Part (FP) or First Part of the Second Part (FS) or Second Part of the Second Part (SS), by Question (Q), Article (A), Objection (Obj) by number, and, where appropriate, Reply to Objection (RObj) by number.

24 Cf Roger Ariew, 'Descartes and Scholasticism: The Intellectual Background to Descartes' Thought.'

25 Whether this constitutes 'direct realism' is a further issue. At best it is the sort of direct realism that is part of the substance tradition, where the form or essence of the thing known is in the mind. If one rejects the Aristotelian substance ontology and the doctrine of knowledge that is part of it (which Descartes does not do), then of course it is no sort of realism at all: it is simply not acceptable as metaphysics, and therefore also not acceptable as an account of knowledge.

26 Kenny, *Descartes*, states, with regard to Descartes's use of the distinction between formal and objective reality, that this 'is a piece of scholastic jargon, but no scholastic theory seems to be involved' (110). This seems profoundly wrong, however. In using the 'jargon,' Descartes is surely locating for his readers how his meditations grow out of, and aim to defend, the traditional metaphysics. See T. Lennon, 'The Inherence Pattern and Descartes' Ideas.' The roots of the Cartesian account of formal and objective reality go back to Aristotle and Aquinas, but the immediate background was provided by Suarez; see in particular Norman J. Wells, 'Material Falsity in Descartes, Arnauld, and Suarez.'

27 See S. Gaukroger, 'Introductory Essay' to his translation of A. Arnauld, *On True and False Ideas*.

28 Cf John Sergeant, *Solid Philosophy Asserted against the Fancies of the Ideists* (London 1696).

29 Unless otherwise noted, references to Scotus are to the *Opus oxoniense*, I, dII, q4, as translated in Duns Scotus, *Philosophical Writings*, ed. and trans. Alan B. Wolter, 96–132. These references will be given as 'Wolter.' Scotus's various works are discussed in this volume, at xvii ff.

   For an important discussion of Scotus's views on knowledge, see Peter C. Vier, *Evidence and Its Function according to John Duns Scotus*. Vier cites a number of Scotus's works beyond the passages in Wolter; references to these sources will be given as 'Vier.' References from Vier to the *Opus oxoniense* that are not in Wolter are referred to as *Oxon*. Other references taken from Vier are to *Quaestiones subtilissime in Metaphysicam Aristotelis*, referred to as *Metaphy*, and to *Quaestiones quodlibetales*, referred to as *Quodl*.

30 Duns Scotus, *Opus*, Wolter, 100.

31 Ibid., 128, 131.

32 Ibid., 100.

33 Ibid., 118.

34 Ibid., 100.

35 Ibid., 128.

36 Ibid., 129.

37 Ibid., 128, 131.

38 Ibid., 110.

39 Ibid., 110–11.

40 Ibid., 109.
41 'Est intellectus hujus propositionis: natura determinatur ad unum, non quidem ad unum producibile, unum inquam numero sive singulare; sed determinatur ad unum determinatum modum producendi, quia non est ibi principium indeterminatum respectu oppositorum sicut est voluntas' (Duns Scotus, *Quodl.* Q2, n10; quoted in Vier, 142).
42 Duns Scotus, *Opus*, Wolter, 109.
43 Ibid., 109.
44 Ibid., 108.
45 Ibid., 103.
46 Ibid., 116.
47 Ibid., 106.
48 Ibid., 108–9.
49 Ibid., 114–15.
50 Ibid., 118.
51 'videt et certus est naturam ut in pluribus uniformiter agere et ordinate' (*Metaphy.* I, q4, n19; quoted in Vier, 143).
52 'Posse habere causatum simpliciter necessarium non est perfectionis in causa secunda; immo et hoc nulli causae secundae convenir ... Simpliciter enim necessario causate includit contradictionem et ideo hoc nulli causae secundae convenit' (Duns Scotus, *Oxon.* II, d1, q3, n12; quoted in Vier, 146).
53 'Causa naturalis, licet ex se terminetur ad effectum, potest tamen impediri ... si ab extrinseco ponatur impedimentum' (Duns Scotus, *Metaphy.* IX, q14, n1; quoted in Vier, 144).
54 *Metaphy.* VI, q2, n7; quoted in Vier, 145.
55 Duns Scotus, *Oxon.* I, d3, q2; quoted in Vier, 145.
56 'Licet aliqua connexio (effectus ad causam) sit secundum quid necessaria, nulla tamen simpliciter est necessaria, quia quaelibet dependet a prima quae contingenter causat ... similitur communitur secundae causae impedibiles, et causa impedibilis quantumcumque non impediatur non est necessaria' (Duns Scotus, *Metaphy.* V, q3, n5; quoted in Vier, 144).
57 Duns Scotus, *Opus*, Wolter, 109–10.
58 Ibid., 118–19.
59 Duns Scotus, *Metaphy.* VI, q2, n7; quoted in Vier, 144.
60 C. Kahn, 'The Greek Verb "To Be" and the Concept of Being.'
61 Ibid., 250.
62 C. Kahn, 'The Thesis of Parmenides,' 712.
63 The importance of the determinable/determinate distinction for understanding Descartes's philosophy has been emphasized by S. Keeling, *Descartes*, 129–30; and Lennon, 'The Inherence Pattern and Descartes' *Ideas*.' See also H. Bracken, 'Some Problems of Substance among the Cartesians,' and 'Substance in Berkeley.'

64 Cf Guéroult, *Descartes' Philosophy Interpreted*, 2:245ff; and also A. Gombay, '"*Cogito ergo Sum*": Inference or Argument?' in R.J. Butler, ed., *Cartesian Studies*, 71–88.

65 Descartes, *Letter* to Clerselier (January 1646), in *Descartes: Philosophical Writings*, trans. E. Anscombe and P. Geach, 299–300.

66 Cf F. Wilson, *Explanation, Causation and Deduction*, and also Study Two, Sec. III. 2, above.

67 As we discussed in Study Two, that one can discern the necessary causal relation in particulars is important to any serious defence of objective necessary connections of the sort defended by Aristotle and Descartes.

68 Cf J. Weinberg, '*Cogito ergo Sum*: Some Reflections on Mr. Hintikka's Article,' 488.

69 J. Hintikka, '*Cogito, ergo Sum*: Inference or Performance?'

70 Ibid., 11.

71 A.J. Ayer, 'Cogito ergo Sum.'

72 J.L. Mackie, 'Self-Refutation – A Formal Analysis.'

73 G. Nahknikian, 'On the Logic of Cogito Propositions.'

74 Hintikka, '*Cogito ergo Sum*,' 12.

75 Ibid., 16.

76 Fred Feldman has argued that Hintikka's 'Fa, ergo $(\exists x)(x = a)$' has essentially the same existential presuppositions as the traditional 'I think, ergo I am,' and that Hintikka's criticisms of the latter apply equally to his own formulation of the *cogito*. Cf F. Feldman, 'On the Performatory Interpretation of the *Cogito*.' But this is to ignore Hintikka's crucial claim that the *cogito* is a performance rather than an inference, and that the 'I think' does not function as a premise but as a statement expressing the awareness that we have of the performance that makes the self-verifying 'I am' or 'I exist' true.

77 Guéroult, *Descartes' Philosophy Interpreted*, 1:61.

78 P.D. Huet, *Censura philosophiae cartesianae*, editio quarta (Paris 1694), Caput I, sec. XI, 54–61.

79 Thomas Heath, *Greek Mathematics*, 2:400–1. See also Pappus of Alexandria (*Bk 7 of the Collection*, Part I: Introduction, Text and Translation, edited with translation and commentary by A. Jones, 82–4.

80 Cf Carl Boyer, *History of Analytic Geometry*, 23.

81 Cf Ibid., ch. 5.

82 Concerning the nature and importance of Cartesian science, see J. Chevalier, *Descartes*, ch. 4.

83 Cf S. Gaukroger, 'The Nature of Abstract Reasoning: Philosophical Aspects of Descartes' Work in Algebra'; and also W. Boyce Gibson, 'La "Géométrie" de Descartes au point de vue de sa méthode.'

84 Descartes, *Discourse on Method, Optics, Geometry, and Meteorology*, trans. P.J. Olscamp, 179.

85 M. Guéroult, *Descartes' Philosophy Interpreted*, 1:229.

86 Ibid., 9.

87 For example, M. Wilson, *Descartes*, 136ff.

88 For an important discussion of this aspect of Descartes's 'rules for the direction of the mind,' see H.H. Joachim, *Descartes's Rules for the Direction of the Mind.*

89 M. Wilson, *Descartes*, 137.

90 H.G. Frankfurt, 'Descartes' Validation of Reason.'
    Frankfurt's discussion is indebted to A. Gewirth, 'Clearness and Distinctness in Descartes.'

91 Concerning Frankfurt's interpretation, see R.A. Watson, 'In Defiance of Demons, Dreamers, and Madmen'; and John H. Bannan, 'Theories of Truth and Methodic Doubt.'

92 B. Williams, *Descartes*, 153.

93 References are to Plato, *Republic*, trans. G.M.A. Grube.

94 For an important discussion of the notion of God in Aristotle, see L.P. Gerson, *God and Greek Philosophy*, ch. 3.

95 F16, R3 (Alexander, apud Simplicius, Commentarius in de Caelo 289.1–15). In Aristotle's 'Fragments,' in vol. 2 of *Aristotle, The Complete Works of Aristotle*, ed. J. Barnes.

96 For a discussion of this Christianization of Aristotle, see Richard Sorabji, 'Infinite Power Impressed: The Transformation of Aristotle's Physics and Theology.'

97 For some other aspects of the ontological argument in Descartes, see Amos Funkenstein, 'Descartes, Eternal Truths and the Divine Omnipotence.'

98 Guéroult, *Descartes' Philosophy Interpreted*, 1:268

99 See R. Descartes, *Œuvres philosophiques*, ed. F. Alquié, 2:432n1.

100 Ibid.

101 R. Descartes, *Descartes' Conversation with Burman*, trans. J. Cottingham.

102 Ibid., 9.

103 Hiram Caton, 'On the Interpretation of the *Meditations.*'

104 Louis Loeb, 'Is There Radical Dissimulation in Descartes' *Meditations?*'

105 Kenneth Dorter, 'Science and Religion in Descartes' *Meditations.*'

106 Guéroult, *Descartes' Philosophy Interpreted*, 1:241.

107 Ibid., 1:243.

108 Ibid., 1:109.

109 Popkin, *History of Scepticism from Erasmus to Descartes*, 187.

110 Ibid., ch. 10.

111 Kenny, *Descartes*, ch. 7.

112 Ibid., 162.

113 M. Wilson, *Descartes*, 176.

114 Kenny, *Descartes*, 121.

115 E.J. Ashworth, 'Descartes' Theory of Clear and Distinct Ideas.'

116 E.B. Allaire, 'The Circle of Ideas and the Circularity of the *Meditations*,' 131.

117 Ibid., 131.

118 Ibid.; Ashworth, 'Descartes' Theory of Clear and Distinct Ideas,' 105.

119 Williams, *Descartes*, 67.

120 Kenny, *Descartes*, 21.

121 S. Foucher, *Critique [of Nicholas Malebranche's] Of the Search after Truth.* For a general discussion of Foucher's significance, see the important account in R.A. Watson, *The Downfall of Cartesianism.*

122 A. Arnauld, *On True and False Ideas*, trans. E.J. Kremer, 20.

123 Ibid., 19.

124 Ibid.

125 S. Foucher, *Critique [of Nicholas Malebranche's] Of the Search after Truth.*

126 Ibid., 21ff.

127 Ibid., 31ff.

128 Ibid., 32–3.

129 P.D. Huet, *An Essay concerning the Weakness of Human Understanding*, trans. E. Combe, 22–3.

130 N. Malebranche, *The Search after Truth*, trans. T. M. Lennon and P.J. Olscamp. There was a direct influence of Suarez on Malebranche; see Desmond Connell, *The Vision in God: Malebranche's Scholastic Sources*, 165–206.

131 Malebranche, *Search after Truth.*, 218.

132 Ibid., 2.

133 Ibid., 217.

134 Ibid., 227.

135 Ibid., Elucidation Ten, 624.

136 Ibid., Elucidation Ten, 614.

137 Ibid.

138 Ibid., 230.

139 Ibid.

140 Ibid., Elucidation Six, 573.

141 Ibid., 574.

142 Ibid.

143 Ibid., 573.

144 Foucher, *Critique [of Nicholas Malebranche's] Of the Search after Truth*, 46.

145 Malebranche, *Search after Truth*, Elucidation Eleven, 633.

146 Ibid., Elucidation Fifteen, 658.

147 Ibid., 660.

148 Ibid., 657.

149 Ibid., 657.

150 See Monte Cook, 'Malebranche versus Arnauld.'

151 See A.O. Lovejoy, '"Representative Ideas" in Malebranche and Arnauld'; Monte

Cook, 'Arnauld's Alleged Representationalism'; Daisy Radner, 'Representationalism in Arnauld's Act Theory of Perception'; and T. Lennon, 'Philosophical Commentary,' in N. Malebranche, *The Search after Truth*, 794–803.

152 J. Locke, *Examination of Malebranche's Opinion*, sect. 20. This was first published in Locke's *Posthumous Works* (London 1706). It appears in J. Locke, *Works*, 10 vols (London 1823). The quoted passage is contained in one of the excerpts published in *The Locke Reader*, ed. J. Yolton.

153 Cf A.A. Luce, *The Dialectic of Immaterialism*; and Charles J. McCracken, *Malebranche and British Philosophy*.

154 Note that the PA is an *ontological* principle. M. Wilson, *Descartes*, 189, misleadingly suggests that an appeal to PA is an 'epistemological argument.'

155 Malebranche, *Search after Truth*, 2.

156 J. Locke, *Essay concerning Human Understanding*, IV.iii.28; see also IV.vi.10.

157 Cf F. Wilson, 'Acquaintance, Ontology and Knowledge.' Also F. Wilson, 'Bradley's Conception of Ideality: Comments on Ferreira's Defence.'

158 D. Hume, *Treatise of Human Nature*. Page references to this text hereafter will be in parentheses, following a 'T.'

159 John Dryden, 'Virgil's Georgics,' in *The Poems of John Dryden*, ed. James Kingsley, 2:955, ll 698–701.

160 Virgil, *The Georgics*, trans. L.P. Wilkinson, 93, ll 491–4.

161 Dryden, 'Virgil's Georgics,' 954, ll. 676–82.

162 Virgil, *Georgics*, 92, ll. 478–82.

163 John Milton, *Paradise Lost*, Bk VII, ll 94–7, in *John Milton: Complete Poems and Major Prose* ed. Merrit Y. Hughes.

164 Ibid., Bk VIII, ll 66–75.

165 Ibid., Bk VII, ll 109–30.

166 Ibid., Bk V, ll 156–9.

167 M. Guéroult, *Descartes' Philosophy Interpreted*, 2:237.

Bibliography

Abelard, Peter (Peter Abailard). *Sic et non*. A critical edition by B. Rover and R. McKeon. Chicago: University of Chicago Press 1976.

Agricola, Rudolphus. *De inventione dialectica libri tres*. Tubingen: M. Niemeyer 1992.

Aguilonius, F. *Opticorum libri sex*. Antwerp, 1613.

Aldrich, Henry. *Artis logicae compendius*. Oxford: e Theatro Sheldoniana, 1691.

– *Artis logica rudimenta*. Ed. with notes by H.L. Mansel, 4th ed. London: Rivington, 1862.

Allaire, E.B. 'The Circle of Ideas and the Circularity of the *Meditations*,' *Dialogue* 5 (1966). 131–53.

A.M. *A Discourse of Local Motion*. London: Moses Pitt 1670.

Anscombe, G.E.M. 'Hume Reconsidered.' *Blackfriars* 43 (1962): 188.

Árdal, P. *Passion and Value in Hume's Treatise*. Edinburgh: University of Edinburgh Press 1966.

– 'Some Implications of the Virtue of Reasonableness in Hume's *Treatise*.' In D. Livingston and J. King, eds, *Hume: A Re-valuation*, 91–106. New York: Fordham University Press 1976.

Ariew, Roger. 'Descartes and Scholasticism: The Intellectual Background to Descartes' Thought.' In J. Cottingham, ed., *Cambridge Companion to Descartes*, 58–90. Cambridge: Cambridge University Press 1992.

Aristotle. *The Basic Works of Aristotle*. Ed. R. McKeon. New York: Random House 1941.

– *The Complete Works of Aristotle*. Rev. Oxford translation. Ed. J. Barnes. Princeton: Princeton University Press 1984.

– *De interpretatione*. Trans. J.L. Ackrill. Oxford: Oxford University Press 1963.

– *The Generation of Animals*. In *The Complete Works of Aristotle*, ed. J. Barnes. Princeton: Princeton University Press 1984.

Armstrong, D.M. *What Is a Law of Nature?* Cambridge: Cambridge University Press 1963.

Arnauld, A. *The Art of Thinking.* Trans. J. Dickoff and P. James. Indianapolis: Bobbs-Merrill 1974.

– *On True and False Ideas.* Trans. E.J. Kremer. Queenston, Ont.: Edwin Mellen Press 1990.

Arnauld, A., and J. Nicole. *La Logique de Port-Royal.* Ed. Ch. Jourdain. Paris: Hachette 1854.

Ashworth, E.J. 'Descartes' Theory of Clear and Distinct Ideas.' In R.J. Butler, ed., *Cartesian Studies,* 89–105.

Aulus Gellius. *Noctes Atticae.* 3 vols. Trans. John C. Rolfe. Cambridge, Mass.: Harvard University Press 1946.

Ayer, A.J. 'Cogito ergo Sum.' *Analysis* 14 (1953–4), 17–33.

– *The Problem of Knowledge.* London: Macmillan 1956.

Bacon, Francis. *New Organon.* Ed. with an Introduction by F.H. Anderson. Indianapolis: Bobbs-Merrill 1960.

– *The Novum Organum ... Epitomiz'd.* London: Thomas Lee 1676.

– *The Philosophical Works.* 3 vols. Trans. with notes by Peter Shaw. London: J.J. and P. Knapton et al. 1733.

Baier, A. *A Progress of the Passions.* Cambridge, Mass.: Harvard University Press 1991.

Bannan, John H. 'Theories of Truth and Methodic Doubt.' *The Modern Schoolman* 58 (1981): 105–12.

Barfoot, Michael. 'Hume and the Culture of Science in the Early Eighteenth Century.' In M.A. Stewart, ed., *Studies in the Scottish Enlightenment,* 151–90. Oxford: Oxford University Press 1990.

Barker, S. *The Elements of Logic.* New York: McGraw-Hill 1965.

Barnes, J., et al., eds. *Articles on Aristotle.* 4 vols. London: Duckworth 1975.

Baron, H. *The Crisis of the Early Italian Renaissance.* 2 vols. Princeton: Princeton University Press 1955; rev. ed. 1966.

Beauchamp, T., and T. Mappes. 'Is Hume Really a Sceptic about Induction?' *American Philosophical Quarterly* 12 (1975): 119–34.

Beauchamp, T. and A. Rosenberg. *Hume and the Problem of Causation.* Oxford: Oxford University Press 1981.

Beck, L.J. *The Metaphysics of Descartes.* Oxford 1965.

Beiser, F.C. *The Sovereignty of Reason: The Defense of Rationality in the Early English Enlightenment.* Princeton: Princeton University Press 1996.

Belsham, Thomas. *Elements of the Philosophy of the Mind, and of Moral Philosophy. To Which is Prefixed a Compendium of Logic.* London: J. Johnson 1801.

Bergmann, G. 'On Non-Perceptual Intuition.' In his *Metaphysics of Logical Positivism,* 228–33. New York: Longmans 1954.

– *Philosophy of Science*. Madison, Wisc.: University of Wisconsin Press 1956.
– 'Stenius on the *Tractatus*.' In his *Logic and Reality*, 242–71. Madison, Wisc.: University of Wisconsin Press 1964.
Berkeley, G. *Philosophical Commentaries*. In his *Works*, vol. 1.
– *Principles of Human Knowledge*. In his *Works*, vol. 2.
– *Works*. 6 vols. Ed. A. Luce and T. Jessop. Nelson: London 1948–57.
Bernard of Clairvaux. *Epistolae*. In *Patrologia Latina*, vol. 182.
Bird, O. 'The Formalizing of the Topics in Mediaeval Logic.' *Notre Dame Journal of Formal Logic* 1 (1960): 138–49.
– *Syllogistic and Its Extensions*. Englewood Cliffs, N.J.: Prentice-Hall 1968.
– 'The Tradition of the Logical Topics: Aristotle to Ockham.' *Journal of the History of Ideas* 23 (1962): 307–23.
Black, D. 'Aristotle's *"Peri hermeneias"* in Medieval Latin and Arabic Philosophy: Logic and the Linguistic Arts.' *Canadian Journal of Philosophy*, Supp. Vol. 17: 25–83.
Blanshard, B. 'The Case for Determinism.' In S. Hook, ed., *Determinism and Freedom*, 19–30. New York: Collier 1961.
Bochenski, I.M. *Ancient Formal Logic*. Amsterdam: North-Holland Publishing Co. 1951.
Boethius. *De topicis differentiis*. Trans. E. Stump. Ithaca, N.Y.: Cornell University Press 1978.
– *In Ciceronis Topica*. Trans. E. Stump. Ithaca, N.Y.: Cornell University Press 1988.
Boole, G. *Laws of Thought*. In vol. 2 of *George Boole's Collected Logical Works*. La Salle, Ill.: Open Court 1952.
Bourgey, L. 'Observation and Experiment in Analogical Explanation.' In J. Barnes et al., eds, *Articles on Aristotle*, 1:175–82. London: Duckworth 1975.
Bouwsma, O.K. 'Descartes' Evil Genius.' *Philosophical Review* 58 (1949): 141–51.
Boyer, Carl. *History of Analytic Geometry*. New York: Scripta Mathematica Studies 1956.
Boyle, Robert. *Disquisition about the Final Causes of Natural Things*. London: H.C. 1688.
– *A Free Enquiry into the Vulgarly Receiv'd Notion of Nature*. London: H. Clark 1686.
Bracken, H. 'Some Problems of Substance among the Cartesians.' *American Philosophical Quarterly* 1 (1964): 129–37.
– 'Substance in Berkeley.' In Warren E. Steinkraus, ed., *New Studies in Berkeley's Philosophy*, 85–97. New York: Holt, Rinehart and Winston 1966.
Bradford, Dennis. 'Hume on Existence.' *International Studies in Philosophy* 15 (1983): 1–12.
Bréhier, Émile. *Chrysippe*. Paris: Felix Alcan 1910.
Breuer, Mordechai. 'Pilpul.' *Encyclopaedia Judaica.*, vol. 13. Jerusalem: Encyclopaedia Judaica 1981.

Broad, C.D. *Examination of McTaggart's Philosophy*. 3 vols. Cambridge: Cambridge University Press 1938.

Brody, B. 'Toward an Aristotelian Theory of Scientific Explanation.' *Philosophy of Science* 39 (1972): 20–31.

Brown, G. Patterson. 'Infinite Causal Regression.' *Philosophical Review* 75 (1966): 510–25.

Brown, J.R. *The Laboratory of the Mind*. London: Routledge 1991.

Brown, R. 'History versus Hacking on Probability.' *History of European Ideas* 8 (1987): 655–73.

Brunet, Pierre. 'La Méthodologie de Mariotte.' *Archives Internationale d'Histoire des Sciences* 1 (1947): 26–59.

Burgersdijck, Franco. *Monitio logica, or an Abstract and Translation of Burgersdiscius His Logick*. By a Gentleman. London: Richard Cumberland 1697.

Buroker, Jill. 'Judgment and Predication in the *Port-Royal Logic*.' In E.J. Kremer, ed., *The Great Arnauld and Some of His Philosophical Correspondents*, 3–27.

Butchvarov, P. *Resemblance and Identity*. Bloomington, Ind.: Indiana University Press 1966.

Butler, R.J., ed. *Cartesian Studies*. Oxford: Blackwell 1972.

Caton, Hiram. 'On the Interpretation of the *Meditations*.' *Man and World* 3 (1970): 224–45.

Chambers, Ephraim. *Cyclopaedia*. London 1728; 2nd ed. 1738.

Chappell, V.C. ed. *Hume*. 1st edn. Garden City, N.Y.: Anchor Books 1966.

Chevalier, J. *Descartes*. Paris: Plon 1921.

Chisholm, R.M. 'Law Statements and Counterfactual Inference.' *Analysis* 15 (1955): 97–105.

Cicero. *De divinatione*. In Cicero, *De senectute, De amicitia, De divinatione*. Trans. W.A. Falconer Cambridge, Mass.: Harvard University Press 1923.

– *De fato*. In Cicero, *De Oratore, together with De fato, Paradoxica stoicorum, De partitione oratoria*. 2 vols. Trans. H. Rackham. Cambridge, Mass.: Harvard University Press 1960.

– *De finibus bonorum et malorum*. Trans. H. Rackham. New York: Macmillan 1914.

– *De oratore*. In Cicero, *De Oratore, together with De fato, Paradoxica stoicorum, De partitione oratoria*. 2 vols. Trans. H. Rackham. Cambridge, Mass.: Harvard University Press 1960.

– *Topica*. In *De inventione, De optimo genere oratorum, Topica*. Trans. H.M. Hubbel. Cambridge, Mass.: Harvard University Press 1960.

Clarke, John. *A Demonstration of Some of the Principal Sections of Sir Isaac Newton's Principles of Natural Philosophy*. London: J. and J. Knapton 1730. Repr. with an Introduction by I.B. Cohen. New York: Johnson Reprint Corporation 1972.

Clarke, Samuel. *A Discourse concerning the Being and Attributes of God.* 4th ed. London: James Knapton 1716.

– *A Discourse concerning the Being and Attributes of God, the Obligations of Natural Religion, and the Truth and Certainty of the Christian Religion.* 6th ed. London: James Knapton 1725.

Coffey, P. *The Science of Logic.* 2 vols. New York: Peter Smith 1938; first published 1912.

Cohen, I.B. *The Birth of a New Physics.* Garden City, N.Y.: Doubleday Anchor 1960.

– 'Newton's Second Law and the Concept of Force in the *Principia.*' *Texas Quarterly* 10 (1967): 127–57.

Coleridge, S.T. *Biographia Literaria.* 2 vols. Ed. J. Shawcross. London: Oxford University Press 1907.

Connell, Desmond. *The Vision in God: Malebranche's Scholastic Sources.* Louvain: Éditions Nauwelaerts 1967.

Cook, Monte. 'Arnauld's Alleged Representationalism.' *Journal of the History of Philosophy* 12 (1974): 53–64.

– 'Descartes' Alleged Representationalism.' *History of Philosophy Quarterly* 4 (1987): 179–85.

– 'Malebranche versus Arnauld.' *Journal of the History of Philosophy* 29 (1991): 183–99.

Copi, I. *Introduction to Logic.* 7th ed. New York: Macmillan 1986.

Copleston, F. *History of Modern Philosophy.* vol. 5 [*Modern Philosophy: The British Philosophers*], Pt 1 ['Hobbes to Paley']. Garden City, N.Y.: Image Books, Doubleday 1964.

Cornford, F.M. *Plato's Theory of Knowledge.* London: Routledge and Kegan Paul 1935.

Costabel, P. 'Newton's and Leibniz's Dynamics.' Trans. J.M. Briggs. *Texas Quarterly* 10 (Autumn 1967): 119–26.

Costello, W.D. *The Scholastic Curriculum at Early Seventeenth-Century Cambridge.* Cambridge, Mass.: Harvard University Press 1958.

Coumet, Ernest. 'Sur L'*Essai de Logique* de Mariotte: L'Établissement des sciences.' In Pierre Costabel, *Mariotte, savante et philosophe*, 277–319. Paris: J. Vrin 1986.

Cronin, T.J. *Objective Being in Descartes and Suarez.* Rome: Gregorian University Press 1966.

Cudworth, Ralph. *True Intellectual System of the Universe.* London: R. Royston 1678. References are to the 2nd ed., London: J. Walthoe et al. 1743.

Cummins, P. 'Hume on the Idea of Existence.' *Hume Studies* 17 (1991): 61–82.

Curtis, Mark H. *Oxford and Cambridge in Transition: 1558–1642.* London: Oxford University Press 1959.

Dear, Peter. *Discipline and Experience.* Chicago: University of Chicago Press 1995.

– 'Marin Mersenne and the Probabilistic Roots of "Mitigated Scepticism,"' *Journal of the History of Philosophy* 22 (1984): 173–205.

Desaguliers, J.T. *Lectures of Experimental Philosophy*. London: W. Mears 1719.

Descartes, R. *Descartes' Conversation with Burman*. Trans. with intro. by J. Cottingham. Oxford: Oxford University Press 1976.

– *Descartes: Philosophical Essays* (*Rules for the Direction of the Mind. Discourse on Method*, and *Meditations on First Philosophy*). Trans. L.J. Lafleur. Indianapolis, Ind.: Bobbs-Merrill 1964.

– *Descartes: Philosophical Writings*. Trans. E. Anscombe and P. Geach. London, 1954.

– *Discourse on Method, Optics, Geometry, and Meteorology*. Trans. P.J. Olscamp. Indianapolis, Ind.: Bobbs-Merrill 1965.

– *Œuvres philosophiques*. 3 vols. Ed. F. Alquié. Paris: Garnier 1963–73.

– *The Philosophical Works of Descartes*. 3 vols. Trans. J. Cottingham, Robert Stoothoff, and Dugald Murdoch. Cambridge: Cambridge University Press 1985.

– *Principles of Philosophy*. Trans. V.R. Miller and R.P. Miller. Dordrecht, Netherlands: D. Reidel 1983.

Des Chene, Dennis. *Physiologia: Natural Philosophy in Late Aristotelian and Cartesian Thought*. Ithaca, N.Y.: Cornell University Press 1996.

Dibon, P. *La Philosophie Néerlandaise au siècle d'or*. Vol. 1. Paris: Elsevier 1954.

Diogenes Laertius. *Lives of the Eminent Philosophers*. Cambridge, Mass.: Harvard University Press 1925.

Dodsley, Robert. *The Preceptor*. 2 vols. London: J. & R. Dodsley 1748.

Doney, W., ed. *Descartes: A Collection of Critical Essays*. Garden City, N.Y.: Doubleday Anchor 1967.

Dorter, Kenneth. 'Science and Religion in Descartes' *Meditations*.' *The Thomist* 37 (1973): 313–40.

Drake, Stillman. *Galileo: Pioneer Scientist*. Toronto: University of Toronto Press 1990.

– 'Impetus Theory Reappraised.' *Journal of the History of Ideas* 36 (1975): 27–46.

Dray, W. *Laws and Explanation in History*. London: Oxford University Press 1957.

Dresden, Sem. *Humanism in the Renaissance*. Trans. M. King. London: Weidenfeld and Nicolson 1968.

Dryden, John. 'Virgil's Georgics.' In *The Poems of John Dryden*. 2 vols. Ed. James Kingsley. Oxford: Oxford University Press 1958.

Duncan, William. *The Elements of Logic*. Repr. Menston, Yorks.: Scolar Press 1970.

Duns Scotus. *Philosophical Writings*. Ed. and trans. Alan B. Wolter. Edinburgh: Nelson 1962.

Dürr, K. *The Propositional Logic of Boethius*. Amsterdam: North-Holland Publishing Co. 1951.

Eagles, C.M. 'David Gregory and Newtonian Science.' *British Journal for the History of Science* 10 (1977): 216–25.

Earman, J. 'Laws of Nature: The Empiricist Challenge.' In R.J. Bogdan, ed., *D.M. Armstrong*, 191–224. Dordrecht, Netherlands: D. Reidel 1984.

Ewing, A.C. *The Fundamental Questions of Philosophy.* London: Routledge and Kegan Paul 1951.

Feldman, F. 'On the Performatory Interpretation of the *Cogito.*' *Philosophical Review* 82 (1973): 345–63.

Fisch, M., and H.W. Jones. 'Bacon's Influence on Sprat's *History of the Royal Society.*' *Modern Language Quarterly* 12 (1951): 399–406.

Fogelin, R. *Hume's Scepticism in the Treatise of Human Nature.* London: Routledge and Kegan Paul 1985.

Foucher, S. *Critique [of Nicholas Malebranche's] Of the Search after Truth.* Trans. R.A. Watson. In R.A. Watson and M. Grene, eds and trans., *Malebranche's First and Last Critics: Simon Foucher and Dortous de Mairan.* Carbondale and Edwardsville, Ill.: Southern Illinois University Press 1995.

Frank, Phillipp. *Philosophy of Science.* Englewood Cliffs, N.J.: Prentice-Hall 1957.

Frankfurt, H. 'Descartes' Validation of Reason.' In W. Doney, ed., *Descartes: A Collection of Critical Essays*, 209–26.

Frede, M. 'Stoic vs. Aristotelian Syllogisms.' *Archiv für Geschichte der Philosophie* 56 (1974): 1–32.

Funkenstein, Amos. 'Descartes, Eternal Truths and the Divine Omnipotence.' In S. Gaukroger, ed., *Descartes: Philosophy, Mathematics and Physics.* Sussex: Harvester Press 1980.

Gabbey, Alan. 'Force and Inertia in Seventeenth-Century Dynamics.' *Studies in History and Philosophy of Science* 2 (1971): 1–67.

– 'Force and Inertia in the Seventeenth Century: Descartes and Newton.' In S. Gaukroger, *Descartes: Philosophy, Mathematics and Physics.* 230–320.

Galen. *Institutio logica.* English translation, Introduction, and Commentary by J.S. Keiffer. Baltimore: Johns Hopkins University Press 1964.

Galilei, Galileo. *Two New Sciences.* Trans. with a new Introduction and Notes by Stillman Drake. Toronto: Wall and Thompson 1989.

Garber, Daniel. *Descartes' Metaphysical Physics.* Chicago: University of Chicago Press 1992.

Gaukroger, S. *Cartesian Logic.* Oxford: Clarendon Press 1989.

– *Descartes: Philosophy, Mathematics and Physics.* Sussex: Harvester Press 1980.

– 'Introductory Essay' to his translation of Arnauld's *On True and False Ideas.* Manchester: Manchester University Press 1990.

– 'The Nature of Abstract Reasoning: Philosophical Aspects of Descartes' Work in Algebra.' In J. Cottingham, ed., *Cambridge Companion to Descartes*, 91–114.

Geach, P. 'Distribution: A Last Word?' In *Logic Matters*, 62–4. Oxford: Blackwell 1972.

- *Reference and Generality.* Ithaca, N.Y.: Cornell University Press 1962.

Gerson, L.P. *God and Greek Philosophy.* London: Routledge 1990.

Gewirth, A. 'Clearness and Distinctness in Descartes.' In W. Doney, ed., *Descartes: A Collection of Critical Essays*, 250–77.

Gibson, Strickland. 'The University of Oxford.' In H.E. Salter and M.D. Lobel, eds, *Victoria History of the County of Oxford*, vol. 3. London: Oxford University Press 1954.

Gibson, W. Boyce. 'La "Géométrie" de Descartes au point de vue de sa méthode.' *Revue de Métaphysique et de Morale* 4 (1896): 386–98.

Gillespie, C.M. 'The Logic of Antisthenes.' *Archiv für Geschichte der Philosophie* 27 (1914): 479–500; 28 (1915): 20–38.

Gilson, E. *The Mystical Theology of Saint Bernard.* Trans. A.H.C. Downes. London: Sheed and Ward 1940.

- *René Descartes, Discours de la méthode: Texte et commentaire.* Paris: Vrin 1962.

Glanvill, Joseph. *Scepsis scientifica.* 2 vols. London: E. Cotes 1665.

- *The Vanity of Dogmatizing.* London: E.C. 1661.

Glucker, J. '*Probabile, Veri Simile.* and Related Terms.' In J.G.F. Powell, ed., *Cicero the Philosopher* 115–43. Oxford: Oxford University Press 1995.

Gombay, A. '"*Cogito ergo Sum*": Inference or Argument?' In R.J. Butler, ed., *Cartesian Studies*, 71–88.

Gotthelf, A. 'Aristotle's Conception of Final Causality.' *Review of Metaphysics* 30 (1976): 226–54.

Goudge, T.A. *The Ascent of Life.* Toronto: University of Toronto Press 1961.

Gould, Josiah B. *The Philosophy of Chrysippus.* Albany: State University of New York Press 1970.

Guéroult, M. *Descartes' Philosophy Interpreted according to the Order of Reasons.* 2 vols. Trans. R. Ariew. Minneapolis, Minn.: University of Minnesota Press 1984.

- 'The Metaphysics and Physics of Force in Descartes.' In S. Gaukroger, ed., *Descartes: Philosophy, Mathematics and Physics*, 196–229.

Hacking, I. *The Emergence of Probability.* Cambridge: Cambridge University Press 1975.

Halley, E. 'A Discourse concerning Gravity.' *Philosophical Transactions of the Royal Society.* (Jan., Feb. 1686).

Hamilton, Sir William. *Lectures on Metaphysics and Logic.* 2 vols. Ed. H.L. Mansel and J. Veitch. Boston: Gould and Lincoln 1860.

Hartley, David. *Observations on Man.* London: S. Richardson 1749. Repr. two vols in one, Delmar, N.Y.: Scholars' Facsimiles and Reprints 1976.

Harvey, W. *The Circulation of the Blood and Other Writings.* Trans. K. Franklin. London: J. Dent, Everyman's Library 1963.

Hatfield, Gary. 'Force (God) in Descartes' Physics.' *Studies in History and Philosophy of Science* 10 (1979): 113–40.

Hausman, Alan. 'Hume's Theory of Relations.' *Noûs* 1 (1967): 55–82.

Hay, W. 'Stoic Use of Logic.' *Archiv für Geschichte der Philosophie* 51 (1969): 145–57.

Heath, Thomas. *Greek Mathematics*. 2 vols. Oxford: Oxford University Press 1921.

Hempel, C.G. and P. Oppenheim. 'Studies in the Logic of Explanation.' In C.G. Hempel, *Aspects of Scientific Explanation and Other Essays*, 245–90. New York: Free Press 1965.

Herival, J. 'Newton's Achievement in Dynamics.' *Texas Quarterly* 10 (1967): 103–18.

Herschell, J.W.F. *Preliminary Discourse on the Study of Natural Philosophy*. London: Longman, Rees, Orme, Brown, Green and Longman, and John Taylor 1832; New York: Johnson Reprint Corp. 1966.

Hintikka, J. '*Cogito, ergo Sum*: Inference or Performance?' *Philosophical Review* 71 (1962): 3–32.

– 'Necessity, Universality, and Time in Aristotle.' In J. Barnes et al., eds, *Articles on Aristotle*, 3:108–24. London: Duckworth 1979.

Hirschfield, John Milton. *The Académie Royale des Sciences 1666–1683*. New York: Arno Press 1981.

Hobbes, Thomas. *Elements of Philosophy, the First Section, Concerning Body*. In W. Molesworth, ed., *The English Works of Thomas Hobbes*. vol. 1. London: John Bohn 1839.

– *Leviathan*. In W. Molesworth, ed., *The English Works of Thomas Hobbes*. Vol. 3. London: John Bohn 1839.

Hochberg, H. 'Natural Necessity and Laws of Nature.' *Philosophy of Science* 48 (1981): 386–99.

– 'Possibilities and Essences in Wittgenstein's *Tractatus*.' In E.D. Klemke, ed., *Essays on Wittgenstein*, 485–533. Urbana, Ill.: University of Illinois Press 1971.

Hocutt, M. 'Aristotle's Four Becauses.' *Philosophy* 49 (1974): 385–99.

Hooke, Robert. 'A General Scheme, or Idea of the General State of Natural Philosophy, and How Its Defects May be Remedied by a Methodical Proceeding in the Making Experiments and Collecting Observations whereby to Compile a Natural History, as the Solid Basis for the Superstructure of True Philosophy.' In his *Posthumous Works*, 1–70. London: Sam. Smith and Benj. Walford 1705

Houghton, Walter E. *The Wellesley Index to Victorian Periodicals*. 5 vols. Toronto: University of Toronto Press 1960.

Howell, W. S. *Eighteenth Century British Logic and Rhetoric*. Princeton: Princeton University Press 1971.

– *Logic and Rhetoric in England, 1560–1700*. Princeton: Princeton University Press 1956.

Huet, Pierre-Daniel. *Censura philosophiae Cartesianae*. Editio quarta. Paris: Anisson 1694.

– *An Essay concerning the Weakness of Human Understanding*. Trans. E. Combe. London: Matthew de Varenne 1725.

– *Traité philosophique de la foiblesse de l'esprit humain.* Amsterdam: Henri du Sauvet 1723.

Hume, D. *Dialogues on Natural Religion.* Ed. N. Kemp Smith. 2nd ed. Indianapolis, Ind.: Bobbs-Merrill 1947.

– *Enquiries concerning Human Understanding and concerning the Principles of Morals.* Ed. L.A. Selby-Bigge. 2nd ed. Oxford: Oxford University Press 1902.

– *Letters.* 2 vols. Ed. J.Y.T. Grieg. Oxford: Oxford University Press 1932.

– *A Treatise of Human Nature.* Ed. L.A. Selby-Bigge. Oxford: Oxford University Press 1888.

Hutcheson, Francis. *Logicae compendium.* Vol. 7 in his *Collected Works.* Hildesheim: Georg Olms 1969.

Imlay, R.A. 'Descartes' Ontological Argument.' *The New Scholasticism* 43 (1969): 440–8.

– 'Descartes' Ontological Argument: A Causal Argument.' *The New Scholasticism* 45 (1971): 348–51.

Jacobs, Louis. *Studies in Talmudic Logic and Methodology.* London: Valentine, Mitchell and Co. 1961.

Jardine, L. *Francis Bacon: Discovery and the Art of Discourse.* London: Cambridge University Press 1974.

– 'Humanistic Logic.' In Charles B. Schmitt et al., eds, *The Cambridge History of Renaissance Philosophy,* 173–98. Cambridge: Cambridge University Press 1988.

– 'Lorenzo Valla and the Intellectual Origins of Humanist Dialectic.' *Journal of the History of Philosophy* 15 (1977): 143–59.

Joachim, H.H. *Descartes's Rules for the Direction of the Mind.* Ed. E.E. Harris. London: Allen and Unwin 1957.

John of Salisbury. *The Metalogicon.* Trans. D.D. McGarry. Berkeley: University of California Press 1955.

Jolivet, Jean. *Abélard, ou la philosophie dans le langage.* Paris: Éditions Seghers 1969.

Jones, P. *Hume's Sentiments.* Edinburgh: Edinburgh University Press 1982.

Kahn, Charles. 'Anaximander's Fragment: The Universe Governed by Law.' In Alexander P.D. Mourelatos, *The Pre-Socratics,* 99–117. Garden City, N.Y.: Doubleday Anchor 1970.

– 'The Greek Verb "To Be" and the Concept of Being.' *Foundations of Language* 2 (1966): 245–65.

– 'Stoic Logic and Stoic LOGOS.' *Archiv für Geschichte der Philosophie* 51 (1969): 158–72.

– 'The Thesis of Parmenides.' *Review of Metaphysics* 22 (1968–69): 700–24.

Kant, I. *Critique of Pure Reason.* Trans. N.K. Smith. London: Macmillan 1929.

Katz, B., and A.P. Martinich. 'The Distribution of Terms.' *Notre Dame Journal of Formal Logic* 17 (1976): 279–83.

Keckermann, Bartolomäus. *Systema physicum*. Hanover: I. Stockelii 1623.

Keeling, S. *Descartes*. Oxford: Oxford University Press 1968; first published London: Ernest Benn 1934.

Kenny, A. *Descartes: A Study of His Philosophy*. New York: Random House 1968.

– *The Five Ways*. New York: Shocken Books 1969.

Keynes, J.N. *Studies and Exercises in Formal Logic*. 2nd ed. London: Macmillan 1887.

Kieffer, J.S. 'Introduction' to Galen, *Institutio logica*.

Kneale, W., and M. Kneale. *The Development of Logic*. London: Oxford University Press 1962.

Kosman, L.A. 'Understanding, Explanation, and Insight in Aristotle's *Posterior Analytics*.' In H.N. Lee et al., eds, *Exegesis and Argument*, 374–92. Assen, Netherlands: Van Gorcum 1973.

Kraye, Jill. 'The Philosophy of the Italian Renaissance.' In G.H.R. Parkinson, ed., *The Renaissance and Seventeenth-Century Rationalism*. Vol. 4 of the *Routledge History of Philosophy*, 16–49. New York: Routledge 1993.

Kremer, E.J., ed. *The Great Arnauld and Some of His Philosophical Correspondents*. Toronto: University of Toronto Press 1994.

Kuhn, Thomas. *The Copernican Revolution*. Cambridge, Mass.: Harvard University Press 1959.

– *The Structure of Scientific Revolutions*. 2nd ed. Chicago: University of Chicago Press 1970.

Legrand, A. *Institutio philosophiae*. London 1672.

Leibniz, G.W.F. 'Monadology.' In *G.W. Leibniz: Philosophical Essays*, trans. R. Ariew and D. Garber. Indianapolis: Hackett 1989.

– *Philosophical Papers and Letters*. 2 vols. Trans. and ed. Leroy E. Leomker. Chicago: University of Chicago Press 1956.

– *Theodicy*. Ed. with Intro. by A. Farrer. Trans. E.M. Huggard. La Salle, Ill.: Open Court 1985.

Lennon, T. *The Battle of the Gods and Giants: The Legacies of Descartes and Gassendi, 1655–1715*. Princeton: Princeton University Press 1993.

– 'The Inherence Pattern and Descartes' Ideas.' *Journal of the History of Philosophy* 12 (1974): 43–52.

– 'Philosophical Commentary.' In N. Malebranche, *The Search after Truth*, 756–848.

Lenz, J. 'Hume's Defense of Causal Inference.' In V.C. Chappell, ed., *Hume*, 169–86.

Lesher, James H. 'The Meaning of *NOUS* in the Posterior Analytics.' *Phronesis* 18 (1973): 44–68.

Lewis, C.I., and C.H. Langford. *Symbolic Logic*. 2nd ed. New York: Dover 1959.

Lewis, D. 'Causation.' *Journal of Philosophy* 70 (1973): 556–67.

Livingston, D. *Hume's Philosophy of Common Life*. Chicago: University of Chicago Press 1984.

Locke, J. *Essay concerning Human Understanding*. Ed. P. Nidditch. London: Oxford University Press 1975.
– *Examination of P. Malebranche's Opinion*. In Locke's *Posthumous Works*. London: J. Churchill 1706; and in Locke, *Works*. 10 vols. London: T. Tegg 1823.
– *The Locke Reader*. Ed. J. Yolton. London: Cambridge University Press 1977.
– *Some Thoughts concerning Education*. In J.T. Axtell, ed., *The Educational Writings of John Locke*, 109–325. Cambridge: Cambridge University Press 1968.
Loeb, L. 'Is There Radical Dissimulation in Descartes' *Meditations?*' In A.O. Rorty, ed., *Essays on Descartes' Meditations*, 243–70. Berkeley: University of California Press 1986.
'Logic.' *Encyclopaedia Britannica*. 1st ed. Vol. 2, 984–1003. Edinburgh: A. Bell and C. MacFarquhar 1771.
'Logic.' *Encyclopaedia Britannica*. 3rd ed. Vol. 10, 189–221. Edinburgh: A. Bell and C. MacFarquhar 1797.
Long, A.A. *Hellenistic Philosophy*. London: Duckworth 1974.
– 'Language and Thought in Stoicism.' In A.A. Long, ed., *Problems in Stoicism*, 75–113. London: Athlone Press 1971.
'Lord Bacon an Imposter?' *Fraser's Magazine* 74 (1866): 718–40.
Lovejoy, A.O. '"Representative Ideas" in Malebranche and Arnauld.' *Mind*. ns 32 (1923): 449–61.
Luce, A.A. *The Dialectic of Immaterialism*. London: Hodder and Stoughton 1963.
Lyall, William Rowe. 'Dissertation Prefixed to the Supplemental Volumes of the Encyclopaedia Britannica by Dugald Stewart.' *Quarterly Review* 17 (1817): 39–71.
McCracken, Charles J. *Malebranche and British Philosophy*. Oxford: Oxford University Press 1983.
Mackie, J.L. 'Causes and Conditions.' In E. Sosa, ed., *Causation and Conditionals*, 15–38.
– 'Self-Refutation – A Formal Analysis.' *Philosophical Quarterly* 14 (1964): 193–203.
Mackintosh, James. '"A General View of the Progress of Metaphysical, Ethical, and Political Philosophy, since the Revival of Letters in Europe" by Dugald Stewart.' *Edinburgh Review* 27 (1816): 180–244.
McLachlan, H. *English Education under the Test Acts*. Manchester: Manchester University Press 1931.
MacLachlan, J. 'Galileo's Experiments with Pendulums: Real and Imaginary.' *Annals of Science* 33 (1976): 173–85.
Malebranche, N. *De la recherche de la vérité*. Ed. G. Rodin-Lewis. Paris: J. Vrin 1962.
– *The Search after Truth*. Trans. T.M. Lennon and P.J. Olscamp. Columbus: Ohio State University Press 1980.
Mariotte, Edme. *Essai de logique*. Paris: Feyard 1992.

Maritain, J. *Formal Logic.* Trans. I. Choquette. New York: Sheed and Ward 1937.

– *An Introduction to Philosophy.* Trans. E.J. Watkin. London: Sheed and Ward 1930.

– *A Preface to Metaphysics.* New York: Mentor 1962.

Martin, C.J. 'William's Machine.' *Journal of Philosophy* 83 (1986): 564–72.

Martin of Dacia, *Martinus de Dacia Quaestiones super librum Periermenius.* In H. Roos, ed., *Marini de Dacia opera, Corpus philosophicorum medii aevi,* vol. 2. Copenhagen: Gad 1961.

Mates, Benson. *Stoic Logic.* Berkeley: University of California Press 1961.

Maxwell, Constantia. *A History of Trinity College Dublin, 1591–1892.* Dublin: Trinity College University Press 1946.

Mersenne, Marin. *La Verite des sciences contre les sceptiques ou Pyrrhoniens.* Paris 1625; facsimile ed., Stutgart-Bad Cannstatt: Friedrich Frommann Verlag 1969.

Mill, James. Review of Thomas Belsham, *Elements of the Philosophy of the Mind, and of Moral Philosophy. To Which Is Prefixed a Compendium of Logic.* In *The Anti-Jacobin Review and Magazine* (May 1802): 1–13.

Mill, John Stuart. *System of Logic.* 8th ed. London: Longmans 1872.

Miller, J.W. *The Structure of Aristotelian Logic.* London: Kegan Paul, Trench, Trubner and Co. 1938.

Milton, John. *Paradise Lost.* In Merrit Y. Hughes, ed., *John Milton: Complete Poems and Major Prose.* Indianapolis, Ind.: Odyssey Press, Bobbs-Merrill 1957.

Montaigne, M. de. *Complete Essays.* Trans. D. Frame. Stanford: Stanford University Press 1958.

Moore, G.E. *Ethics.* London: Oxford University Press 1912.

Mossner, E.C. 'Hume's Early Memoranda 1729–40: The Complete Text.' *Journal of the History of Ideas* 9 (1948): 492–518.

Mueller, I. 'Stoic and Peripatetic Logic.' *Archiv für Geschichte der Philosophie* 51 (1969): 173–87.

Mullin, Amy. 'Masks and Accusations: Descartes and Bourdin.' Presented at the 'Descartes 400 Symposium.' Toronto 1996.

Nadler, S. *Arnauld and the Cartesian Philosophy of Ideas.* Manchester: Manchester University Press 1989.

Nahknikian, G. 'On the Logic of Cogito Propositions.' *Noûs* 3 (1969): 197–220.

Napier, Macvey. 'Remarks, Illustrative of the Scope and Influence of the Philosophical Writings of Lord Bacon.' *Transactions of the Royal Society of Edinburgh* 8 (1818): 373–426.

Naylor, R.H. 'Galileo: The Search for the Parabolic Trajectory.' *Annals of Science* 33 (1976): 153–72.

Newton, Isaac. *The Correspondence of Isaac Newton.* 7 vols. Ed. H.W. Turnbull. Cambridge: Cambridge University Press 1959.

– *Mathematical Principles of Natural Philosophy and His System of the World.* Trans.

Andrew Motte 1729. Revised by Florian Cajori. Berkeley: University of California Press 1947.

Normore, Calvin. 'Meaning and Objective Being: Descartes and His Sources.' In A.O. Rorty, ed., *Essays in Descartes' Meditations*, 223–42. Berkeley: University of California Press 1986.

– 'Medieval Connectives, Hellenistic Connections: The Strange Case of Propositional Logic.' In M.J. Osler, ed., *Atoms, Pneuma, and Tranquillity*. 25–38. Cambridge: Cambridge University Press 1991.

Ong, Walter J. *Ramus and Talon Inventory*. Cambridge, Mass.: Harvard University Press 1958.

– *Ramus, Method and the Decay of Dialogue*. Cambridge, Mass.: Harvard University Press 1958.

Pappus of Alexandria, *Book 7 of the Collection*. Part I: Introduction, Text and Translation, ed. with trans. and commentary by A. Jones. New York: Springer Verlag 1986.

Pariente, Jean-Claude. *L'Analyse du langage à Port-Royal*. Paris: Les Éditions de Minuit 1985.

Parry, W.T. 'Quantification of the Predicate and Many-sorted Logic.' *Philosophy and Phenomenological Research* 26 (1966): 342–59.

Pears, D. *Hume's System*. Oxford: Oxford University Press 1990.

Peghaire, J. *Intellectus et ratio selon S. Thomas D'Aquin*. Ottawa: Inst. d'Études Médiévales 1936.

Peirce, C.S. *Collected Papers*. 8 vols. Vols. 1–6 ed. C. Hartshorne and P. Weiss; vols 7–8 ed. A.W. Burks. Cambridge, Mass.: Harvard University Press 1958–60.

Pemberton, Henry. *A View of Sir Isaac Newton's Philosophy*. London 1728; repr. with an Introduction by I.B. Cohen. New York: Johnson Reprint Corp. 1972.

Penelhum, T. *Hume*. New York: St Martin's Press 1975.

Peter of Spain, *Summulae logicales*. Ed. I.M. Bochenski. Turin: Marietti 1947.

Piltz, Anders. *The World of Medieval Learning*. Trans. D. Jones. Oxford: Blackwell 1981.

Pitt, J.C. 'Galileo: Causation and the Use of Geometry.' In R.E. Butts and J.C. Pitt, eds, *New Perspectives on Galileo*, 181–95. Dordrecht, Netherlands: Kluwer Publishers 1978.

Plato, *Gorgias*. Trans. R. Waterfield. Oxford: Oxford University Press 1994.

– *Republic*. Trans. G.M.A. Grube. Indianapolis, Ind.: Hackett 1974.

– *Sophist*. Trans. In F.M. Cornford, *Plato's Theory of Knowledge*.

– *Theatetus*. Trans. In F.M. Cornford, *Plato's Theory of Knowledge*.

– *Timaeus*. Trans. In F.M. Cornford, *Plato's Cosmology*. London: Routledge and Kegan Paul 1956.

Playfair, J. 'Dissertation Second.' *Encyclopaedia Britannica*. 5th ed., Supp. vol. 2. Edinburgh: Constable 1817.

Popkin, R.H. 'The High Road to Pyrrhonism.' *American Philosophical Quarterly* 2 (1965): 18–32.

– *The History of Scepticism from Erasmus to Descartes*. Assen: Van Gorcum 1960.

– *The History of Scepticism from Erasmus to Spinoza*. Berkeley: University of California Press 1979.

– 'Joseph Glanvill: A Precursor of Hume.' *Journal of the History of Ideas* 14 (1953): 292–303.

– 'So, Hume Did Read Berkeley.' *Journal of Philosophy* 61 (1964): pp. 74–5.

Prichard, H.A. *Knowledge and Perception*. London: Oxford University Press 1950.

Prior, A.N. 'Determinables, Determinates, and Determinants.' *Mind* ns, 58 (1949): 1–20, 178–94.

– *Formal Logic*. 2nd ed. London: Oxford University Press 1960.

Quine, W.V.O. Review of P. Geach, *Reference and Generality*. *Philosophical Review* 73 (1964): 100–4.

Radner, Daisy. 'Representationalism in Arnauld's Act Theory of Perception.' *Journal of the History of Philosophy* 14 (1976): 96–8.

Ramée, Pierre de la. *Dialectique*. 1555. Edition critique avec introduction, notes et commentaires de Michel Dassonville. Geneva: Librairie Droz 1964.

Randall, J.H., Jr. 'The Development of Scientific Method in the School of Padua.' *Journal of the History of Ideas* 1 (1940): 177–206.

Rapin, R. *Reflexions sur la philosophie ancienne et moderne*. Paris 1676.

Raymond, P. 'La Théorie de l'induction – Duns Scot précurseur de Bacon.' *Études Franciscaines* 21 (1909): 113–26, 270–9.

Reif, P. 'The Textbook Tradition in Natural Philosophy 1600–1650.' *Journal of the History of Ideas* 30 (1969): 17–32.

Richards, Thomas J. 'Hume's Two Definitions of "Cause,"' in V.C. Chappell, ed., *Hume*, 148–61.

Rist, J. 'Zeno and the Origins of Stoic Logic.' In J. Braunschweig, ed., *Les Stoiciens et leur logique*, 387–400. Paris: Vrin 1978.

Robertson, J. Charles. 'A Bacon-Facing Generation: Scottish Philosophy in the Early Nineteenth Century.' *Journal of the History of Philosophy* 14 (1976): 37–49.

Robison, J.A. 'Hume's Two Definitions of "Cause,"' in V.C. Chappell, ed., *Hume*, 129–47.

– 'Hume's Two Definitions of "Cause" Reconsidered.' In V.C. Chappell, ed., *Hume*, 162–8.

Robison, J. 'Philosophy.' *Encyclopaedia Britannica*. 3rd ed. 14:573–600. Edinburgh: A. Bell and C. MacFarquhar 1797.

Rochot, B. 'Roberval, Mariotte et la logique.' *Archives Internationales d'Histoire des Sciences* 6 (1965): 38–43.

Rohault, J. *Rohault's System of Natural Philosophy*. 2 vols. Trans. J. Clarke, with notes by Samuel Clarke. London: J. Knapton etc. 1723. Repr. with a New Introduction by L. Laudan, New York: Johnson Reprint Corp. 1969.

Rose, L. *Aristotle's Syllogistic.* Springfield, Ill.: Charles C. Thomas 1968.

Ross, W.D. *Aristotle.* 5th ed. London: Methuen 1964.

Russell, B. 'Lectures on the Philosophy of Logical Atomism.' In his *Logic and Knowledge*, ed. R.C. Marsh, 175–282.

– *Logic and Knowledge.* Ed. R.C. Marsh. London: Allen and Unwin 1956.

– 'On the Relations of Universals and Particulars.' In his *Logic and Knowledge*, ed., R.C. Marsh, 103–24.

Sambursky, S. *The Physics of the Stoics.* London: Routledge and Kegan Paul 1959.

Sanderson, Robert. *Logicae artis compendium.* Oxford: Josephus Barnesius 1615. 2nd ed., Oxford: Iohannes Lichfield and Jacobus Short 1618; and *Editio nova emendata*, Oxford: Oxford University Press 1841.

– *Logicae artis compendium.* In vol. 6 of his *Works*. 6 vols. Ed. William Jacobson. Oxford: Oxford University Press 1854.

Schmitt, Charles B. 'Experience and Experiment: A Comparison of Zabarella's View with Galileo's in *De Motu.' Studies in the Renaissance* 16 (1969): 80–138.

Seigel, J.E. '"Civic Humanism" or Ciceronian Rhetoric?' *Past and Present* 34 (1966): 3–48.

Sergeant, John. *Method to Science.* London: W. Redmayne 1696.

– *Solid Philosophy Asserted against the Fancies of the Ideists.* London: W. Redmayne 1696; 2nd ed. London: R. Clavel 1698.

Sextus Empiricus, *Against the Logicians.* Books I and II of this volume are Books VII and VIII of *Against the Mathematicians (Adversos mathematicos)*, in Sextus Empiricus, *Works*, vol. 2. Trans. R.G. Bury. Cambridge, Mass.: Harvard University Press 1935.

– *Outlines of Pyrrhonism.* Trans. R.G. Bury. Cambridge, Mass.: Loeb Classical Library, Harvard University Press 1920.

'sGravesande, W.J. *Mathematical Elments of Physics.* London: G. Strahan et al. 1720.

Shapin, S. *The Scientific Revolution.* Chicago: University of Chicago Press 1996.

Shine, Hill, and Helen Chadwick Shine. *The Quarterly Review under Gifford.* Chapel Hill: University of North Carolina Press 1949.

Simon of Faversham, *Quaestiones super libro Perihermenias.* In Pasquale Mazzarella, ed., *Magistri Simonis Anglici sive de Faversham Opera omnia, Opera logica.* Vol. 1. Padua: C.E.D.A.M. 1957.

Siraisi, Nancy. *Arts and Sciences at Padua; The 'Studium' of Padua before 1350.* Toronto: Pontifical Institute of Mediaeval Studies 1973.

– *Medieval and Early Renaissance Medicine.* Chicago: University of Chicago Press 1990.

Smith, Norman Kemp. *The Philosophy of David Hume.* London: Macmillan 1941.

Sonntag, Otto. 'Leibig on Francis Bacon and the Utility of Science.' *Annals of Science* 31 (1974): 373–86.

Sorabji, R. 'Infinite Power Impressed: The Transformation of Aristotle's Physics and Theology.' In Richard Sorabji, ed., *Aristotle Transformed: The Ancient Commentators and Their Influence*, 181–98. London: Duckworth 1990.

– *Necessity, Cause, and Blame*. London: Duckworth,1980.

Sosa, E., ed. *Causation and Conditionals*. London: Oxford University Press 1975.

Sprat, Thomas. *History of the Royal Society*. Ed. J.I. Cope and H.W. Jones. St. Louis, Miss.: Washington University Studies 1958.

Stelling-Michaud, Sven. *Les Universités Européennes du XIVe au XVIIIe siècle: Actes du Colloque internationale à l'occasion du VIe centenaire de l'Université Jagellonne de Cracovie*. Geneva: Droz 1967.

Stewart, Dugald. 'Dissertation First: [prefixed to the Supplemental Volumes of the Encyclopaedia Britannica] General View of the Progress of Metaphysical, Ethical, and Political Philosophy, since the Revival of Letters in Europe.' In *Encyclopaedia Britannica*. 5th ed., Supp. vol. 1. Edinburgh: Constable 1817.

– *Philosophy of the Active and Moral Powers of Man*. Vol. 2. Vol 7 of his *The Collected Works of Dugald Stewart*. Ed. Sir W. Hamilton. Edinburgh: Constable 1854.

Stewart, M.A. 'Hume and the "Metaphysical Argument *A Priori*."' In A.J. Holland, ed., *Philosophy: Its History and Historiography*. 243–70. Dordrecht, Netherlands: Reidel 1985.

Stone, L. *The University in Society*. 2 vols. Princeton: Princeton University Press 1974.

Strawson, P.F. *Introduction to Logical Theory*. London: Methuen 1952.

Stubbs, J.W. *The History of the University of Dublin*. Dublin: Hodges, Figgis and Co. 1889.

Stump, E. *Dialectic and Its Place in the Development of Medieval Logic*. Ithaca, N.Y.: Cornell University Press 1989.

– 'Logic in the Early Twelfth Century.' In N. Kretzmann, ed., *Meaning and Inference in Medieval Philosophy,* 31–56. Dordrecht, Netherlands: Kluwer 1988.

– 'Topics: Their Development and Absorption into Consequences.' In N. Kretzman, A. Kenny, and J. Pinborg, eds, *The Cambridge History of Later Medieval Philosophy*. 273–99. Cambridge: Cambridge University Press 1982.

Suarez, F. *Disputationes metaphysicae*. 2 vols. Hildesheim: George Olms 1965.

– *On the Essence of Finite Being as Such, On the Existence of That Essence and Their Distinction*. Trans. from *Disputationes metaphysicae, Disputatio XXXI*, by N.J. Wells. Milwaukee, Wisc.: Marquette University Press 1983.

– *On the Various Kinds of Distinctions*. Trans. from *Disputationes metaphysicae, Disputatio VII*. by C. Vollert. Milwaukee, Wisc.: Marquette University Press 947.

Tatham, Edward. *The Chart and Scale of Truth*. Oxford: J. Fether et al. 1793.

Taylor, A.E. *Thomas Hobbes*. London: Archibald Constable and Co. 1908.

Thieberger, Frederic. *The Great Rabbi Loew of Prague*. London: Farrar, Strauss and Young 1955.

Thomas Aquinas, *Summa theologica*. Trans. by the Fathers of the Dominican Province. London: Benziger Bros. 1947.

Thomason, R. 'Species, Determinates and Natural Kinds.' *Noûs* 3 (1969): 95–101.

Thorndyke, L. 'The Cursus Philosophicus before Descartes.' *Archive Internationales d'Histoire des Sciences* 4 (1951): 16–24.

Tooley, Michael. 'The Nature of Laws.' *Canadian Journal of Philosophy* 7 (1977): 667–98.

Turnbull, R.G. 'Aristotle's Debt to the "Natural Philosophy" of the *Phaedo*.' *Philosophical Quarterly* 8 (1958); 131–43.

Tweyman, S. 'The Reliability of Reason.' In R.J. Butler, ed., *Cartesian Studies*, 123–36.

Urbach, P. *Francis Bacon's Philosophy of Science: An Account and a Reappraisal*. La Salle, Ill.: Open Court 1987.

van Leeuwen, H.G. *The Problem of Certainty in English Thought from Chillingworth to Locke*. The Hague: Nijhoff 1963.

Versfeld, M. *An Essay on the Metaphysics of Descartes*. London 1940.

Vier, Peter C. *Evidence and Its Function according to John Duns Scotus*. St Bonaventure, N.Y.: Franciscan Institute 1951.

Virgil. *The Georgics*. Trans. L.P. Wilkinson. Harmondsworth, Middlesex: Penguin 1982.

Vlastos, G. 'Equality and Justice in Early Greek Cosmology. In D.J. Furley and R.E. Allen, eds, *Studies in Presocratic Philosophy*. 2 vols. 1:56–91. London: Routledge and Kegan Paul 1970.

– 'Reasons and Causes in the *Phaedo*.' *Philosophical Review* 78 (1969): 291–325.

von Leibig, J. 'Induction and Deduction.' *Cornhill Magazine* 12 (1865): 296–305.

– 'Lord Bacon as a Natural Philosopher.' *Macmillan's Magazine* 8 (1863): 237–49, 257–67.

– 'Was Lord Bacon an Imposter?' *Fraser's Magazine* 75 (1867): 482–95.

von Tschirnhaus, E.W. *Médecin de l'esprit, ou préceptes généraux de l'art de découvrir*. Ed. with Introduction by J.-P. Wurtz. Paris: Editions Ophyrs 1980.

von Wright, G.H. *The Logical Problem of Induction*. 2nd ed. Oxford: Blackwell 1957.

Wallis, John. *Institutio logicae*. Oxford: Sheldoniano 1687. References are to the edition in the *Opuscula quaedam miscellanea* appended to Wallis's *Operiaem mathematicorum*. Vol. 3. Oxford: Sheldoniano 1699. The pagination of *Opuscula* recommences at page 1, following 708 pages numbering the works that preceded. The *Institutio logica* begins at page 81 of this new series.

Watkins, J.W.N. *Hobbes's System of Ideas*. 2nd ed. London: Hutchinson 1973.

Watson, R.A. *The Downfall of Cartesianism*. The Hague: Nijhoff 1966.

– 'In Defiance of Demons, Dreamers, and Madmen.' *Journal of the History of Philosophy* 14 (1976): 342–53.

Watts, Isaac. *Logick*. 2nd ed. London: John Clark and Richard Hett, Emmanuel Matthews and Richard Ford 1726.

Weinberg, J. *Abstraction, Relation and Induction*. Madison, Wisc.: University of Wisconsin Press 1965.

– 'Cogito *ergo Sum*: Some Reflections on Mr. Hintikka's Article.' *Philosophical Review* 71 (1962): 483–91.

– 'Induction.' In his *Abstraction, Relation and Induction*, 121–53.

– 'The Nominalism of Berkeley and Hume.' In his *Abstraction, Relation and Induction*, 3–60.

– *Ockham, Descartes, Hume*. Madison, Wisc.: University of Wisconsin Press 1977.

Wells, Norman J. 'Material Falsity in Descartes, Arnauld, and Suarez.' *Journal of the History of Philosophy* 22 (1984): 25–50.

Westfall, R.S. *Force in Newton's Physics*. New York: American Elsevier 1971.

Wheelwright, P. *The Presocratics*. New York: Odyssey Press 1966.

Whiston, William. *Praelectiones physico-mathematicae*. 2nd ed. London: Benjamin Motte 1726.

– *Sir Isaac Newton's Mathematick Philosophy More Easily Demonstrated*. London: J. Senex and W. Taylor 1716.

White, T. *An Exclusion of Sceptics from All Title to Dispute: Being an Answer to the Vanity of Dogmatizing*. London: John Williams 1665.

William of Ockham, *Summa logicae*. In *William Ockham opera philosophica, et theologica*. 17 vols. *Opera philosphica*, vol. 1, ed. P. Boehner et al. St Bonaventure, N.Y.: Franciscan Institute 1974.

Williams, B. *Descartes: The Project of Pure Enquiry*. Harmondsworth, Middlesex: Penguin 1978.

Wilson, F. 'Abstract Ideas and Other Rules of Language in Hume.' Presented to the Conference on Ideas: Sensory Experience, Thought, Knowledge and their Objects in 17th and 18th Century Philosophy. University of Iowa, Iowa City, Apr. 1989. Portions of this presentation were subsequently published as 'Association, Ideas, and Images in Hume,' in P.D. Cummins and G. Zoeller, eds, *Minds, Ideas, and Objects*. North American Kant Society Studies in Philosophy, 2:255–74. Atascadero, Calif.: Ridgeview Publishing Co. 1992.

– 'Acquaintance, Ontology and Knowledge.' *The New Scholasticism* 54 (1970): 1–48.

– 'Association, Ideas and Images in Hume.' In P. Cummins and G. Zoeller, eds, *Minds, Ideas, and Objects*, 255–74. Atascadero, Calif.: Ridgeview Publishing 1992.

– 'Bradley's Conception of Ideality: Comments on Ferreira's Defence.' *Bradley Studies*, 1 (1995): 139–52.

– 'Bradley's Contribution to Empiricism.' In J. Bradley, ed., *Philosophy after F.H. Bradley*, 251–82. Bristol: Thoemmes Press 1996).

- 'Critical Notice' of I. Hacking, *The Emergence of Probability. Canadian Journal of Philosophy* 8 (1978): 587–97.
- Critical Review of P. Jones, *Hume's Sentiments. Noûs* 20 (1988): 274–81.
- *Empiricism and Darwin's Science.* Dordrecht, Netherlands: Kluwer 1991.
- *Explanation, Causation and Deduction.* Dordrecht, Netherlands: D. Reidel 1986.
- 'Explanation in Aristotle, Newton and Toulmin.' *Philosophy of Science* 36 (1969): 291–310 and 400–28.
- 'The History of Relations from Burgersdijck to Bradley.' Paper presented to the Conference on the Ontology of Relations, State University of New York at Buffalo, Spring 1995. Manuscript available from the author.
- 'Hume and Derrida on Language and Meaning.' *Hume Studies* 12 (1986): 99–121.
- 'Hume on the Abstract Idea of Existence.' *Hume Studies* 17 (1991): 167–201.
- 'Hume's Cognitive Stoicism.' *Hume Studies*, Supplement (1985): 521–68.
- 'Hume's Defence of Causal Inference.' *Dialogue* 22 (1983): 661–94.
- *Hume's Defence of Causal Inference.* Toronto: University of Toronto Press 1997.
- 'Hume's Defence of Science.' *Dialogue* 25 (1986): 611–28.
- 'Is Hume a Sceptic with regard to Reason?' *Philosophy Research Archives* 10 (1984): 275–320.
- *Laws and Other Worlds.* Dordrecht, Netherlands: Kluwer 1986.
- 'The Lockean Revolution in the Theory of Science.' In S. Tweyman and G. Moyal, eds., *Early Modern Philosophy*, 65–97. Delmar, N.Y.: Caravan Books 1986.
- 'Logical Necessity in Carnap's Later Philosophy.' In A. Hausman and F. Wilson, *Carnap and Goodman: Two Formalists*, 97–225. The Hague: Martinus Nijhoff 1967.
- 'Mill's Proof That Happiness Is the Criterion of Morality.' *Journal of Business Ethics* 1 (1982): 59–72.
- 'On the Hausman's "New Approach."' In R. Muehlmann, ed., *Berkeley's Metaphysics*, 67–88. University Park, Penn.: Pennsylvania State University Press 1995.
- *Psychological Analysis and the Philosophy of John Stuart Mill.* Toronto: University of Toronto Press 1990.
- Review of M. Mandelbaum, *The Anatomy of Historical Knowledge. Philosophical Review* 88 (1979): 663–8.
- 'The Ultimate Unifying Principle of Coleridge's Metaphysics of Relations and of Our Knowledge of Them.' *Ultimate Reality and Meaning* 21 (1998): 273–300.
- 'Was Hume a Sceptic with regard to the Senses?' *Journal of the History of Philosophy* 27 (1989): 49–73.
- 'Was Hume a Subjectivist?' *Philosophy Research Archives* 14 (1989): 247–82.
- 'Weinberg's Refutation of Nominalism.' *Dialogue* 8 (1969): 460–74.
- 'The World and Reality in the *Tractatus*.' *Southern Journal of Philosophy* 5 (1967): 253–60.

Wilson, Margaret. *Descartes*. London: Routledge and Kegan Paul 1978.

Winkler, K. *Berkeley: An Interpretation*. Oxford: Oxford University Press 1989.

Wittgenstein, L. 'On Logic, and How Not to Do It.' Review of P. Coffey, *The Science of Logic*. In Eric Homberger, William Janeway, and Simon Schama, eds, *The Cambridge Mind*, 127–9. London: Jonathan Cape 1970.

Wolff, Michael. 'Philoponous and the Rise of Preclassical Dynamics.' In R. Sorabji, ed., *Philoponous and the Rejection of Aristotelian Science*, 84–120. London: Duckworth 1987.

Wood, P.B. 'Methodology and Apologetics: Thomas Sprat's *History of the Royal Society*.' *British Journal for the History of Science* 45 (1986): 1–26.

Woods, J. 'On Species and Determinates.' *Noûs* 1 (1967): 243–54.

Worster, Benjamin. *A Compendious and Methodical Account of the Principles of Natural Philosophy*. London: W. Innys 1722.

Yolton, J. 'Locke's Unpublished Marginal Replies to John Sergeant.' *Journal of the History of Ideas* 12 (1951): 528–59.

– *Perceptual Acquaintance from Descartes to Reid*. Minneapolis: University of Minnesota Press 1984.

Index